AN INTRODUCTORY TEXT TO BIOENGINEERING

ADVANCED SERIES IN BIOMECHANICS

Editor: Y C Fung *(University of California, San Diego)*

Vol. 1: Selected Works on Biomechanics and Aeroelasticity
Parts A & B
by Y C Fung

Vol. 2: Introduction to Bioengineering
Ed. Y C Fung

Vol. 3: Basic and New Aspects of Gastrointestinal Ultrasonography
Eds. S Ødegaard, O H Gilja & H Gregersen

Advanced Series in Biomechanics – Vol. 4

AN INTRODUCTORY TEXT TO BIOENGINEERING

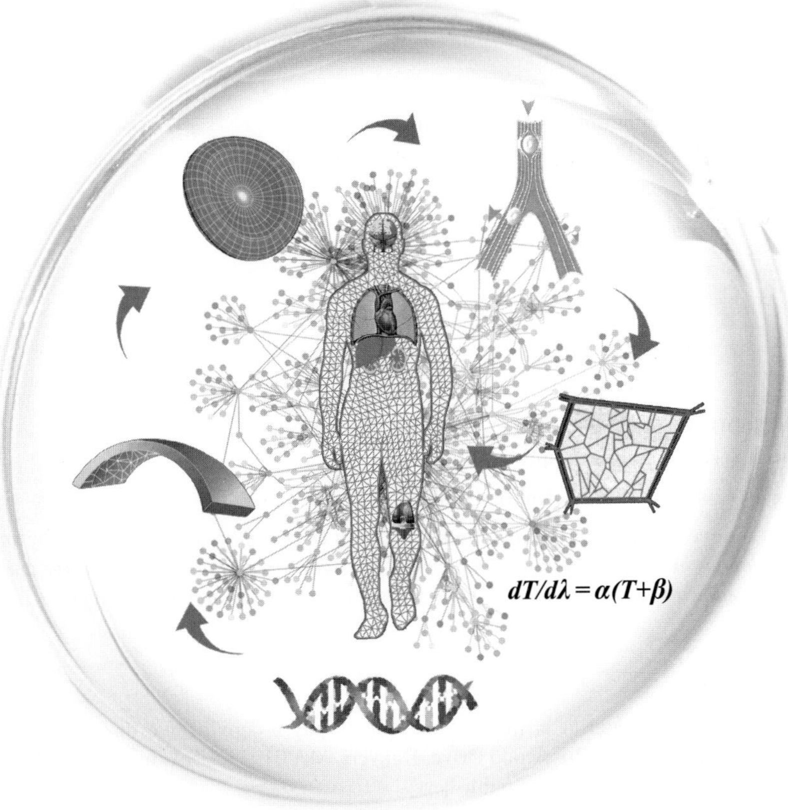

Editors **Shu Chien Peter C Y Chen Y C Fung**

University of California, San Diego, USA

World Scientific

NEW JERSEY · LONDON · SINGAPORE · BEIJING · SHANGHAI · HONG KONG · TAIPEI · CHENNAI

Published by

World Scientific Publishing Co. Pte. Ltd.
5 Toh Tuck Link, Singapore 596224
USA office: 27 Warren Street, Suite 401-402, Hackensack, NJ 07601
UK office: 57 Shelton Street, Covent Garden, London WC2H 9HE

British Library Cataloguing-in-Publication Data
A catalogue record for this book is available from the British Library.

AN INTRODUCTORY TEXT TO BIOENGINEERING
Advanced Series in Biomechanics — Vol. 4

ISBN-13 978-981-270-793-2
ISBN-10 981-270-793-X
ISBN-13 978-981-270-794-9 (pbk)
ISBN-10 981-270-794-8 (pbk)

Typeset by Stallion Press
Email: enquiries@stallionpress.com

Printed by Fulsland Offset Printing (S) Pte Ltd, Singapore

CONTENTS

Contributors ix

Preface xvii

Acknowledgments xix

SECTION I: INTRODUCTORY CHAPTERS

Chapter 1: Overview 3
Shu Chien

Chapter 2: Perspectives of Biomechanics 13
Yuan-Cheng B. Fung and Wei Huang

SECTION II: CARDIOVASCULAR BIOENGINEERING

Chapter 3: Cardiac Electromechanics in the Healthy Heart 37
Roy C. P. Kerckhoffs and Andrew D. McCulloch

Chapter 4: Cardiac Biomechanics and Disease 53
Jeffrey H. Omens

Chapter 5: Bioengineering Solutions for the Treatment of Heart Failure 69
John T. Watson and Shu Chien

Chapter 6: Molecular Basis of Modulation of Vascular Functions by Mechanical Forces 79
Shu Chien

Chapter 7: Autoregulation of Blood Flow: Examining the Process of Scientific Discovery 99
Paul C. Johnson

SECTION III: BLOOD CELL BIOENGINEERING

Chapter 8: Molecular Basis of Cell and Membrane Mechanics 117
Lanping Amy Sung

Chapter 9: Cell Activation in the Circulation: The Auto-Digestion Hypothesis 131
Geert W. Schmid-Schönbein

Chapter 10: Blood Substitutes and the Design of Oxygen Non-Carrying and Carrying Fluids 149

Marcos Intaglietta

SECTION IV: RESPIRATORY-RENAL BIOENGINEERING

Chapter 11: Analysis of Human Pulmonary Circulation: A Bioengineering Approach 163

Wei Huang, Michael R. T. Yen and Qinlian Zhou

Chapter 12: Pulmonary Gas Exchange 181

Peter D. Wagner

Chapter 13: Engineering Approaches to Understanding the Kidney 209

Scott C. Thomson

SECTION V: TISSUE ENGINEERING AND REGENERATIVE MEDICINE

Chapter 14: Skeletal Muscle Tissue Bioengineering 225

Richard L. Lieber and Samuel R. Ward

Chapter 15: Multi-Scale Biomechanics of Articular Cartilage 243

Won C. Bae and Robert L. Sah

Chapter 16: Design and Development of an *In Vivo* Force-Sensing Knee Prosthesis 261

Darryl D. D'Lima and Peter C. Y. Chen

Chapter 17: The Implantable Glucose Sensor in Diabetes: A Bioengineering Case Study 279

David A. Gough

Chapter 18: Stem Cells in Regenerative Medicine 291

Shu Chien and Lawrence S. B. Goldstein

SECTION VI: NANOSCIENCE AND NANOTECHNOLOGY

Chapter 19: Engineering Compounds Targeted to Vascular Zip Codes 313

Erkki Ruoslahti

Chapter 20: The Structure of the Central Nervous System and Nanoengineering Approaches for Studying and Repairing It 327

Gabriel A. Silva

Chapter 21: Cellular Biophotonics: Laser Scissors (Ablation) 353

Michael W. Berns

Chapter 22: Microelectronic Arrays: Applications from DNA Hybridization Diagnostics to Directed Self-Assembly Nanofabrication 369

Michael J. Heller and Dietrich Dehlinger

SECTION VII: GENOMIC ENGINEERING AND SYSTEMS BIOLOGY

Chapter 23: Systems Biology: A Four-Step Process 387

Jennifer L. Reed and Bernhard O. Palsson

Chapter 24: Bioinformatics and Systems Biology: Obtaining the Design Principles of Living Systems 401

Shankar Subramaniam

Chapter 25: Synthetic Biology: Bioengineering at the Genomic Level 427

Natalie Ostroff, Mike Ferry, Scott Cookson, Tracy Johnson and Jeff Hasty

Chapter 26: Network Genomics 453

Trey Ideker

Chapter 27: Genomes, Genomic Technologies and Medicine 473

Xiaohua Huang

SECTION VIII: SOCIO-ECONOMICAL ASPECTS OF BIOENGINEERING

Chapter 28: Ethics for Bioengineers 489

Michael Kalichman

Chapter 29: Opportunities and Challenges in Bioengineering Entrepreneurship 507

Jen-Shih Lee

Chapter 30: How to Move Medical Devices from Bench to Bedside 521

Paul Citron

Index 533

CONTRIBUTORS

Won C. Bae, PhD
Postdoctoral Researcher
Department of Bioengineering
University of California, San Diego
9500 Gilman Drive
La Jolla, CA 92093-0412, USA

Michael W. Berns, PhD
The Arnold and Mabel Beckman Professor and
Professor of Biomedical Engineering, Surgery, and Cell Biology
The Beckman Laser Institute
University of California, Irvine
1002 Health Sciences Road East
Irvine, CA 92612
Adjunct Professor of Bioengineering
Whitaker Institute for Biomedical Engineering
University of California, San Diego
La Jolla, CA 92093, USA

Peter C. Y. Chen, PhD
Project Scientist
Department of Bioengineering
University of California, San Diego
9500 Gilman Drive
La Jolla, CA 92093-0412, USA

Shu Chien
University of Professor and Y.C. Fung Professor
Departments of Bioengineering and Medicine, and
Director, The Whitaker Institute for Biomedical Engineering
University of California, San Diego
9500 Gilman Drive
Powell-Focht Bioengineering Hall, Room 134
La Jolla, CA 92093-0412, USA

Paul Citron, MSEE
Adjunct Professor
Department of Bioengineering
University of California, San Diego
9500 Gilman Drive
La Jolla, CA 92093-0412, USA

Scott Cookson, MS
Graduate Researcher
Department of Bioengineering
University of California, San Diego
9500 Gilman Drive
La Jolla, CA 92093-0412, USA

Dietrich Dehlinger, MS
Graduate Researcher
Departments of Electrical and Computer Engineering
University of California, San Diego
9500 Gilman Drive
La Jolla, CA 92093-0407, USA

Darryl D. D'Lima, MD, PhD
Director
Shiley Center for Orthopaedic Research and Education at Scripps Clinic
11025 North Torrey Pines Road, Suite 140
La Jolla, CA 92037-1030
Department of Bioengineering
University of California, San Diego
La Jolla, CA 92093-0412, USA

Mike Ferry, MS
Graduate Researcher
Department of Bioengineering
University of California, San Diego
9500 Gilman Drive
La Jolla, CA 92093-0412, USA

Yuan-Cheng B. Fung, PhD
Professor Emeritus
Department of Bioengineering
University of California, San Diego
9500 Gilman Drive
La Jolla, CA 92093-0412, USA

Lawrence S. B. Goldstein, PhD
Investigator, Howard Hughes Medical Institute
Professor, Department of Cellular and Molecular Medicine
Director, UC San Diego Stem Cell Program
University of California, San Diego
9500 Gilman Drive
La Jolla, CA 92093-0683, USA

David A. Gough, PhD
Professor
Department of Bioengineering
University of California, San Diego
9500 Gilman Drive
La Jolla, CA 92093-0412, USA

Jeff Hasty, PhD
Associate Professor
Department of Bioengineering and
Institute of Nonlinear Science
University of California, San Diego
9500 Gilman Drive
La Jolla, CA 92093-0412, USA

Michael J. Heller, PhD
Professor
Departments of Nanoengineering and Bioengineering
University of California, San Diego
9500 Gilman Drive
La Jolla, CA 92093-0412, USA

Wei Huang, PhD
Associate Project Scientist
Department of Bioengineering
University of California, San Diego
9500 Gilman Drive
La Jolla, CA 92093-0412, USA

Xiaohua Huang, PhD
Assistant Professor
Department of Bioengineering
University of California, San Diego
9500 Gilman Drive
La Jolla, CA 92093-0412, USA

Trey Ideker, PhD
Associate Professor
Department of Bioengineering
University of California, San Diego
9500 Gilman Drive
La Jolla, CA 92093-0412, USA

Marcos Intaglietta, PhD
Distinguished Professor
Department of Bioengineering
University of California, San Diego
9500 Gilman Drive
La Jolla, CA 92093-0412, USA

Paul C. Johnson, PhD, Dr Med (Hon)
Adjunct Professor
Department of Bioengineering
University of California, San Diego
9500 Gilman Drive
La Jolla, CA 92093-0412, USA

Tracy Johnson, PhD
Assistant Professor
Division of Biological Sciences
Department of Molecular Biology
University of California, San Diego
9500 Gilman Drive
La Jolla, CA 92093-0412, USA

Michael Kalichman, PhD
Director, Research Ethics Program
University of California, San Diego
9500 Gilman Drive
La Jolla, CA 92093-0612, USA

Roy C. P. Kerckhoffs, PhD
Postdoctoral Researcher
Department of Bioengineering
University of California, San Diego
9500 Gilman Drive
La Jolla, CA 92093-0412, USA

Jen-Shih Lee, PhD, MS, BS
Chief Executive Officer, Global Monitors, Inc.
PO Box 675772
Rancho Sante Fe, CA 92067
Adjunct Professor of Bioengineering
University of California, San Diego
La Jolla, CA 92093-0412
Professor Emeritus of Biomedical Engineering
University of Virginia
Charlottesville, VA 22908, USA

Richard L. Lieber, PhD
Professor
Vice Chair, Departments of Orthopedic Surgery and Bioengineering
University of California and V.A. Medical Centers, San Diego
9500 Gilman Drive
La Jolla, CA 92093-9151, USA

Andrew D. McCulloch, PhD
Professor and Chair
Department of Bioengineering
University of California, San Diego
9500 Gilman Drive
La Jolla, CA 92093-0412, USA

Jeffrey H. Omens, PhD
Adjunct Professor
Departments of Medicine and Bioengineering
University of California, San Diego
9500 Gilman Drive
La Jolla, CA 92093-0613J, USA

Natalie Ostroff, MS
Graduate Researcher
Department of Bioengineering
University of California, San Diego
9500 Gilman Drive
La Jolla, CA 92093-0412, USA

Bernhard O. Palsson
Galetti Professor of Bioengineering
Adjunct Professor of Medicine
Department of Bioengineering
University of California, San Diego
9500 Gilman Drive
La Jolla, CA 92093-0412, USA

Jennifer L. Reed
Assistant Professor
Department of Chemical and Biological Engineering
University of Wisconsin-Madison
3639 Engineering Hall, 1415 Engineering Drive Madison
WI 53706-1607, USA

Erkki Ruoslahti, MD, PhD
Distinguished Professor
Burnham Institute for Medical Research (at UCSB)
1105 Life Sciences Technology Bldg
University of California Santa Barbara
Santa Barbara, CA 93106-9610, USA

Robert L. Sah, MD, ScD
Professor and Vice Chair
Department of Bioengineering,
Stein Institute for Research on Aging, and
The Whitaker Institute for Biomedical Engineering
University of California, San Diego
9500 Gilman Drive
La Jolla, CA 92093-0412
Professor
Howard Hughes Medical Institute, USA

Geert W. Schmid-Schönbein, PhD, FAHA
Professor, Department of Bioengineering and
The Whitaker Institute for Biomedical Engineering
University of California, San Diego
9500 Gilman Drive
La Jolla, CA 92093-0412, USA

Gabriel A. Silva, MSc, PhD
Assistant Professor
Departments of Bioengineering and
Ophthalmology, and Neurosciences Program
University of California, San Diego
UCSD Jacobs Retina Center
9415 Campus Point Drive
La Jolla, CA 92037-0946, USA

Shankar Subramaniam, PhD
Professor
Department of Bioengineering
University of California, San Diego
9500 Gilman Drive
La Jolla, CA 92093-0412, USA

Lanping Amy Sung, PhD
Associate Professor
Department of Bioengineering
University of California, San Diego
9500 Gilman Drive
La Jolla, CA 92093-0412, USA

Scott C. Thomson, MD, FAHA
Professor of Medicine
Division of Nephrology-Hypertension
University of California, San Diego and
VA San Diego Healthcare System
3350 La Jolla Village Drive
San Diego, CA 92161-9151, USA

Peter D. Wagner, MD
Distinguished Professor of Medicine and Bioengineering
Chief, Division of Physiology, Department of Medicine
University of California, San Diego
9500 Gilman Drive
La Jolla, CA 92093-0623A, USA

Samuel R. Ward, PT, PhD
Assistant Professor
Departments of Orthopedic Surgery and Radiology
University of California and V.A. Medical Centers, San Diego
9500 Gilman Drive
La Jolla, CA 92093-9151, USA

John T. Watson, PhD
Associate Director, William J. von Liebig Center and
Professor & Vice Chair, Department of Bioengineering
University of California, San Diego
9500 Gilman Drive
La Jolla, CA 92093-0433, USA

Michael R. T. Yen, PhD
Professor
Department of Biomedical Engineering
University of Memphis
Memphis, TN 38152, USA

Qinlian Zhou, PhD
Research Scientist
Department of Biomedical Engineering
University of Memphis
Memphis, TN 38152, USA

PREFACE

This book is written to provide an introduction to the rapidly advancing, interdisciplinary field of Bioengineering. The theme of the book is integrative bioengineering, which brings together the fundamental concepts and techniques in engineering and biomedical sciences and demonstrates their interplays.

This book evolved from the book *Introduction to Bioengineering* edited by Y. C. Fung and published in 2000 also by World Scientific Publishing Co. which was written primarily as a textbook for bioengineering undergraduate students. The current book is still well suited as an introductory text for bioengineering undergraduate students, but it can also be used for bioengineering graduate students. Furthermore, it would be very valuable as a reference book or introductory text for scientists and students in other disciplines in engineering or in the life sciences who would like to learn the fundamentals of bioengineering and to invent.

This book covers bioengineering of several body systems, organs, tissues, and cell types, integrating physiology with engineering concepts and approaches. It presents novel developments in tissue engineering, regenerative medicine, nanoscience and nanotechnology. It addresses the state-of-the-art of genomic engineering and systems biology. Furthermore, it discusses socio-economic aspects of bioengineering.

Thus, this book integrates (1) biology, medicine and engineering, (2) different levels of the biological hierarchy, (3) basic knowledge with applications, and (4) science/technology with ethics/entrepreneurism/translation. The chapters are written by authors who have outstanding accomplishments in their fields, with extensive use of diagrams and graphics. The book is organized and edited in a cohesive manner to facilitate learning and to stimulate innovation.

Shu Chien, Peter C. Y. Chen, and Y. C. Fung
Editors

ACKNOWLEDGMENTS

The Editors wish to express their sincere thanks to Rowella Garcia of the Department of Bioengineering at UCSD and Judy Blake of the Shiley Center for Orthopaedic Research and Education at The Scripps Clinic for their excellent editorial work in the preparation of this book.

SECTION I

INTRODUCTORY CHAPTERS

CHAPTER 1

OVERVIEW

Shu Chien

Abstract

As indicated in the Preface, the theme of this book is *Integrative Bioengineering*, which brings together the fundamental concepts and techniques in engineering and biomedical sciences and demonstrates their interplays. It is organized in a cohesive manner to facilitate learning and to stimulate innovation. There are eight sections that are composed of 30 chapters. Section I is introductory in nature, and it is followed by Secs. II to IV that cover the bioengineering of several organ systems, with a close relation to physiology, i.e. Cardiovascular Bioengineering, Blood Cell Bioengineering, and Respiratory-Renal Bioengineering. Sections V to VII present three new areas of developments in bioengineering, viz. Tissue Engineering & Regenerative Medicine, Nanoscience & Nanotechnology, and Genomic Engineering & Systems Biology. The last section (VIII) addresses the important socio-economical aspects of bioengineering.

Section I on **Introductory Chapters** provides an Overview of the book in this Chapter 1 and Perspectives of Biomechanics in Chapter 2.

Chapter 2. *Perspectives of Biomechanics*, by Y. C. Fung and Wei Huang: Biomechanics, which is mechanics applied to biology, covers a wide territory. In this chapter, the use of biomechanics in bioengineering is illustrated with several examples on tissue remodeling in blood vessels. These examples include the determination of zero stress state of a blood vessel from its opening angle, which changes after hypertension and following tissue remodeling, the usage of the intrinsic modes approach for the analysis continuous, the elucidation of the dynamics of tissue remodeling from long-term blood pressure recording, and the correlation of gene expression and physiological changes to understand how forces trigger gene actions for tissue remodeling. The overall objective is to demonstrate how theory and experiment can be coupled, and how design and science are linked in biomechanics.

Section II on **Cardiovascular Bioengineering** is composed of five chapters (3 to 7), which present the functions of the heart and blood vessels across the entire biological hierarchy, from molecules/cells to tissues/organs, as well as the cardiovascular system as a whole, under normal conditions and in disease states. These chapters illustrate how bioengineering approaches are used to study cardiovascular physiology and pathophysiology in health and disease.

Chapter 3. *Cardiac Electromechanics in the Healthy Heart*, by Roy Kerckhoffs and Andrew McCulloch. This chapter provides a multi-scale description of cardiac structure and electrical and mechanical functions, from the organ level to tissue and cell scales. Electrical function in the heart is intimately linked to mechanical function. Muscle contraction at the cellular level is triggered by electrical activity via the flux of calcium ions, while cell electrophysiology is modified by feedback from mechanical alterations (mechano-electric feedback). These electromechanical interactions have important medical implications, e.g. their roles in regional wall remodeling induced by clinical therapy such as chronic pacing and cardiac resynchronization therapy. Examples are given to illustrate the iterative interactions between bioengineering experiments and simulations that help in understanding cardiac function.

Chapter 4. *Cardiac Biomechanics and Disease*, by Jeffrey H. Omens: Biomechanics plays a central role in research on cardiac structure and function, and diseases of the heart are often the direct consequence of impaired mechanics. There is a need to understand the structure and function of the heart in quantifying cardiac function via experimental procedures at multi-scales in health and disease. Determinations of cardiac function range from the measurement of cardiac output in the intact system to experimental procedures on cardiac muscle cells, as well as roles of individual proteins within the cells and the extracellular matrix. It is important to determine how the heart responds to external loads and whether it can transduce the mechanical signals to provide adequate compensatory growth; the inadequacy of these responses can lead to failure of the mechanical pump.

Chapter 5. *Bioengineering Solutions for the Treatment of Heart Failure*, by John T. Watson and Shu Chien. In heart failure, the heart cannot pump a sufficient blood flow to meet the body's metabolic requirements; it can result from coronary artery disease, high blood pressure, diabetes, and several other conditions. The incidence of heart failure has steadily increased over the last 50 years and presents a major public health problem that needs a bioengineering solution. Design principles are presented for mechanical circulatory support systems, including ventricular assist devices and heart replacements or total artificial hearts. These systems have provided safe and successful alternatives in treatment for cardiac transplantation in end-stage heart failure patients. This is still a nascent field that will benefit greatly from advances in bioengineering and related fields.

Chapter 6. *Molecular Basis of Modulation of Vascular Functions by Mechanical Forces*, by Shu Chien: Mechanical forces such as shear stress can modulate cellular functions. This chapter discusses the participation of a multitude of mechano-sensors in initiating mechanotransduction by activating signaling pathways to modulate gene and protein expressions in endothelial cells, and hence their functions. Sustained laminar shear stress with a definite direction, as seen in straight part of the aorta, is protective against atherosclerosis by down-regulating the expression of chemotactic and up-regulating growth-arrest molecules. In contrast, the disturbed flow (without clear direction) observed at branch points up-regulates molecules that enhance monocyte entry, endothelial turnover, and lipid permeability, and are hence atherogenic. Coupling of mechanics and biology helps to elucidate vascular functions in health and disease.

Chapter 7. *Autoregulation of Blood Flow*, by Paul C. Johnson: Autoregulation is the tendency for blood flow to remain constant in an organ during changes in arterial pressure. It results from local regulatory mechanisms in the organ and is independent of the control by the central nervous system. This chapter examines the evidence for and against the two possible explanations, i.e. the metabolic hypothesis (with metabolic mediators as the stimulus) and the myogenic hypotheses (with intravascular pressure as the stimulus). Autoregulation may be important in clinical situations such as atherosclerosis and hemorrhage. Reviewing the development of experimental evidence for the two hypotheses provides insight to the manner in which concepts arise and are tested experimentally.

Section III on **Blood Cell Bioengineering** is composed of three chapters (8 to 10), which present the molecular basis of cell and membrane mechanics (with a focus on the red blood cell membrane and its role in blood rheology), the role of white blood cell activation and inflammation in the induction of cardiovascular diseases, and the design of blood substitutes based on bioengineering considerations of oxygen delivery and rheology of blood flow.

Chapter 8. *Molecular Basis of Cell and Membrane Mechanics*, by Lanping Amy Sung: This chapter discusses the interrelation and integration between biomechanics and molecular biology, as exemplified by the author's research on the red blood cell (RBC) membrane. The interdisciplinary work involves the cloning of cDNAs that encode RBC membrane skeletal proteins, characterization of the genomic organization, creation of gene-knockout mouse model by disrupting a target gene in an embryonic stem cell, expression of recombinant proteins and mapping their binding sites, mechanical testing of genetically engineered erythrocytes, construction of the first 3-D model for a junction complex, and establishment of the 3-D nano-mechanics of RBC membrane skeleton. This chapter provides a valuable example for the new generation of bioengineers on how to combine engineering principles and techniques with molecular genetics.

Chapter 9. *The Auto-Digestion Hypothesis for Cell Activation in the Circulation*, by Geert W. Schmid-Schönbein. Studies on a variety of acute and chronic disease conditions have identified a common feature in terms of *inflammation*, which can be triggered by many abnormal states to lead to tissue injury. Inflammation can be detected by the presence of enhanced concentrations of certain proteins in the plasma. There is evidence that the pancreas can release a digestive enzyme into the blood under low-flow condition to activate white blood cells in the blood to cause inflammation. The study of white blood cell activation and inflammation may serve as a key entry point to develop a systematic model of cardiovascular disease. This chapter provides an example of using rigorous engineering analysis to develop novel medical intervention.

Chapter 10. *Blood Substitutes and the Design of Oxygen Non-carrying and Carrying Fluids*, by Marcos Intaglietta. Analysis of oxygen transport at the level of microscopic blood vessels has resulted in the development of blood substitutes with novel properties different from blood, including a high oxygen affinity that allows the targeting of oxygen delivery to regions with low pO_2 and a high viscosity that insures the maintenance of functional capillary density in extreme anemia. The design of the blood substitutes is based on the conjugation of polyethylene glycol with molecular hemoglobin. These novel substitutes have been shown to be effective in treating conditions such as hemorrhage and extreme hemodilution, thus providing a realistic re-deployment of existing resources of human blood.

Section IV on **Respiratory-Renal Bioengineering** is composed of three chapters (11 to 13), which present the bioengineering approaches to study the pulmonary circulation, gas exchange in the lung, and functions of the kidney. These chapters emphasize the importance of biological structure, biomechanics, and transport processes in the regulation of pulmonary and renal functions in health and disease.

Chapter 11. *Analysis of Human Pulmonary Circulation with a Bioengineering Approach*, by Wei Huang, Michael R. T. Yen, and Qinlian Zhou: This chapter presents a bioengineering approach to study human pulmonary circulation based on the principles of continuum mechanics in conjunction with detailed measurements of pulmonary vascular geometry, vascular elasticity, and blood rheology. Experimental data are used to construct a mathematical model of pulsatile flow in the human lung. Input impedance of every order of pulmonary blood vessels is calculated under physiological condition, and pressure-flow relation of the whole lung is predicted theoretically. The influence of variations in vessel geometry and elasticity on impedance spectra is analyzed. The goal is to understand the detailed pulmonary blood pressure-flow relationship in the human lung for clinical application.

Chapter 12. *Pulmonary Gas Exchange*, by Peter D. Wagner: This chapter discusses a linked series of transport functions that employ both convective and diffusive movements of gases. Ventilation brings O_2 from the air to the 300 million alveoli during inspiration by convection. Gas diffusion in the alveoli leads to the elimination of CO_2 during expiration. Pulmonary blood flow, again being convective, moves the blood out of the alveolar capillaries and back to the left heart for distribution to the tissues. These three transport functions (ventilation, diffusion and perfusion) can be modeled mathematically with remarkable accuracy using simple mass conservation principles. Such interdisciplinary approaches illustrate the application of engineering principles of transport process for the understanding of respiratory functions.

Chapter 13. *Engineering Approaches to Understanding the Kidney*, by Scott C. Thomson. Engineering models have been essential for the understanding of the various functions of the kidney, including the mechanisms, control and regulation of glomerular filtration, tubular reabsorption, as well as the elaboration of a concentrated urine. Following a brief overview, this chapter discusses how engineering approaches can be used to understand these kidney functions and how they are stabilized by the internal negative feedback mechanisms. The application of the core attributes of engineering, viz. imagination, judgment, and mathematical reasoning, will remain essential to the comprehension of kidney physiology, including the explanation of the properties and functions that arise from the interacting parts in this complex organ.

Section V on **Tissue Engineering and Regenerative Medicine** is composed of five chapters (14 to 18), which present research and development of skeletal muscle tissue engineering, articular cartilage biomechanics, *in vivo* force-sensing knee prosthesis, implantable glucose sensors for diabetes, and stem cells in regenerative medicine. These chapters illustrate how engineering principles and techniques can be applied to the regeneration of tissues and organs, including the use of stem cells.

Chapter 14. *Skeletal Muscle Tissue Engineering*, by Richard L. Lieber and Samuel R. Ward: Skeletal muscle represents a classic example of structure-function relationship in the biological system. Skeletal muscle anatomy can be determined by using a combination of direct tissue dissection and magnetic resonance imaging (MRI). A major determinant of whole muscle force is the skeletal muscle architecture, including the orientation and number of fibers within the muscle. Within each muscle fiber, the arrangement of the sarcomeres is a major determinant of the muscle fiber force generated. The sarcomeres are sensitive to length and velocity. Virtually every aspect of skeletal muscle architecture and function can be studied using the modern tools of bioengineering that include imaging, mechanics, molecular biology, cell biology and bioinformatics.

Chapter 15. *Multi-Scale Biomechanics of Articular Cartilage*, by Won C. Bae and Robert L. Sah. Articular cartilage is the connective tissue that covers the ends of bones in the body, bearing and transmitting load while allowing low-friction and low-wear joint articulation. The biomechanical functions of articular cartilage have been examined at multiple length scales, ranging from intact joints to cellular and molecular components. This chapter provides an introduction to (1) the composition, structure, and function of articular cartilage, (2) biomechanical tests of articular cartilage, and (3) mathematical analysis of tissue deformation and strain. The knowledge gained from the multi-scale biomechanical studies facilitates the understanding of cartilage function in growth, aging, health and disease.

Chapter 16. *Design and Development of an In Vivo Force-Sensing Knee Prosthesis*, by Darry L. D'Lima and Peter C. Y. Chen: A total knee replacement tibial tray component with four embedded force transducers and a telemetry system has been developed to measure directly tibiofemoral compressive forces *in vivo*. After extensive laboratory testing to determine performance, accuracy and safety, trial surgical implantation has been performed to demonstrate feasibility and to assess the utility of the prosthesis as a dynamic ligament-balancing device for proper alignment and success of the operation. Knee forces are monitored during recovery and rehabilitation, as well as in daily living and exercise. These measurements are supplemented by video motion analysis, electromyography, and ground reaction force measurement to provide the information needed for the orthopedic scientific community to improve knee prosthesis design.

Chapter 17. *The Implantable Glucose Sensor in Diabetes: A Bioengineering Case Study*, by David A. Gough: Regulation of blood glucose by the body is a fundamentally important process that is impaired in diabetes, which has become increasing more prevalent as a serious health problem. Since the basis of all therapies for diabetes is the restoration of normal blood glucose control, glucose monitoring is of central importance. This chapter focuses on the recent advances made in the author's laboratory to develop implantable glucose sensor technology and the efforts to close the loop to maintain glucose homeostasis. These developments provide an excellent example for the combination of engineering principles and technology with biomedical experimentation in solving an important health problem.

Chapter 18. *Stem Cells in Regenerative Medicine*, by Shu Chien and Larry Goldstein: Stem cells can either self-renew for long periods without differentiation or can become differentiated under specific conditions into specialized cells. They have great potential to treat disease someday by regenerating the dysfunctional cells or by providing novel ways to develop either drugs or other therapies. Embryonic stem cells are pluripotent in that they can differentiate into all types of cells in any organ/tissue. Adult stem cells are multipotent in that they differentiate only into the types of cells that exist in the organ/tissue in which they reside. There is the possibiliity that adult stem cells may become pluripotent and differentiate into cells of other organs/tissues. Bioengineers can play a significant role in fostering the advance of stem cell research and its eventual clinical applications.

Section VI on **Nanoscience and Nanotechnology** is composed of four chapters (19 to 22), which present the new developments in this cutting-edge field, including the engineering of compounds targeting to vascular zip codes, use of nanoengineering approaches to study and repair the central nervous system, utilization of cellular biophotonics to solve bioengineering and medical problems, and application of microelectronic arrays for DNA hybridization diagnostics and self-assembly nanofabrication. The application of these novel developments has important medical implications in improving the diagnosis and therapy of a variety of diseases.

Chapter 19. *Engineering Compounds Targeted to Vascular Zip Codes*, by Erkki Ruoslahti: The blood vessels in different tissues carry specific molecular markers. Various disease processes, such as cancer, inflammation or atherosclerosis, express their own molecular markers on the vasculature to create a "zip code" system of vascular addresses. The vascular addresses for tissues and disease processes reside in the endothelium lining the blood vessels, and are thus readily accessible from the blood stream. The screening of phage-displayed peptide libraries *in vivo* has been an effective way of identifying vascular zip codes. Targeting a drug to these addresses can enhance the efficacy of the drug while reducing its side effects. The greatest potential of the vascular targeting technology may be the construction of smart nanodevices.

Chapter 20. *The Structure of the Central Nervous System and Nanoengineering Approaches for Studying and Repairing It*, by Gabriel A. Silva: Nanotechnologies involve materials and devices with an engineered functional organization at the nanometer scale. Applications of nanotechnology to biology and physiology provide targeted interactions at a fundamental molecular level. In neuroscience, this entails specific interactions with neurons and glial cells. Examples include technologies designed with improved interactions with neural cells, advanced molecular imaging technologies, applications of materials and hybrid molecules for neural regeneration and neuroprotection, and targeted delivery of drugs and small molecules across the blood-brain barrier.

Chapter 21. *Cellular Biophotonics*: *Laser Scissors (Ablation)*, by Michael W. Berns: This chapter examines the use of light (photons) at tissue and cellular levels with a focus on the understanding and application of light to solve bioengineering and medical problems. The presentations in this chapter include the mechanisms of photon interaction at the tissue and cellular levels, the use of light and light-sensitive photochemical agents for the diagnosis and treatment of cancer, and the use of a laser microbeam at cellular and subcellular levels. Also discussed are the use of laser microbeam to manipulate the organelles of dividing cells, the use of GFP gene-fusion proteins in facilitating visualization and targeting of subcellular structures, and the combined use of cell tracking and robotics in developing an internet-based laser microscope.

Chapter 22. *Microelectronic Arrays: Applications from DNA Hybridization Diagnostics to Directed Self-Assembly Nanofabrication*, by Michael J. Heller and Dietrich Dehlinger: This chapter describes the microelectronic array devices that have been developed for DNA hybridization analysis, clinical genotyping diagnostics, and layer-by-layer directed self-assembly of molecular and nanoparticle entities into higher order structures. Such devices can produce electric fields on their surfaces to transport or bind charged molecules and nanostructures, including DNA, RNA, peptides, proteins, nanoparticles, cells, etc. and also have the ability to carry out directed self-assembly of nanoparticles into multilayer structures, thus combining "top-down" and "bottom-up" technologies for the assembly and integration of nanocomponents into higher order structures.

Section VII on **Genomic Engineering and Systems Biology** is composed of five chapters (23 to 27), which present the state-of-the-art of these frontier fields. These chapters cover the steps involved in systems biology research, use of bioinformatics and systems biology to obtain the design principles of living systems, application of synthetic biology to study bioengineering at the genomic level, modeling the influences of genes and genomics in the context of a larger biomolecular system or network, and development of novel genomic technologies for personalized medicine. This section provides an outstanding collection of cohesive and mutually reinforcing chapters that introduce this important new field of genomic engineering and systems biology that transects life sciences, medicine and engineering.

Chapter 23. *Systems Biology: A Four-Step Process*, by Jennifer L. Reed and Bernhard O. Palsson: Systems biology focuses on the study of biological networks through the processes of (1) network reconstruction, (2) computer model formulation, (3) hypothesis generation, and (4) experimental validation. The first two steps involve identification of components and their interactions, and the result is network reconstruction: an accounting of all components and their interactions comprising the network. From this network reconstruction, an *in silico* model can be generated (step 3) to be used to predict and analyze the behavior of biological systems (step 4). The chapter concludes with a discussion of systems biology applications to address specific biological and industrial questions.

Chapter 24. *Bioinformatics and Systems Biology: Obtaining the Design Principles of Living Systems*, by Shankar Subramaniam: Because of the need for living systems to self-organize and self-evolve, the coupling between multiple time or length scales are non-hierarchical. Events that happen in seconds and minutes give rise to processes that occur in days to weeks. Processes that are traceable to a single cell can result in a systemic response spanning the entire physiology. The blueprint for continuous adaptation, error-checking and optimization of the system is built into living systems such that they sample infinite number of different states. Yet the end-point physiology is often similar or even identical. This implies that multiple solutions lead to nearly the same optimality and behavior of the system. This chapter explores the features of living systems from an engineering and design perspective and attempts to identify methods that can be used to decipher the rules that govern living systems.

Chapter 25. *Synthetic Biology: Bioengineering at the Genomic Level*, by Natalie Ostro, Mike Ferry, Scott Cookson, Tracy Johnson, and Jeff Hasty: The developing discipline of synthetic biology attempts to recreate in artificial systems the emergent properties found in natural biology. Because the genetic networks in cells are complex, redesigning simpler synthetic systems for study is a valuable approach at the genome and gene network levels. Recent activities on designing synthetic gene networks that mimic the functionality of natural systems are not only easier to construct, but also make them more amenable to tractable experimentation and mathematical modeling. The construction and testing of artificial systems resembling natural systems promise to advance our understanding of how biological systems function by providing information about cellular processes that cannot be obtained by studying intact native systems.

Chapter 26. *Network Genomics*, by Trey Ideker: Network genomics models the influence of genes and genomics in the context of a larger biomolecular system or network, which is a comprehensive collection of molecules and molecular interactions that regulate cellular function. This approach allows us to experimentally measure and define biomolecular interactions at large scale. Once all of the interactions present in a network have been catalogued, it would be possible to ask critical questions on topological structure of the network, signal transmission through the network, and evolutionary conservation of parts of the network. Most importantly, the interaction network can be used as a storehouse of information from which to extract and construct computer-based models of cellular processes in health and disease.

Chapter 27. *Genomes, Genomic Technologies and Medicine*, by Xiaohua Huang: The sequencing of the human genome was one of the greatest breakthroughs in the last century. Genome sequencing with currently available technologies, however, remains slow and expensive for routine sequencing of individual human genomes for biomedical applications. This chapter describes the author's strategies and recent progresses in engineering the next-generation technologies for genome sequencing and for digital enumerations of the molecular components in the cells. Micro- and nano-technologies are used to engineer fully automated, miniaturized "lab-on-a-chip" devices to enable massive parallel manipulations and analyses of biological molecules on an unprecedented scale so that each individual human genome can be sequenced for as little as US$1000.

Section VIII. Socio-economical Aspects of Bioengineering is composed of three chapters (28 to 30), which present several important non-technical subjects on ethics for bioengineers, bioengineering entrepreneurship, and bioengineering translation. These are very valuable topics for the bioengineers in translating their scientific knowledge to practical applications.

Chapter 28. *Ethics For Bioengineers*, by Michael Kalichman: Adherence to ethical principles is essential in bioengineering research and practice. Although news media reports on ethical violations, misconduct, etc. are typically in high profile areas such as politics, sports, and business, scientists and engineers are not immune from lapses of good judgment. The primary goal of science and engineering is to generate new knowledge for the benefit of humankind. Scientists and engineers should be particularly concerned about the integrity of their disciplines and resist the temptation for misrepresentation. The challenges are to identify the ethical dimensions of the work, to be aware of how to avoid the unethical pitfalls, to develop the skills for ethical decision-making, and to be clear about the obligation to act responsibly. Integrity is central to what it means to be an outstanding bioengineer.

Chapter 29. *Opportunities and Challenges in Bioengineering Entrepreneurship*, by Jen-Shih Lee: The contributions made by biomedical engineering entrepreneurs have improved the understanding of biomedical sciences and also health care for people. This chapter reviews some of the innovations made by the biomedical engineering industry and viewed by physicians as significance in improving the health of their patients. Two innovations, cardiac pacemakers and hemodialysis, are highlighted to elaborate on their entrepreneurial growth into billion-dollar industries. Pointers are offered for readers to evaluate the chance of success for their inventions and to gain insights on the commitment required for entrepreneurship. The chapter concludes with the heading

"Biomedical Engineers Mean Business" to encourage bioengineering students to consider entrepreneurs as their career option when opportunities arise.

Chapter 30. *How to Move Medical Devices from Bench to Bedside*, by Paul Citron: Although there are similarities in how any new technology in any industry migrates from Research and Development (R&D) to the ultimate customer, the medical device industry has certain elements that are unique to it. Understanding these factors and accommodating them as an integral part of the business development plan can make the difference between a mere laboratory curiosity and an innovation that serves the needs of seriously ill patients, while producing financial returns for the industry at the same time. This chapter addresses some of these factors that are unique to the medical device industry and how they relate to the success of the innovation process.

In summary, the 30 Chapters in eight sections of this book provide a rather comprehensive and up-to-date coverage of the current state of bioengineering, showing the rapid advances in the field and the great excitement and tremendous promise it holds for the future. There is no limit in the development of bioengineering, and it is all up to the imagination and innovation of the scientists and students.

CHAPTER 2

PERSPECTIVES OF BIOMECHANICS

Yuan-Cheng B. Fung and Wei Huang

Abstract

Biomechanics is mechanics applied to biology. It covers a very wide territory. In this chapter, the use of biomechanics in bioengineering is illustrated with a few examples on tissue remodeling in blood vessels. Our objective is to demonstrate how theory and experiment couple together, and how design and science do link in biomechanics.

1. Introduction

Biomechanics is the study of the effects of forces in a living organism. Force and motion obey Newton's law, which says that force is equal to mass × acceleration. Forces cause stress and strain. Stress and strain are related by constitutive equations. The constitutive equation of a living material is affected by tissue remodeling. Biomechanics must clarify the mass distribution in a biological system, the forces that exist, and the motion that is permitted and occurs in life. Ultimately, we try to understand the mechanics of the DNA as it affects a living organism's life, and *vice versa*.

Today, biomedical engineering has given us the following:

- better understanding of physiology,
- mathematical models, computational methods,
- artificial internal organs,
- artificial limbs, joints, and prostheses,
- better understanding of hearing, vision, speaking,
- a number of implantable materials,
- new biosensors,
- new imaging techniques: CT, NMR, PET, etc.,
- clinical devices, instruments, techniques,

- remote sensors, virtual reality,
- minimally invasive surgery techniques, and many more.

Most of the successes listed above are still limited in scope and have not achieved their full potential. Real progress awaits further research.

It is easy to see the role played by mechanics in the items listed above. Artificial hearts and heart assist devices are fluid mechanical devices. Artificial kidney is a device for mass transport. Pacemaker must be structurally sound. Implantable devices must have proper mechanical properties. Design of implantable biosensors requires fluid mechanical and mass transport analyses. Imaging and clinical devices and virtual reality, and the minimally invasive surgery techniques such as angioplasty rely on biomechanics. Other well-known contributions include sports and sports medicine, orthopedics, treatment of automobile injuries, burn injuries, and equipment for rehabilitation. Biomechanics of hearing, seeing, and speaking are extremely important subjects. Lengthening of bone to solve gait problems can be done with proper understanding of biomechanics.

To illustrate the use of biomechanics in bioengineering, let us consider a couple of examples below.

2. The Zero-Stress State of a Blood Vessel

First, let us describe a very simple experiment. In Fig. 1, an aorta is sketched. If we cut an aorta twice by cross-sections perpendicular to the longitudinal axis of the vessel, we obtain a ring. If we cut the ring radially, it will open up into a sector.[1,2] By using equations of static equilibrium, we know that the stress resultants and stress moments are zero in the open sector. Whatever stress remains in the vessel wall must be locally in equilibrium. If we cut the open sector further, and can show that no additional strain results, then we can say that the sector is in zero-stress state.

We did a simple experiment illustrated in Fig. 2 (from Fung and Liu[3]). Five consecutive segments (rings), each 1 mm long, were cut from a rat aorta. The first four segments were then cut radially and successively at the positions indicated in Fig. 1, namely, inside, outside, anterior, posterior; designated as I, O, A, P, respectively. The fifth segment was cut in all four positions, resulting in four pieces designated a, b, c, d. The open sectors of the first four rings are shown in the upper row on the right. When the four pieces of the fifth ring were reassembled in the order of abcd, bcda, cdab, bcda, with tangents matched at successive ends, we obtain four configurations shown in the lower row of Fig. 2. They resemble the shape of the four cut segments of the first row quite well. This tells us that the arterial wall is not axisymmetric, that different parts of the circumference are different, and that one cut is almost as good as four cuts in relieving the residual stress. Hence, we may say that one cut of the ring reduces the ring into zero-strain state within the first order of infinitesimals.

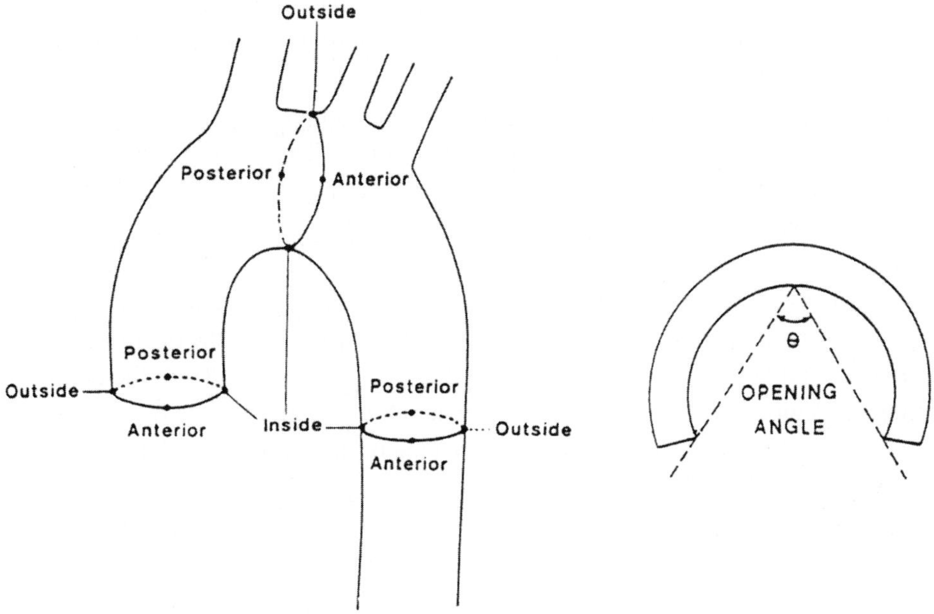

Fig. 1. Sketch of an aorta with an indication of the cutting positions. Right: schematic cross-section of a cut vessel segment at zero-stress, defining the opening angle θ.

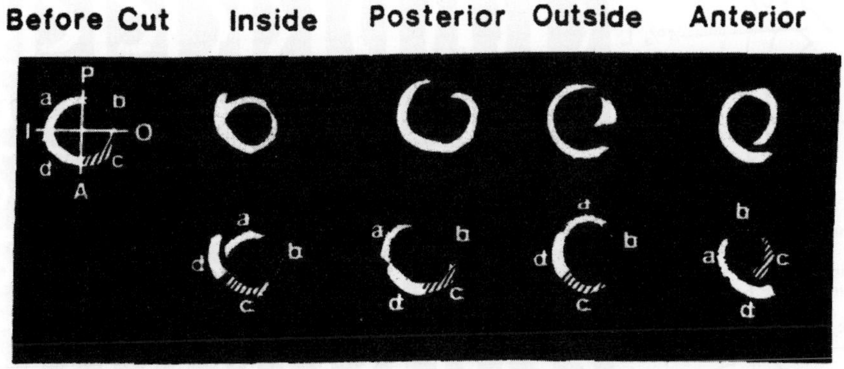

Fig. 2. The figures in the upper row show an arterial segment of a rat before cut and after cutting at four positions. The lower row shows the same vessel cut into four pieces and reassembled in four ways. It appears that one cut is sufficient to reduce an arterial segment at no-load to the zero-stress state.[3]

Having been assured that the open sector represents zero-stress state of a blood vessel, we conclude that the zero-stress state of an artery is not a tube. It is a series of open sectors. To characterize the open sectors, we define an *opening angle* as the angle subtended by two radii drawn from the midpoint of the inner wall (endothelium) to the tips of the inner wall of the open sections (see Fig. 1).

Although the opening angle is a convenient simple measure of the zero-stress state of an artery, it is not a unique measure of the artery, because its value depends on where the cut was made. This is clearly illustrated in Fig. 2. The four segments were not identical. The blood vessel shown in Fig. 2 was not uniform circumferentially. Hence in stating an opening angle, one must explain where the cut was made. This requirement is a limitation to the usefulness of the opening angle as a simple measure of a complex phenomenon.

The photographs in the first column of Fig. 3 show a more complete picture of the zero-stress state of a normal young rat aorta.[3] The entire aorta was cut successively into many segments of approximately one diameter long. Each segment was then cut radially at the "outside" position indicated in Fig. 1. It was found that the opening angle varied along the rat aorta: it was about 160° in the ascending aorta, 90° in the arch, 60° in the thoracic region, 5° at the diaphragm level, and 80° toward the iliac bifurcation point.

Following the common iliac artery down a leg of the rat, we found that the opening angle was in the 100° level in the iliac artery, dropped down in the popliteal artery region to 50°, then rose again to the 100° level in the tibial artery. In the medial plantar artery of the rat, the micro arterial vessel 50 μm diameter had an opening angle of the order of 100°.[4]

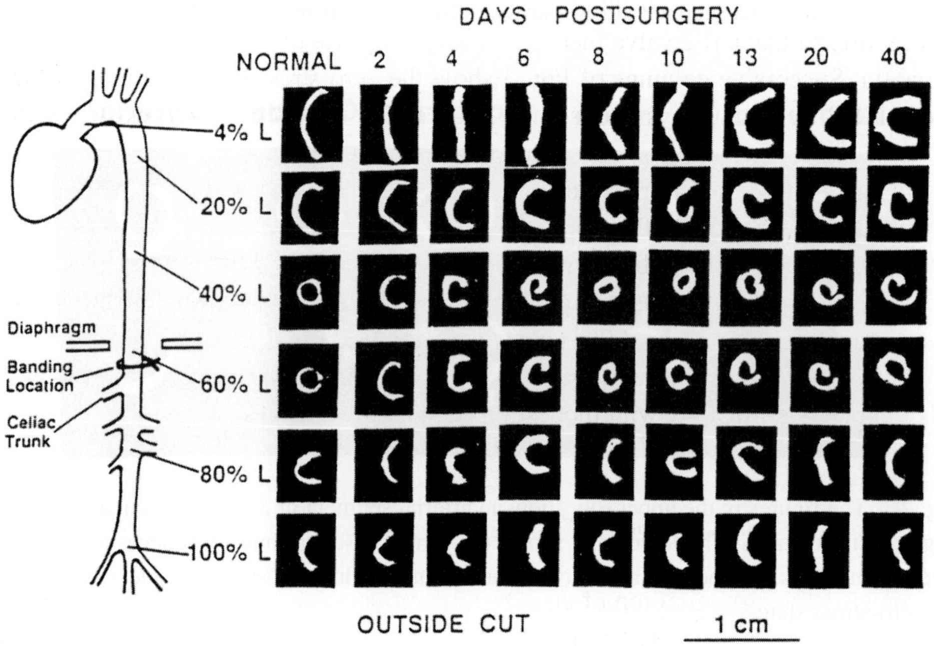

Fig. 3. Photographs of the cross-sections of a rat aorta cut along "outside" line shown in Fig. 1. The first column shows zero-stress state of normal aorta. Other columns show changed zero-stress states a number of days after a sudden onset of hypertension. Successive rows correlate with locations on the aorta expressed in percentage of total length of aorta, L, from the aortic valve. The aortic cross-sectional area was clamped 97% by a metal band below the diaphragm to induce the hypertension.[3]

There are similar spatial variations of opening angles in the aorta of the pig and in pulmonary arteries and veins and trachea of the dog.

3. Hypertension Causes Changes of the Opening Angle of Aorta

We created hypertension in rats by constricting the abdominal aorta with a metal clip placed right above the celiac trunk.[3] The clip severely constricted the aorta locally, reducing the cross-sectional area of the lumen by 97%, with only about 3% of the normal area remaining. This causes a 20% step-increase of blood pressure in the upper body, and a 55% step-decrease of blood pressure in the lower body immediately following the surgery. Later, the blood pressure increased gradually, following a time course shown in Fig. 4.

It is seen that in the upper body the blood pressure rose rapidly at first, then more gradually, tending to an asymptote at about 75% above normal. In the lower body, the blood pressure rose to normal value in about four days, then gradually increased further to an asymptotic value of 25% above normal. Parallel with these changes of blood pressure, the zero-stress state of the aorta changed. The changes are illustrated in Fig. 3 in which the location of any section on the aorta is indicated by the percentage distance of that section to the aortic valve measured along the aorta divided by the total length of the aorta. Successive columns of Fig. 3 show the zero-stress configurations of the rat aorta at 0, 2, 4, … , 40 days after surgery. Successive rows refer to successive locations on the aorta.

Figure 5 shows the course of change of the opening angle of the rat aorta in greater detail. Figures 3 and 5 together show that the blood vessel changed its opening angle in a few days following the blood pressure change. We found similar changes in pulmonary arteries after the onset of pulmonary hypertension by exposing rats to hypoxic gas containing 10% oxygen, 90% nitrogen, at atmospheric pressure.[5]

Since opening angle changes reflect structural changes, we conclude that blood vessels remodel significantly with modest blood pressure changes.

4. What Does the Change of Opening Angle Mean?

The open sector configuration of an artery at zero-stress looks like a curved beam and mechanically can be analyzed as a curved beam (see Fig. 1). A beam can change its curvature only if one side of the beam lengthens while the other side of the beam shortens. If the opening angle increases due to tissue remodeling, then the endothelial side of the blood vessel wall must have an increase in circumferential strain, while the adventitial side of the blood vessel wall must have a decrease of circumferential strain (see Fig. 6).[6] Since these increases and decreases are not due to external loads, but are due directly to growth and resorption of the tissue in remodeling, we can conclude without much ado

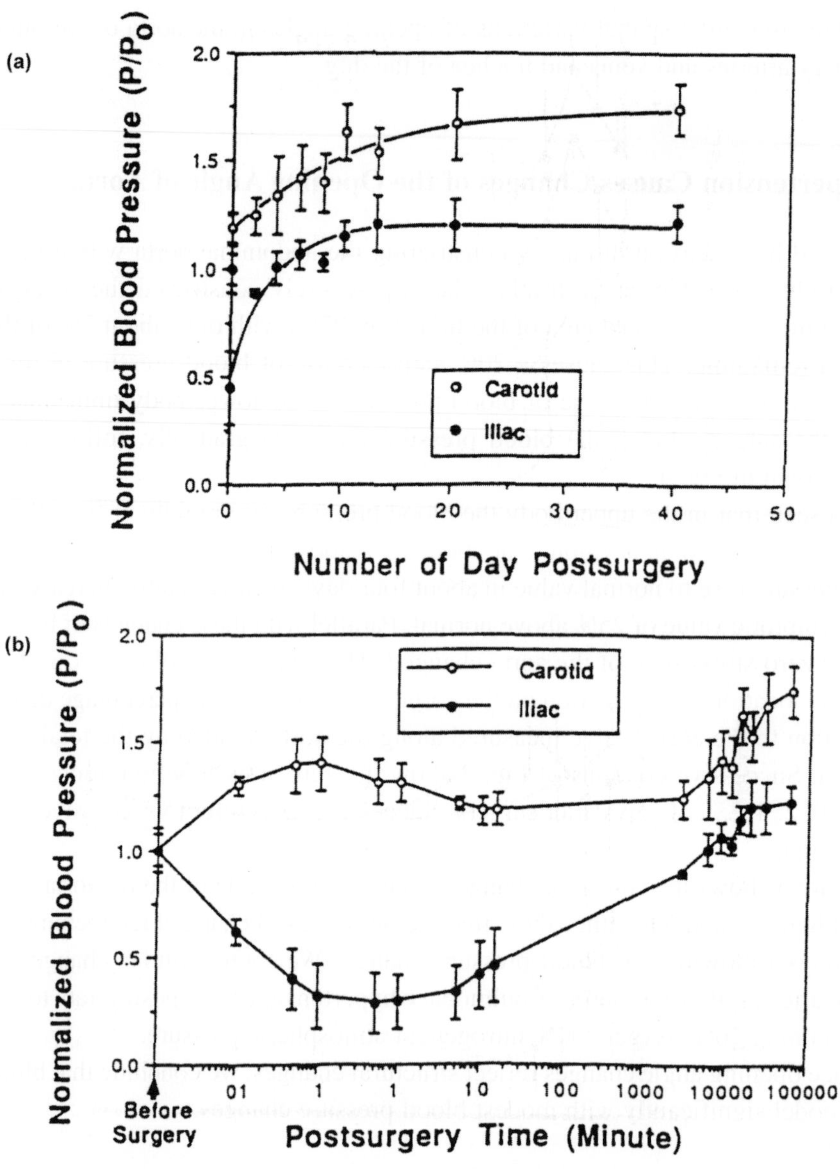

Fig. 4. The course of change of blood pressure (normalized with respect to that before surgery) after banding the aorta.[3]

that the change of opening angle of blood vessels due to change of blood pressure is due to *nonuniform* remodeling throughout the thickness of the vessel wall.

From the point of view of studying tissue remodeling, the zero-stress state is significant because it reveals the configuration of the vessel in the most basic way, without being complicated by elastic deformation. If cellular or extracellular growth or resorption occurs in the blood vessel due to any physical, chemical and biological stimuli, they will be revealed by the change of zero-stress state.

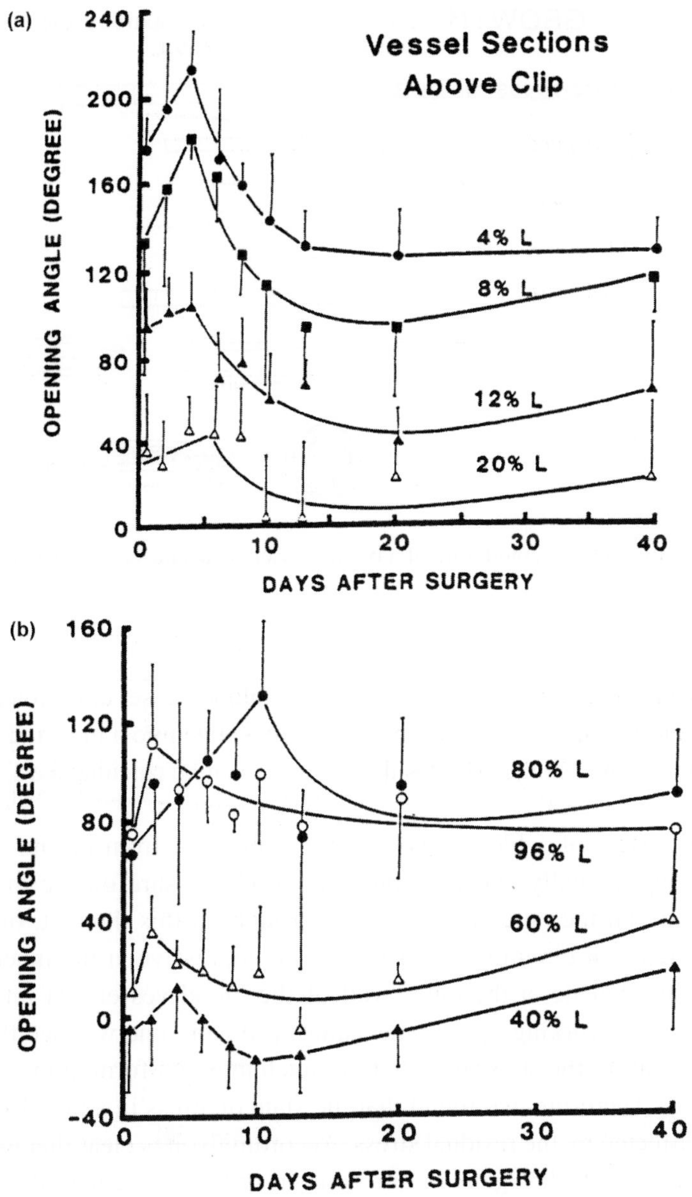

Fig. 5. The course of change of the opening angle of the rat aorta at the zero-stress state following aortic banding to change blood pressure. L is the length of aorta. $\%L$ indicates location of section from aortic valve. **(a)** Locations above the constriction. **(b)** Locations below the constriction.[3]

5. What is the Residual Stress Doing There?

The state of a body on which all the external loads are removed is called a *no-load state*. The internal stress existing in a body at the no-load state is called the *residual stress*. For a blood vessel, if there is no longitudinal tension and no transmural pressure, then

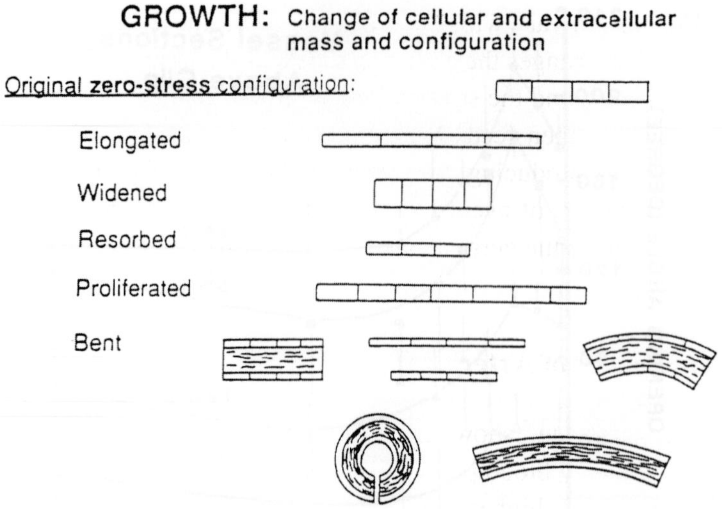

Fig. 6. Illustration that the remodeling of a blood vessel is best described by change of its zero-stress state.[6]

it is at a no-load state. From Fig. 6, we see that a blood vessel at no-load state can be obtained by bending the vessel wall in its zero-stress state into a tube and then welding the edges into a seamless tube. The residual stresses can be calculated according to this point of view.

The *in vivo* state of a blood vessel can be obtained from the no-load state by stretching it longitudinally and then put the blood pressure on the inner wall and external pressure on the outer wall. Follow through on this thought, one can easily show that the residual circumferential stress is compressive at the inner wall of the blood vessel, and tensile at the outer wall of the blood vessel. On introducing the blood pressure, and working out the mechanics problem, one will see that the residual stresses will make the stress distribution much more uniform in the vessel wall at the *in vivo* state. Therefore, we found that the state of stress in a blood vessel *in vivo* is very much affected by the residual stress. Accordingly, it is clear that we must know the zero-stress state of all organs in our body in order to calculate the stress in our body *in vivo*. Thus, the very simple experiment illustrated in Fig. 1 is indeed fundamental and far-reaching.

6. Tissue Remodeling Revealed by Change of Zero-Stress State

The stress and strain in our body change normally or pathologically. In principle, the reason for these changes is very simple. By molecular mechanism, living cells can make proteins to enlarge themselves or build up intercellular matrix. Hence, new materials are made from a molecular pool according to the laws of molecular biomechanics. Hence, the mass of the tissue can vary with time. It follows that the structure of the tissue can change with time. This changes the zero-stress state of the tissue. The stress-strain laws of the tissue will change, and the stress distribution will change.[7,8]

The changing stress then feeds back onto the cells, causing them to react to the changing stress field by producing new materials, or move, or proliferate by cell division. This logical sequence of events has been demonstrated. In the following, a brief sketch is given to indicate the current status of our knowledge.

7. Tissue Remodeling of Arteries Under Stress

Figure 7, from Fung and Liu,[5] shows the history of tissue remodeling of rat pulmonary artery when the pulmonary blood pressure was raised from the normal value by breathing normal sea level air to a higher value obtaining by breathing a gas with 10% O_2 and 90% N_2 at atmospheric pressure. The left hand side figure is a sketch of the pulmonary artery of the left lung. The first row shows the morphological changes of the cross-section of the arterial wall in the arch region. It is seen that significant changes occurred already in two hours. The intima changed first, followed by the media in which the vascular smooth muscle resides. The adventitia changed slower. The succeeding rows show the changes in the vessel walls of smaller arteries.

Figure 8, from Liu and Fung,[9] shows how the stress-strain relationship of rat thoracic aorta is changed by tissue remodeling during the development of diabetes following an

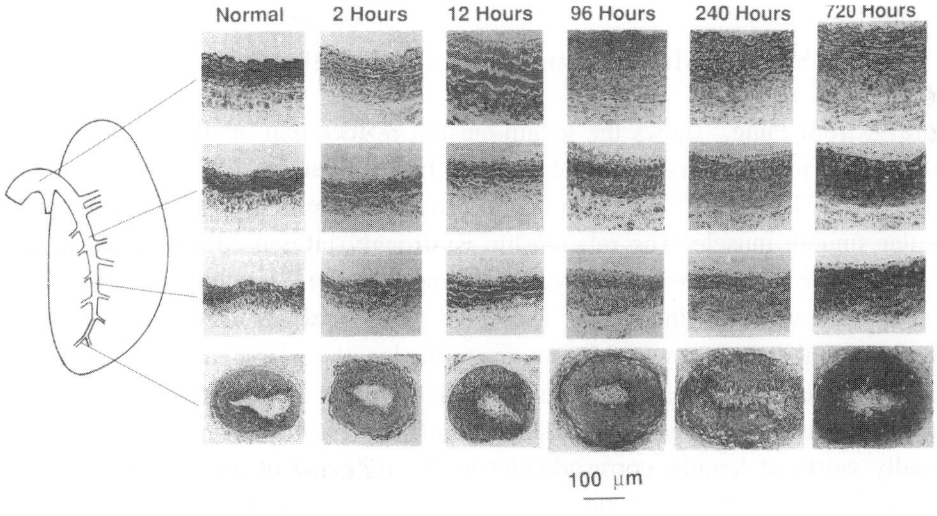

Fig. 7. Remodeling of rat pulmonary arteries when the animal is subjected to pulmonary hypoxic hypertension by breathing hypoxic gas to the length of time shown in the figure. Photographs of histological slides from four regions of main pulmonary artery of a normal rat and hypertensive rats with different periods of hypoxia. Specimens were fixed at no-load condition.[5]

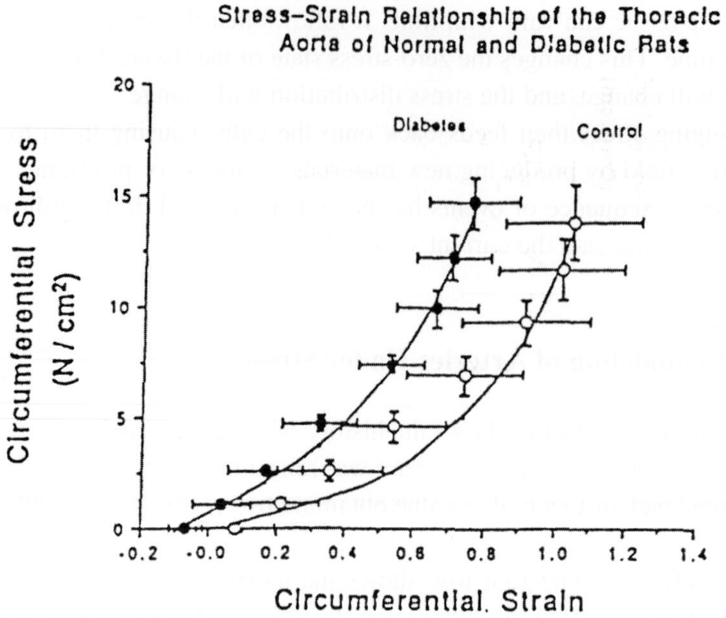

Fig. 8. Change of the stress-strain relationship of rat thoracic aorta during the development of diabetes 20 days after an injection of AZT.[9]

injection of streptozocin. The general trend remains the same, but the elastic constants at any given strains are changed.

8. Vascular Smooth Muscle Length-Tension Relationship

Figure 9, from Fung,[6] shows the length-tension relationship of the vascular smooth muscle in the pig coronary arterioles deduced by the present author[6] from the experimental data of Kuo *et al.*[10–12] This is the homeostatic length-tension relationship of the vascular smooth muscle. The relationship is strongly influenced by the shear stress imposed on the blood vessel wall by the flowing blood, τ. The upper panel shows the relationship in vessels without flow. The lower panel shows the effect of flow-induced shear, τ. Note the similarity of these curves to the length-tension curve of the tetanized skeletal muscle, and to the curve of muscle length versus the peak tension in isometric twitch of heart muscle. The major distinction of the vascular smooth muscle is that it normally works at lengths corresponding to the right leg of an arch shaped curve, whereas the heart and skeletal muscles normally work at lengths corresponding to the left leg of the arch shaped curve.

Knowing the curves of Fig. 9, we can now understand the autoregulation of the blood flow; and hyperemia, the increased blood flow in a vessel upon release from compression; and the Bayliss phenomena. Autoregulation is the tendency for blood

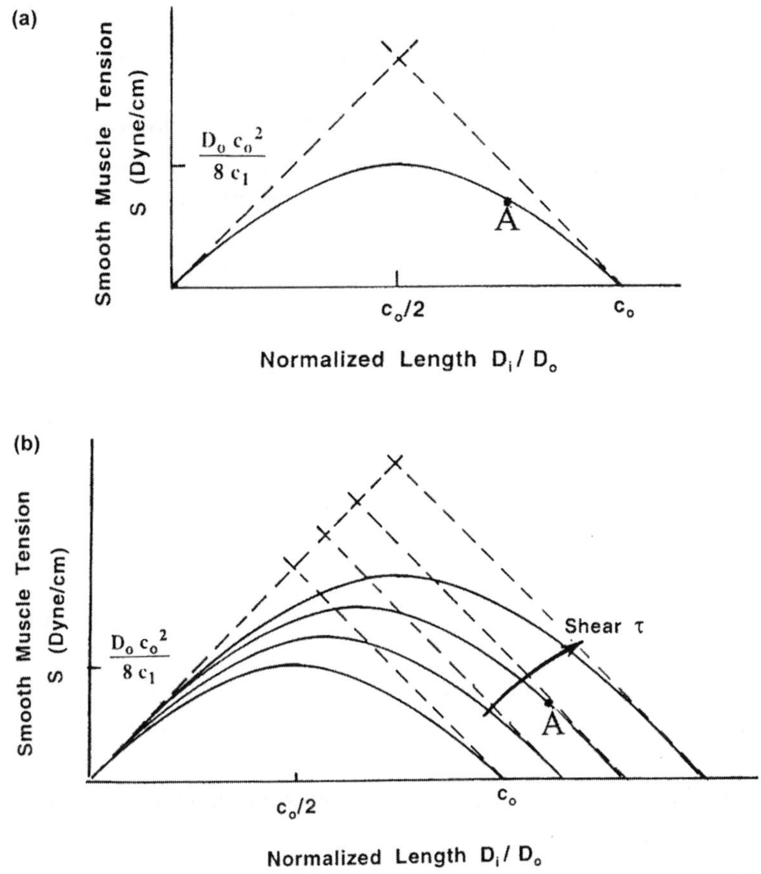

Fig. 9. **(a)** Relationship between the vascular smooth muscle tension and length derived in Ref. 5 from the experimental results given in Refs. 10 to 12. **(b)** The influence of shear stress τ acting on the endothelium.[6]

flow to remain constant in the face of changes in arterial pressure to the organ. It is seen in virtually all organs of the body. It is most pronounced in the brain and kidney. (See Johnson[13] for detailed discussion and literature.) Each organ has a steady state flow-and-pressure relationship. Suppose in an autoregulated region, say at a point A in Fig. 9, you made a sudden increase of arterial blood pressure. By the elasticity of the artery the arterial diameter will be increased in response to the sudden increase of pressure, thereby, the resistance to blood flow is transiently reduced and the flow increases. Then the tension in the vascular smooth muscle is increased to a new value above that of the point A in Fig. 9. The curve in Fig. 9 shows that at equilibrium, such an upward movement of tension will have a muscle length that is shorter than that of point A. The smooth muscle shortens to reach the new equilibrium state. The vessel diameter is reduced, the flow resistance increases, and the flow falls back toward the normal value. This is autoregulation, which used to be considered mysterious. With Fig. 9, we see it as a revelation of the basic feature of the smooth muscle property, and the fact that the normal working condition of the artery corresponds to a point lying on the right leg of the arch-like curve of Fig. 9.

Reactive hyperemia is the period of elevated flow that follows a period of circulatory arrest, e.g. by inflating a cuff on an arm as we use a manometer to measure our blood pressure. On releasing the cuff pressure, the blood flow burst forth far above the normal value. (See Johnson[13] for details.) The explanation based on Fig. 9 is that under compression the tension in the smooth muscle is reduced: the condition is represented by a point lower than the normal condition of point A. Hence, the smooth muscle lengthened and the vessel circumference enlarged while the vessel was compressed. Upon release of the compression and the return of the blood flow, the enlarged blood vessel causes a large flow. Later, the muscle length adjustments take place to return the flow to the normal condition. Reactive hyperemia is often used to detect if a vessel has an arteriosclerotic plaque, because the plaque abolishes reactive hyperemia.

9. Analysis of Blood Pressure Signals for Physiology and Medicine

Blood pressure reading is an important physiological indicator in medicine. In clinical practice today, the pumped-cuff-Korotkoff sound method of determining blood pressure is convenient, but it is not continuous, and is a reference number only, not precise enough for computational analysis. For detailed information, physicians and researchers measure blood pressure with a dwelling catheter.

Figure 10a shows a continuous record of the pulmonary arterial blood pressure in a free-ranging rat with a dwelling catheter in our laboratory. The dwelling catheter was

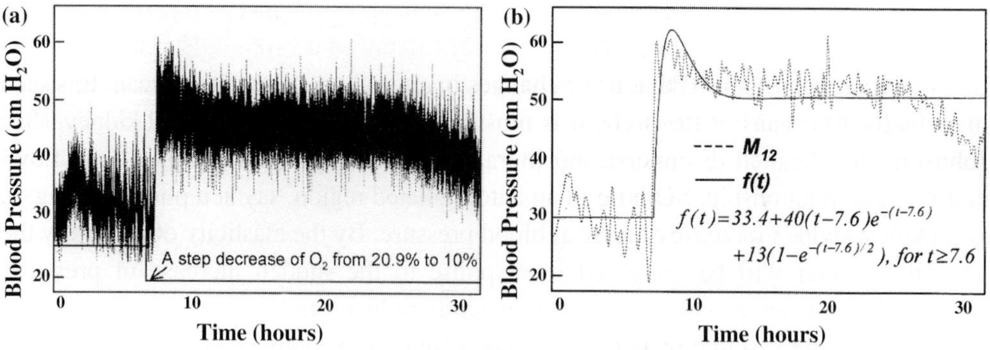

Fig. 10. **(a)** A continuous record of the pulmonary arterial blood pressure of a rat subjected to step lowering of oxygen concentration in breathing gas. **(b)** A typical mean trend of the blood pressure and its best-fit formula $f(t)$.[19] As it is explained in Huang *et al.*, the IMF method, means of various orders can be defined rigorously, the ones with the higher order have fewer zero crossings. In this figure, the order of the mean is 12.[19]

implanted in the pulmonary arterial trunk through the right jugular vein before pressure measurement. After recovering from anesthesia, the rat was placed in a hypoxic chamber. After recovery from the anesthesia, the rat was put in the hypoxic chamber for a rest of six hours in the normal lab air. The oxygen concentration of the breathing gas in the chamber was reduced to 10% in 1.5 ± 0.5 minute, and then the oxygen level was maintained at 10% for 24 hours.

After 24 hours, the oxygen tension in the chamber was quickly returned to normal by opening the door of the chamber. Pressure data were collected during these 30-hour period, which include the changeovers from 20.9% to 10% oxygen level and *vice versa*. The pressure was recorded continuously by a computer at a sampling rate of 100 points/second over a 30-hour period with a time lag of two seconds in every 60 seconds for computer processing. As shown in Fig. 10a, the blood pressure is a variable with nonstationary, nonlinear, and stochastic features. How complex are the pressure records! How do we understand these records?

Looking at the actual recording as shown in Fig. 10a, we find that the task is not as easy as it appears. Well known methods are associated with such mathematical terms are Fourier analysis, Fourier spectrum, Hilbert spectrum, wavelets, etc. We found that the intrinsic modes approach of Huang *et al.*[14] most appropriate. A software package is developed for the practical computation by Huang *et al.*[14] The software can analyze the digital signals into intrinsic modes of oscillations, each mode has the character that the local oscillations about the modal curve has an average of zero. In general, any given signal of nature has a finite number of modes: successive modes have fewer and fewer zero crossings; the last mode has no zero-crossing at all, it represents a trend. In our study, a pulmonary arterial pulse pressure record has typically ten to 14 intrinsic modes. Each complete record is resolved into a complete set of intrinsic modes. Thus we can define an average signal, or mean signal as the sum of a certain number of last intrinsic modes depending on an arbitrarily chosen number of zero-crossings you may wish to allow. The total signal is the sum of the mean signal and a signal of oscillations about the mean. Figure 10b illustrates the process. The details are given in Huang *et al.*[15–17]

We have demonstrated that it is possible to get a much better understanding of the mean signals and oscillations about the mean signals and arrythmia by the intrinsic mode approach.[15–17] Such understanding of signals will allow us to study tissue remodeling as a dynamic process. Our objective is to distinguish tissue remodeling in response to slowly varying mean stresses, from that see response to oscillations about the mean.

10. Correlation Between the Trends of Gene Expressions and Physiological Changes

Blood pressure produces stresses and strains in the cells of the blood vessel wall. Vascular cells, namely, vascular endothelial cells, vascular smooth muscle cells, and fibroblasts, are subjected to the stress and strain system induced by the blood pressure, and are able to convert mechanical stimuli into intracellular signals that provoke genes into action to produce mRNAs which make proteins.

To study the relationship between the action of the genes and the physiological changes in tissue remodeling, we measured also the activities of the genes in the cells of pulmonary arterial wall by a *cDNA microarray with colorimetric detection method*,[18] while determining the indicial response functions of tissue remodeling of the pulmonary arteries in response to hypoxic hypertension.[19–21] We used the same hypoxic hypertension rat model described in Sec. 9, took the arterial specimens at scheduled times, extracted the total RNA in each specimen.[19–21] Then the mRNA was isolated, polymerized cDNA, colored with biotin, and prepared for hybridization. The detailed procedures and commercial kits were described in Refs. 22 to 24. The solution of each specimen containing all the **targets**, which are the genes in each of these processed specimen solutions, was used to measure gene expression in a microarray test equipment designed and constructed by Konen Peck.[18] In Peck's array, 9600 selected genes were deposited as ***probes*** on a nylon membrane in a rectangular matrix pattern. They were PCR products of human cDNA clones rearrayed from the *IMAGE Consortium cDNA libraries*[25] based on the Unigene clustering.[26] Hybridization of the probes and targets occurred when the specimen solution was spread over the probe array membrane. Colorimetry was used to scan genes on the nylon membrane after hybridization. The changes of gene expression were expressed in terms of the changes of the intensity of color. The colorimetric readings were normalized against sample size by dividing each reading with the sum of all significant readings of that sample. The normalized readings were considered as measures of gene expression. The ratio of the gene expression in active state divided by that of the *in vivo* state minus one is defined as the **action of the gene**.

The correlation coefficients of the gene actions and the physiological changes can be computed. The definition of the correlation coefficients is as follows: Let $y_j(t)$, $j = 1, 2, 3, \ldots$, represent the history of the jth physiological feature, at a series of instants of time t_m, $m = 1, 2, \ldots$. Let $z_k(t)$ be the action of the gene number k, $k = 1, 2, 3, \ldots$ at time t_m, $m = 1, 2, \ldots$. Then we define the correlation coefficient by the following formula:

$$R(y_j, z_k) = \left[\sum y_j(t_m) z_k(t_m) \Delta t_m\right] \left[\sum y_j^2(t_m) \Delta t_m \cdot \sum z_k^2(t_m) \Delta t_m \right. \tag{1}$$

where t_m, with $m = 1, 2, \ldots$, are the instants of time at which y_j, z_k are measured. Here Δt_m are the spacing of time intervals, $\Delta t_m = t_{m+1} - t_m$.

We computed the changes of blood pressure, zero-stress state, arterial vessel wall thickness, Young's elastic moduli of pulmonary arteries from the *in vivo* value, and the correlation coefficients with the action of every gene whose expression reading was significant. We lined up the genes in a row according to the size of the correlation coefficient. In a row, there is the first one, we call it the top gene. Plotting the action of the top gene and the physiological features together, we obtained the curves shown in Fig. 11.

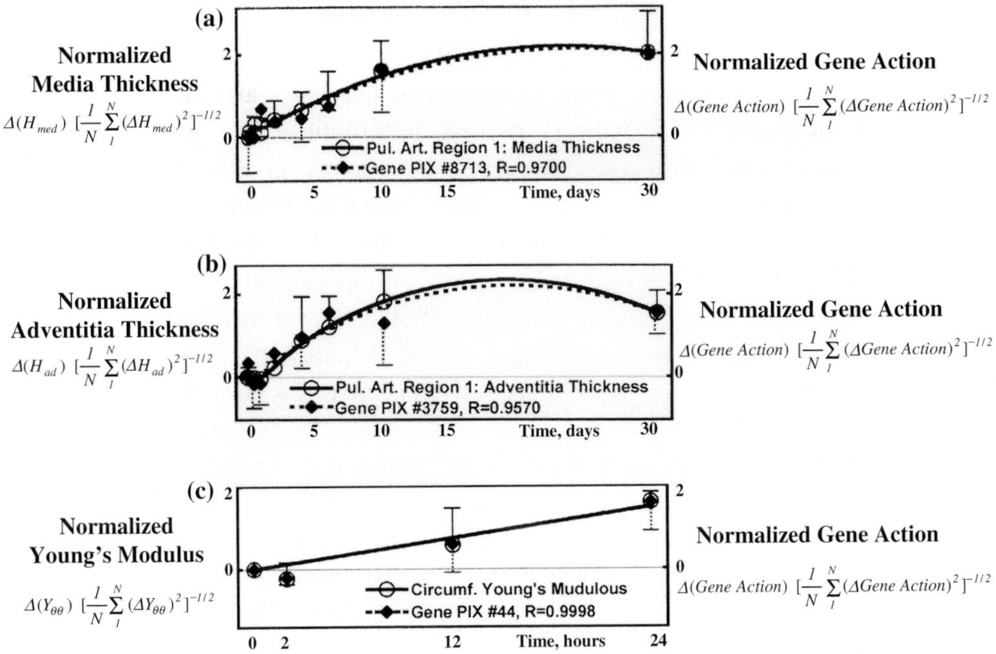

Fig. 11. Matching gene activities with physiological changes. **(a)** Action of gene PIX #8713 (*inorganic pyrophosphatase gene*), vs. the change of media thickness (H_{med}) of pulmonary arterial trunk. **(b)** Action of gene PIX #3759 (*osteoblast-specific factor 2, fasciclin I-like gene*) vs. the change of adventitia thickness (H_{ad}) of pulmonary arterial trunk. **(c)** Action of gene PIX #44 (*dynein, cytoplasmic light polypeptide gene*), vs. the change of Young's modulus of elasticity of the pulmonary artery, $Y_{\theta\theta}$, relating circumferential stress and strain. Symbols: small circle, O, normalized physiological changes, 1 SEM flag up; ♦, normalized gene action, 1 SEM flag down.[21]

The story told by these figures is that the genes work all the time. The physiological changes have roots in the genes' activity, and *vice versa*. However, how forces trigger the gene actions for tissue remodeling is barely known. It is our mission to find out this unknown.

11. Tensorial Description of the Fibrous Proteins in Tissue Remodeling of Blood Vessels

Blood vessels are constantly subjected to blood pressure, and remodel in response to changes of blood pressure in health and disease. In vascular remodeling, the changes of proteins in the vascular cells are related to the stresses and strains in the cells in the vessel wall. If the relationship between the changes of proteins in the cells and the stress and strain in the cells can be described by a valid constitutive equation, then every term

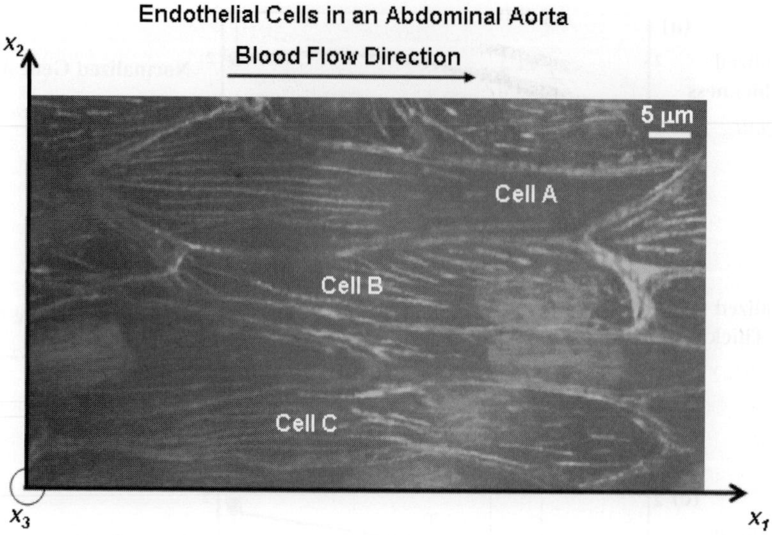

Fig. 12. A plan view of the endothelial cells in a rat's abdominal aorta in a plane perpendicular to the endothelium (x_3 axis). The F-actin fibers and nuclei in the cells were stained. The thick bright lines are the boundaries of the endothelial cells. The thin lines inside the cells are the F-actins. The F-actin fibers appear complex. Three projected cells are enclosed with cell boundaries, and labeled as Cell A, Cell B, and Cell C. A Cartesian coordinate system with x_1-axis parallel to the direction of blood flow, x_3-axis perpendicular to the basal lamina, and x_2-axis normal to x_1 and x_3 is used as references to label the endothelium.[27]

in the equation must be tensors of the same rank. Since stress and strain are tensors, the protein changes must also be stated in tensor language. We find it useful to define a second order ***molecular configuration tensor*** P_{ij}^{g} for fibrous proteins in tissue remodeling of blood vessels to be used in such equations, where i, j refers to coordinate axes x_1, x_2, x_3, and g refers to the kind of fibrous proteins.[27]

Figure 12 is a photograph to show the complex distributions of phalloidin-stained F-actin fibers in the endothelial cells of a rat's abdominal aorta. The thick bright lines are the boundaries of the endothelial cells. The thin lines inside the cells are the F-actins. Three projected cells are enclosed with cell boundaries, and labeled as Cell A, Cell B, and Cell C. These actin fibers have influence on the stress and strain in the artery, which in turn affects the blood flow and physiology of the entire organ and the whole animal. The actin fibers remodel when the blood vessel remodels under stresses. Therefore they have influence on the constitutive equation of the blood vessel wall. Many forms of constitutive equations have been presented in biomechanics. However, so far, none of the existing constitutive equations pays special attention to the complex features of fibrous proteins as shown in Fig. 12.

By the central dogma of molecular biology, DNA makes RNA which makes protein. If the newly synthesized proteins can be stained and photographed, then we know their geometry in three dimensions. Quantization of the geometry of these protein fibers can be done as illustrated in Fig. 13.

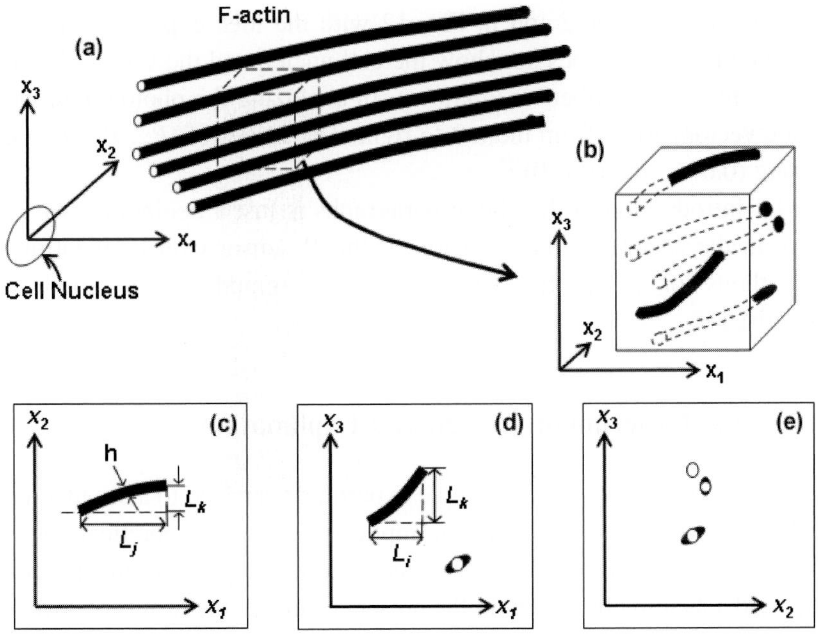

Fig. 13. A scheme to describe quantitatively what we see in Fig. 12.[27]

Figure 13 illustrates the steps taken by microscopists to describe the geometry of the protein fibers in tissues.[27] Assume that the fibers are as shown in Fig. 13a. We consider a unit cube as shown in Fig. 13b, superpose it on the fibers of Fig. 13a, and photograph the fibers cut by the six faces of the cube. We use a rectangular Cartesian frame of reference with coordinates (x_1, x_2, x_3), i.e. x_i, with $i = 1, 2$, or 3. A plane normal to the x_i axis cuts the cells in the x_j-x_k plane, $j, k = 1, 2$, or 3. Considering a fiber which appears as a circular spot in the x_j-x_k plane is interpreted as perpendicular to the x_j-x_k plane, i.e. lying in the direction of the x_i axis. A spot that is elliptical is interpreted as a fiber that cuts the x_j-x_k plane at an angle: in that case we count a circle with radius equal to the minor axis of the ellipse as the normal component. In the x_j-x_k plane shown in Figs. 13c to 13e, we resolve the length of a fiber into rectangular components L_j, L_k. Let the thickness of the fiber be h. The diameter of the dots is also h. For an elongated dot, we treat a circle of diameter h as a dot and the length of (major – minor) diameter as L_j, L_k. On summing over all segments in an area A (of the rectangle of the figures), we can compute the P_{ij}, P_{ik}, and P_{ii}. If these numbers were known for all $i, j, k = 1, 2, 3$, we may write the following matrix as P_{ij}^g:

$$\begin{bmatrix} P_{11} & P_{12} & P_{13} \\ P_{21} & P_{22} & P_{23} \\ P_{31} & P_{32} & P_{33} \end{bmatrix} = P_{ij}^g . \tag{2}$$

By comparing the photograph in Fig. 12 with the idea expressed in Fig. 13, we notice that the photograph does not show the cell nuclei, and the vectorial sense of the mRNA which made these fibers is unknown. In analyzing the photograph, we assume all fibers are vectors going from the left to right. The values of (P_{31}, P_{32}, P_{33}) in Cell A in Fig. 12 are (0.074, −0.0017, 0).[27]

The brief introduction in the above paragraphs is just a beginning. We anticipate an increase in the accuracy and precision of the P^g_{ij} tensor over conventional image analysis methods for quantitative studies on tissue remodeling of blood vessels in health and disease.

12. Theory and Experiment, Design and Explanation

The example above illustrates how a simple experiment can lead to a broader and deeper theoretical investigation with fundamental implications. These theoretical investigations lead to new experiments; new experiments lead to new theory. Hopefully, the spiral leads to a greater understanding of nature. In the present example, the results of several rounds of theory and experiment have led to the concept of tissue engineering, because if tissues remodel under stress, then the stress can be used as a tool to control or "engineer" the tissue. We believe that this concept has an impact on medical advances.

The simple experiment shown in Fig. 1 was done in 1982 because new papers appeared which showed that there is strong stress concentration in heart and blood vessels due to internal pressure. These stress concentrations were demonstrated by the theory of linear elasticity. When people improved the theory with the nonlinear stress-strain law, and taking the nonlinear finite strains-deformation gradient relationship into account, the stress concentrations did not decrease. Not only did they not decrease, they became much worse. On the other hand, there is no biological reason why such stress concentration should be there. In this situation, we thought the fault might lie in the assumption that the heart and blood vessels are stress free when the pressure loading is removed. To test this suspicion, we cut up the specimen to see if there were residual strains. There were, as the evidence presented in Fung[1] showed. Paul Patitucci, a graduate student at that time, made lots of contributions. Almost simultaneously, without us knowing it, Vaishnav and Vossoughi made similar investigations and obtained the same results.[28,29] Following this, other important papers followed. We have listed most of them in the list of References at the end of this chapter.[1–9,22–24,30–55] New investigations have turned to cellular and molecular details to explain the macroscopic observations. The better we know how our body works, the better we feel.

13. Conclusion

What impact does this kind of biomechanics have on the classical subject of continuum mechanics? The impact lies in calling for vigorous renewal. Biomechanics differs from other branches of applied mechanics by (1) non-conservation of mass, (2) non-conservation of structure, (3) nonlinearity of the equations describing the mechanical properties, (4) the stress-strain relationship changes with the changing state of stress or disease, (5) the stress-growth laws define the outcome of the external or internal loading, and (6) by probing more deeply into the basic reasons of the features named above, one is led to the study of biomechanics of the cells and biomolecules. One can expect a vigorous new development of applied mechanics to answer this calling!

In this brief introduction, we refrain from describing the contributions made by biomechanics to the broad field of mechanics. Classical fluid and solid mechanics are based on Newtonian viscosity and Hooke's law. Biomechanics deal with Non-Newtonian fluids and constitutive equations that reflect the molecular remodeling of living cells under stress. The molecular remodeling is governed by DNA, and is done by a sequence of genes. The mechanics of gene action lies at the focus of bioengineering.

References

1. Y. C. Fung, What principle governs the stress distribution in living organs? In: *Biomechanics in China, Japan, and USA* (Science Press, Beijing, 1983), pp. 1–13.
2. S. Q. Liu and Y. C. Fung, Relationship between hypertension, hypertrophy, and opening angle of zero-stress state of arteries following aortic constriction, *J. Biomech. Eng.* **111**: 325–335 (1989).
3. Y. C. Fung and S. Q. Liu, Change of residual strains in arteries due to hypertrophy caused by aortic constriction, *Circ. Res.* **65**: 1340–1349 (1989).
4. Y. C. Fung and S. Q. Liu, Strain distribution in small blood vessels with zero-stress state taken into consideration, *Am. J. Physiol.* **262**: H544–552 (1992).
5. Y. C. Fung and S. Q. Liu, Changes of zero-stress state of rat pulmonary arteries in hypoxic hypertension, *J. Appl. Physiol.* **70**: 2455–2470 (1991).
6. Y. C. Fung, *Biodynamics: Circulation*, 2nd ed. (Springer-Verlag, New York, 1984) changed title to *Biomechanics: Circulation* (Springer-Verlag, New York, 1996).
7. Y. C. Fung, *Biomechanics: Motion, Flow, Stress and Growth* (Springer-Verlag, New York, 1990).
8. Y. C. Fung, *Biomechanics: Mechanical Properties of Living Tissues*, 2nd ed. (Springer-Verlag, New York, 1981, 1993).
9. S. Q. Liu and Y. C. Fung, Changes in the rheological properties of blood vessel tissue remodeling in the course of development of diabetes, *Biorheology* **29**: 443–457 (1992).
10. L. Kuo, M. J. Davis and W. M. Chilian, Myogenic activity in isolated subepicardial and subendocardial coronary arterioles, *Am. J. Physiol.* **255**: H1558–1562 (1988).
11. L. Kuo, M. J. Davis and W. M. Chilian, Endothelium-dependent, flow-induced dilation of isolated coronary arterioles, *Am. J. Physiol.* **259**: H1063–1070 (1990).
12. L. Kuo, W. M. Chilian and M. J. Davis, Interaction of pressure- and flow-induced responses in porcine coronary resistance vessels, *Am. J. Physiol.* **261**: H1706–1715 (1991).
13. P. C. Johnson, *Peripheral Circulation* (Wiley, New York, 1978).
14. N. E. Huang, Z. Shen, S. R. Long, M. L. Wu, H. H. Shih, Q. Zheng, N. C. Yen, C. C. Tung and H. H. Liu, The empirical mode decomposition and the Hilbert spectrum for nonlinear and non-stationary time series analysis, *Proc. R. Soc. Lond. A* **454**: 903–995 (1998).
15. W. Huang, Z. Shen, N. E. Huang and Y. C. Fung, Engineering analysis of biological variables: an example of blood pressure over 1 day, *Proc. Natl. Acad. Sci. USA* **95**: 4816–4821 (1998).

16. W. Huang, Z. Shen, N. E. Huang and Y. C. Fung, Use of intrinsic modes in biology: examples of indicial response of pulmonary blood pressure to +/− step hypoxia, *Proc. Natl. Acad. Sci. USA* **95**: 12766–12771 (1998).
17. W. Huang, Z. Shen, N. E. Huang and Y. C. Fung, Nonlinear indicial response of complex nonstationary oscillations as pulmonary hypertension responding to step hypoxia, *Proc. Natl. Acad. Sci. USA* **96**: 1834–1839 (1999).
18. J. J. Chen, R. Wu, P. C. Yang, J. Y. Huang, Y. P. Sher, M. H. Han, W. C. Kao, P. J. Lee, T. F. Chiu, F. Chang *et al.*, Profiling expression patterns and isolating differentially expressed genes by cDNA microarray system with colorimetry detection, *Genomics* **51**: 313–324 (1998).
19. W. Huang, Y. P. Sher, D. Delgado-West, J. T. Wu, K. Peck and Y. C. Fung, Tissue remodeling of rat pulmonary artery in hypoxic breathing. I. Changes of morphology, zero-stress state, and gene expression, *Ann. Biomed. Eng.* **29**: 535–551 (2001).
20. Y. C. Fung and W. Huang, The physics of hypoxic pulmonary hypertension and its connection with gene actions. In: *Hypoxic Pulmonary Vasoconstriction: Cellular and Molecular Mechanisms*, ed. J. X.-J. Yuan (Kluwer Academic Publishers, Boston, 2004), pp. 35–50.
21. W. Huang, Y. P. Sher, K. Peck and Y. C. Fung, Matching gene activity with physiological functions, *Proc. Natl. Acad. Sci. USA* **99**: 2603–2608 (2002).
22. S. Q. Liu, Influence of tensile strain on smooth muscle cell orientation in rat blood vessels, *J. Biomech. Eng.* **120**: 313–320 (1998).
23. S. Q. Liu and Y. C. Fung, Changes in the organization of the smooth muscle cells in rat vein grafts, *Ann. Biomed. Eng.* **26**: 86–95 (1998).
24. R. J. Price and T. C. Skalak, Distribution of cellular proliferation in skeletal muscle transverse arterioles during maturation, *Microcirculation* **5**: 39–47 (1998).
25. G. Lennon, C. Auffray, M. Polymeropoulos and M. B. Soares, The I.M.A.G.E. Consortium: an integrated molecular analysis of genomes and their expression, *Genomics* **33**: 151–152 (1996).
26. G. D. Schuler, M. S. Boguski, E. A. Stewart, L. D. Stein, G. Gyapay, K. Rice, R. E. White, P. Rodriguez-Tome, A. Aggarwal, E. Bajorek *et al.*, A gene map of the human genome, *Science* **274**: 540–546 (1996).
27. W. Huang, Tensorial description of the geometrical arrangement of the fibrous molecules in vascular endothelial cells, *Mol. Cell. Biomech.* **8**: 1–13 (2008).
28. R. N. Vaishnav and J. Vossoughi, Estimation of residual strains in aortic segments. In: *Biomedical Engineering, II. Recent Developments*, ed. C. W. Hall (Pergamon Press, New York, 1983), pp. 330–333.
29. R. N. Vaishnav and J. Vossoughi, Residual stress and strain in aortic segments, *J. Biomech.* **20**: 235–239 (1987).
30. Y. C. Fung and S. Q. Liu, Determination of the mechanical properties of the different layers of blood vessels *in vivo*, *Proc. Natl. Acad. Sci. USA* **92**: 2169–2173 (1995).
31. J. P. Xie, S. Q. Liu, R. F. Yang and Y. C. Fung, The zero-stress state of rat veins and vena cava, *J. Biomech. Eng.* **113**: 36–41 (1991).
32. J. P. Xie, J. Zhou, and Y. C. Fung, Bending of blood vessel wall: stress-strain laws of the intima-media and adventitial layers, *J. Biomech. Eng.* **117:** 136–145 (1995).
33. H. C. Han and Y. C. Fung, Longitudinal strain of canine and porcine aortas, *J. Biomech.* **28**: 637–641 (1995).
34. J. C. Debes and Y. C. Fung, Biaxial mechanics of excised canine pulmonary arteries, *Am. J. Physiol.* **269**: H433–442 (1995).
35. Y. C. Fung, Stress, strain, growth, and remodeling of living organisms, *Z. Angew. Math. Phys.* (Special Issue) **46**: S469–S482 (1995).
36. H. C. Han and Y. C. Fung, Direct measurement of transverse residual strains in aorta, *Am. J. Physiol.* **270**: H750–759 (1996).

37. Y. C. Fung, New trends in biomechanics. Keynote lecture. In: *Proceedings of the 11th Conference on Engineering Mechanics*, eds. Y. K. Lin and T. C. Su (American Society of Civil Engineers, 1996), pp. 1–15.
38. S. Q. Liu and Y. C. Fung, Indicial functions of arterial remodeling in response to locally altered blood pressure, *Am. J. Physiol.* **270**: H1323–1333 (1996).
39. H. Gregersen, G. Kassab, E. Pallencaoe, C. Lee, S. Chien, R. Skalak and Y. C. Fung, Morphometry and strain distribution in guinea pig duodenum with reference to the zero-stress state, *Am. J. Physiol.* **273**: G865–874 (1997).
40. J. Zhou and Y. C. Fung, The degree of nonlinearity and anisotropy of blood vessel elasticity, *Proc. Natl. Acad. Sci. USA* **94**: 14255–14260 (1997).
41. S. Q. Liu and Y. C. Fung, Changes in the organization of the smooth muscle cells in rat vein grafts, *Ann. Biomed. Eng.* **26**: 86–95 (1998).
42. R. Skalak and C. F. Fox, eds., *Tissue Engineering* (Alan R. Liss, Inc. New York, 1988).
43. S. L. Y. Woo and Y. Seguchi, eds., *Tissue Engineering — 1989*, ASME Publications No. BED–Vol. 14, (American Society of Mechanical Engineers, New York, 1989).
44. E. Bell, ed., *Tissue Engineering: Current Perspectives* (Birkhäuser, Boston, 1993).
45. H. Abè, K. Hayashi and M. Sato, eds., *Data Book on Mechanical Properties of Living Cells, Tissues, and Organs* (Springer, Tokyo, 1996).
46. J. Berry, A. Rachev, J. E. Moore, and J. J. Meister, Analysis of the effect of a non-circular two-layer stress-free state on arterial wall stresses, *Proc. 14th Ann. Intern. Conf. IEEE Eng. Med. Bio. Sec.* (1992), pp. 65–66.
47. A. Rachev, Theoretical study of the effect of stress-dependent remodeling on arterial geometry under hypertension condition. *13th Southern Biomed. Eng. Conf. U. District Columbia* (Washington, D.C., 1994).
48. T. Adachi, M. Tanaka and Y. Tomita, Uniform stress state in bone structures with residual stress, *J. Biomech. Eng.* **120**: 342–348 (1998).
49. H. C. Han, L. Zhao, M. Huang, L. S. Hou, Y. T. Huang and Z. B. Kuang, Postsurgical changes of the opening angle of canine autogenous vein graft, *J. Biomech. Eng.* **120**: 211–216 (1998).
50. L. E. Niklason, J. Gao, W. M. Abbott, K. K. Hirschi, S. Houser, R. Marini and R. Langer, Functional arteries grown *in vitro*, *Science* **284**: 489–493 (1999).
51. A. Rachev, N. Stergiopulos and J. J. Meister, A model for geometric and mechanical adaptation of arteries to sustained hypertension, *J. Biomech. Eng.* **120**: 9–17 (1998).
52. J. Vossoughi, Z. Hedjazi and F. S. Boriss, II, Intimal residual stress and strain in large arteries, *BED-24, Bioeng. Conf. ASME* (1993), pp. 4394–4437.
53. L. A. Taber, A model for aortic growth based on fluid shear and fiber stresses, *J. Biomech. Eng.* **120**: 348–354 (1998).
54. S. Q. Liu, Biomechanical basis of vascular tissue engineering, *Crit. Rev. Biomed. Eng.* **27**: 75–148 (1999).
55. B. L. Langille, Arterial remodeling: relation to hemodynamics, *Can. J. Physiol. Pharmacol.* **74**: 834–841 (1996).

SECTION II

CARDIOVASCULAR BIOENGINEERING

CHAPTER 3

CARDIAC ELECTROMECHANICS IN THE HEALTHY HEART

Roy C. P. Kerckhoffs and Andrew D. McCulloch

Abstract

In this chapter, cardiac function is discussed in the context of the healthy heart as part of a system. Anatomy and physiology will be described for each scale, starting with the organ level, and continuing to tissue and cell scales. For each scale, we will have a closer look at both electrical and mechanical functions.

Electrical function in the heart is intimately linked to mechanical function: muscle contraction is generated at the cellular level, and triggered by electrical activity via the flux of calcium ions. Furthermore, cell electrophysiology is modified by feedback from mechanical alterations (mechano-electric feedback or MEF). From a clinical standpoint, understanding these electromechanical interactions is becoming increasingly important because, for example, of regional wall remodeling induced by chronic pacing and the rapidly growing interest in cardiac resynchronization therapy (CRT).

As electrophysiology and mechanics in the heart are virtually inseparable, so are experiment and mathematical models of cardiac bioengineering. Through iterative interactions between experiment and simulation, insight in cardiac physiology is growing. Hence, in this chapter we will refer to examples of bioengineering experiments and simulations that helped in understanding normal cardiac function. The role of mechanics in cardiac disease can be found in Chapter 4.

1. Organ Level

1.1 *Gross anatomy*

The primary function of the heart is to pump blood through the circulatory system. In fact, the heart consists of two pumps, located side-by-side, one on the left, and one on the right. Both sides contain an atrium and a ventricle. The heart is surrounded by the pericardium: a thin protecting membrane of connective tissue. The right atrium (RA) collects venous blood that returns to the heart from the rest of the body, while the left atrium

(LA) collects blood that returns from the lungs. In addition, both atria facilitate filling of the ventricles. The inter-atrial septum separates the LA and the RA. The tricuspid valve separates the right atrium from the right ventricle (RV). The RV is crescent-shaped in cross-section, wrapped partially around the (ellipsoid-shaped) left ventricle (LV). The mitral valve separates the left atrium from the left ventricle. The interventricular septum, which anatomically belongs to the LV, separates the LV and RV and can transmit forces from one to the other. The surfaces of the LV and RV cavities are called the endocardium; the outer surface of the heart is called the epicardium, while the middle layers of the ventricles are referred to as myocardium (the cardiac muscle). The RV is connected to the pulmonary artery through the pulmonary valve. The RV maintains the pulmonary circulation, where blood receives oxygen and disposes of carbon dioxide in the lungs. The LV is connected to the aorta through the aortic valve. The LV maintains the systemic circulation, where blood delivers oxygen and nutrients to the rest of the body and receives carbon dioxide and waste products. The contracting cardiac walls pump blood through the (lower pressurized) pulmonary and (higher pressurized) systemic circulation, hence the wall of the RV is thinner and less powerful than that of the LV (Fig. 1).

1.2 *Global heart function*

From the opened or closed states of the valves, four cardiac phases can be distinguished for the ventricles (the isovolumic relaxation and filling phases, which together constitute diastole, and the isovolumic contraction and ejection phases that comprise systole). A single cardiac cycle in the normal human LV at rest is summarized in Fig. 2 with typical orders of magnitude for time and cavity pressure. During the filling phase, the mitral

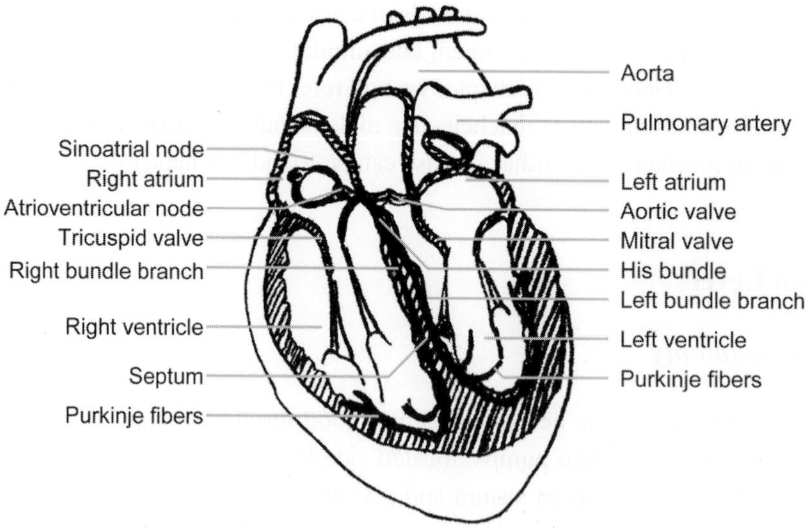

Fig. 1. Gross anatomy, also showing the conduction system.

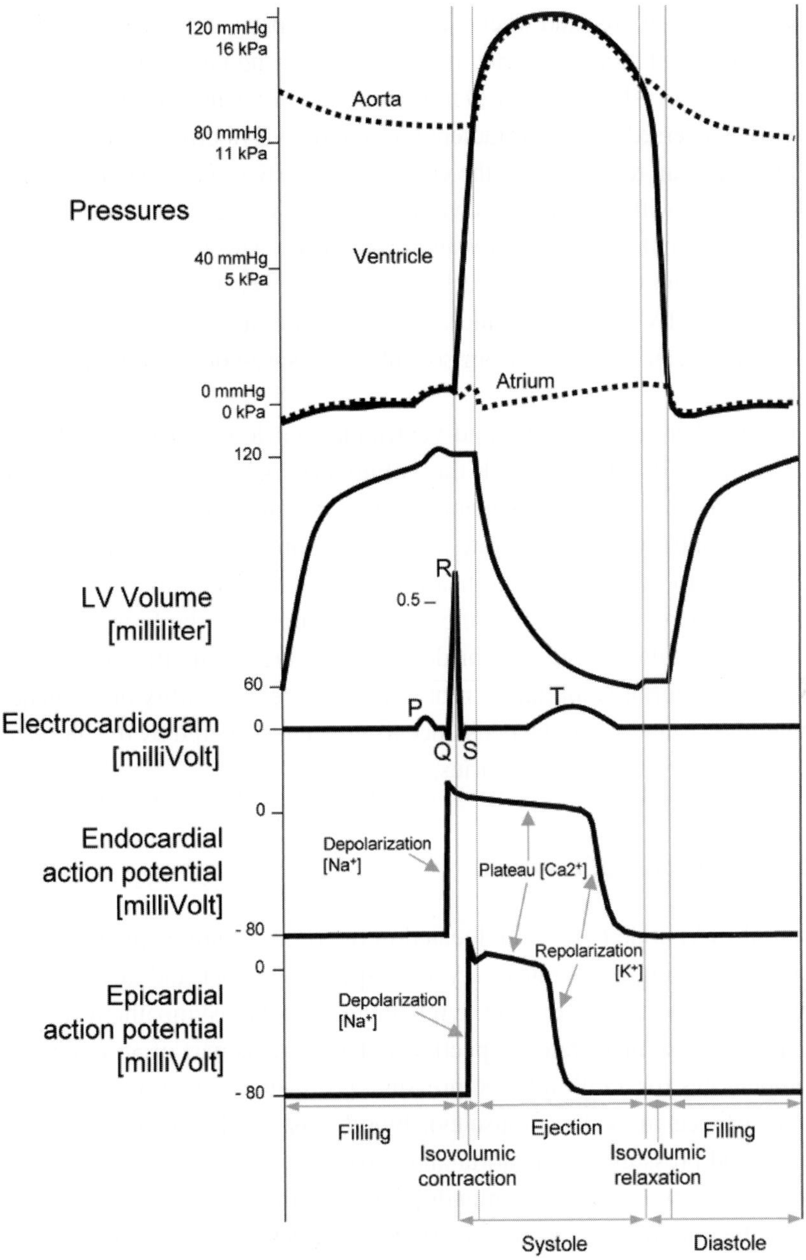

Fig. 2. Pressures, LV cavity volume, electrocardiogram, and two ventricular action potentials as a function of and synchronized in time. Modified from Katz.[2]

valve is open and the aortic valve is closed. About 500 ms after the mitral valve opens, active myofiber stress (force per unit area) in the LV wall increases the pressure in both ventricles and closes the atrioventricular valves (~1.5 kPa, ~11 mmHg). Now, all valves are closed and the isovolumic contraction phase is entered. LV pressure rises quickly during isovolumic systole (~50 ms) and when it exceeds aortic pressure (~10 kPa, ~75 mmHg) the aortic valve opens and blood is ejected into the aorta during the ejection phase (reaching a maximum pressure of ~16 kPa or ~120 mmHg). After ~250 ms, LV pressure becomes low enough and the reversed pressure-gradient through the valve orifice during flow deceleration provides the main force to close the valve which is effective in producing almost complete valve closure, with only a very small reversed flow. Then, the isovolumic relaxation phase begins (~80 ms). When LV pressure drops below left atrial pressure (~0 kPa), the cycle starts over again with the rapid early filling phase. Diastolic pressures for the RV are in the same order of LV pressures, while systolic pressures are typically 20%–25% as large. Additional insight can be gained from so-called pressure-volume loops, as described in Chapter 4.

The relation between pressure and ventricular outflow is determined by the interaction between the RV and LV with the pulmonary and systemic circulation, respectively. Major determinants from the cardiac side are contractile stress generation in the cardiac walls (contractility), heart rate and the amount of ventricular filling (preload). The latter relates to the fact that the heart is able to generate higher systolic pressures as a function of increasing end-diastolic volumes, known as the Frank-Starling mechanism. From the circulation side, major determinants are the inertia of the blood; aortic valve resistance; arterial compliance, representing the extensibility of the major arteries and peripheral resistance, and the resistance to flow encountered by the blood as it flows from the major arteries to the capillaries.

1.3 *The heart as part of a system*

The sinoatrial (SA) node (Sec. 2.2), located in the right atrium, is the heart's own natural pacemaker. Resting heart rates vary with species, gender and age. However, heart rate is under autonomic control by sympathetic and parasympathetic nerves, which increase and decrease heart rate, respectively. The heart is part of an integrated system which compensates for a fall in blood pressure as a result of exercise (due to decreased arteriolar resistance in skeletal muscles), blood loss,[1] and also during heart failure.[2] Decreased blood pressures in the aorta and carotids will stimulate the baroreceptors in those vessels. That will increase sympathetic stimulation on the heart, which in its turn will increase beating frequency and contractility and hence increase total cardiac output in liter/minute. Water and salt will be retained in the kidneys to increase total blood volume and therefore preload, and blood supply will be redistributed throughout the body.

2. Tissue Level

2.1 *Fiber and sheet architecture*

The myocardium is largely composed of a few billion cardiac muscle cells (known as myocytes or myofibers), which have a complex three-dimensional architecture.[3]

Although the myocytes are relatively short, they are rod-shaped and connected such that at any point in the normal heart wall there is a clear predominant muscle fiber direction. Each ventricular myocyte is connected via gap junctions at intercalated disks to an average of 11.3 neighbors, 5.3 on the sides and 6.0 at the ends.[4] About the mean, myofiber angle dispersion is typically 10°–15°[5] except in certain pathologies. Similar patterns have been described for various mammals, from humans to rats. In the human or dog left ventricle, the muscle fiber angle typically varies continuously from about –60° (i.e. 60° clockwise from the circumferential direction) at the epicardium to about +70° at the endocardium. The rate of change of fiber angle is usually greatest at the epicardium, so that circumferential (0°) fibers are found in the outer half of the wall. Figure 3 shows the direction of myofibers in the rabbit heart.

Furthermore, cardiac myofibers are organized into laminar sheets about four–six cells thick.[6] Sheets are extensive cleavage planes running approximately radially from endocardium toward epicardium in transmural section. Like the fibers, the sheets also have a branching pattern with the number of branches varying considerably through the wall thickness.

Myofiber orientations have been reconstructed from histological measurements[7–9] and diffusion weighted MRI.[10,11] Sheet orientations also have been obtained by histological measurements,[12] but there is recent evidence that sheets can also be measured with diffusion weighted MRI.[13] It has been shown[14] that the distributions of sheet orientations measured within the left ventricular wall of the dog heart coincided closely with those predicted from observed three-dimensional wall strains (regional length change) using the assumption that laminae are oriented in planes that contain the muscle fibers and maximize interlaminar shearing. This assumption also leads to the conclusion that two families of sheet orientations may be expected. Indeed, a retrospective analysis of the histology supported this prediction and more recent observations confirm the presence of two distinct populations of sheet plane in the inner half of the ventricular wall.[14]

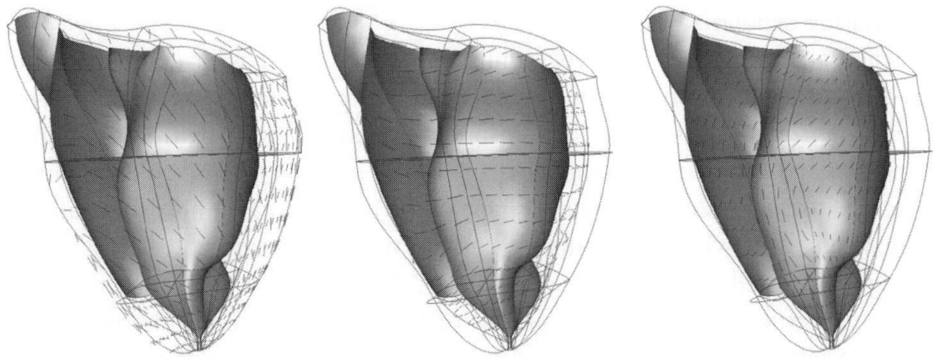

Fig. 3. Direction of myofibers shown on the epicardium (left), midwall (mid), and endocardium (right) for a finite element model of the rabbit ventricular anatomy.[7] These fiber directions were fitted to measurements.

2.2 *Electrical propagation*

The electrocardiogram (ECG or EKG) is a recording of electrical differences in the heart, measured by electrodes on the body surface (Fig. 2). Einthoven was the first to measure the ECG accurately at the beginning of the 20th century.

In a normal heart, an electrical wave starts in the sinoatrial node by spontaneous depolarization. This node is a cluster of modified myocytes located at the endocardium of the right atrium. The resulting depolarization wave propagates over neighboring cells, thus depolarizing the atria. This is reflected by the P-wave in the ECG. The atrioventricular (AV) node, located near the septum between the atria and the ventricles, slows down the depolarization wave. Distal to the AV node, the wave is conducted fast through the bundle of His, splitting in a left and right branch. These branches bifurcate into many small branches, thus forming the fast conducting Purkinje system, located in the subendocardium of the left and right ventricle. These Purkinje fibers are insulated with connective tissue. Purkinje fibers enable a fast propagation of depolarization, typically 3–4 m/s.[15]

The Purkinje system ends in the Purkinje-muscular junctions. Here, the depolarization wave enters the ventricular myocardium. In the ECG, the period between the P-wave and the time of ventricular myocardial depolarization, is referred to as the P-R interval, reflecting the time between atrial and ventricular activation. In the myocardium, propagation is much slower,[16] where the wave propagates mainly from endocardium to epicardium. In the ECG, ventricular activation is reflected by the QRS complex. Myocardial propagation velocities of 0.6–1.0 m/s and 0.2–0.5 m/s have been reported parallel and perpendicular to myofibers.[16–18] From the moment of leaving the Purkinje fibers, the whole human ventricular myocardium is depolarized within ~50 ms,[19] which is reflected by the narrow QRS width. Since it takes cells at the endocardium longer to repolarize (going back to the electrical rest-state) compared with epicardial cells, repolarization occurs in the opposite direction — from epicardium to endocardium. This is reflected by an upward T-wave in the ECG, because electrical gradients in the tissue have the same direction during repolarization as during depolarization.

Electrical propagation in myocardium can be modeled with computer simulations, in which the action potential within a cell is predicted by sets of differential equations. Difficulty in solving for the spread of activation arises from the inherent nonlinearity and geometrical complexity of cardiac tissue. The finite element method can be used to discretize the irregular shape of the heart into a number of smaller regular elements, over which the spatially varying transmembrane potential is continuously approximated (Fig. 4). The geometry, local muscle fiber orientation and material properties (e.g. tissue resistance) of the myocardium are interpolated. The resulting system of equations is solved to give model predictions of electrical activation in a portion of the heart.

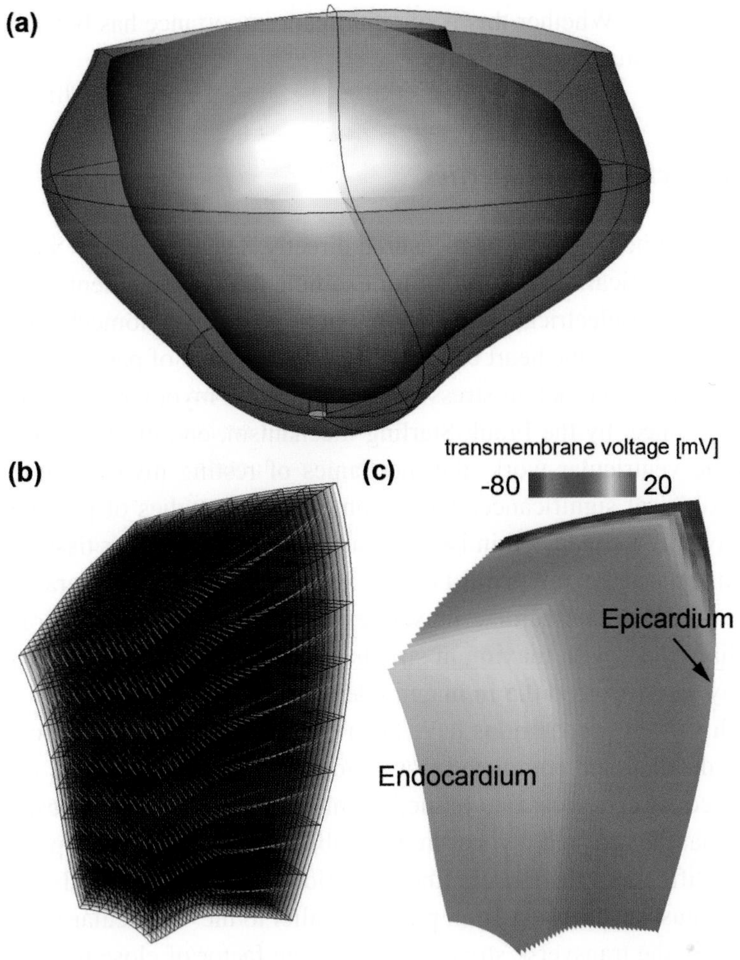

Fig. 4. Computational model of electrical activation in a wedge of myocardium. (**a**) A coarse finite element model of the dog ventricular anatomy with semi-transparent epicardium; (**b**) a wedge segment of the left ventricle, refined to 2048 elements displaying average myocyte orientation; and (**c**) the simulated spread of activation after it has traversed most of the ventricular wall and is nearing the epicardium.

With a two-dimensional computer model of cardiac electromechanics, it has been shown that contraction affects the ECG (a result of mechano-electrical feedback, see Sec. 3.4): the ECG exhibited a shorter QT-interval when contraction was included in the simulation as opposed to computing electrophysiology only. In this model, the heart was embedded in a torso model, which enabled the calculation of an ECG.

Using another computer model, Hooks and colleagues[20] have shown that different conductivities result locally in three perpendicular directions (orthotropy) from sheet structure as opposed to different conductivities parallel and transverse to the myofiber (transverse isotropy). Whether this is of significant importance has been questioned by another simulation study.[21]

2.3 *Passive mechanical properties*

Stress in the intact heart cannot be measured directly. Therefore stress values are calculated from mathematical models (using for example the finite element method, as mentioned in Sec. 2.2 for electrical propagation, but now to solve biomechanics equations). Generally, total stress in the heart can be considered the sum of passive stress, when the tissue is still at rest, and active stress generated by the myocytes after depolarization (see Sec. 3.3). Since, by the Frank-Starling mechanism, end-diastolic volume directly affects systolic ventricular work, the mechanics of resting myocardium have fundamental physiological significance. Most biomechanical studies of passive myocardial properties have been conducted in isolated, arrested whole heart or tissue preparations. Passive cardiac muscle exhibits most of the mechanical properties characteristic of soft tissues in general.[22] Intact cardiac muscle experiences finite deformations during the normal cardiac cycle, with maximum strains (which are generally radial and endocardial) that may easily exceed 0.5 in magnitude.

Although various preparations have been used to study resting myocardial elasticity, the most detailed and complete information has come from biaxial and multiaxial tests of isolated sheets of cardiac tissue, mainly from the dog.[23–25] These experiments have shown that the arrested myocardium exhibits significant anisotropy with substantially greater stiffness in the muscle fiber direction than transversely (Fig. 5). In equibiaxial tests of muscle sheets cut from planes parallel to the ventricular wall, fiber stress was greater than the transverse stress by an average factor of close to 2.0.[26] Moreover, as suggested by the structural organization of the myocardium, sheet structure seems to be more important for mechanics than for electrophysiology.[27] Strain distributions were in better agreement with experiments when orthotropic passive properties were

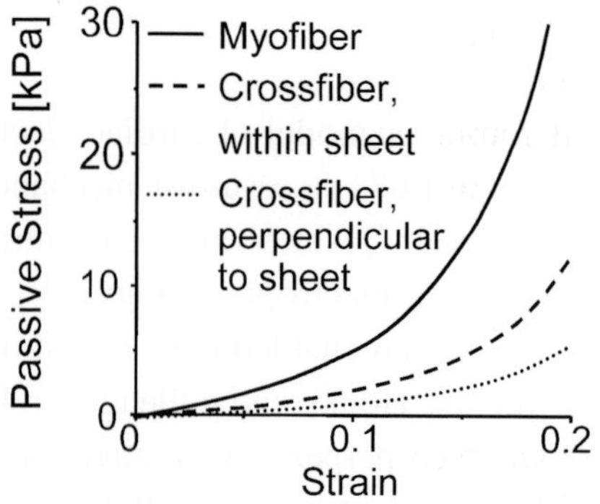

Fig. 5. Passive stress of myocardium, showing nonlinearity and orthotropy.

included in a computer model of cardiac mechanics as opposed to transversely isotropic properties.[28]

The biaxial stress-strain properties of passive myocardium display some heterogeneity in the cardiac walls.[29] Tissue specimens in the dog from the inner and outer thirds of the left ventricular free wall were stiffer than those from the midwall and interventricular septum, but the degree of anisotropy was similar in each region. Biaxial testing of the pericardium and epicardium have shown that these tissues have distinctly different properties than the myocardium being very compliant and isotropic at low biaxial strains (< 0.1–0.15) but rapidly becoming very stiff and anisotropic as the strain is increased.[24,30]

Various constitutive models have been proposed for the elasticity of passive cardiac tissues. Because of the large deformations and nonlinearity of these materials, the most useful framework has been provided by the pseudostrain-energy formulation for hyperelasticity. For a detailed review of the material properties of passive myocardium and approaches to constitutive modeling, the reader is referred to Chapters 1–6 of Glass *et al.*[31] In hyperelasticity, the components of the stress are obtained from a strain energy W as a function of the Lagrangian (Green's) strain components.

The myocardium is generally assumed to be an incompressible material, which is a good approximation in the isolated tissue, although in the intact heart there can be significant redistribution of tissue volume associated with phasic changes in regional coronary blood volume.

3. Cell Level

3.1 *Anatomy*

Myocytes are approximately rod-shaped with a length of ~100 μm and a diameter of ~25 μm. They contain numerous myofibrils, which are long cylindrical organelles, composed of regular three-dimensional arrays of thick (myosin) and thin (actin) myofilaments (Fig. 6). Neighboring myofibrils have their sarcomeres aligned, which gives myofibers a striated appearance. The sarcomere is considered to be the basic contractile unit of cardiac and skeletal muscle. Myofibrils and mitochondria (the cell's energy providers) make up almost 85% of the myocyte volume.[2]

3.2 *Electrophysiology*

During diastole, a negative resting electrical potential of about –80 mV exists in the cell (Fig. 2). This is caused by a gradient of different ions (e.g. K^+, Ca^{2+}, Na^+) inside and outside the cell. There is more potassium (K^+) within than outside the cell and

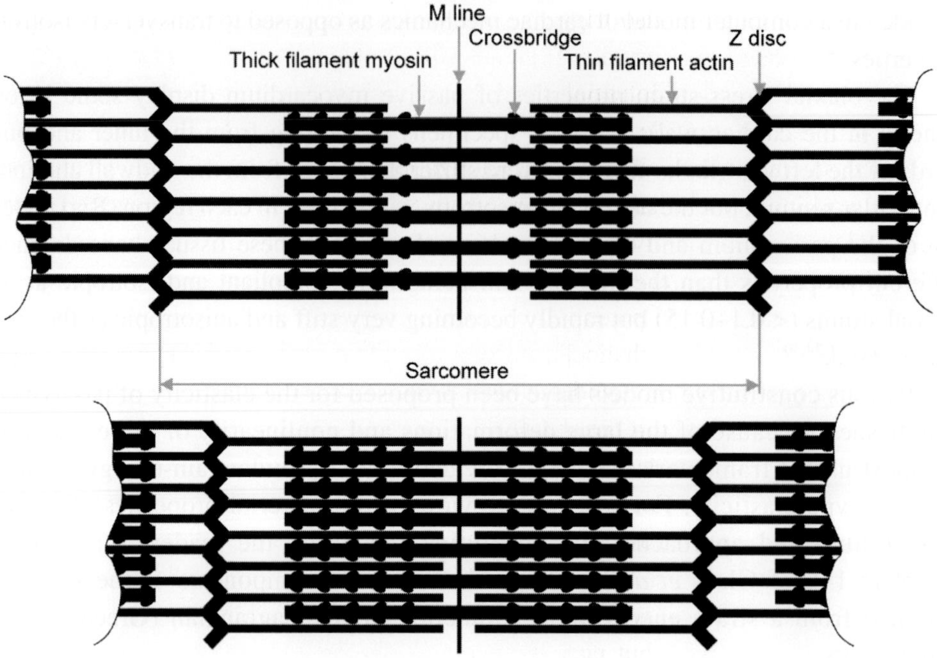

Fig. 6. Schematic of the arrangement of the myofilaments myosin and actin. Myosin filaments are connected at the M line; actin filaments are connected through the Z discs. The region between two Z discs is the sarcomere. In the bottom part of the figure, the sarcomere is contracted, showing the sliding of the myofilaments.

more calcium (Ca^{2+}) and sodium (Na^{+}) present extracellular than intracellular. An action potential is generated in a resting myocyte when the polarity over the cell membrane changes from negative to positive (depolarization), caused by adjacent depolarizing cells.

A large inward current of sodium is responsible for the initial upstroke of the transmembrane action potential (see the endocardial and epicardial action potentials in Fig. 2). Next, calcium enters the cell through L-type calcium channels, located in the cell's outer membrane, which contributes to the action potential plateau.[32] Calcium entry triggers calcium release through the ryanodine receptors in the membrane of the sarcoplasmatic reticulum (which stores most of the intracellular calcium). The combination of calcium influx and release raises the calcium concentration in the cell's myoplasm, allowing it to bind to the regulatory protein troponin C, located on the actin filament. The resulting conformational change of the regulatory proteins tropomyosin and troponin on the actin filament exposes myosin-binding sites. A force-generating bond between actin and myosin, a crossbridge, is formed when myosin heads attach to the actin binding sites. Depending on the loading of the myocyte, actin filaments slide along the myosin filaments, causing shortening of the sarcomere (see Fig. 6 and Sec. 3.3).

Intracellular calcium also triggers calcium-activated potassium channels that generate outward potassium currents. Together with voltage-activated potassium channels, these will end the action potential. Pumps and exchangers actively restore concentrations of ions during diastole.

Based on electrophysiological characterization, the presence of three different subtypes of working myocardial cells in the ventricular walls have been suggested, named after their location in the myocardium: endocardial (closest to the cavity), midmyocardial (or just simply M), and epicardial (closest to the outer surface of the heart) cells.[33,34] These cells produce action potentials of different shape and length (Fig. 2), due to different expressions of ion channels, different calcium fluxes,[35,36] and hence different active mechanical behavior.[36] Midmyocardial cells exhibit the longest action potentials, while epicardial cells exhibit the shortest.

However, in intact tissue, transmural differences in for example action potentials are much smaller due to electrotonic coupling.[37] With a numerical tissue model of cardiac electrophysiology it has been shown that electrotonic coupling has a diminishing effect on heterogeneity in the intact myocardium which strongly depends on how strongly cells are coupled with each other.[38] Coupled, epicardial cells still exhibit the shortest action potentials, but now the endocardial cells exhibit the longest (see also Sec. 2.2).

3.3 *Active mechanical properties*

The amount of active stress generated depends on the length of the sarcomere and on the amount of calcium in the cell (Fig. 7). The relationship between free calcium concentration and isometric muscle tension has mostly been investigated in muscle preparations in which the outer cell membrane has been chemically permeabilized. Because there is evidence that this chemical "skinning" alters the calcium sensitivity of myofilament interaction, recent studies have also investigated myofilament calcium sensitivity in intact muscles tetanized by high frequency stimulation in the presence of a compound such as ryanodine that open calcium release sites in the sarcoplasmic reticulum. Intracellular calcium concentration was estimated using calcium-sensitive optical indicators such as Fura. The myofilaments are activated in a graded manner by small concentrations of calcium, which binds to troponin-C according to a sigmoidal relation.[39] Half-maximal tension in cardiac muscle is developed at intracellular calcium concentrations of 10^{-6} to 10^{-5} M depending on factors such as species and temperature.[40]

In the normal heart, sarcomere lengths range from 1.6 μm (generating zero active stress) to 2.4 μm (generating maximum active stress). This is the basis for the Frank-Starling mechanism. Early evidence for a negative dependence of cardiac muscle isometric stress on length at higher sarcomere lengths was found to be caused by shortening in the central region of the isolated muscle at the expense of

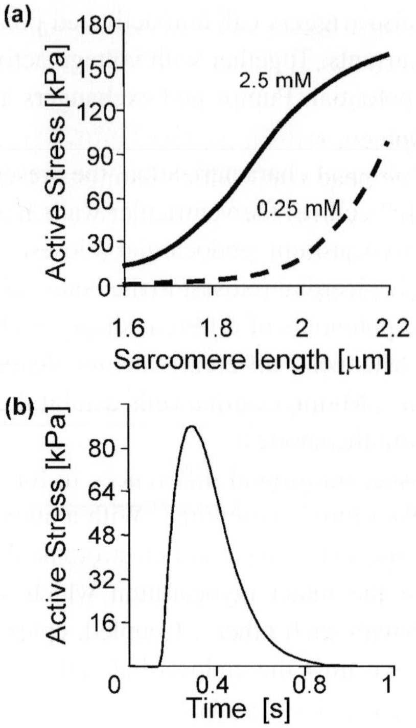

Fig. 7. Active stress of myocardium. **(a)** Active stress as a function of sarcomere length and extracellular calcium concentration, shown here for 0.25 and 2.5 mM. **(b)** Active stress as a function of time at a sarcomere length of 1.9 μm.

stretching at the damaged ends where specimen was attached to the test apparatus. If muscle length is controlled so that sarcomere length in the undamaged part of the muscle is indeed constant, or if the developed tension is plotted against the instantaneous sarcomere length rather than the muscle length, the negative dependence is eliminated.[41]

The isotonic force-velocity relation of cardiac muscle is similar to that of skeletal muscle, and A.V. Hill's well-known hyperbolic relation is a good approximation except at larger forces greater than about 85% of the isometric value. The maximal (unloaded) velocity of shortening is essentially independent of preload, but does change with time during the cardiac twitch. de Tombe and ter Keurs[42] using sarcomere length-controlled isovelocity release experiments found that viscous forces imposes a significant internal load opposing sarcomere shortening. If the isotonic shortening response is adjusted for the effects of passive viscoelasticity, the underlying cross-bridge force-velocity relation is found to be linear.

In an interesting and important development, biaxial testing of ventricular myocardium has shown that developed active stress also has a large component in directions transverse to the mean myofiber axis that can exceed 50% of the axial fiber component.[43] The magnitude of this transverse active stress depended significantly on the biaxial loading conditions. Moreover, evidence from osmotic swelling and other studies suggests that transverse strain can affect contractile tension development along the fiber axis by altering distances between myofibrils.[44] The mechanisms of transverse active stress development remain unclear but two possible contributors are the geometry of the crossbridge head itself which is oriented oblique to the myofilament axis,[45] and the dispersion of myofiber orientation.[5] Strain distributions as computed by computer models of cardiac mechanics showed better agreement with experiments when this transverse active behavior was included,[45,46] but the role of sheets here remains unclear.

The three types of cells, exhibiting different electrophysiological properties, have also been found to exhibit different active mechanical properties.[36] In the study by Cordeiro *et al.* unloaded cell shortening, calcium transients, and inward L-type calcium current characteristics were examined of epicardial, endocardial, and midmyocardial isolated cells from the canine left ventricle. Time to peak and latency to onset of contraction were shortest in epicardial and longest in endocardial cells; midmyocardial cells displayed an intermediate time to peak. Sarcoplasmic reticulum calcium content was largest in epicardial cells and contributed to a faster time to peak. Differences in active mechanical properties are caused partly by the different electrophysiological properties of these cells, but also partly by intrinsic differences in excitation-contraction coupling and mechano-electrical feedback.

3.4 *Electromechanics*

It has been known for over a hundred years that heart rhythm can be disturbed by mechanical chest impacts, possibly leading to sudden cardiac death. Thus, mechanics affect electrophysiology. Electrophysiology of the cell experiences feedback from mechanical alterations (mechano-electric feedback or MEF[46,47]) through the dependence of myofilament calcium sensitivity on sarcomere length,[48] and through ion channels (chloride and potassium) that are activated when cells are stretched (stretch activated channels or SACs[49]). Stretch also alters calcium handling and action potential generation.

Kohl *et al.*[50] have investigated the influence of strains on electrophysiology in the whole heart through stretch activated channels. In the latter study, a local region in the left ventricular wall with a higher compliance (aneurysm) generated a propagating action potential, showing that mechano-electric feedback can induce cardiac arrhythmias.

A computer model of electrophysiology of rabbit cell electrophysiology was recently extended to include transmurally varying SACs. The model predicted that endocardial and midmyocardial cells are the most sensitive to stretch-induced changes in action potential duration. Inclusion of a potassium SAC may reduce repolarization gradients in intact myocardium caused by intrinsic ion channel densities, nonuniform strains and electrotonic effects.[51]

Acknowledgments

The authors were supported by the National Biomedical Computation Resource (P41 RR08605) (to A.D.M), the National Science Foundation (BES-0096492 and BES-0506252) (to A.D.M), and NIH grant HL32583. A.D.M is a co-founder of Insilicomed Inc., a licensee of UCSD-owned software used in this research. We are grateful to Sarah Healy for providing the results shown in Fig. 4.

References

1. K. Mackway-Jones, B. A. Foex, E. Kirkman and R. A. Little, Modification of the cardiovascular response to hemorrhage by somatic afferent nerve stimulation with special reference to gut and skeletal muscle blood flow, *J. Trauma* **47**: 481–485 (1999).
2. A. M. Katz, *Physiology of the Heart*, 3rd ed. (Lippincott Williams & Wilkins, Philadelphia, PA, 2001).
3. D. D. Streeter Jr., Gross morphology and fiber geometry of the heart. In: *Handbook of Physiology*, Section 2: The Cardiovascular System, Chapter 4, eds. R. M. Berne and N. Sperelakis (American Physiological Society, Bethesda, MD, 1979), pp. 61–112.
4. J. E. Saffitz, H. L. Kanter, K. G. Green, T. K. Tolley and E. C. Beyer, Tissue-specific determinants of anisotropic conduction velocity in canine atrial and ventricular myocardium, *Circ. Res.* **74**: 1065–1070 (1994).
5. W. J. Karlon, J. W. Covell, A. D. McCulloch, J. J. Hunter and J. H. Omens, Automated measurement of myofiber disarray in transgenic mice with ventricular expression of ras, *Anat. Rec.* **252**: 612–625 (1998).
6. I. J. LeGrice, B. H. Smaill, L. Z. Chai, S. G. Edgar, J. B. Gavin and P. J. Hunter, Laminar structure of the heart: ventricular myocyte arrangement and connective tissue architecture in the dog, *Am. J. Physiol.* **269**: H571–582 (1995).
7. F. J. Vetter and A. D. McCulloch, Three-dimensional analysis of regional cardiac function: a model of rabbit ventricular anatomy, *Prog. Biophys. Mol. Biol.* **69**: 157–183 (1998).
8. C. Stevens and P. J. Hunter, Sarcomere length changes in a 3D mathematical model of the pig ventricles, *Prog. Biophys. Mol. Biol.* **82**: 229–241 (2003).
9. I. J. Legrice, P. J. Hunter and B. H. Smaill, Laminar structure of the heart: a mathematical model, *Am. J. Physiol.* **272**: H2466–2476 (1997).
10. D. F. Scollan, A. Holmes, R. Winslow and J. Forder, Histological validation of myocardial microstructure obtained from diffusion tensor magnetic resonance imaging, *Am. J. Physiol.* **275**: H2308–2318 (1998).
11. T. G. Reese, R. M. Weisskoff, R. N. Smith, B. R. Rosen, R. E. Dinsmore and V. J. Wedeen, Imaging myocardial fiber architecture *in vivo* with magnetic resonance, *Magn. Reson. Med.* **34**: 786–791 (1995).
12. I. J. LeGrice, B. H. Smaill, L. Z. Chai, S. G. Edgar, J. B. Gavin and P. J. Hunter, Laminar structure of the heart: ventricular myocyte arrangement and connective tissue architecture in the dog, *Am. J. Physiol.* **269**: H571–582 (1995).
13. W. Y. Tseng, V. J. Wedeen, T. G. Reese, R. N. Smith and E. F. Halpern, Diffusion tensor MRI of myocardial fibers and sheets: correspondence with visible cut-face texture, *J. Magn. Reson. Imaging* **17**: 31–42 (2003).
14. T. Arts, K. D. Costa, J. W. Covell and A. D. McCulloch, Relating myocardial laminar architecture to shear strain and muscle fiber orientation, *Am. J. Physiol. Heart Circ. Physiol.* **280**: H2222–2229 (2001).

15. R. J. Myerburg, K. Nilsson and H. Gelband, Physiology of canine intraventricular conduction and endocardial excitation, *Circ. Res.* **30**: 217–243 (1972).
16. D. E. Roberts and A. M. Scher, Effect of tissue anisotropy on extracellular potential fields in canine myocardium *in situ*, *Circ. Res.* **50**: 342–351 (1982).
17. D. W. Frazier, W. Krassowska, P. S. Chen, P. D. Wolf, N. D. Danieley, W. M. Smith and R. E. Ideker, Transmural activations and stimulus potentials in three-dimensional anisotropic canine myocardium, *Circ. Res.* **63**: 135–146 (1988).
18. B. Taccardi, E. Macchi, R. L. Lux, P. R. Ershler, S. Spaggiari, S. Baruffi and Y. Vyhmeister, Effect of myocardial fiber direction on epicardial potentials, *Circulation* **90**: 3076–3090 (1994).
19. D. Durrer, R. T. van Dam, G. E. Freud, M. J. Janse, F. L. Meijler and R. C. Arzbaecher, Total excitation of the isolated human heart, *Circulation* **41**: 899–912 (1970).
20. D. A. Hooks, K. A. Tomlinson, S. G. Marsden, I. J. LeGrice, B. H. Smaill, A. J. Pullan and P. J. Hunter, Cardiac microstructure: implications for electrical propagation and defibrillation in the heart, *Circ. Res.* **91**: 331–338 (2002).
21. P. Colli-Franzone, L. Guerri and B. Taccardi, Modeling ventricular excitation: axial and orthotropic anisotropy effects on wavefronts and potentials, *Math. Biosci.* **188**: 191–205 (2004).
22. Y. C. Fung, *Biomechanics: Mechanical Properties of Living Tissues*, 2nd ed. (Springer-Verlag Inc., New York, 1993).
23. H. R. Halperin, P. H. Chew, M. L. Weisfeldt, K. Sagawa, J. D. Humphrey and F. C. Yin, Transverse stiffness: a method for estimation of myocardial wall stress, *Circ. Res.* **61**: 695–703 (1987).
24. J. D. Humphrey, R. K. Strumpf and F. C. Yin, Biaxial mechanical behavior of excised ventricular epicardium, *Am. J. Physiol.* **259**: H101–108 (1990).
25. L. L. Demer and F. C. Yin, Passive biaxial mechanical properties of isolated canine myocardium, *J. Physiol.* **339**: 615–630 (1983).
26. F. C. Yin, R. K. Strumpf, P. H. Chew and S. L. Zeger, Quantification of the mechanical properties of noncontracting canine myocardium under simultaneous biaxial loading, *J. Biomech.* **20**: 577–589 (1987).
27. P. J. Hunter, A. D. McCulloch and H. E. ter Keurs, Modelling the mechanical properties of cardiac muscle, *Prog. Biophys. Mol. Biol.* **69**: 289–331 (1998).
28. T. P. Usyk, R. Mazhari and A. D. McCulloch, Effect of laminar orthotropic myofiber architecture on regional stress and strain in the canine left ventricle, *J. Elast.* **61**: 143–164 (2000).
29. V. P. Novak, F. C. Yin and J. D. Humphrey, Regional mechanical properties of passive myocardium, *J. Biomech.* **27**: 403–412 (1994).
30. M. C. Lee, Y. C. Fung, R. Shabetai and M. M. LeWinter, Biaxial mechanical properties of human pericardium and canine comparisons, *Am. J. Physiol.* **253**: H75–82 (1987).
31. L. Glass, P. Hunter and A. D. McCulloch, *Theory of Heart: Biomechanics, Biophysics and Nonlinear Dynamics of Cardiac Function*, Institute for Nonlinear Science, ed. H. Abarbanel (Springer-Verlag, New York, 1991).
32. D. M. Bers, Cardiac excitation-contraction coupling, *Nature* **415**: 198–205 (2002).
33. C. Antzelevitch, S. Sicouri, S. H. Litovsky, A. Lukas, S. C. Krishnan, J. M. Di Diego, G. A. Gintant and D. W. Liu, Heterogeneity within the ventricular wall. Electrophysiology and pharmacology of epicardial, endocardial, and M cells, *Circ. Res.* **69**: 1427–1449 (1991).
34. D. Fedida and W. R. Giles, Regional variations in action potentials and transient outward current in myocytes isolated from rabbit left ventricle, *J. Physiol.* **442**: 191–209 (1991).
35. K. R. Laurita, R. Katra, B. Wible, X. Wan and M. H. Koo, Transmural heterogeneity of calcium handling in canine, *Circ. Res.* **92**: 668–675 (2003).
36. J. M. Cordeiro, L. Greene, C. Heilmann, D. Antzelevitch and C. Antzelevitch, Transmural heterogeneity of calcium activity and mechanical function in the canine left ventricle, *Am. J. Physiol. Heart Circ. Physiol.* **286**: H1471–1479 (2004).

37. P. Taggart, P. Sutton, T. Opthof, R. Coronel and P. Kallis, Electrotonic cancellation of transmural electrical gradients in the left ventricle in man, *Prog. Biophys. Mol. Biol.* **82**: 243–254 (2003).
38. C. E. Conrath, R. Wilders, R. Coronel, J. M. de Bakker, P. Taggart, J. R. de Groot and T. Opthof, Intercellular coupling through gap junctions masks M cells in the human heart, *Cardiovasc. Res.* **62**: 407–414 (2004).
39. J. C. Rüegg, *Calcium in Muscle Activation: A Comparative Approach*, 2nd ed., Zoophysiology, Vol. 19 (Springer-Verlag, Berlin, 1988).
40. D. M. Bers, *Excitation-Contraction Coupling and Cardiac Contractile Force* (Dordrecht, Kluwer, 1991).
41. H. E. ter Keurs, W. H. Rijnsburger, R. van Heuningen and M. J. Nagelsmit, Tension development and sarcomere length in rat cardiac trabeculae. Evidence of length-dependent activation, *Circ. Res.* **46**: 703–714 (1980).
42. P. P. de Tombe and H. E. ter Keurs, An internal viscous element limits unloaded velocity of sarcomere shortening in rat myocardium, *J. Physiol.* **454**: 619–642 (1992).
43. D. H. Lin and F. C. Yin, A multiaxial constitutive law for mammalian left ventricular myocardium in steady-state barium contracture or tetanus, *J. Biomech. Eng.* **120**: 504–517 (1998).
44. M. Schoenberg, Geometrical factors influencing muscle force development. I. The effect of filament spacing upon axial forces, *Biophys. J.* **30**: 51–67 (1980).
45. M. Schoenberg, Geometrical factors influencing muscle force development. II. Radial forces, *Biophys. J.* **30**: 69–77 (1980).
46. P. Kohl and U. Ravens, Cardiac mechano-electric feedback: past, present, and prospect, *Prog. Biophys. Mol. Biol.* **82**: 3–9 (2003).
47. M. J. Lab, P. Taggart and F. Sachs, Mechano-electric feedback, *Cardiovasc. Res.* **32**: 1–2 (1996).
48. D. G. Allen and S. Kurihara, The effects of muscle length on intracellular calcium transients in mammalian cardiac muscle, *J. Physiol.* **327**: 79–94 (1982).
49. W. Craelius, Stretch-activation of rat cardiac myocytes, *Exp. Physiol.* **78**: 411–423 (1993).
50. P. Kohl, P. Hunter and D. Noble, Stretch-induced changes in heart rate and rhythm: clinical observations, experiments and mathematical models, *Prog. Biophys. Mol. Biol.* **71**: 91–138 (1999).
51. S. N. Healy and A. D. McCulloch, An ionic model of stretch-activated and stretch-modulated currents in rabbit ventricular myocytes, *Europace* **7**(Suppl. 2): 128–134 (2005).

CHAPTER 4

CARDIAC BIOMECHANICS AND DISEASE

Jeffrey H. Omens

Abstract

Biomechanics plays a central role in research of the heart and its function, fundamentally because the main role of the heart as an organ is the mechanical pumping of blood through the circulation of the body. Diseases of the heart are many times a direct consequence of impaired mechanics. Although quantifying cardiac function in the normal and diseased heart can be as simple as determining the cardiac output, i.e. the amount of blood pumped per unit time, the underlying structure and function of this organ are very complex, and scientific investigations encompass experimental procedures on the intact organ, muscle tissue and cardiac muscle cells, as well as roles of individual proteins within the cells and extracellular matrix. Cardiac biomechanics examines not only the structural basis of tissue stress development and global function, but also how cells and tissue respond to external loads. The normal response of the heart to increases in stress is hypertrophy (growth) of the cells and tissue, but defects in the transduction of the mechanical signals can lead to inadequate compensatory growth, and in many cases failure of the mechanical pump.

1. Introduction

The heart is a mechanical pump, and its main function is simple: to provide adequate oxygenated blood to the organs and tissues of the body, including itself. From this perspective, we can characterize the function of the heart with simple parameters such as cardiac output, i.e. the amount of blood that the heart is able to eject over a certain amount of time. Indeed many clinical parameters used to assess cardiac function and dysfunction are straightforward such as peak pressures developed by the contracting ventricles or the fraction of blood volume ejected with each heart beat (ejection fraction). But to accomplish this pumping action a very complex structure exists which not only can alter its function during acute stress (i.e. running up stairs), but has the ability to remodel itself in the long term to adapt to adverse environmental conditions, in an attempt to continue to deliver the needed blood supply to the body. Understanding of this remodeling requires biomechanical experiments and modeling at various scales from tissue to cell. When an insult is too great for a compensated response, the pump can fail. Heart failure is becoming a more common disease as the population ages and more people survive diseases such as myocardial infarction. It is still not clear why heart failure occurs, and how mechanical forces and structural changes are related to heart failure. Biomechanical function, cardiac remodeling and the biomechanical events associated with heart failure will be the focus of this chapter. The relationship between the electrical activity and mechanical function in the normal heart are given in Chapter 3.

2. The Heart as a Mechanical Pump

Like other organs of the body, the heart is composed of cells surrounded by an extracellular matrix which structurally joins the cells together and can transmit force. In muscle tissue, when the contractile cells (myocytes) generate forces by contracting, these forces need to be transmitted between cells and layers of cells in a synchronous fashion in order to develop pressure inside the ventricular chambers and pump blood throughout the body. Mammalian hearts have evolved with a complex structure in order to optimize the pumping capability of these chambers. The structure of mammalian hearts is remarkably conserved between species: the hearts from mice, rats, rabbits, dogs, pigs and humans all look very much the same, except for their size. They all develop about the same amount of pressure. Thus experimental work in animal hearts has direct implications for man. This is especially important for the mouse heart, since genetically engineered mice have been made to mimic many human diseases, and the role of individual proteins can be more easily examined in the mouse heart.

2.1 *Anatomy and structure of the heart*

The structure and anatomy of mechanical organs such as the heart determine how the organ will function. The mammalian heart consists of four chambers, two atria and two ventricles. The left side of the heart, the left atrium and ventricle, pump oxygenated blood to the entire periphery, while the right side collects deoxygenated blood from the body and pumps it to the lungs to be oxygenated. The flow of blood is directed by four valves at the "base" the heart: they are all one-way valves, two of which permit flow from the atria to the ventricles, and the other two are at the outflow area of each ventricle to prevent backflow from the aorta on the left side and the pulmonary artery on the right side. The working muscle of the heart is supplied with oxygen by a system of arteries and veins within the myocardium, known as coronary vessels. The coronary arteries originate near the base of the aorta, and blood flow through these arteries occurs mostly during the passive part of the cardiac cycle, diastole. Blockage of these coronary arteries leads to local tissue damage and cell death — a myocardial infarction. The muscular walls of the ventricles are typically divided into three layers, the endocardium at the inner surface, the midwall myocardium, and the epicardium at the outer surface. The contractile cells of the myocardium, the myocytes, are arranged in distinct bundles that wrap around in a helical pattern to form the continuum of the myocardium. At any particular location in the heart, the myocytes are arranged tangent to the epicardial surface plane, and their angle with respect to a circumferential axis varies from about –90° to +90° from epicardial to endocardium, and are roughly circumferential near the midwall.[1] The myocytes are also arranged in a laminar "sheet" structure.[2] These sheets are about three to four cells thick, and run roughly radial through the wall. The relative motion of these sheets during cardiac contraction plays an important role in cardiac mechanics.[3] The myocytes are connected together by an extensive extracellular matrix synthesized by the cardiac fibroblasts, composed mostly of collagen types I and III.[4] Although the extracellular matrix only composes up to 5% of the tissue volume, this complicated matrix connects cells to each other, and also helps form the sheet structure and the overall geometry of the heart, and is a primary determinant of the passive material properties of the tissue.[5]

2.2 *Electrical activation*

Cardiac contraction starts with an electrical signal being propagated throughout the tissue. When this electrical signal reaches the individual myocytes, they contract to produce the mechanical pumping action of the ventricles. Details on electrical activation can be found in Chapter 3, including the electrocardiogram. The electrical activity of the heart originates in a small area in the right atrium called the sino-atrial (S-A) node. Specialized cells (pacemaker cells) control the electrical activity of this region, which in turn control the contraction rate of the entire heart. Once the S-A node starts an action potential, this electrical signal then propagates though the specialized conduction system of the myocardium. The signal travels through the atria, through the atrioventricular node, delaying the signal, before it continues to the ventricles. The electrical signal travels quickly through the conduction system, essentially activating the entire ventricular muscle in synchrony. Defects in electrical activation may have severe consequences for cardiac function. Small areas of tissue may depolarize spontaneously and initiate premature contractions and other arrhythmias. Transmission of the electrical signal may be blocked in certain areas due to disease or drugs, leading to conditions such as high-frequency contractions (flutter) and non-coordinated contractions (fibrillation).

2.3 *Cellular contraction*

When the myocytes are electrically stimulated, their membranes depolarize, setting off a chain of events within the cell leading to contraction of each individual myocyte.

The main ionic player in the process of muscle contraction is calcium. The initial step in muscle cell contraction is an influx of calcium into the cell due to membrane depolarization. This ionic flow, through the L-type calcium channels, occurs primarily in t-tubules, which are internal extensions of the cell membrane that help distribute this calcium flux throughout the interior of the cell. These calcium ions do not go far: they diffuse across a small subspace and reach the sarcoplasmic reticulum, where they induce further rapid calcium ion release, in high concentrations. It is the calcium from the sarcoplasmic reticulum which diffuses into the vicinity of the sarcomere, the force generating unit of the muscle cell. The sarcomere of cardiac muscle is about 2.2 μm long, and is composed of overlapping thin (actin) and thick (myosin) filament proteins. Transverse arms or cross-bridges are activated in the presence of calcium to pull the myosin past the actin, shortening the distance between the Z-discs (ends of the sarcomeres). This results in sarcomere shortening, which in turn produces muscle cell shortening and stress. Since sarcomeres are arranged in a regular, aligned pattern, it would be expected that muscle force is developed along the axis of the muscle (and sarcomere). In general this is the case, but there may also be a substantial "cross-fiber" component of active muscle stress.[6,7] Indeed in the intact heart, most models of active stress show substantial cross-fiber stress.[8] It may be that this transverse stress is due to active tension development in that direction, or in whole muscle, due to tethering effects between muscles and layers of muscle.

To examine force development directly, we have developed a system to examine forces in individual myocytes. Termed traction force microscopy, this method has been used in several other cell types to examine patterns of developed forces.[9] Neonatal myocytes are isolated and cultured in a polyacrylamide gel (Fig. 1). The gel has an elastic modulus of 1.2 kPa and contains 1 μm fluorescent beads. The cell is stimulated with a square-wave electrical pulse, contracting with the myocyte, and the displacement of the embedded markers can be used to estimate the local traction forces.

2.4 *Global and regional function*

At the organ level, we think of the heart as a pump, capable of generating internal pressure and ejecting blood. At the cellular level, individual myocytes are structurally connected together to form the myocardial tissue. During each cardiac cycle, the cells are electrically stimulated, contract and relax. When examined at the tissue (regional) level, the myocardium is an anisotropic material (different material properties in different directions), and is also heterogeneous (properties vary with location) and viscoelastic (properties of both fluids and solids combined).

2.4.1 *Global cardiac function*

At the global level, cardiac function can be expressed in terms of ventricular blood pressure (P) and chamber volume (V). For the most part, P-V mechanics have been

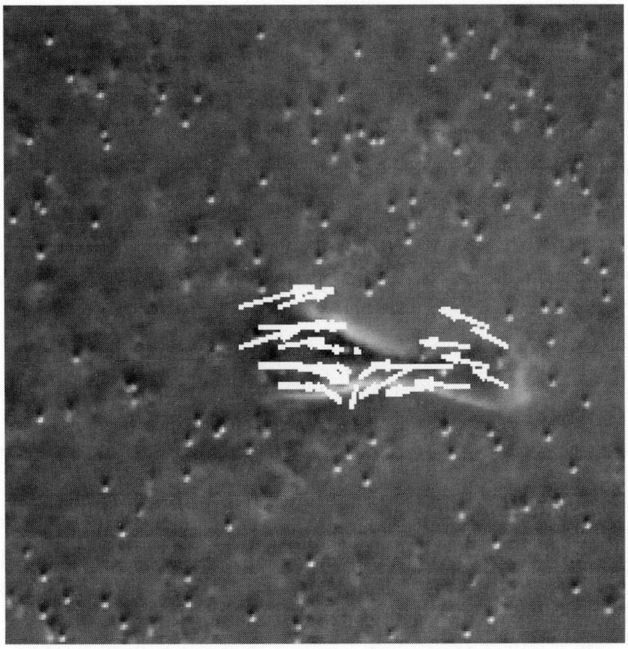

Fig. 1. Deformation of a polyacrylamide gel by a contracting neonatal rat cardiac myocyte. Traction force microscopy measures the displacements of fiduciary fluorescent beads embedded in the gel, and fits these to a map of most likely traction forces (represented by arrows).

determined for the left ventricle, but the general principles apply to the right ventricle also. In terms of function, the atria act as primer pumps, adding inflow to the ventricles before the inflow valves to the ventricles close. The cardiac cycle has two phases, systole (contracted) and diastole (relaxed). The spread of the action potential and contraction of the myocytes defines the start of systole. The initial phase of systole is isovolumic, during which both valves of the ventricle are closed, the volume in the chamber is constant, and the pressure rises due to the muscle contraction. When the pressure in the ventricle becomes higher than in the aorta, the aortic (outflow) valve opens, and blood is ejected from the ventricle. Systole ends when the aortic valve again closes, signaling the end of ejection (although there is still some active stress in the myocytes well into diastole). The initial phase of diastole is also isovolumic, as pressure falls. When the chamber pressure is low enough, the mitral (inflow) valve opens, and the blood begins to fill the ventricular chamber, quickly at first, then slowing as the chamber fills. Finally, atrial contraction is again initiated, and the next cycle begins.

The main mechanical function of the heart is to develop internal pressure. Simple biomechanics help to see how the heart accomplishes this task. If P is plotted against V for a cardiac cycle, one obtains a pressure-volume loop (Fig. 2a). A theoretical P-V loop has two vertical lines corresponding to isovolumic phases. It is

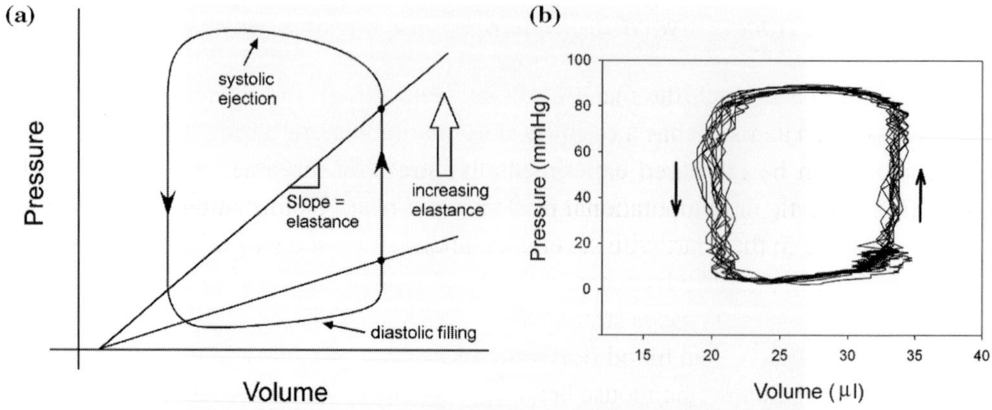

Fig. 2. **(a)** Theoretical pressure-volume loop for the left ventricle. Note isovolumic phases (vertical portions), and between these isovolumic phases are roughly the systolic ejection at the top, and diastolic filling phases at the bottom. At any point on the loop, and instantaneous "elastance" of the ventricle is found from the slope of the line through that P-V point. During isovolumic contraction, the elastance increases (myocytes develop internal stress), leading to the pressure development during this phase. **(b)** Actual pressure-volume data recorded in a mouse heart, for several cardiac cycles. Notice volume scale is in microliters — the entire heart is about 5–6 mm in diameter. Also notice the "isovolumic" phases can show some change in volume (usually due to measurement inaccuracies).

difficult to measure ventricular volumes accurately, especially in small animals; an example of experimental pressure volume loops are shown in Fig. 2b. An interesting way to look at ventricular function is to determine the instantaneous stiffness (elastance) of the chamber.[10] This is the slope of the line passing through the various points of the P-V relation (Fig. 2a). If we look at the elastance through a cardiac cycle, we see it is lowest near the bottom of the P-V loop (diastolic filling) and highest at the end of ejection (end-systole). The elastance, or chamber stiffness, rises and falls with each cardiac cycle, roughly corresponding to the state of muscle mechanical activation. How does the heart develop pressure during isovolumic contraction, i.e. without changes in the chamber volume? We know that the muscle cells contract to squeeze the chamber and develop pressure. During this phase of the cycle, the chamber volume is constant, so even though cells are contracting and developing force, they are probably close to isometric (length is constant). A simple Laplace relationship for a sphere, $P = 2T/r$, shows how the internal pressure balances the wall tension,[11] where P is the chamber pressure, T is the wall stress (tension) and r is the radius. During this phase, r is constant, and the pressure and wall stress both increase. As the wall stress goes up (due to cross-bridge interactions), the elastance is also increasing, meaning that the tissue is getting stiffer. Thus the chamber stiffness and developed stress increase during isovolumic contraction to increase the internal pressure.

2.4.2 *Regional (local) myocardial function*

On a local, regional level, the material of the heart stiffens during systole and relaxes during diastole. The tissue has a complex deformation pattern during the cardiac cycle. Deformation can be measured experimentally; stress on the other hand, can only be found from analytic or computational models of the heart. Deformation, or finite strain, can be measured in the heart with several techniques, and is a way of quantifying function in the tissue. Recent advancements with echocardiograpic methods have allowed for measurement of local tissue strains and strain rate,[12] in addition to classic measures of chamber dimensions and blood flows. Magnetic resonance imaging is used clinically and in animals, including the mouse heart (Fig. 3). If the data acquisition is gated to the cardiac cycle (via the ECG signal), then 2D images can be obtained for geometric reconstruction and function measures such as stroke volume and ejection fraction.

Another technique used for examining detailed deformation in the myocardium is with implanted markers in animal experiments, and visualization of these markers with X-ray imaging. This technique has been used extensively for structure-function studies of the myocardium. Small metal markers are inserted in or on the heart muscle and imaged with X-ray. A calibration grid is used to reconstruct the three-dimensional locations of the implanted markers, and their motion is recorded via computer acquisition of the video images. This technique had been used to examine regional passive and active mechanics,[13,14] tissue mechanics during hypertrophy,[15] relationships of coronary flow to regional mechanics[16,17] and mechanics of regional scar formation after infarction.[18] Recently, we have used this technique to examine the regional mechanics of the myocardium with different ventricular pacing sites. Ventricular pacing is used clinically as a therapy for dysynchronous electrical activation, and it has been shown that pacing the heart on the ventricle results in altered global function. As seen in Fig. 4, normal

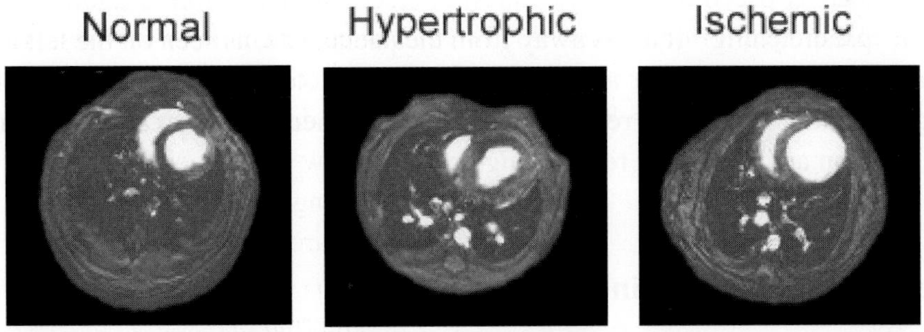

Fig. 3. Cardiac imaging performed with a 7T small animal magnetic resonance scanner. Left (circular) and right (crescent) ventricular chambers are white. These axial, equatorial slices are from a normal adult mouse, age 36 weeks; Age-matched mouse 11 days after aortic banding, which produces a pressure overload hypertrophy and extensive wall thickening; the effect of coronary artery ligation (ischemia) is septal bulging and ventricular dilation as seen after 30 days of infarction.

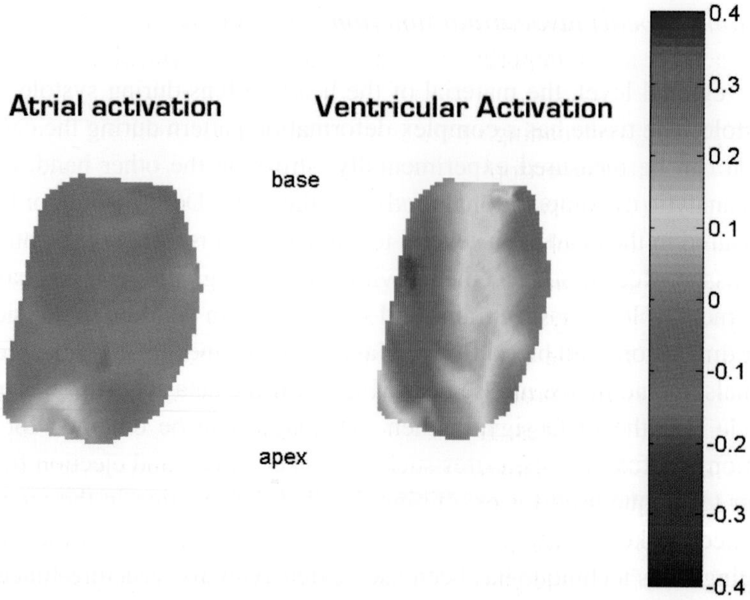

Fig. 4. Local deformation patterns with and without ventricular pacing. Color scale shows local circumferential strain on a surface near the midwall of the left ventricular anterior wall, at a time in early systole when ventricular pressure is rising rapidly. Apex and base directions are labeled. With atrial activation (normal sinus rhythm), the tissue is starting to shorten (slightly negative strain) as expected, in a uniform fashion. With ventricular pacing (opposite the measurement site), this area undergoes "prestretch", as seen by the large positive strains even after pressure in the chamber has started to rise.

activation from the atrium results in mostly small, negative circumferential strains at the beginning of systole (which become more negative as systole continues, not shown). If the same heart is paced at a ventricular site, electrical activation is altered, and this results in "prestretching" of areas away from the pacing site, as seen on the left as areas of positive strain which occur at the beginning of systole when the tissue is normally starting the shorten.[19] These regional variations in mechanics have significance for global function and long term remodeling processes.

3. Mechanotransduction in Cardiac Myocytes

The main function of muscle cells is to develop stress internally, and have this stress transmitted from cell to cell in a tissue, resulting in mechanical motion, local force development and pumping of blood. Dysfunction of the force development machinery within a myocyte will have obvious consequences for contractile function. But stress in cardiac tissue has another very important effect: cell function and remodeling are determined to a large extent by external tissue stress. Of course chemical factors such as cytokines, hormones and growth factors control growth and remodeling of cells, but mechanical factors play an important role in maintaining cell function and initiating cell remodeling.

External mechanical loading of myocytes can directly increase protein synthesis rates and cause cellular hypertrophy.[20] How the cells actually sense these loads and initiate nuclear protein synthesis is not known.[21] Several hypotheses have been proposed for the mechanisms of mechanotransduction in myocytes.[22] Mechanical stress may be directly transmitted to the nucleus via cytoskeletal connections, or indirectly through a variety of signaling pathways.[23] Stretch activated ion channels may be involved.[24] It has also been proposed that the extracellular matrix transmits the stimulus for myocyte hypertrophy via membrane proteins called integrins.[25] The sarcomere itself may also be a "length sensor" in myocytes. If a structure such as the sarcomere is the sensor, it remains unclear whether the sarcomeric proteins would sense changes in length (strain), or somehow transduce forces (stress). Since a direct "stress sensor" is not known to exist,[26] it is likely that the cellular transducer is actually a strain transducer. This is certainly the case for man-made gauges: strain or displacement is measured directly, but force is measured indirectly through the deformation of an elastic structure. Thus, the distinction between transducing stress and strain at the sensor level is probably related to the stiffness of the deformation-sensing element.

In order to directly evaluate the response of myocytes to external loads, cells are isolated from the tissue and cultured on deformable substrates.[20,27] Gopalan *et al.*[28] have examined the response of cultured myocytes to different patterns of stretch. Microfabrication techniques were used to pattern collagen on deformable elastomers, producing elongated aligned myocytes. In addition, the substrate itself may be grooved to produce these elongated cells (Figs. 5a and 5b) which enable the application of

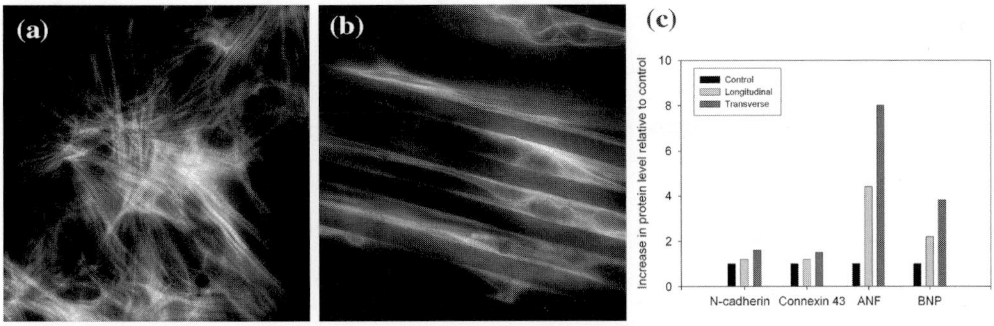

Fig. 5. Cell culture of myocytes with unpatterned **(a)** and patterned **(b)** substrates, both stained with phalloidin for actin visualization. The patterned cells are on a 10 μm grooved silicone substrate, leading to an aligned and elongated cell phenotype. **(c)** Upregulation of proteins with stretch in aligned cultures. After 24 hours, cellular proteins are upregulated more by transverse-dominated stretch (10% transverse, 5% longitudinal), compared to an opposite longitudinal-dominated stretch.

different strain patterns relative to a local cell axis. An elliptical cell stretcher applied 2:1 anisotropic strain statically to the elastic substrate and the attached cells. After one day of static stretch, principal strain parallel to myocytes did not significantly alter protein expression. In contrast, 10% strain transverse to the long axis of the cell resulted in upregulation of protein signal intensity of several hypertrophic markers (Fig. 5c). Thus myocytes are able to differentiate the direction of external loads and show differential hypertrophy responses depending on the pattern of loading.

3.1 *The cytoskeleton and its role in load transduction*

The internal cytoskeleton forms the structure responsible for maintaining cell shape, structural integrity, scaffolding for internal transport and cell division, and several putative signaling cascades.[29] The coupling between the extracellular matrix and the cytoskeleton at the cell membrane suggests that external physical signals may mechanically propagate via the cytoskeleton[30,31] Cytoskeletal proteins have been implicated in several load-sensing pathways (Fig. 6), and cytoskeletal defects are associated with diseases such as dilated cardiomyopathy and heart failure.[32,33] Thus there is evidence that cytoskeletal proteins play a critical role in biomechanical signaling, and may be involved with ventricular dilation and heart failure.[34]

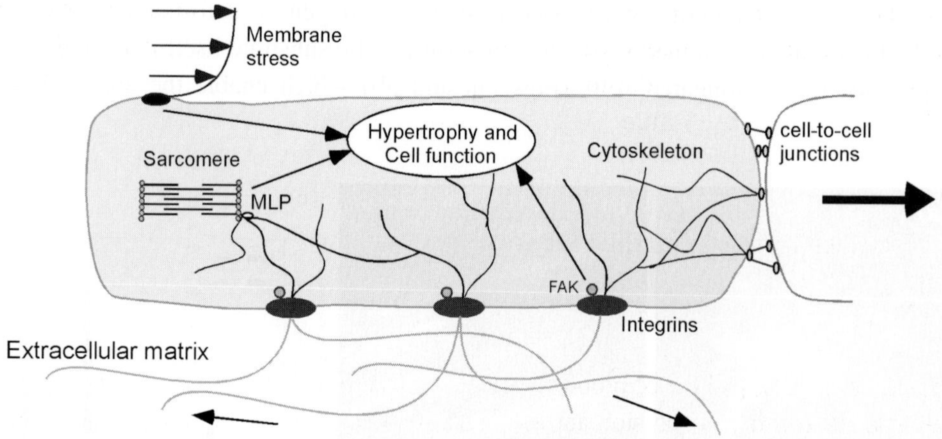

Fig. 6. Cytoskeletal components which may be involved with mechanotransduction, regulating hypertrophy and cell function. External stresses are transmitted to the cell membrane. They may be transduced at membrane-associated integrins via the focal adhesion kinase (FAK) pathway, or directly through the cytoskeleton to the sarcomere, for example via the MLP complex at the Z-disc, or directly to the nucleus. Stresses from neighboring cells are transmitted through cell-to-cell junctions, probably through the cytoskeleton. Other external stresses, for example fluid induced shear on the membrane,[35] may also regulate function via unknown receptor/signaling pathways.

3.2 *Cytoskeletal LIM proteins and cardiomyopathies*

A particular set of cytoskeletal proteins called LIM proteins have been implicated in the stress-sensing pathways in myocytes.[36] LIM is an acronym for the three genes in which this type of domain was first described.[37] Several LIM proteins are Z disc associated, and the Z disc of the cardiac myocyte is known to be an anchoring site for many structural proteins such as actin, α-actinin and titin.[29] The LIM domain is known to be a protein binding interface[38] capable of anchoring different proteins together, thus it would be expected that proteins with a LIM domain may be part of the stress transmitting pathways within the cell.

Defects in muscle LIM protein, MLP, have been associated with dilated cardiomyopathy in humans.[39] MLP deficient mice develop cardiac dilation and eventual heart failure,[40] implicating this part of the cytoskeleton in the transition to heart failure. Even though mice with defective MLP develop heart failure, it is not known how the cytoskeletal defect acts in the mechanosensing pathways. To investigate whether MLP plays a role in force transmission, isolated left ventricular papillary muscles from two-week-old MLP knockout and wildtype mice were mechanically tested.[39] Echocardiography in two-week-old knockout mice did not reveal signs of impaired cardiac performance, thus at this early age the heart appeared to be functioning normally even though the molecular defect was present. Mechanical tests of papillary muscles and trabeculae from these young animals showed mostly normal systolic function, but impaired diastolic function in terms of a decrease in stiffness of the muscle tissue, decrease in the minimum rate of change of force, and impaired relaxation (Fig. 7).[41] Calcium handling appeared to be normal, thus implying that a defect in this cytoskeletal protein results in early diastolic dysfunction which may be a factor leading to cardiac dilation and heart failure.

4. Biomechanics of Cardiac Disease

4.1 *Myocardial ischemia*

Cardiac ischemia, or lack of blood flow, is the most common cause of death: there can be a sudden coronary occlusion and tissue infarction (localized area of cell death), or a slower vessel occlusion over longer periods (weeks to years) that weakens the heart muscle eventually leading to some form of heart failure. Atherosclerosis is usually the cause of the coronary artery plugging. Infarcted myocardium is a completely different material compared to normal tissue: a healed infarct is composed mainly of collagen, similar to a scar on the skin. The myocytes disappear, leaving only a relatively stiff, non-contracting, thinned area of tissue. Depending on the size of the infarct, the heart may be able to functionally compensate with growth of the non-infarcted tissue. As expected, the stress and strain in the infarcted region will differ significantly from the

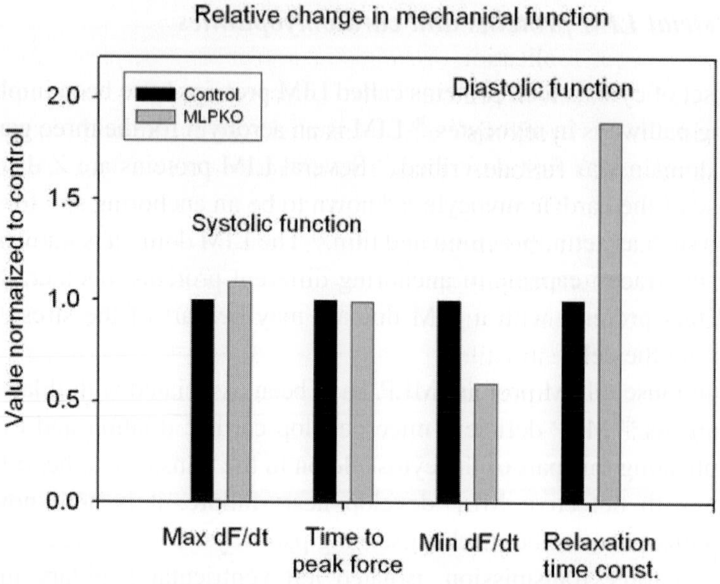

Fig. 7. Changes in systolic and diastolic mechanics in myocardial tissue from young animals with cytoskeletal defects (MLP knockout), before signs of heart failure are present. Data were acquired from trabeculae samples contracting at 1 Hz, with uniaxial force measured during each twitch.[41] Systolic parameters in the MLP knockout tissue are unchanged, but diastolic parameters (minimum rate of change of force and relaxation time constant) show deficiencies, indicating that diastolic dysfunction is present before overt signs of heart failure in this model, hence the changes in diastolic properties are a possible mechanism of cardiac dilation and eventual heart failure.

contracting tissue,[42] and the contracting tissue will have alterations in regional mechanics. Other factors such as decreased chamber distensibility and slowed myocardial relaxation will impede global function of the infarcted left ventricle, eventually contributing to failure of the pump.

4.2 *Cardiac hypertrophy*

The heart attempts to compensate for increased pumping demands (more pressure or volume) by geometric remodeling. At the cellular level, myocyte hypertrophy is a response of myocytes to increased loading conditions. Mature myocytes cannot divide, so growth of these cells is a mechanism for the heart to increase its pumping capability in the long term. Myocyte hypertrophy is considered a pathologic process when the muscle becomes structurally abnormal and functionally impaired. It is interesting that two distinct “types” of cardiac hypertrophy have been described: pressure and volume overload hypertrophy. Each is different at the cellular and organ levels. In volume overload or eccentric hypertrophy, myocytes respond to an increase in diastolic pressure and volume with series replication of sarcomeres, increased axial myocyte length and chamber dilatation.[43] This is in contrast to the increased myocyte cross-sectional area by the parallel addition of myofibrils[44] and substantial wall thickening (concentric hypertrophy) during pressure overload.[45] These two distinct modes of hypertrophy imply that the myocytes can sense different types of loads and activate different molecular signaling pathways,[46] although the molecular mechanisms driving these differences are unknown. Ventricular hypertrophy has implications for cardiac mechanics: for example, thickening of the ventricular wall tends to reduce the wall stress at the same loading pressure. But filling and contractile mechanics are also affected by geometric alterations in hypertrophy, and many times the material properties of the tissue are altered during hypertrophic remodeling.

4.3 *Heart failure*

Both cardiac hypertrophy and myocardial ischemia in many cases lead to heart failure, or an inability of the ventricles to supply the required cardiac output. As discussed above, the inability of the heart to compensate to changes in external loads, i.e. a defect in mechanotransduction, can be a factor in the transition to heart failure. Heart failure is typically associated with defective calcium handling within the myocyte,[47] as well as programmed cell death, or apoptosis.[48] Dilation of the ventricle is typically associated with heart failure, thus tissue stress and geometric remodeling play important roles in the progression to failure. Although failure of systolic function is by definition directly correlated with heart failure, there has been much investigation into the role of diastolic properties in the genesis of this disorder.[49] Factors such as diastolic chamber compliance and myocardial relaxation have been implicated in the development of heart failure. In fact, "diastolic" heart failure can exist, in which systolic function of the heart is mostly normal, but the diastolic dysfunction and clinical signs of heart failure are present.

5. Summary

The mechanical pumping function of the heart is straightforward: the heart contracts to provide the blood necessary for the organs and tissues of the body. But the structure and function of the tissue components are highly complex, and the heart has the ability to change function not only acutely in response to extra demands, but can also remodel in the long term to compensate for changes in loading conditions and other environmental factors. Biomechanical analysis of tissue function and remodeling in diseases such as heart failure allow for the detailed understanding of mechanisms of disease in this mechanical pump.

Acknowledgments

Several members of the UCSD Cardiac Mechanics Research Group were responsible for data and figures in this chapter, including Ilka Lorenzen-Schmidt, John Gallager, Ben Coppola, Peter Costandi and Jeff Jacot. This work was supported by NIH Grants HL43026, HL64321 and HL32583.

References

1. D. D. Streeter, Jr, H. M. Spotnitz, D. P. Patel, J. Ross, Jr. and E. H. Sonnenblick, Fiber orientation in the canine left ventricle during diastole and systole, *Circ. Res.* **24**: 339–347 (1969).
2. I. J. LeGrice, B. H. Smaill, L. Z. Chai, S. G. Edgar, J. B. Gavin and P. J. Hunter, Laminar structure of the heart: ventricular myocyte arrangement and connective tissue architecture in the dog, *Am. J. Physiol.* **269**: H571–582 (1995).
3. I. J. LeGrice, Y. Takayama and J. W. Covell, Transverse shear along myocardial cleavage planes provides a mechanism for normal systolic wall thickening, *Circ. Res.* **77**: 182–193 (1995).
4. J. B. Caulfield and T. K. Borg, The collagen network of the heart, *Lab. Invest.* **40**: 364–371 (1979).
5. T. F. Robinson, L. Cohen-Gould, S. M. Factor, M. Eghbali and O. O. Blumenfeld, Structure and function of connective tissue in cardiac muscle: collagen types I and III in endomysial struts and pericellular fibers, *Scanning Microsc.* **2**: 1005–1015 (1988).
6. D. H. Lin and F. C. Yin, A multiaxial constitutive law for mammalian left ventricular myocardium in steady-state barium contracture or tetanus, *J. Biomech. Eng.* **120**: 504–517 (1998).
7. G. I. Zahalak, V. de Laborderie and J. M. Guccione, The effects of cross-fiber deformation on axial fiber stress in myocardium, *J. Biomech.* **121**: 376–385 (1999).
8. J. M. Guccione, K. D. Costa and A. D. McCulloch, Finite element stress analysis of left ventricular mechanics in the beating dog heart, *J. Biomech.* **28**: 1167–1177 (1995).
9. S. Munevar, Y. Wang and M. Dembo, Traction force microscopy of migrating normal and H-ras transformed 3T3 fibroblasts, *Biophys. J* **80**: 1744–1757 (2001).
10. K. Sagawa, *Cardiac Contraction and the Pressure-Volume Relationship* (Oxford University Press, New York, 1988).
11. Y. C. Fung, *Biomechanics: Mechanical Properties of Living Tissues* (Springer-Verlag Inc., New York, 1981).
12. J. D'hooge, B. Bijnens, J. Thoen, F. van de Werf, G. R. Sutherland and P. Suetens, Echocardiographic strain and strain-rate imaging: a new tool to study regional myocardial function, *IEEE Trans. Med. Imaging* **21**: 1022–1030 (2002).
13. J. H. Omens, K. D. May and A. D. McCulloch, Transmural distribution of three-dimensional strain in the isolated arrested canine left ventricle, *Am. J. Physiol.* **261**: H918–928 (1991).
14. L. K. Waldman, *In vivo* measurement of regional strains in myocardium. In: *Frontiers in Biomechanics*, eds. G. W. Schmid-Schönbein, S. L. Woo and B. W. Zweifach (Springer-Verlag, New York, 1986), pp. 99–116.
15. J. H. Omens and J. W. Covell, Transmural distribution of myocardial tissue growth induced by volume overload hypertrophy in the dog, *Circulation* **84**: 1235–1245 (1991).
16. K. D. May-Newman, J. H. Omens, R. S. Pavelec and A. D. McCulloch, Three-dimensional transmural mechanical interaction between the coronary vasculature and passive myocardium in the dog, *Circ. Res.* **74**: 1166–1178 (1994).
17. R. Mazhari, J. H. Omens, J. W. Covell and A. D. McCulloch, Structural basis of regional dysfunction in acutely ischemic myocardium, *Cardiovasc. Res.* **47**: 284–293 (2000).

18. S. D. Zimmerman, W. Karlon, J. Holmes, J. Omens and J. Covell, Structural and mechanical factors influencing infarct scar collagen organization, *Am. J. Physiol. Heart Circ. Physiol.* **278**: H194–200 (2000).
19. B. A. Coppola, J. W. Covell, A. D. McCulloch and J. H. Omens, Asynchrony of ventricular activation affects maginitude and timing of fiber stretch in late-activated regions of the canine heart, *Am. J. Physiol. Heart Circ. Physiol.* **293**: H754–761 (2007).
20. D. L. Mann, R. L. Kent and G. Cooper, IV, Load regulation of the properties of adult feline cardiocytes: growth induction by cellular deformation, *Circ. Res.* **64**: 1079–1090 (1989).
21. P. Tavi, M. Laine, M. Weckstrom and H. Ruskoaho, Cardiac mechanotransduction: from sensing to disease and treatment, *Trends Pharmacol. Sci.* **22**: 254–260 (2001).
22. J. Lammerding, R. D. Kamm and R. T. Lee, Mechanotransduction in cardiac myocytes, *Ann. N. Y. Acad. Sci.* **1015**: 53–70 (2004).
23. R. S. Reneman, T. Arts, M. van Bilsen, L. H. E. H. Snoeckx and G. J. van der Vusse, Mechanoperception and mechanotransduction in cardiac adaptation: mechanical and molecular aspects. In: *Molecular and Subcellular Cardiology: Effects of Structure and Function*, eds. S. Sideman and R. Beyar (Plenum Press, New York, 1995), pp. 185–194.
24. J. O. Bustamante, A. Ruknudin and F. Sachs, Stretch-activated channels in heart cells: relevance to cardiac hypertrophy, *J. Cardiovasc. Pharmacol.* **17**: S110–113 (1991).
25. R. L. Juliano and S. Haskill, Signal transduction from the extracellular matrix, *J. Cell. Biol.* **120**: 577–585 (1993).
26. T. Arts, F. W. Prinzen, L. H. E. H. Snoeckx and R. S. Reneman, A model approach to the adaptation of cardiac structure by mechanical feedback in the environment of the cell. In: *Molecular and Subcellular Cardiology: Effects of Structure and Function*, eds. S. Sideman and R. Beyar (Plenum Press, New York, 1995), pp. 217–228.
27. J. Sadoshima, L. Jahn, T. Takahashi, T. J. Kulik and S. Izumo, Molecular characterization of the stretch-induced adaptation of cultured cardiac cells, *J. Biol. Chem.* **267**: 10551–10560 (1992).
28. S. M. Gopalan, C. Flaim, S. Bhatia, M. Hoshijima, R. Knöll, K. R. Chien, J. H. Omens and A. D. McCulloch, Anisotropic stretch-induced hypertrophy in neonatal ventricular myocytes micropatterned on deformable elastomers, *Biotechnol. Bioeng.* **81**: 578–587 (2003).
29. K. A. Clark, A. S. McElhinny, M. C. Beckerle and C. C. Gregorio, Striated muscle cytoarchitecture: an intricate web of form and function, *Annu. Rev. Cell. Dev. Biol.* **18**: 637–706 (2002).
30. D. G. Simpson, L. Terracio, M. Terracio, R. L. Price, D. C. Turner and T. K. Borg, Modulation of cardiac myocyte phenotype *in vitro* by the composition and orientation of the extracellular matrix, *J. Cell. Physiol.* **161**: 89–105 (1994).
31. P. Ruiz and W. Birchmeier, The plakoglobin knock-out mouse: a paradigm for the molecular analysis of cardiac cell junction formation, *Trends Cardiovasc. Med.* **8**: 97–101 (1998).
32. S. Hein, D. Scholz, N. Fujitani, H. Rennollet, T. Brandt, A. Friedl and A. Schaper, Altered expression of titin and contractile proteins in failing human myocardium, *J. Mol. Cell. Cardiol.* **26**: 1291–1306 (1994).
33. K. R. Chien, Stress pathways and heart failure, *Cell* **98**: 555–558 (1999).
34. J. A. Towbin, The role of cytoskeletal proteins in cardiomyopathies, *Curr. Opin. Cell. Biol.* **10**: 131–139 (1998).
35. I. Lorenzen-Schmidt, G. W. Schmid-Schönbein, W. R. Giles, A. D. McCulloch, S. Chien and J. H. Omens, Chronotropic response of cultured neonatal rat ventricular myocytes to short-term fluid shear, *Cell Biochem. Biophys.* **46**: 113–122 (2006).
36. A. M. Katz, Cytoskeletal abnormalities in the failing heart: out on a LIM?, *Circulation* **101**: 2672–2673 (2000).
37. Q. Zhou, P. Ruiz-Lozano, M. E. Martone and J. Chen, Cypher, a striated muscle-restricted PDZ and LIM domain-containing protein, binds to alpha-actinin-2 and protein kinase C, *J. Biol. Chem.* **274**: 19807–19813 (1999).

38. K. L. Schmeichel and M. C. Beckerle, The LIM domain is a modular protein-binding interface, *Cell* **79**: 211–219 (1994).
39. R. Knöll, M. Hoshijima, H. M. Hoffman, V. Person, I. Lorenzen-Schmidt, M. L. Bang, T. Hayashi, N. Shiga, H. Yasukawa, W. Schaper, W. McKenna, M. Yokoyama, N. J. Schork, J. H. Omens, A. D. McCulloch, A. Kimura, C. C. Gregorio, W. Poller, J. Schaper, H. P. Schultheiss and K. R. Chien, The cardiac mechanical stretch sensor machinery involves a Z disk complex that is defective in a subset of human dilated cardiomyopathy, *Cell* **111**: 943–956 (2002).
40. S. Arber, J. J. Hunter, J. Ross, Jr., M. Hongo, G. Sansig, J. Borg, J.-C. Perriard, K. R. Chien and P. Caroni, MLP-deficient mice exhibit a disruption of cardiac cytoarchitectural organization, dilated cardiomyopathy, and heart failure, *Cell* **88**: 393–403 (1997).
41. I. Lorenzen-Schmidt, B. D. Stuyvers, H. E. D. J. ter Keurs, M. Date, M. Hoshijima, K. Chien, A. D. McCulloch and J. H. Omens, Young MLP deficient mice show diastolic dysfunction before the onset of dilated cardiomyopathy, *J. Mol. Cell. Cardiol.* **39**: 241–250 (2005).
42. J. W. Holmes, H. Yamashita, L. K. Waldman and J. W. Covell, Scar remodeling and transmural deformation after infarction in the pig, *Circulation* **90**: 411–420 (1994).
43. P. Anversa, R. Ricci and G. Olivetti, Quantitative structural analysis of the myocardium during physiological growth and induced cardiac hypertrophy: a review, *J. Am. Coll. Cardiol.* **7**: 1140–1149 (1986).
44. W. Grossman, D. Jones and L. P. McLaurin, Wall stress and patterns of hypertrophy in the human left ventricle, *J. Clin. Invest.* **56**: 56–64 (1975).
45. S. H. Smith and S. P. Bishop, Regional myocyte size in compensated right ventricular hypertrophy in the ferret, *J. Mol. Cell. Cardiol.* **17**: 1005–1011 (1985).
46. S. Sopontammarak, A. Aliharoob, C. Ocampo, R. A. Arcilla, M. P. Gupta and M. Gupta, Mitogen-activated protein kinases (p38 and c-Jun NH2-terminal kinase) are differentially regulated during cardiac volume and pressure overload hypertrophy, *Cell Biochem. Biophys.* **43**: 61–76 (2005).
47. J. A. Birkeland, O. M. Sejersted, T. Taraldsen and I. Sjaastad, EC-coupling in normal and failing hearts, *Scand. Cardiovasc. J.* **39**: 13–23 (2005).
48. H. Sabbah and V. Sharov, Apoptosis in heart failure, *Prog. Cardiovasc. Dis.* **40**: 549–562 (1998).
49. W. H. Gaasch and M. R. Zile, Left ventricular diastolic dysfunction and diastolic heart failure, *Annu. Rev. Med.* **55**: 373–394 (2004).

CHAPTER 5

BIOENGINEERING SOLUTIONS FOR THE TREATMENT OF HEART FAILURE

John T. Watson and Shu Chien

Abstract

For the last 50 years, deaths from cardiovascular disease have steadily declined. The major exception is heart failure (HF), which is a disease condition where the heart cannot pump enough blood to meet the metabolic requirements of the patient's body. HF has steadily increased over the last 50 years for both men and women of all races. This chapter discusses HF as a public health problem and a bioengineering solution for reversing this condition and restoring the patients' cardiac function. Design principles are presented for mechanical circulatory support systems, including ventricular assist devices and heart replacements or total artificial hearts.

1. Heart Failure is a Public Health Epidemic

Deaths from cardiovascular disease have steadily declined during the last 50 years, but a major exception is heart failure (HF), which is a disease condition where the heart is not able to fill properly (insufficient venous return) or pump with enough force (insufficient cardiac output), or both.[1] The steady increase in HF involves both men and women of all races,[2] and it can affect the left or the right ventricle, or both.[3] It is the most costly health condition for the United States healthcare system.[4] This chapter discusses HF as a public health problem and a bioengineering solution for reversing this condition and restoring the patients' cardiac function.[5]

Coronary artery disease, high blood pressure and diabetes are the leading causes of HF. Other causative or contributory factors to HF are cardiomyopathy (heart muscle disease), heart valve diseases, conduction arrhythmias, and congenital malformations. Factors that affect heart muscle may also contribute to HF, e.g. HIV/AIDS, cancer treatments, thyroid disorders, and drug and alcohol abuse.

HF includes multiple symptoms such as:

- shortness of breath
- swollen ankles and legs

- fatigue
- angina
- loss of appetite
- weight gain or loss.

Shortness of breath and swollen ankles and legs result from fluid retention in the lungs and lower extremities, respectively. Fatigue and shortness of breath are hallmarks of HF.

HF affects an estimated five million US citizens, with 550,000 new cases each year, and contributes to about 300,000 deaths per year, according to data collected by the National Heart, Lung, and Blood Institute (NHLBI) of the National Institutes of Health (NIH). It costs about US$26 billion annually. The prognosis is a 50% mortality rate at five years.

End-stage HF patients (American Heart Association Class IV) have an even poorer prognosis with a 50% mortality rate at six months. Heart transplantation is the effective treatment for these patients, with survival rates of about 75% at three years. There are tens-of-thousands of patients living with heart transplants. However, the utilization of this therapy is limited by the donor supply, which remains plateaued at around 2000 per year (US) and may be trending downward.

Now the bioengineering solutions: mechanical circulatory support systems (MCSS) are starting to fill the gap that exists for treating the estimated 50,000 patients that could benefit from permanent approaches to restore cardiac-like function and reverse their HF condition.[6] The following sections describe the design philosophy, systems, clinical outcomes and adverse biological effects of MCSS and ventricular assist devices (VAD) or total artificial hearts (TAH). Many aspects of these nascent technologies will benefit from further clinical and basic research and development by teams of clinicians, engineers and scientists.

2. A National Program

A new clinical therapy typically takes 15–20 years from laboratory concept, through animal experiments and clinical trials, before it reaches initial clinical use. Several years of additional experience and improved practice are needed before the therapy is in general clinical application. The innovation phases (up to and including clinical trials) usually require government funding because the timeline is much too long and the outcome too financially risky for private investment. In the early 1960s, as the US public developed confidence in our engineering ability to create almost anything that was humanly conceivable, Congress passed legislation mandating the development of an artificial kidney and artificial heart, together with travel to the moon. Thus, the NIH was charged to implement programs to meet the public health needs for treating kidney and heart failure.

Without a Congressional mandate, it would not have been possible to pursue such targeted therapies that require a critical mass of research teams working collaboratively on the same objective (HF) with a programmatic integration of different approaches, because the individual NIH investigator-initiated grants system does not provide the leadership and review monitoring process that would facilitate research with a 15-year horizon.

In 1964, NHLBI began a peer-review program for research and development of systems to treat acute and chronic HF. Acute treatment (days to weeks) is needed for patients who have reversible HF conditions that require temporary ventricular assistance.[7] Chronic HF was the most challenging problem that requires creation of a multi-year VAD or TAH function.

NHLBI supported the building of a research capacity by soliciting bioengineering teams, comprised of engineers, scientists and clinicians, to demonstrate the feasibility of individual VAS components, to advance the state-of-the-art in sub-program areas such as blood pumps, electrical energy converters, control methods, compliance chambers, and energy transmission techniques, and to enhance knowledge of relevant biomaterials and their biocompatibility.

In 1980, based on peer-review, five contract research programs were established to integrate the most promising concepts into a complete ventricular assist system (VAS) and, subsequently, to demonstrate performance and reliability in both bench tests and animal studies. Government support led to increased investments of funds through industry and university partnerships. Currently, several VAD designs are available for different clinical applications, predominantly for use as bridge to heart transplant, and the research focus of NHLBI today is on the development of VAD and TAH systems for the chronic treatment of pediatric and adult advanced HF.[8,9]

3. Design Concept Based on Clinical Indications

The design of bioengineered clinical treatment systems should be based on clinical indications and patients needs. Top priority is given to the patient, considering safety, function and reliability. In the 1970s, the clinical goal was that MCSS must demonstrate a high level of reliability for two years.[10] This is equivalent to 80–100 million heart beats or VAS cycles. In the 1990s, based on early clinical MCSS outcomes and new technological advances, the NHLBI increased the mission target to five years (200–250 million cycles).

In order to convert clinical "indications" to engineering specifications, the NHLBI defined a number of physiologic and engineering parameters for the first-generation MCSS based on clinical outcomes suggested by clinicians. In the 1960s there was an epidemic of heart attack-induced deaths for middle-aged men, and this led to the suggestion of a primary physiological design point of 10 L/min output at a rate of 120 beats per minute, with a mean blood pressure of 100 mm Hg. These requirements then define a stroke volume (SV = cardiac output/rate, or ~83 ml). This further determines the physical volume of the blood pumping chamber and, when the activation mechanism (energy converter) and system controller are taken into account, the volume of the device.

The power needed at the level of heart is equal to $\Delta P \times$ flow. For a left ventricular assist device (LVAD) to operate with an output of 10 L/min and a pressure of 100 mm Hg, the power can be calculated as 100 mm Hg × 10 L/min × 2.2 × 10^{-3} (a conversion factor to generate the unit of power in watts) to be 2.2 watts. For a TAH, the right ventricle should be able to develop a 40 mm Hg mean pressure in the pulmonary artery. Thus, the power requirements for both ventricles of a TAH can be calculated to be 2.2 w + [40 mm Hg × 10 L/min × 2.2 × 10^{-3}], or 3.08 watts. This is a tremendous amount of power that must be continuously supplied.

Present MCSS are powered by secondary batteries, and patients carry as much battery power as they wish for their daily activities. The more efficient the MCSS and the higher the density of the batteries, the less cumbersome will be the rechargeable battery pack for the patient. A desirable overall efficiency for a MCSS would be around 20%–25%. This is the efficiency from the battery supply to the level of blood flow. Current MCSS have about half the desirable efficiency. For example, the power needed for the AbioCor (Abiomed, Inc.; Danvers, MA) TAH to deliver a 10 L/min output at the pressure levels mentioned above is about 25 watts, which corresponds to an overall efficiency of only 12% (100% × 3.08/25).

Another clinical need is for "transparency" or "biocompatibility" of the MCSS to the patient, especially for permanent implants. Two approaches are under development for MCSS. One approach is to use surfaces that are so smooth that they have no imperfections larger than the size of blood platelets (~2 μm), based on the concept is that "smooth" surfaces should be invisible and thus biologically acceptable.[11] In practice, however, "smooth" surfaces are detected by the body's defense systems and patients need chronic immunological and anti-coagulation treatments to reduce the chances of a severe immune response, thrombus formation and embolization.

The second approach is to work with textured surfaces.[12] The texture produces a programmed biological response. In the blood, the textured surface forms a blood coagulum which is replaced over a few weeks by a pseudo-intima surface that is basically biologically acceptable. The immune response to a textured surface is usually mild, but it is amplified by blood transfusions, which result in the capture of foreign blood cells by the textured surface. When this occurs, the ensuing immunological response can usually be controlled with modest immunosuppression. The rate of thrombosis also seems to be lower with textured surfaces.[13]

Overall, our understanding of biology and engineering has not been sufficient to create the original concept of "transparency" for the blood contacting surfaces. Although, clinical results for the majority of patients are acceptable, biomaterials and clinical advances are needed for reducing the severity and frequency of adverse events associated with blood contacting surfaces.

4. Systems Approach to Design

Many medical therapies are based on treating a particular "target" which is scientifically associated with a clinical symptom. It happens sometimes that multiple pharmaceuticals may be prescribed in accordance with the clinical indications without considering possible interactions. Modern pharmaceuticals are good examples of this approach. The broader view of systems biology allows the alleviation of multiple symptoms with a treatment directed to the root mechanism for restoration of function. By practice, engineers have taken a systems approach in developing mechanical circulatory support systems.

Pharmaceutical treatment of the symptoms of HF might include multiple drugs such as diuretics for swollen ankles and legs, ACE inhibitors to reduce peripheral vascular resistance and myocardial oxygen consumption, beta blockers to reduce myocardial wall tension, digoxin to improve myocardial contractility, and spironalactone to partially reverse myocardial remodeling associated with hypertrophy of the heart. The interactions of these drugs are unknown. Implantable defibrillators are used for HF patients at high risk of sudden cardiac death due to arrthymias and cardiac resynchronization pacemakers are implanted in some patients to improve cardiac function.

MCSS treats HF from a systems rather than a symptom-based approach by defining a primary design point that would ensure restoration of cardiac function to meet metabolic requirements during rest, daily basic activities and moderate exercise. The systems concept is that by restoring cardiac output, this may eliminate or alleviate the reversible HF symptoms. A specification of 10 L/min cardiac output was chosen as the systems design point to optimize for load margins and efficiency. While the MCSS is not expected to operate at this load continuously, this design goal would provide considerable bandwidth of operation to meet the patient's routine cardiac output needs, reverse most clinical symptoms, while reducing the power requirements across most of the operating range of daily living (3–10 L/min).

5. Clinical Indications

Several forms of VAD have been commonly employed in treating the post-cardiotomy syndrome characterized by a failure to wean the patient from cardiopulmonary bypass and as a bridge for the HF patient who qualifies for a cardiac transplant. According to the guidelines for using a VAD as a bridge to transplant, it is indicated when, despite optimal medical therapy, the patient continues to suffer one or more of the following conditions:

- failing hemodynamics
  - low cardiac output
  - low systolic blood pressure requiring the intra-aortic balloon pump
  - elevated pulmonary capillary wedge pressure

- persistent pulmonary edema off ventilator
- neurologic or renal failures due to low flow that is potentially reversible
- fluid and electrolyte imbalance
- episodes of severe arrhythmias refractory to therapy.

There are many patients who need but do not qualify for cardiac transplant because of their age and/or co-morbidities such as diabetes and end-stage kidney disease. Such patients with end-stage failure, were shown to be candidates for permanent MCSS by the multi-center study called the REMATCH trial (Randomized Evaluation of Mechanical Assistance for the Treatment of Congestive Heart Failure) supported by the NHLBI.[14] This study compared the outcomes of end-stage HF patients receiving long-term implantation of a MCSS versus optimal medical management. Device patients have a clear benefit for survival and quality of life compared to these medial patients.[15] A MCSS may also serve as a bridge to recovery in some circumstances, e.g. some cases of post-cardiotomy syndrome, acute myocarditis, and acute myocardial infarction treated with a VAS have spontaneous recovery of natural cardiac function.[16]

6. Clinical Outcomes and Secondary Prevention

Thousands of adult HF patients have been supported by one of the several FDA approved MCSS. The number of implanted devices has been estimated to be 5600 Heartmate, 1700 Novacor, 3500 Thoratec, and ~3000 others. The feasibility of clinical translation of these technologies is shown by the implants performance by multiple surgical teams at over 200 medical centers in 23 countries. There are many multi-year patients. The longest continuous implant has been a rotary device for six years, a patient who lives in England and travels the world.[17]

REMATCH patients have a clear survival benefit compared to the medial patients, but less than the benefit expected from a cardiac transplant. The REMATCH study carefully documented major clinical adverse events of bleeding, infection, neurological events and device failures. In 2001, adverse events occurred twice as often in patients with MCSS as those on optimal medical therapy. Since 2001, MCSS patient survival has improved greatly and the adverse events rate has been reduced.[18,19] However, there are still many opportunities for bioengineering improvements in all aspects of MCSS design.

The improving clinical outcomes suggest that MCSS may have become a useful secondary prevention strategy for patients with advanced HF. To answer this and other questions, a national registry was established (INTERMACS) for all MCSS patients — designed as a joint effort of the National Heart, Lung and Blood Institute (NHLBI), the

Centers for Medicare and Medicaid Services (CMS), and the Food and Drug Administration (FDA).[20] Actual clinical use will depend on registry results for clinical indications, function, patient quality of life, and cost.

7. Regenerating Cardiac Function

A small percentage of MCSS patients spontaneously recover normal cardiac function after six to 18 months of mechanical circulatory support,[21,22] thus allowing surgical removal of the device. Further research can lead to combination therapies using MCSS for cardiac stability and adjunctive procedures to regenerate cardiac function. The latter can include protein, gene and cell therapies as well as surgical reshaping of the natural ventricles. Combination therapy to restore cardiac function will have great appeal for patients that are otherwise healthy except for their HF.

8. Summary

Heart failure may be the world's costliest medical condition. While cardiovascular death and disability have declined substantially over the last few decades, HF morbidity and death continue to increase. Bioengineering of MCSS has provided the first successful clinical treatment alternative for end-stage HF since cardiac transplantation was introduced in 1967.

MCSS has proved a safe and beneficial clinical strategy for secondary prevention for HF, both by chronic application and enabling adjunctive procedures that restore cardiac function. These patients are extending the natural history of HF and should provide new insights into its basic mechanisms. This is a very nascent field that will benefit greatly from advances in bioengineering and related fields.

References

1. T. Thom, N. Haase, W. Rosamond, V. J. Howard, J. Rumsfeld, T. Manolio, Z. J. Zheng, K. Flegal, C. O'Donnell, S. Kittner *et al.*, Heart disease and stroke statistics — 2006 update: a report from the American Heart Association Statistics Committee and Stroke Statistics Subcommittee, *Circulation* **113**: e85–151 (2006).
2. Vital Statistics of the United States, National Center for Health Statistics [http://www.cdc.gov/mmwr/preview/mmwrhtml/00024847.htm].
3. S. A. Hunt, ACC/AHA 2005 guideline update for the diagnosis and management of chronic heart failure in the adult: a report of the American College of Cardiology/American Heart Association Task Force on Practice Guidelines (Writing Committee to Update the 2001 Guidelines for the Evaluation and Management of Heart Failure), *J. Am. Coll. Cardiol.* **46**: e1–82 (2005).

4. Centers for Medicare and Medicaid Services. NCA Tracking Sheet for Ventricular Assist Devices as Destination Therapy (CAG-00119R). http://www.cms.hhs.gov/mcd/viewtrackingsheet.asp?id=187 (2006).
5. O. H. Frazier, The development, evolution, and clinical utilization of artificial heart technology, *Eur. J. Cardiothorac. Surg.* **11** (Suppl.): S29–31 (1997).
6. J. R. Hogness and M. van Antwerp (eds.), *The Artificial Heart: Prototypes, Policies, and Patients* (Institute of Medicine, National Academy Press Washington, D.C., USA, 1991)
7. National Institutes of Health, US Department of Health and Human Services. The Artificial Heart and the NHLBI Report of the Workshop on the Artificial Heart: Planning for Evolving Technologies (1994).
8. O. H. Frazier, Prologue: ventricular assist devices and total artificial hearts. A historical perspective, *Cardiol. Clin.* **21**: 1–13 (2003).
9. D. B. Olsen, Rotary blood pumps: a new horizon, *Artif. Organs* **23**: 695–696 (1999).
10. G. M. Pantalos, F. Altieri, A. Berson, H. Borovetz, K. Butler, G. Byrd, A. A. Ciarkowski, R. Dunn, O. H. Frazier, B. Griffith *et al.*, Long-term mechanical circulatory support system reliability recommendation: American Society for Artificial Internal Organs and The Society of Thoracic Surgeons: long-term mechanical circulatory support system reliability recommendation, *Ann. Thorac. Surg.* **66**: 1852–1859 (1998).
11. F. B. Russell, D. M. Lederman, P. I. Singh, R. D. Cumming, R. A. Morgan, F. H. Levine, W. G. Austen and M. J. Buckley, Development of seamless tri-leaflet valves, *Trans. Am. Soc. Artif. Intern. Organs* **26**: 66–71 (1980).
12. O. H. Frazier, R. T. Baldwin, S. G. Eskin and J. M. Duncan, Immunochemical identification of human endothelial cells on the lining of a ventricular assist device, *Tex. Heart Inst. J.* **20**: 78–82 (1993).
13. R. M. Lazar, P. A. Shapiro, B. E. Jaski, M. K. Parides, R. C. Bourge, J. T. Watson, L. Damme, W. Dembitsky, J. D. Hosenpud, L. Gupta *et al.*, Neurological events during long-term mechanical circulatory support for heart failure: the Randomized Evaluation of Mechanical Assistance for the Treatment of Congestive Heart Failure (REMATCH) experience, *Circulation* **109**: 2423–2427 (2004).
14. E. A. Rose, A. J. Moskowitz, M. Packer, J. A. Sollano, D. L. Williams, A. R. Tierney, D. F. Heitjan, P. Meier, D. D. Ascheim, R. G. Levitan *et al.*, The REMATCH trial: rationale, design, and end points. Randomized Evaluation of Mechanical Assistance for the Treatment of Congestive Heart Failure, *Ann. Thorac. Surg.* **67**: 723–730 (1999).
15. M. A. Dew, R. L. Kormos, S. Winowich, E. A. Stanford, L. Carozza, H. S. Borovetz and B. P. Griffith, Human factors issues in ventricular assist device recipients and their family caregivers, *ASAIO J.* **46**: 367–373 (2000).
16. D. J. Goldstein, N. Moazami, J. A. Seldomridge, H. Laio, R. C. Ashton, Jr., Y. Naka, D. J. Pinsky and M. C. Oz, Circulatory resuscitation with left ventricular assist device support reduces interleukins 6 and 8 levels, *Ann. Thorac. Surg.* **63**: 971–974 (1997).
17. S. Westaby, O. H. Frazier and A. Banning, Six years of continuous mechanical circulatory support, *N. Engl. J. Med.* **355**: 325–327 (2006).
18. J. W. Long, A. G. Kfoury, M. S. Slaughter, M. Silver, C. Milano, J. Rogers, R. Delgado and O. H. Frazier, Long-term destination therapy with the HeartMate XVE left ventricular assist device: improved outcomes since the REMATCH study, *Congest. Heart Fail.* **11**: 133–138 (2005).
19. G. H. Mudge, Jr., J. C. Fang, C. Smith and G. Couper, The physiologic basis for the management of ventricular assist devices, *Clin. Cardiol.* **29**: 285–289 (2006).
20. Request for Proposals (NHLBI-HV-05–08), Interagency Registry of Mechanical Circulatory Support for End-Stage Heart Failure (2004).

21. C. M. Terracciano, J. Hardy, E. J. Birks, A. Khaghani, N. R. Banner and M. H. Yacoub, Clinical recovery from end-stage heart failure using left-ventricular assist device and pharmacological therapy correlates with increased sarcoplasmic reticulum calcium content but not with regression of cellular hypertrophy, *Circulation* **109**: 2263–2265 (2004).
22. S. Maybaum, P. Stockwell, Y. Naka, K. Catanese, M. Flannery, P. Fisher, M. Oz and D. Mancini, Assessment of myocardial recovery in a patient with acute myocarditis supported with a left ventricular assist device: a case report, *J. Heart Lung Transplant* **22**: 202–209 (2003).

CHAPTER 6

MOLECULAR BASIS OF MODULATION OF VASCULAR FUNCTIONS BY MECHANICAL FORCES

Shu Chien

Abstract

Mechanical forces such as shear stress can modulate gene and protein expressions and hence cellular functions by activating membrane sensors and intracellular signaling. Studies on cultured endothelial cells have shown that laminar shear stress causes a transient increase in MCP-1 expression, which involves the Ras-MAP kinase signaling pathway. We have demonstrated that integrins and the vascular endothelial growth factor receptor Flk-1 can sense shear stress, with integrins being upstream to Flk-1. Other possible membrane components involved in the sensing of shear stress include G-protein coupled receptors, ion channels, intercellular junction proteins, membrane glycocalyx, and the lipid bilayer. Mechanotransduction involves the participation of a multitude of sensors, signaling molecules, and genes. Microarray analysis has demonstrated that shear stress can up- and down-regulate different genes. Sustained shear stress down-regulates atherogenic genes (e.g. MCP-1) and up-regulates growth arrest genes. In contrast, the disturbed flow observed at branch points and induced in step-flow channels causes sustained activation of MCP-1 and the genes that enhance cell turnover and lipid permeability. These findings provide a molecular basis for the explanation of the preferential localization of atherosclerotic lesions at regions of disturbed flow, such as the arterial branch points, and the sparing of the straight part of the arterial tree. The combination of mechanics and biology (from molecules-cells to organs-systems), can help to elucidate the physiological processes of mechano-chemical transduction and improve the management of important clinical conditions such as coronary artery disease.

1. Introduction

Endothelial cells (ECs) form the interface of the vessel wall with the circulating blood. In addition to being a permeability barrier, ECs perform many other functions such as the production, secretion, and metabolism of biochemical substances and the modulation of contractility of vascular smooth muscle cells (SMCs). Besides their modulation by chemical ligands (e.g. growth factors and hormones), ECs and SMCs also respond to mechanical factors such as pressure and flow, and these responses play a significant role in regulating vascular performance in health and disease.

In atherosclerosis, the accumulation of atheromatous materials in the artery wall causes a narrowing of the vessel lumen and hence a reduction of blood flow and the consequent clinical problems. The two major elements in atherosclerosis are the low density lipoprotein (LDL) and monocytes. Monocytes are transformed in the artery wall into macrophages, which can engulf oxidized LDL to form foam cells (Fig. 1). The accumulation of foam cells, smooth muscle cells and extracellular matrix in the subendothelial intima is the basis for atheroma formation.[1]

The preferential localization of atherosclerotic lesions in branch points and curved regions[2] suggests that local hemodynamic forces play a significant role in atherogenesis.[1,3] There is increasing evidence that hemodynamic factors modulate both monocyte entry and lipid accumulation in the artery wall. The hemodynamic forces exerted on the vessel wall include the shear stress, which acts tangentially on the luminal surface of the vessel as a result of flow and the normal stress and circumferential stress resulting from the action of pressure. Investigations on the

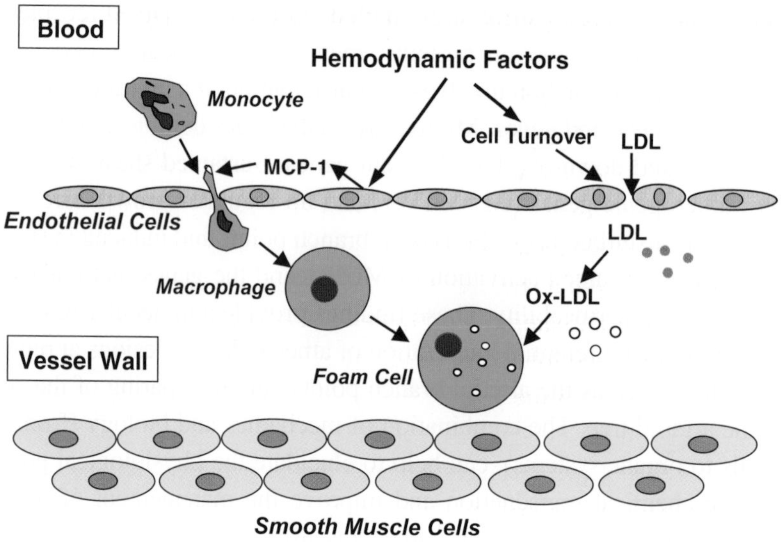

Fig. 1. Schematic drawing to show the roles of hemodynamic factors in (1) enhancing cell turnover to increase LDL permeability and (2) causing monocyte chemotactic protein-1 (MCP-1) secretion to induce monocyte entry across the endothelial cells (ECs). In the subendothelial intima, monocytes are transformed into macrophages, which ingest the oxidized LDL (ox-LDL) to form foam cells. The foam cells and the smooth muscle cells (SMCs) that migrate into the neointima are the main cellular elements in the atheroma.

role of hemodynamic forces in regulating EC function can help to elucidate (1) the fundamental mechanism of mechano-chemical transduction, and (2) the biomechanical and molecular bases of the preferential localization of atherosclerosis in the arterial tree.

2. Effects of Shear Stress and Flow Patterns on Mechanosensing, Signal Transduction, and Gene Expression in Endothelial Cells

2.1 *Effects of shear stress on monocyte chemotactic protein-1 (MCP-1) gene expression*

Cells can respond to chemical and mechanical stimuli by modulating their gene expression, and hence protein expression and functions such as secretion, migration, proliferation, and apoptosis. There are many studies on the effects of shear stress on EC gene expression.[4,5] The understanding of the mechanisms by which cells sense mechanical stimuli, modify their signaling pathway, and modulate gene expression is of fundamental importance in bioengineering and medicine.

It is well known that arterial branch points, in comparison to the straight segments, have a high LDL permeability[6] and are sites of preferential localization of macrophages underneath the ECs,[7] these results suggest that local hemodynamic forces in such lesion-prone areas can enhance the entry of LDL and monocytes. The entry of monocytes into the blood vessel wall is induced by a number of chemotactic agents, especially the monocyte chemotactic protein-1 (MCP-1).[8] MCP-1 is secreted by the EC. Over-expression of the MCP-1 gene and the consequent secretion of MCP-1 protein can attract monocyte entry into the vessel wall.[9]

The effects of shear stress on MCP-1 expression by ECs have been studied in cultured human umbilical vein endothelial cells (HUVECs) with the use of a flow channel system.[10] The application of a shear stress of 12 dynes/cm^2 (which is within the normal range in arteries of approximately 10–20 dynes/cm^2) causes an increase in MCP-1 gene expression to more than two folds in 1.5 hr (Fig. 2). Sustained application of shear stress beyond 4 hr, however, causes the gene expression to decrease below the pre-shear level.

2.2 *Effects of shear stress on MAPK signaling pathways*

The promoter region of MCP-1 gene contains the TPA responsive element (TRE), which is the *cis*-element for the shear stress activation of MCP-1.[11] The transcription factor for TRE is activator protein-1 (AP-1), which is a dimer composed of cJun-cFos or cJun-cJun. The activation of c-Jun and c-Fos is mediated by the mitogen activated protein kinase (MAPK) signaling pathways.[4,12] The MAPK pathways are also involved in cell mitosis and apoptosis. Therefore, the understanding of these pathways is important

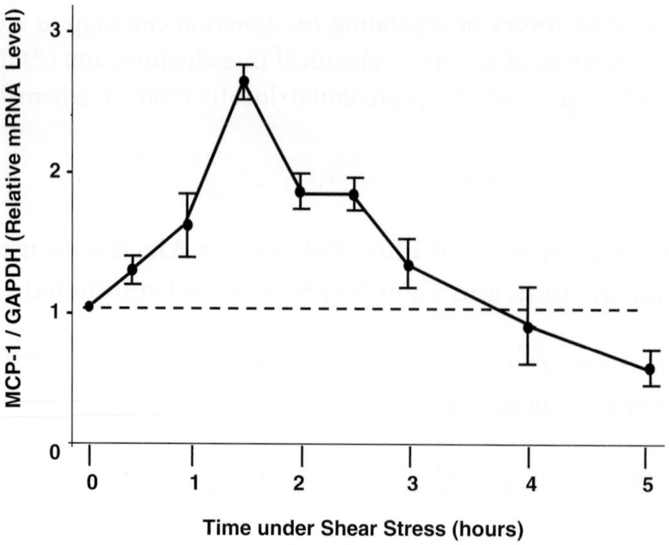

Fig. 2. Time course of the gene expression of monocyte chemotactic protein-1 (MCP-1) in confluent cultured human umbilical vein endothelial cells (HUVECs) in response to a sustained shear stress of 12 dynes/cm^2. The Northern blot results of MCP-1 mRNA were divided by those of the housekeeping gene GAPDH for normalization. Note that the MCP-1 gene expression in response to a step shear stress is transient, reaching a peak at 1.5 hr and declined to below the static control level even though the shear stress was continuously applied for 5 hr or longer. Modified from Shyy *et al.*[10]

for the elucidation of both the chemoattraction of monocytes by MCP-1 and the modulation of LDL permeability by changes in cell turnover.

The MAPK signaling pathways involve a group of protein kinases that are activated sequentially to result in a phosphorylation cascade. There are several parallel pathways in this system (e.g. JNK and ERK shown in Fig. 3), with the molecule Ras being a common upstream molecule. When Ras is activated, it binds to GTP. The application of shear stress (12 dynes/cm^2) to cultured bovine aortic endothelial cells (BAECs) in a flow channel activates Ras to cause it to bind GTP instead of GDP and the ensuing activation of both ERK and JNK.[5,13,14] The Ras-GTP binding occurs in seconds after the beginning of shearing, and the activation of the downstream signaling molecules and gene expression follow sequentially (Fig. 4). The response of Ras is very transient, and the early onset and short duration of its activation reflects its upstream location relative to the MAPKs in the signaling pathways.

Ras N17, which is a negative mutant molecule of Ras, markedly suppresses the shear-activation of TRE and MCP-1.[12] Upstream to Ras, focal adhesion proteins (e.g. focal adhesion kinase, p60Src), and adapter molecules (e.g. Shc, Grb2 and Sos) have been found to be important, because uses of their negative mutants suppress the shear-activation of TRE.[15]

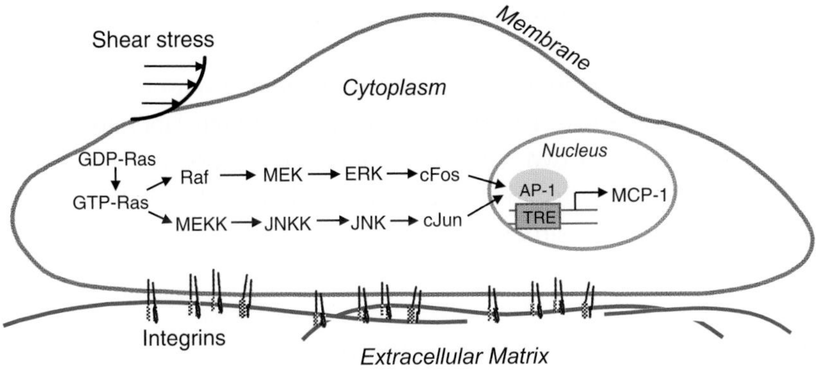

Fig. 3. Schematic diagram showing the Ras-MAP kinase signaling pathways mediating the mechano-chemical transduction in endothelial cells to cause an increase in MCP-1 gene transcription in response to shear stress. The application of shear stress activates Ras by changing it from the GDP-bound state to become GTP bound. This activates the phosphorylation cascades of the MAP kinase pathway, leading to the activation of cFos and cJun and hence the transcription factor activator protein-1 (AP-1). AP-1 binds to the TPA responsive element (TRE) to cause the activation of MCP-1 transcription.

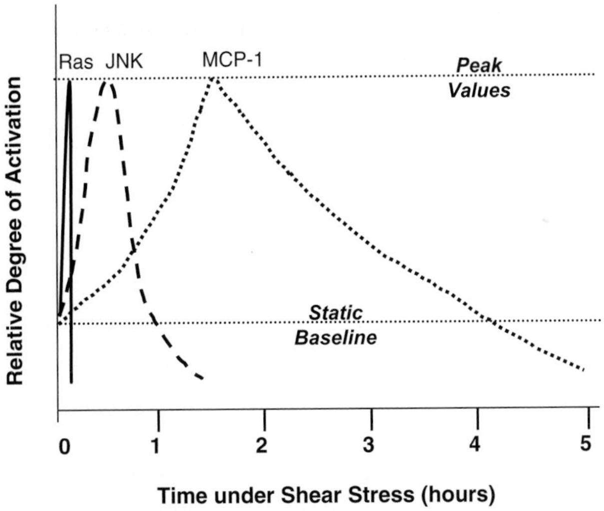

Fig. 4. Time courses of the sequential activation and down-regulation of Ras, JNK, and MCP-1. The activation of Ras reaches a peak in less than 1 min, followed by an increase in the kinase activity of JNK with a peak at 30 min. The induction of the MCP-1 gene occurs later than the activation of these signaling molecules, and peaks at 90 min. Following prolonged shearing, the activities of the signaling molecules and the MCP-1 gene expression become lower than those in the static controls. The lower horizontal line represents the static control level, and the upper horizontal line represents the peak values for the three parameters measured. Data from Li *et al.*[12] and Shyy *et al.*[10]

2.3 *Mechano-sensors for shear stress*

There is evidence that there are several types of membrane sensors for the mechanical stimulation due to shear stress. Experiments conducted in our laboratory have demonstrated the responsiveness of two types of membrane receptors, viz. a receptor tyrosine kinase (Flk-1) and the integrins, to mechanical stimulations. In addition, the cell membrane lipid bilayer may also participate in mechanosensing.

2.3.1 *Membrane receptors — receptor tyrosine kinase*

Receptor tyrosine kinases (RTKs) are membrane receptors that can undergo phosphorylation on their tyrosine residues in response to appropriate stimuli. Such tyrosine phosphorylation activates the receptor such that it becomes associated with the adaptor molecules (e.g. Shc), and this in turn activates the downstream signaling pathway such as Ras-JNK. An example of RTK is the vascular endothelial growth factor (VEGF) receptor Flk-1, which is known to be activated by its chemical ligand VEGF. We found that the application of shear stress activates Flk-1 in the same manner as the chemical stimulation due to VEGF, i.e. Flk-1 undergoes oligomerization and phosphorylation, binds to Shc and other adaptor molecules, and subsequently activates Ras and its downstream molecules.[16] The time courses of the responses of Flk-1 to these two modes of stimuli are similar: starting rapidly in less than 1 min and subsiding totally in 30 min. The effect of shear stress on Flk-1 activation is not altered by the blockade of binding of VEGF to Flk-1, indicating that the mechanical activation of Flk-1 is not the result of a paracrine or autocrine release of VEGF.

2.3.2 *Membrane integrins*

Integrins are transmembrane receptors that link the intracellular cytoskeletal proteins with the proteins in the ECM to provide two-way communications between the cell and its ECM.[17] There are more than 20 types of integrins, each of which is composed of α and β subunits. One of the major integrins in vascular ECs is $\alpha_v\beta_3$, which interacts specifically with the ECM proteins vitronectin and fibronectin. Shear stress causes the association of $\alpha_v\beta_3$ with Shc and the subsequent activation of the Ras pathway, only when the ECs are cultured on ECM composed of vitronectin or fibronectin.[18] In such systems, anti-$\alpha_v\beta_3$ antibodies markedly attenuate the shear-induced signaling. Another EC integrin is $\alpha_6\beta_1$, which has laminin as its cognate ECM molecule. Shear stress can activate $\alpha_6\beta_1$ in ECs only when they are cultured on laminin. These findings indicate that integrins play a significant role in the initiation of signaling in response to shear stress and that this integrin-mediated signaling involves specific interaction of the integrins with their cognate ECM proteins. Furthermore, the integrin activation by shear stress, as in the case of the activation by EC adhesion, occurs only when new integrin-ECM bonds are formed and it subsides with continued binding. Thus, integrin-mediated signaling is dynamic as well as specific.[18]

2.3.3 *Interplay between Flk-1 and integrins in mechanosignaling*

Experiments have been performed on cultured BAECs to determine whether the mechano-activations of Flk-1 and integrin are parallel or serial processes, and if the latter what is the sequence.[19] These experiments involve assessing the effects of integrin blockade on mechano-activation of Flk-1 and the effects of Flk-1 blockade on mechano-activation of integrin. The shear activation of Flk-1, as assessed by its tyrosine phosphorylation and its association with Cbl, was abolished by using integrin-blocking antibodies such as LM609 and 6S6. In contrast, the shear-activation of integrin $\alpha_v\beta_3$, as assessed by its association with Shc, was not significantly affected by using the Flk-1 blocking agent SU-1498. These results suggest that integrins are upstream of Flk-1 in their sequential shear-activations. The shear-activation of Flk-1 was abolished by the actin-disrupting agent cytochalasin D, suggesting that the effect of integrin on Flk-1 is mediated by the actin cytoskeleton. It is possible that the tension induced by shear stress on the luminal side of the EC membrane can be transmitted along the plane of the membrane or via the cytoskeleton to the integrins, which in turn activates Flk-1.

2.3.4 *Effects of shear stress on membrane fluidity*

The effects of shear stress on membrane lipid fluidity of BAECs have been determined in a flow chamber by using the fluorescence recovery after photobleaching (FRAP) method with DiI as the lipid stain.[20] The DiI-diffusion coefficient (*D*) on the up- and downstream sides of the cell was determined by using a confocal microscope. After a step shear stress of 10 dynes/cm^2, *D* shows an upstream increase and downstream decrease in <10 sec, and both changes disappear rapidly. There is a secondary, larger increase in upstream *D*, which reaches a peak at 7 min and decreases thereafter despite continuation of shearing. Downstream *D* shows little secondary changes throughout the 10-min shearing. When the 10-dynes/cm^2 shear stress is attained with a ramp rate of 20 dynes/cm^2 per min,[21] both the up- and downstream portions show a rapid decrease of *D* (within 5 sec). While step-shear of 10 dynes/cm^2 activates ERK and JNK, ramping shear stress to the same level does not. Benzyl alcohol, which increases *D*, enhances the activities of both MAPKs; cholesterol, which reduces *D*, diminishes these activities. The results indicate that the EC membrane lipid bilayer can sense the temporal features of the applied shear stress with spatial discrimination, and that the shear-induced membrane perturbations can be transduced into MAPK activation. It is possible that the changes in

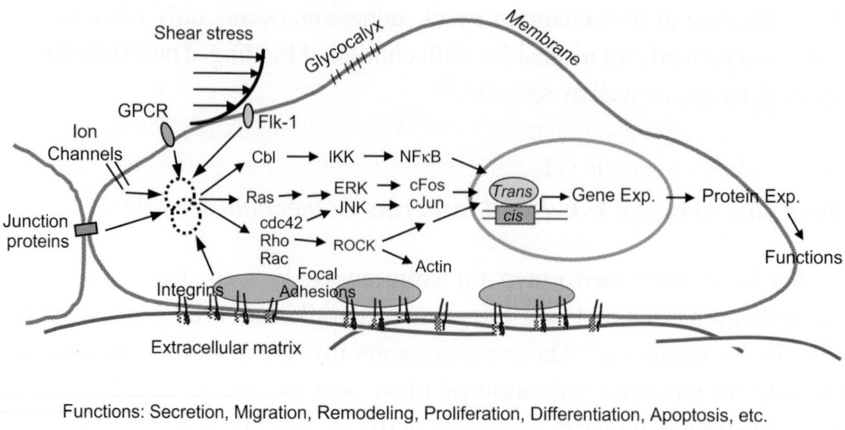

Fig. 5. Schematic diagram showing the multiple mechanosensors that may play a role in mechanotransduction in endothelial cells. The potential sensors shown in this diagram include Flk-1, G-protein coupled receptor, ion channels, junction proteins, and integrins (which link the extracellular matrix and focal adhesion proteins), as well as the cell membrane and glycocalyx. The shear-activation of the sensors can activate the adapter molecules (shown as two dashed circles) and a myriad of signaling pathways. In this diagram only a few pathways are shown. The activation of the signaling pathways leads to modulation of gene expression and hence protein expression and cellular functions. The mechanotransduction can also lead to cell remodeling through the modification of the organization of cytoskeleton, including actin. The cytoskeletal elements can also participate in the modulation of signaling pathways.

membrane fluidity may play a significant role in the shear-induced modulation of the conformation and/or interaction of membrane proteins.

2.3.5 *Summary on mechanosensing*

In addition to the sensors mentioned above, many other membrane elements may also sense mechanical forces to initiate intracellular signaling (Fig. 5). These include G-protein coupled receptors,[22] ion channels,[23] and intercellular junction proteins,[24] as well as interaction between the glycocalyx and the membrane.[25]

3. Effects of Shear Stress on Endothelial Cell Proliferation and Survival

3.1 *DNA microarray studies on shear stress modulation of gene expression*

Using the DNA microarray technology, Chen *et al.*[26] have investigated gene expression profiles in cultured human aortic endothelial cells (HAECs) in response to 24 hr of laminar shear stress at 12 dynes/cm^2. This relatively long-term shearing of cultured HAECs modulates the expression of a number of genes. Several genes related to inflammation and EC proliferation are down-regulated, suggesting that 24-hr shearing may keep ECs in a relatively non-inflammatory and non-proliferative state in comparison to static cells. The genes significantly up-regulated by the 24-hr shear stress include those involved in EC survival and angiogenesis (Tie2 and Flk-1) and vascular remodeling (matrix metalloproteinase 1). These results provide information on the profile of gene expression in shear-adapted ECs, which is the case for the native ECs in the straight part of the aorta *in vivo*. Effects of shear stress on EC gene expression profile have been reported by other groups,[27–29] and we have developed a method to analyze the time course of gene expression profile following the application of laminar shear stress to HAECs.[30]

3.2 *Roles of p53, p21^{cip1}, GADD45, and Rb dephosphorylation in the decrease in cell proliferation in response to long-term laminar shearing*

Based on the finding that relatively long-term shearing tends to keep ECs in a low-proliferative state, Lin *et al.*[31] have studied the mechanism underlying the inhibition of endothelial cell growth by laminar shear stress. The tumor suppressor gene p53 is found to increase in BAECs subjected to 24 hr of laminar shear stress at 3 dynes/cm^2 or higher. One of the mechanisms of the shear-induced increase in p53 is its stabilization following its phosphorylation by c-Jun N-terminal kinase (JNK). Twenty-four hour laminar shear stress also increases the growth arrest proteins GADD45 (growth arrest and DNA damage inducible protein 45) and p21^{cip1} and a decrease in phosphorylation of the retinoblastoma gene product (Rb). These results suggest that prolonged laminar shear stress causes a sustained p53 activation, which induces the up-regulation of GADD45 and p21^{cip1}. The resulting Rb hypophosphorylation, probably through the down-regulation of cyclin dependent kinase cdk4, leads to the cell cycle arrest. This inhibition of EC proliferation by laminar shear stress may serve an important homeostatic function by preventing atherogenesis in the straight part of the arterial tree, which is constantly subjected to high levels of laminar shearing.

3.3 *Effects of flow patterns on cell proliferation and survival*

The proliferation of vascular ECs has been assessed in the step flow channel[32] by determining the incorporation of the labeled nucleotide BrdU (Fig. 6). Under static condition, BrdU incorporation is low and randomly distributed throughout the channel. Following 24 hr laminar shear at 12 dynes/cm^2, BrdU incorporation is markedly enhanced in the re-attachment area and its vicinity (which simulates the flow condition at branch points), and is much lower elsewhere, including the downstream laminar flow region[5] (Fig. 6b). The same distribution pattern is seen for the expression of genes related to cell cycle control

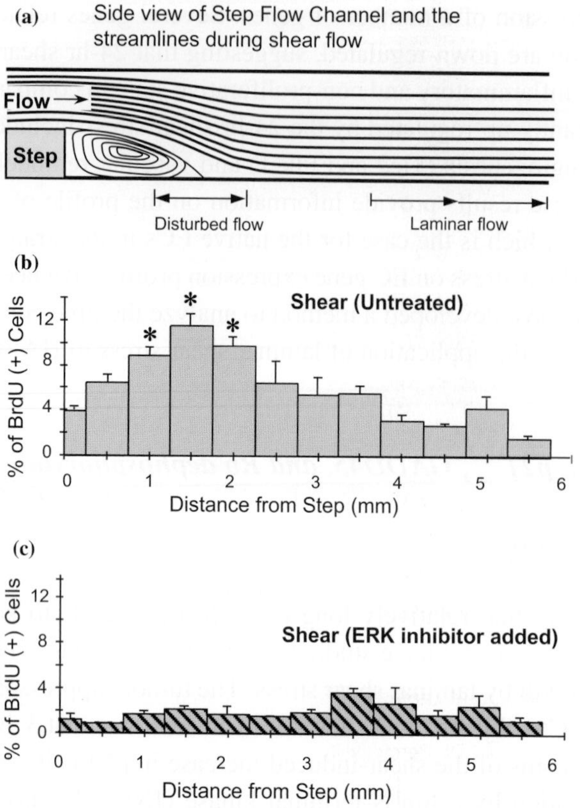

Fig. 6. The proliferation of cultured bovine aortic endothelial cells (BAECs), as assessed by the BrdU incorporation assay, is markedly elevated in the disturbed flow region near the re-attachment point in the step flow channel. Top panel **(a)** shows the side view of the step flow channel and the flow patterns in the disturbed flow region and the downstream laminar flow region. Middle panel **(b)** shows BrdU incorporation into endothelial cells (four experiments) without treatment to block signaling pathways. Bars are mean ± SEM. Asterisks indicate significantly higher BrdU incorporation in the region of disturbed flow near the re-attachment point, as compared with static control. Bottom panel **(c)** shows BrdU incorporation into endothelial cells treated with the inhibitor for ERK (PD98059). Note the blockade of the increase in BrdU incorporation in the disturbed flow region by this ERK inhibitor.

for proliferation and of signaling molecules such as ERK. Inhibition of ERK with PD98059 abolished the increase in BrdU incorporation in the disturbed flow region (Fig. 6c).

These results indicate that the disturbed flow pattern in the re-attachment area at branch points, via ERK activation, enhances cell proliferation. In contrast, the laminar flow region has a low cell proliferation rate (Fig. 6b) and, as mentioned above, the MAPK signaling pathway and MCP-1 gene expression are down-regulated following sustained shearing. This down-regulation of MCP-1 does not occur in the re-attachment region, e.g. vessel branches, where there is a lack of sustained laminar shear (Fig. 7).[33]

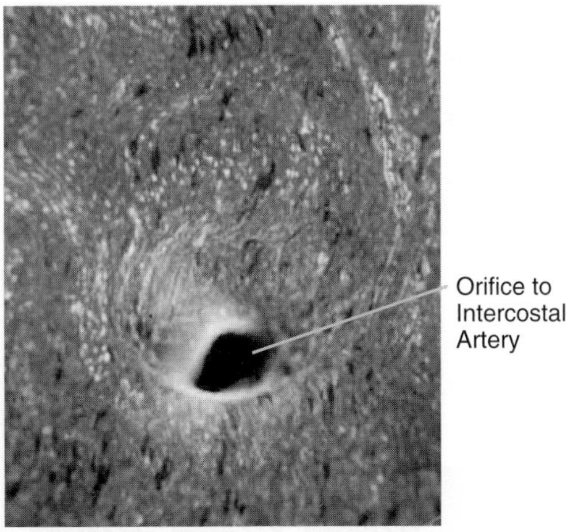

Fig. 7. Histochemical staining for MCP-1 in an *en face* preparation of rat descending aorta. Rat aorta was perfused and fixed under 100 mmHg. The dissected vessel was stained with anti-rat MCP-1 (Serotec) and a secondary antibody conjugated with R-phycoerythin. Note the increased staining around the intercostal orifice. Modified from Chien.[33]

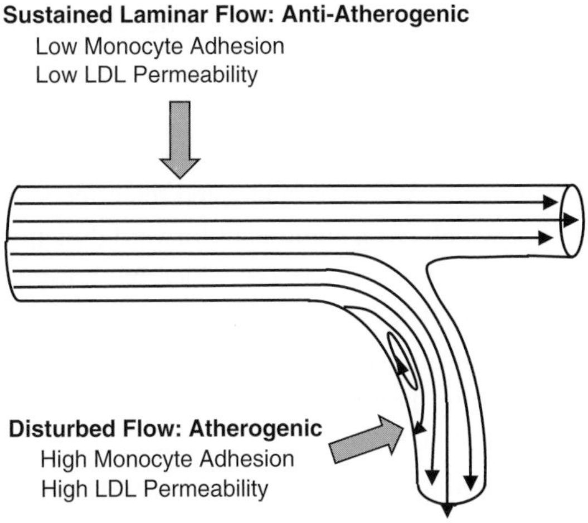

Fig. 8. Diagram showing the atheroprotective mechanisms in the straight part of the aorta where the sustained laminar flow causes low monocyte adhesion (due to down-regulation of MCP-1) and low LDL permeability (due to growth arrest). In contrast, the complex flow pattern at branch point is atherogenic by causing opposite changes.

Thus, our findings indicate that the complex flow pattern near the branch points increases EC proliferation and does not down-regulate the expression of MCP-1 gene, thus placing these regions at risk for atherogenesis. In contrast, the sustained shearing in the laminar flow region induces EC cell arrest and down-regulates MCP-1 expression, thus providing a protective action against atherogenesis (Fig. 8).

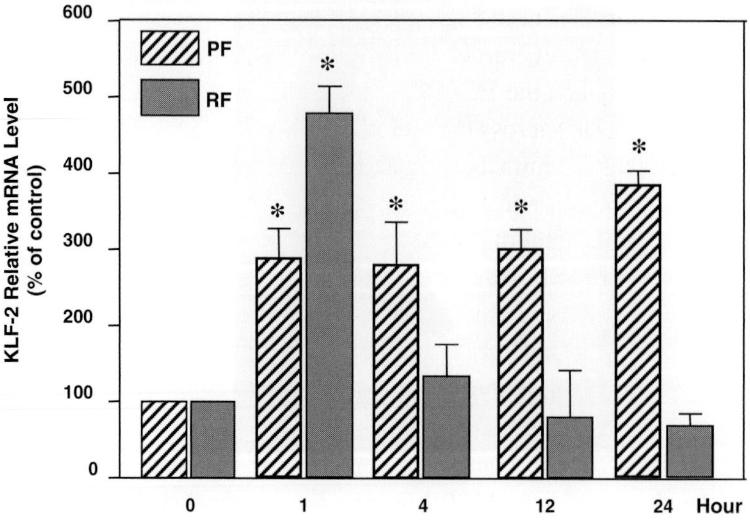

Fig. 9. Effects of pulsatile flow (PF, shear stress 12 ± 4 dynes/cm^2) and reciprocating flow (OF, 0.5 ± 4 dynes/cm^2) on KLF2 expression in confluent HUVECs over 24 h. The mRNA levels of KLF2 were determined by real-time RT-PCR and normalized to GAPDH. The results were compared with the concurrent "static" control (0.5 dyn/cm^2); $*p < 0.05$ vs. control. PF induced a sustained expression of the KLF2 whereas RF caused a transient induction with an ensuing suppression. Modified from Wang *et al.*[34]

Krüppel-like factor-2 (KLF-2) is a transcription factor abundantly expressed in ECs and is beneficial to EC survival.[34] We have contrasted the effects of pulsatile shear stress (12 ± 4 dynes/cm^2 at 1 Hz), which has a significant forward direction, with those of reciprocating shear stress (0.5 ± 4 dynes/cm^2 at 1 Hz), which has a minimal forward direction, on the expression of KLF-2 in HUVECs.[34] The mRNA level of KLF2 increases significantly after exposure to both pulsatile and reciprocating shear stresses at 1 hr (Fig. 9). With continued reciprocating shearing, however, KLF2 gene expression decreases below the basal level at 4 hr and remains low throughout the 24 hr of shearing. In contrast, pulsatile shear stress results in a sustained up-regulation of KLF2 for as long as 24 hr under continuous shearing (Fig. 9). Blockade of KLF-2 causes a significant decrease in KLF2 gene expression. These results indicate that sustained application of reciprocating shear stress, which lacks a significant forward component, inhibits KLF2 expression and decreases the ability of ECs to survive against oxidative stress. Pulsatile shear stress, which has a significant forward component, results in continued expression of KLF2 and is beneficial to EC survival.

4. Role of Hemodynamic Factors in Transendothelial Permeability of Macromolecules

At the branch points and curved regions of the arterial tree, which have a predilection for atherosclerosis, blood flow is unsteady and the shear stress shows marked spatial and temporal variations.[35] These findings led to the hypothesis that complex flow patterns cause an accelerated EC turnover (including cell mitosis and death), such that the resulting leakiness between the ECs undergoing turnover increases the permeability of large molecules (e.g. LDL) across the endothelial layer.[36] Our experimental studies have provided evidence that EC mitosis and death are associated with the leakage of macromolecules such as albumin and LDL on individual cell basis. Studies performed in a number of laboratories, including our own have shown that these events of accelerated EC turnover occur primarily in areas with disturbed blood flow, e.g. arterial branch points.[37] Electron microscopic studies have identified the widening of the intercellular junctions around the ECs undergoing mitosis or dying.[38]

The potential role of hemodynamic factors and EC turnover in the focal nature of lipid infiltration has been studied in rabbit thoracic aorta.[39] Three and a half minutes after intravenous injection of Evans Blue albumin (EBA), the animal was sacrificed, and the formalin-fixed aortic tree was first stained with Harris' hematoxylin, followed by bisbenzimide after its bleaching. The luminal surface of the entire thoracic aorta was examined by light microscopy for mitosis, nuclear shape index [S.I. = 4π (Area)/(Perimeter)^2], and EBA leakage. S.I. varies from 1 for a sphere to 0 for a line.

At intercostal orifices, EC nuclei on the upstream side are orientated toward the center of the orifices with low S.I. (more spherical) but on lateral sides they are oriented diagonally with high S.I. (more elongated) At the downstream apex of the orifices, the nuclei are small, round and without a distinct orientation, but those immediately distal to the apex are oriented nearly parallel to the inter-orifice line and have very low S.I. The mitotic cells are found primarily in areas with a high S.I. The nuclear orientation suggests that flow pattern is rather complex even in the straight portion of aorta and that small branches have significant influences on the flow pattern, as found in large branches.[40]

Mitotic cells and EBA leaky spots are distributed primarily around the orifices of intercostal arteries (Figs. 10 and 11), with the number densities decreasing laterally away from

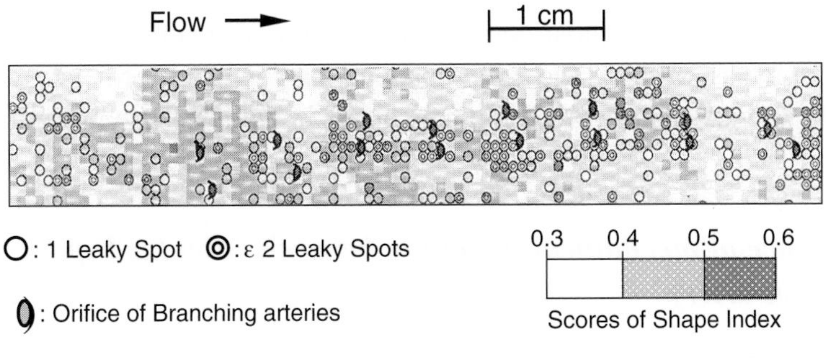

Fig. 10. Drawing of an *en face* preparation of the rabbit thoracic aorta showing the shape index [S.I. = 4π (Area)/(Perimeter)^2] and the presence of albumin leaky spots in various regions. Note that the S.I. values and the number of leaky spots are higher near the intercostal orifices. Modified from Chien.[33]

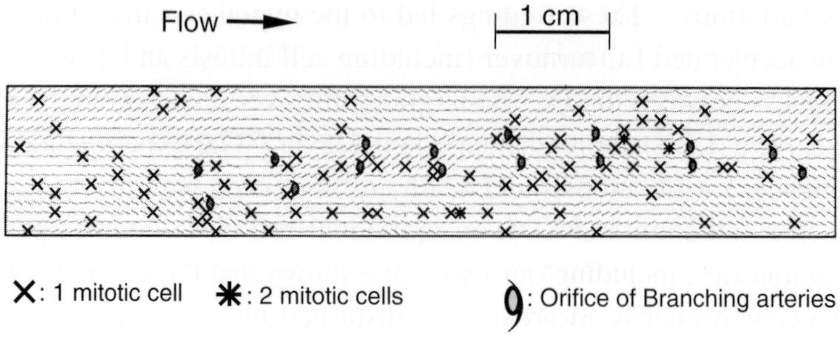

Fig. 11. Drawing of an *en face* preparation of the rabbit thoracic aorta showing the distribution of mitotic cells per unit area. Note that the number of mitotic cells, similar to the leaky spots shown in Fig. 10, is higher near the intercostal orifices. Modified from Chien.[33]

the orifices. The distribution of mitosis and EBA-leaky spots is similar to the pattern of lipid accumulation in cholesterol-fed rabbits.[41] Our results provide evidence in support of the roles of disturbed flow and shear gradient in promoting EC turnover. Most of the EBA leaky spots, with or without cell mitosis, are localized in regions with secondary flows.

The results of experimental and theoretical studies indicate the following sequence of events, in addition to those mediated by SREBP, contribute to the focal nature of atherosclerosis.

Local hemodynamic factors → ↑EC turnover → ↑Local LDL → Focal lipid
(complex flow pattern) (mitosis & death) permeability accumulation

5. Implications in Clinical Conditions

The results outlined above have considerable implications in the prevention and treatment of cardiovascular diseases. Two examples are given below to illustrate how such knowledge is relevant to coronary artery disease, which is a major cause of death in the United States. The two most common modalities of treatment of coronary artery narrowing are bypass surgery and balloon angioplasty. The molecular responses of the vascular cells to mechanical forces can play a role in both of these procedures.

5.1 *Importance of mechanical matching between vascular bypass and native artery*

In vascular bypass surgery, an artificial vascular graft or a vessel segment from the patient is used to provide the bypass between the aorta and the coronary artery distal to the site of obstruction. One of the most commonly used bypass vessels is the saphenous vein from the patient's leg. While the saphenous vein graft has had its successes, recurrence of obstruction (re-stenosis) is a major complication. Veins are normally exposed to much lower pressures (<10 mmHg) than arteries (~100 mmHg), and their structural features and hence mechanical properties are quite different. The insertion of a segment of the saphenous vein into the arterial system exposes the thin-walled and compliant vein to a pressure much higher than what the venous wall is prepared to accommodate. This mechanical mismatch causes the saphenous vein segment to bulge, and this sudden enlargement leads to a geometric mismatch. Thus, the junction area provides a condition similar to that in the step flow channel[42] to result in eddy flow and flow re-attachment, and becomes vulnerable to atherogenesis because of the activation of events such as cell proliferation and MCP-1 expression. Therefore, it is beneficial to reinforce the saphenous vein prior to grafting to prevent the occurrence of geometric mismatch.[42] In the development of tissue-engineered vascular graft, whether cell-based and/or biomaterial-based, it is important to match the mechanical properties of the graft with those of the native vessel.

5.2 *Use of a Ras negative mutant to prevent re-stenosis following balloon angioplasty*

In treating coronary artery disease by balloon angioplasty, a catheter with a balloon near its tip is advanced to the site of stenosis under X-ray visualization. Inflation of the balloon presses on the plaque and opens the obstructed vessel lumen. Although the procedure is usually successful, vessel wall thickening recurs in about one-third of the cases within a few months. The re-stenosis is principally due to the proliferation of smooth muscle cells, as a result of stimulation by the mechanical injury incurred by the ballooning and the chemical factors in the blood following endothelial denudation. Ras plays a pivotal role in the regulation of the intracellular signaling events for many functions in a variety of cells, including the proliferation of vascular smooth muscle cells. Therefore, we have tested the possibility of using the negative mutant RasN17 to prevent the vascular stenosis induced by balloon injury in animal experiments.[43,44] In order to provide an efficient mode of transfection of the RasN17 into the vessel wall, it was packaged into non-replicating adenovirus to produce AdRasN17. As a control, a non-therapeutic molecule LacZ was similarly packaged to produce AdLacZ.

Under general anesthesia, the common carotid artery of the rat was subjected to balloon injury with a balloon catheter,[43] similar to the procedure used clinically in balloon angioplasty. By the use of temporary clamps, the injured artery was divided into cranial and caudal segments. AdRasN17 was introduced into one segment and AdLacZ into the other. After 5 min, the lumen contents were rinsed out, flow was re-established after removal of the clamps, the neck wound was closed, and the animal was allowed to recover. Six weeks later, the common carotid artery was removed for histological examination. The control artery without any balloon injury had a thin wall and a wide lumen. In the absence of RasN17 treatment, balloon injury (with or without AdLacZ) caused

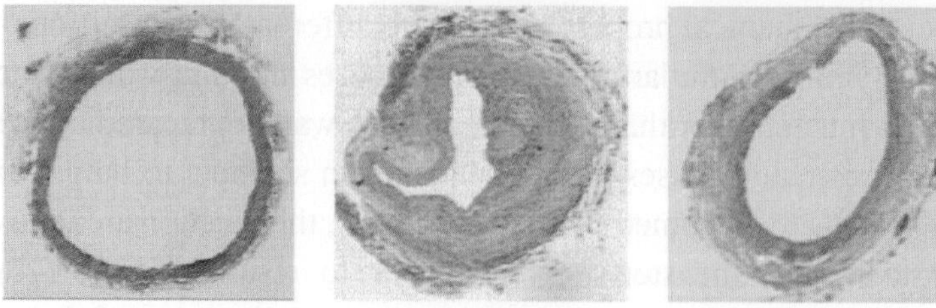

Fig. 12. RasN17 prevented re-stenosis of rat common carotid artery after balloon injury. From the study by Jin *et al.*[43] Note the wide lumen and thin wall in the control artery, the severe stenosis following balloon injury without AdRasN17 (with only control AdLacZ), and the marked attenuation of stenosis and increase in lumen patency with AdRasN17 treatment following balloon injury. Modified from Chien.[33]

a marked thickening of the vessel wall and a severe narrowing of the lumen. AdRasN17 treatment greatly attenuated the wall thickening following balloon injury, and the lumen was sufficiently wide to allow an essentially normal flow (Fig. 12). These results and similar findings on the coronary artery of the pig suggest the potential value of using RasN17 as an agent to prevent the re-stenosis resulting from balloon angioplasty and illustrate how basic research can generate new approaches for clinical management of disease.

6. Summary and Conclusions

Hemodynamic forces can modulate the structure and function of ECs and smooth muscle cells in blood vessels. Under normal conditions, these modulating influences allow the vascular wall to adapt to changes in pressure and flow for optimizing its functional performance. In disease states, however, the abnormal responses can disturb the homeostasis and initiate or aggravate pathological processes.

Changes in flow conditions can activate EC membrane receptors (e.g. VEGFR, integrins, and others) to initiate the signal transduction involving upstream molecules such as Ras and downstream molecules such as JNK and ERK, as well as MCP-1. Sustained laminar flow encountered in the straight part of the arterial tree down-regulates such activation and hence minimizes monocyte entry into the vascular wall. Such laminar flow is also associated with a reduced lipid accumulation in ECs because of the lower LDL permeability due to the up-regulation of growth-arrest genes. The reductions in monocyte entry and lipid accumulation induced by long-term laminar shear stress are protective against atherogenesis.

Atherosclerosis occurs primarily in arterial branch points and curved regions. The complex flow pattern, especially the flow re-attachment, in these lesion-prone regions is associated with an increased LDL permeability due to a high cell turnover rate and the widening of the intercellular junction. There is also an increased tendency for monocyte accumulation due to the lack of MCP-1 down-regulation. The greater accumulations of both LDL and monocytes are important atherogenic factors.

Consideration of the results on mechano-chemical transduction can lead to the improvement of methods of prevention and treatment of coronary artery disease. In vascular bypass surgery, it is important to consider mechanical matching between the graft and the vessel in order to prevent the creation of complex flow patterns due to geometric mismatch. Adenovirus-mediated transfer of RasN17, e.g. by using coated stent, has the potential of being used as a prophylactic procedure against re-stenosis in balloon angioplasty by suppressing smooth muscle cell proliferation.

The combination of mechanics and biology (from molecules-cells to organs-systems), can play a major role in enhancing our understanding of the physiological processes of mechano-chemical transduction and improving the methods of the management of important clinical conditions such as coronary artery disease.

Although there have been significant advances in our knowledge of the roles of mechanical factors in modulating lipid dynamics in the artery wall, additional research is needed to further understand the molecular and mechanical bases. Thus, we need to elucidate the initial event of mechano-sensing, the interplays of mechano-sensors and signaling pathways, the mechano-modulations of gene expression, and the mechano-regulation of vascular functions *in vivo* in health and disease. Such studies will enhance our understanding of the mechanisms of mechanotransduction and help to improve the diagnosis and treatment of cardiovascular diseases.

Acknowledgments

This work was supported in part by grants HL043026, HL080518 and HL085159 from the National Heart, Lung, and Blood Institute and a Development Award from the Whitaker Foundation. The author would like to acknowledge the valuable collaboration of many excellent colleagues in these studies.

References

1. D. Steinberg, Role of oxidized LDL and antioxidants in atherosclerosis, *Adv. Exp. Med. Biol.* **369**: 39–48 (1995).
2. J. F. Cornhill, E. E. Herderick and H. C. Stary, Topography of human aortic sudanophilic lesions, *Monogr. Atheroscler.* **15**: 13–19 (1990).
3. R. M. Nerem, Hemodynamics and the vascular endothelium, *J. Biomech. Eng.* **115**: 510–514 (1993).
4. P. F. Davies, Flow-mediated endothelial mechanotransduction, *Physiol. Rev.* **75**: 519–560 (1995).
5. Y. S. Li, J. H. Haga and S. Chien, Molecular basis of the effects of shear stress on vascular endothelial cells, *J. Biomech.* **38**: 1949–1971 (2005).
6. R. G. Gerrity and C. J. Schwartz, Structural correlates of arterial endothelial permeability in the Evans blue model, *Prog. Biochem. Pharmacol.* **13**: 134–137 (1977).

7. R. A. Malinauskas, R. A. Herrmann and G. A. Truskey, The distribution of intimal white blood cells in the normal rabbit aorta, *Atherosclerosis* **115**: 147–163 (1995).
8. W. S. Shin, A. Szuba and S. G. Rockson, The role of chemokines in human cardiovascular pathology: enhanced biological insights, *Atherosclerosis* **160**: 91–102 (2002).
9. M. Namiki, S. Kawashima, T. Yamashita, M. Ozaki, T. Hirase, T. Ishida, N. Inoue, K. Hirata, A. Matsukawa, R. Morishita *et al.*, Local overexpression of monocyte chemoattractant protein-1 at vessel wall induces infiltration of macrophages and formation of atherosclerotic lesion: synergism with hypercholesterolemia, *Arterioscler. Thromb. Vasc. Biol.* **22**: 115–120 (2002).
10. Y. J. Shyy, H. J. Hsieh, S. Usami and S. Chien, Fluid shear stress induces a biphasic response of human monocyte chemotactic protein 1 gene expression in vascular endothelium, *Proc. Natl. Acad. Sci. USA* **91**: 4678–4682 (1994).
11. J. Y. Shyy, M. C. Lin, J. Han, Y. Lu, M. Petrime and S. Chien, The cis-acting phorbol ester "12-O-tetradecanoylphorbol 13-acetate"-responsive element is involved in shear stress-induced monocyte chemotactic protein 1 gene expression, *Proc. Natl. Acad. Sci. USA* **92**: 8069–8073 (1995).
12. Y. S. Li, J. Y. Shyy, S. Li, J. Lee, B. Su, M. Karin and S. Chien, The Ras-JNK pathway is involved in shear-induced gene expression, *Mol. Cell. Biol.* **16**: 5947–5954 (1996).
13. H. Jo, K. Sipos, Y. M. Go, R. Law, J. Rong and J. M. McDonald, Differential effect of shear stress on extracellular signal-regulated kinase and N-terminal Jun kinase in endothelial cells. Gi2- and G beta/gamma-dependent signaling pathways, *J. Biol. Chem.* **272**: 1395–1401 (1997).
14. M. Takahashi and B. C. Berk, Mitogen-activated protein kinase (ERK1/2) activation by shear stress and adhesion in endothelial cells. Essential role for a herbimycin-sensitive kinase, *J. Clin. Invest.* **98**: 2623–2631 (1996).
15. S. Li, M. Kim, Y. L. Hu, S. Jalali, D. D. Schlaepfer, T. Hunter, S. Chien and J. Y. Shyy, Fluid shear stress activation of focal adhesion kinase. Linking to mitogen-activated protein kinases, *J. Biol. Chem.* **272**: 30455–30462 (1997).
16. K. D. Chen, Y. S. Li, M. Kim, S. Li, S. Yuan, S. Chien and J. Y. Shyy, Mechanotransduction in response to shear stress. Roles of receptor tyrosine kinases, integrins, and Shc, *J. Biol. Chem.* **274**: 18393–18400 (1999).
17. M. A. Schwartz, Integrin signaling revisited, *Trends. Cell Biol.* **11**: 466–470 (2001).
18. S. Jalali, M. A. del Pozo, K. Chen, H. Miao, Y. Li, M. A. Schwartz, J. Y. Shyy and S. Chien, Integrin-mediated mechanotransduction requires its dynamic interaction with specific extracellular matrix (ECM) ligands, *Proc. Natl. Acad. Sci. USA* **98**: 1042–1046 (2001).
19. Y. Wang, H. Miao, S. Li, K. D. Chen, Y. S. Li, S. Yuan, J. Y. Shyy and S. Chien, Interplay between integrins and FLK-1 in shear stress-induced signaling, *Am. J. Physiol. Cell Physiol.* **283**: C1540–1547 (2002).
20. P. J. Butler, G. Norwich, S. Weinbaum and S. Chien, Shear stress induces a time- and position-dependent increase in endothelial cell membrane fluidity, *Am. J. Physiol. Cell Physiol.* **280**: C962–969 (2001).
21. P. J. Butler, T. C. Tsou, J. Y. Li, S. Usami and S. Chien, Rate sensitivity of shear-induced changes in the lateral diffusion of endothelial cell membrane lipids: a role for membrane perturbation in shear-induced MAPK activation, *FASEB J.* **16**: 216–218 (2002).
22. M. J. Kuchan, H. Jo and J. A. Frangos, Role of G proteins in shear stress-mediated nitric oxide production by endothelial cells, *Am. J. Physiol.* **267**: C753–758 (1994).
23. V. G. Romanenko, P. F. Davies and I. Levitan, Dual effect of fluid shear stress on volume-regulated anion current in bovine aortic endothelial cells, *Am. J. Physiol. Cell Physiol.* **282**: C708–718 (2002).
24. E. Tzima, M. Irani-Tehrani, W. B. Kiosses, E. Dejana, D. A. Schultz, B. Engelhardt, G. Cao, H. DeLisser and M. A. Schwartz, A mechanosensory complex that mediates the endothelial cell response to fluid shear stress, *Nature* **437**: 426–431 (2005).

25. S. Weinbaum, X. Zhang, Y. Han, H. Vink and S. C. Cowin, Mechanotransduction and flow across the endothelial glycocalyx, *Proc. Natl. Acad. Sci. USA* **100**: 7988–7995 (2003).
26. P. Chen, Y. S. Li, Y. Zhao, K. D. Chen, S. Li, J. Lao, S. Yuan, J. Y. Shyy and S. Chien, DNA microarray analysis of gene expression in endothelial cells in response to 24-h shear stress, *Physiol. Genomics* **7**: 55–63 (2001).
27. G. Garcia-Cardena, J. Comander, K. R. Anderson, B. R. Blackman and M. A. Gimbrone, Jr., Biomechanical activation of vascular endothelium as a determinant of its functional phenotype, *Proc. Natl. Acad. Sci. USA* **98**: 4478–4485 (2001).
28. S. M. McCormick, S. G. Eskin, L. V. McIntire, C. L. Teng, C. M. Lu, C. G. Russell and K. K. Chittur, DNA microarray reveals changes in gene expression of shear stressed human umbilical vein endothelial cells, *Proc. Natl. Acad. Sci. USA* **98**: 8955–8960 (2001).
29. A. R. Brooks, P. I. Lelkes and G. M. Rubanyi, Gene expression profiling of human aortic endothelial cells exposed to disturbed flow and steady laminar flow, *Physiol. Genomics* **9**: 27–41 (2002).
30. Y. Zhao, B. P. Chen, H. Miao, S. Yuan, Y. S. Li, Y. Hu, D. M. Rocke and S. Chien, Improved significance test for DNA microarray data: temporal effects of shear stress on endothelial genes, *Physiol. Genomics* **12**: 1–11 (2002).
31. K. Lin, P. P. Hsu, B. P. Chen, S. Yuan, S. Usami, J. Y. Shyy, Y. S. Li and S. Chien, Molecular mechanism of endothelial growth arrest by laminar shear stress, *Proc. Natl. Acad. Sci. USA* **97**: 9385–9389 (2000).
32. J. J. Chiu, D. L. Wang, S. Chien, R. Skalak and S. Usami, Effects of disturbed flow on endothelial cells, *J. Biomech. Eng.* **120**: 2–8 (1998).
33. S. Chien, Molecular and mechanical bases of focal lipid accumulation in arterial wall, *Prog. Biophys. Mol. Biol.* **83**: 131–151 (2003).
34. N. Wang, H. Miao, Y. S. Li, P. Zhang, J. H. Haga, Y. Hu, A. Young, S. Yuan, P. Nguyen, C. C. Wu *et al.*, Shear stress regulation of Krüppel-like factor 2 expression is flow pattern-specific, *Biochem. Biophys. Res. Commun.* **341**: 1244–1251 (2006).
35. S. Glagov, C. Zarins, D. P. Giddens and D. N. Ku, Hemodynamics and atherosclerosis. Insights and perspectives gained from studies of human arteries, *Arch. Pathol. Lab. Med.* **112**: 1018–1031 (1988).
36. S. Weinbaum, G. Tzeghai, P. Ganatos, R. Pfeffer and S. Chien, Effect of cell turnover and leaky junctions on arterial macromolecular transport, *Am. J. Physiol.* **248**: H945–960 (1985).
37. S. Chien, Molecular basis of rheological modulation of endothelial functions: importance of stress direction, *Biorheology* **43**: 95–116 (2006).
38. Y. L. Chen, K. M. Jan, H. S. Lin and S. Chien, Ultrastructural studies on macromolecular permeability in relation to endothelial cell turnover, *Atherosclerosis* **118**: 89–104 (1995).
39. S. Chien, Mechanotransduction and endothelial cell homeostasis: the wisdom of the cell, *Am. J. Physiol. Heart Circ. Physiol.* **292**: H1209–1224 (2007).
40. T. Karino, Microscopic structure of disturbed flows in the arterial and venous systems, and its implication in the localization of vascular diseases, *Int. Angiol.* **5**: 297–313 (1986).
41. C. Schwenke and T. E. Carew, Initiation of atherosclerotic lesions in cholesterol-fed rabbits. I. Focal increases in arterial LDL concentration precede development of fatty streak lesions, *Arteriosclerosis* **9**: 895–907 (1989).
42. S. Q. Liu, Prevention of focal intimal hyperplasia in rat vein grafts by using a tissue engineering approach, *Atherosclerosis* **140**: 365–377 (1998).
43. G. Jin, J. C.-H. Wu, Y. S. Li, Y. L. Hu, J. Y. Shyy and S. Chien, Effects of active and negative mutants of Ras on rat arterial neointima formation, *J. Surg. Res.* **94**: 124–132 (2000).
44. C. H. Wu, C. S. Lin, J. S. Hung, C. J. Wu, P. H. Lo, G. Jin, Y. J. Shyy, S. J. Mao and S. Chien, Inhibition of neointimal formation in porcine coronary artery by a Ras mutant, *J. Surg. Res.* **99**: 100–106 (2001).

CHAPTER 7

AUTOREGULATION OF BLOOD FLOW: EXAMINING THE PROCESS OF SCIENTIFIC DISCOVERY

Paul C. Johnson

Abstract

Autoregulation of blood flow is the tendency for blood flow to remain constant in an organ during changes in arterial perfusion pressure. It is due to local mechanisms in the organ and is independent of the regulatory mechanisms of the central nervous system. The possible explanations that have received the most attention are the metabolic and myogenic hypotheses. We review the experimental evidence for and against these hypotheses both to develop our understanding of autoregulation and as examples of the process of scientific inquiry. Evidence is presented in support of the metabolic hypothesis that autoregulation is due to a link between blood flow and energy metabolism but the lack of definitive evidence for most proposed mediators is noted. The myogenic response has been more controversial since it proposes that the arterioles regulate flow indirectly by constricting in response to intravascular pressure as a stimulus. The possible involvement of integrins and stretch sensitive channels in this response is described. Autoregulation may be important in clinical situations such as atherosclerosis and blood loss. Reviewing the development of our understanding of autoregulation provides insight to the manner in which concepts arise and are tested experimentally.

1. Introduction

The main purpose of this chapter is to examine mechanisms that are involved in a phenomenon called *autoregulation of blood flow*. Autoregulation is defined as "the tendency for blood flow in an organ to remain constant despite changes in arterial perfusion pressure."[1] It is seen in virtually all organs and tissues of the body and is therefore of fundamental importance to understanding the regulation of the cardiovascular system. It is a local phenomenon and is not related to mechanisms acting through the central nervous system.

1.1 *The strategy of investigation*

Examination of research on the cause of autoregulation of blood flow also provides an opportunity to follow the strategy used by scientists as they seek to understand biological phenomena. A common strategy is to develop a hypothesis to explain a phenomenon of interest and devise an experiment to test that hypothesis. An experimental test often involves introducing a perturbation in the system and measuring the response. In the process of developing a hypothesis there is a natural human tendency to select the simplest mechanism that might explain the phenomenon, based on existing knowledge. That is, faced with several possible mechanisms that might explain the phenomenon and the response to the perturbation, it is often tacitly assumed that the simplest explanation is most likely to be correct.

However, applying this rule to biological systems poses difficulties. Biological systems of remarkable complexity and diversity have developed in the course of evolution. As noted by Francois Jacob,[2] "evolution is a tinkerer." Thus, when the organism is faced with a changing environment, small changes that are advantageous to survival are selectively conserved. Over a long period of time the accumulation of these small changes constitutes a drastic change. As a consequence, in studying biological phenomena, it is important to bear in mind that the process of evolution may not produce the simplest mechanism.

2. The Cardiovascular System

Multicellular organisms have developed specialized systems to maintain a constant internal environment for the parenchymal cells. One of the most important and complex of these is the cardiovascular system, which delivers nutrients and removes waste products, maintains a constant hydration of the tissues, provides a communication system among organs and in many other ways maintains a constant internal environment of the body. Many of these functions depend on upon controlling the rate at which blood flows through the tissue, which is determined both by the pumping action of the heart and the caliber of the small blood vessels that regulate the flow through the organ. These vessels are called *arterioles* and are the terminal branches of the arterial network. The pressure profile of the cardiovascular system is shown in Fig. 1.

Note that the pressure drop is greatest in the arterioles, as would be expected since these vessels are the primary regulators of blood flow. The wall of the arteriole is invested with a layer of vascular smooth muscle that is relatively thick in relation to its diameter; and this feature enables the arteriole to change its diameter greatly under different conditions.

3. Autoregulatory Flow Patterns

Two examples of autoregulation are shown in Fig. 2. This figure shows the flow immediately following step changes in the pressure (open circles) and the steady state flow

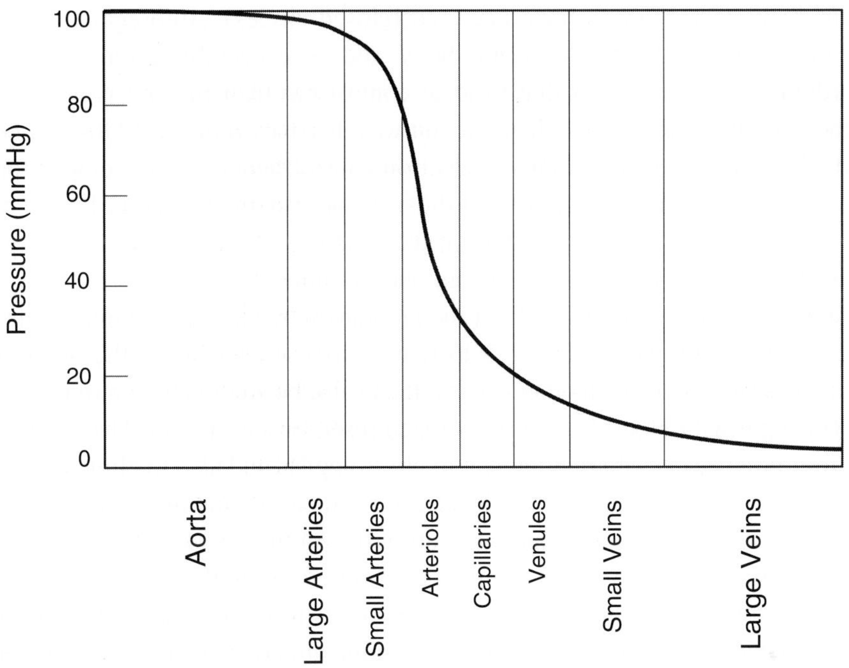

Fig. 1. Mean pressure distribution in the peripheral circulation. Note the steep pressure drop in the arterioles.

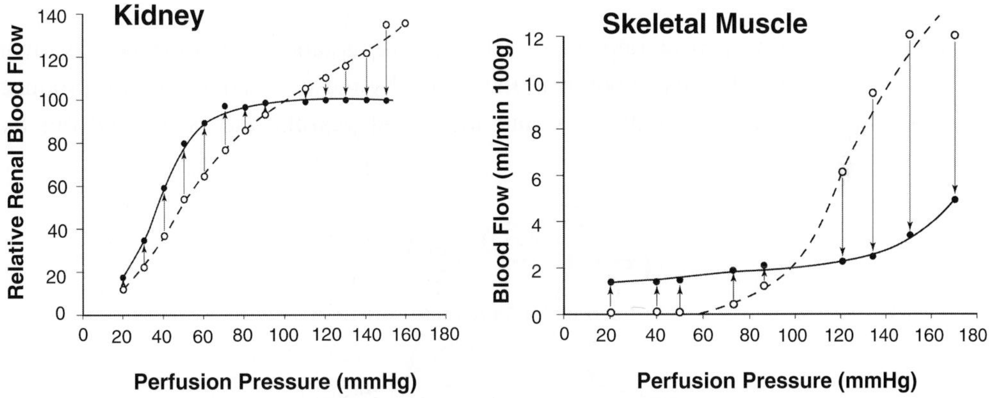

Fig. 2. Blood flows in kidney and skeletal muscle with alteration of arterial perfusion pressure. In the kidney, pressure was decreased in steps of 10 to 80 mmHg and increased in steps of 10 to 60 mmHg from an initial pressure of 100 mmHg. The open circles represent the initial flow value immediately after the step changes in pressure. The closed circles represent the final flow value at the same pressure when the autoregulatory response was complete. Note that steady state flow is essentially independent of pressure in the kidney over the range 80 to 160 mmHg. A similar experiment in skeletal muscle yields similar results but flow is not as closely regulated in the muscle as in the kidney. Modified from Johnson.[6]

after autoregulatory changes have occurred (closed circles). Autoregulation has been found in virtually all organs of the body and is especially prominent in brain, myocardium and kidney. It is due to local control mechanisms and is not dependent upon the nervous system. This phenomenon was discovered in the 1930s and began to be studied in detail in the late 1940s as electronic instruments were developed for measurement of pressure and flow in the cardiovascular system. The contribution of engineering advances to our understanding of the function of the cardiovascular system is an interesting story in itself and this is but one example.

Autoregulation can be observed in the laboratory by placing a clamp on the artery feeding an organ and partly closing it to reduce arterial pressure to the organ without affecting pressure and flow in other organs. It can also be studied by removing the organ and perfusing it with blood from a pressurized reservoir or a pump. The mechanism(s) of this response became of great interest to physiologists and also to clinicians as it provided some of the first evidence for local control mechanisms in the peripheral blood vessels. Note in Fig. 2 that after an initial drop, flow returns toward the previous value. Since the arterioles are the site of the major pressure drop in the peripheral circulation, it was inferred that the arterioles dilate as pressure is reduced and constrict as pressure is elevated. Since flow initially changes in the same direction as pressure and then tends to return to the initial level, the explanations most commonly advanced involved a flow-dependent mechanism acting on the vascular smooth muscle cells in the arterioles.

4. The Metabolic Hypothesis

A cartoon illustrating a flow-dependent scheme, the metabolic hypothesis, is shown in Fig. 3. According to this hypothesis the parenchymal cells continuously produce a vasodilator substance that diffuses into the blood vessels and is washed out by

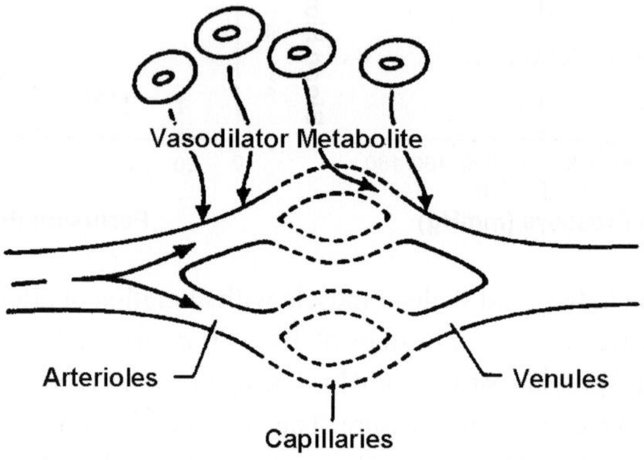

Fig. 3. A depiction of the metabolic hypothesis showing tissue distribution and washout of a hypothetical vasodilator metabolite produced by the parenchymal cells. Modified from Johnson.[1]

the blood stream.[1] Carbon dioxide is an example of such a vasodilator metabolite and is produced by the oxidative metabolism of the organ. In this case the concentration of CO_2 in the tissues will be a function of both the rate of production and of washout by the blood stream. If flow is reduced, the metabolite would accumulate in the tissue, which would lead to dilation of the arterioles and a tendency for flow to return toward the initial level. In addition, when blood flow is reduced oxygen delivery to the tissues would fall, leading to a reduction in oxygen levels in the organ. If the drop in flow were large or if some tissue areas were not well supplied with oxygen to begin with, flow reduction would cause a shift from aerobic to anaerobic metabolism in the tissues. The anaerobic metabolism would cause production of lactic acid, which is also a vasodilator metabolite. Conversely, if flow is increased above normal levels the CO_2 concentration in the tissues would decrease and the vasodilator stimulus to the arterioles would decrease. Evidence obtained with a variety of techniques indicates that the importance of the metabolic mechanism in autoregulation varies among organs. The arterioles in the cerebral circulation are very sensitive to increases in CO_2 and H^+ ions. However, arterioles in many other organs that show autoregulation are rather insensitive to these two mediators.

Another mediator which has been suggested for a flow-dependent feedback mechanism is adenosine. Adenosine is a very potent vasodilator and is produced by the breakdown of ATP when oxygen levels in the tissue are insufficient to maintain normal oxidative metabolism. The key question is whether adenosine is produced during autoregulatory responses. To date it has been shown only in the brain that reduction of arterial pressure causes an increase in adenosine in the tissue.[3] It has not been demonstrated in other organs such as the myocardium with normally high metabolic activity.

Recent studies have demonstrated that arterioles dilate when the oxygen level in the vicinity of the vessel decreases. This is believed to involve the production of a vasodilator (nitric oxide) by the endothelial cells of the arteriole.[4] Nitric oxide is also released by the endothelial cells in response to other stimuli. The contribution of this mechanism in autoregulation of blood flow has not been established.

In considering the potential of the metabolic hypothesis to explain autoregulation of blood flow it is clear that such a mechanism has a significant limitation. It cannot return flow exactly to its previous level after a step increase or decrease in pressure since that would also return the level of the vasodilator metabolite in the tissue to its previous level and there would be no continuing stimulus or error signal for the altered state of the arteriole.

5. The Myogenic Hypothesis

5.1 *Development of the concept*

An alternative explanation for autoregulation of blood flow is the *myogenic hypothesis*. This hypothesis was proposed over 100 years ago based on very indirect evidence.

At that time there were no methods for measuring vascular responses directly as we have today and volume changes in organs were used to infer vascular changes. William Bayliss, a British physiologist, noted that a short period of occlusion of the abdominal aorta (usually lasting about 30 seconds) of the dog caused the volume of the hind limb to decrease during the occlusion and then increase above the normal level after the occlusion was released.[5] The limb then gradually returned to the initial volume. He proposed a novel explanation: namely that the blood pressure in the arterioles provided a stimulus to the smooth muscle in the vascular wall causing the arterioles to be partially constricted by force of the internal pressure. A cartoon of this phenomenon is shown in Fig. 4. Bayliss reasoned that when he occluded the aorta, the pressure in the arterioles in the hind limb fell, decreasing this stimulus and causing the arterioles to relax. When the pressure was rapidly restored, the relaxed smooth muscle in the arterioles allowed the vessels to dilate beyond their normal diameter, leading to the increased volume of the hind limb for a short period of time until the return of the normal pressure led to arteriolar constriction again.

This explanation was widely criticized for several reasons.[6] First, Bayliss did not account for accumulation of vasodilator metabolites (the metabolic hypothesis) in his experiments. Second, the site of the aortic occlusion also stopped flow to other organs and tissue such as the adrenal glands and could lead to release of vasodilator hormones from those areas. Third, if this response to pressure were a property of arterioles generally a rise in arterial pressure, for whatever reason, would lead to a vicious cycle leading to greater and greater constriction of the arterioles and elevation of arterial pressure until an artery burst. Bayliss, however, had pointed out that pressure sensors in the carotid artery and aorta provide a reflex mechanism acting through the autonomic

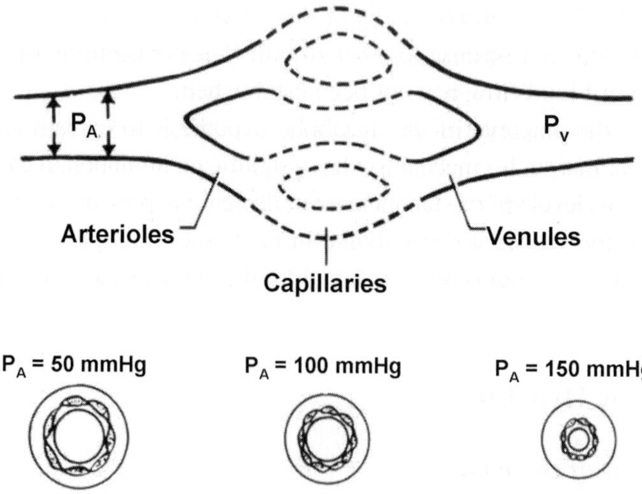

Fig. 4. Depiction of the diameter of an arteriole at several levels of arterial pressure according to the myogenic hypothesis.

nervous system to regulate systemic arterial pressure. This reflex would prevent a vicious cycle from developing. In a monograph written 20 years later, Bayliss (cited in Johnson[6]) distanced himself from the myogenic hypothesis and said "we must regard this issue as largely undecided." The myogenic hypothesis was largely forgotten for many years.

5.2 *Rediscovery of the myogenic hypothesis*

As noted above, advances in methodology for measuring blood flow, pressure and other variables beginning in the late 1940s made it possible to examine the relation between blood flow and the driving head of pressure in individual organs. In this period a Swedish investigator, Bjorn Folkow, suggested that the myogenic hypothesis might explain the changes he observed in blood flow in a surgically isolated skeletal muscle with brief changes in arterial pressure.[7] This observation led to renewed interest in the myogenic hypothesis. Other investigators became interested and evidence from such whole organ studies began to accumulate, some of which supported the myogenic hypothesis and some did not. Some of the findings supported the metabolic hypothesis. Since these studies measured the response of flow to a pressure change, they essentially used a "black box" approach. With this approach it was difficult to develop critical tests of any one hypothesis that excluded all other possibilities. It became necessary to open the black box.

5.3 *Contribution of advances in microcirculation*

Researchers began to observe the microcirculation shortly after the microscope was developed. In fact the first observations were made by Malpighi and Van Leevenhoek in the decades following William Harvey's postulate of the existence of the blood circulation in 1628. Their observations of red cells flowing from arterial to venous vessels provided crucial evidence in support of his hypothesis.[8] However, obtaining detailed quantitative measurements of the microcirculation proved to be difficult. It was only in the early 20th century that measurements began and then on a very limited basis. This changed in the 1960s with the development of photometric instrumentation for measuring red cell velocity, micropressure techniques and compact video cameras and recorders for following dimensional changes in microcirculatory vessels.[9] Researchers interested in the microcirculation, led to a very significant degree by the bioengineering group at the University of California, San Diego,[10,11] developed such special purpose instrumentation for microcirculatory studies.

The new methods allowed detailed studies of the regulation of blood flow in the microcirculation. Studies on the arterioles in an isolated thin transparent tissue called the mesentery verified that the arterioles dilated at reduced arterial pressure as shown in

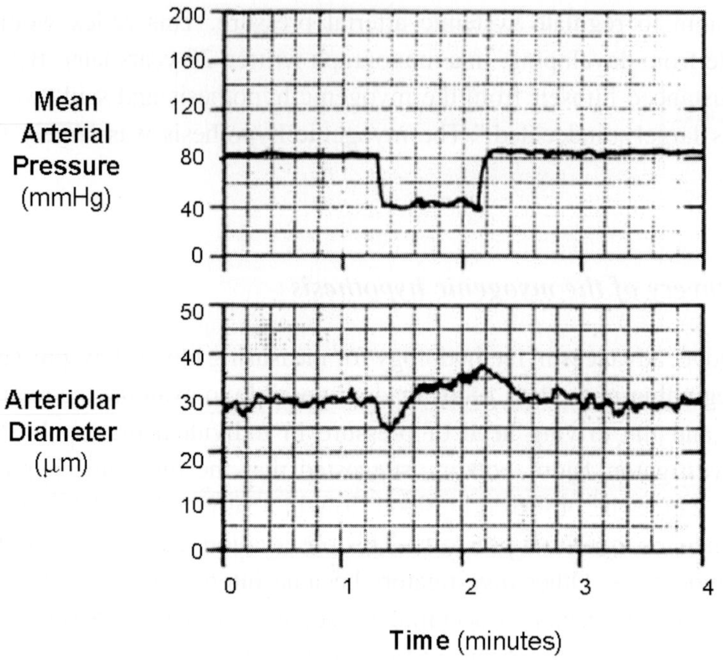

Fig. 5. Internal diameter of an arteriole in the mesentery during a step reduction of arterial pressure from 80 to 40 mmHg. A step reduction in pressure leads first to a small decrease in diameter followed by a gradual increase above the initial value. Sudden restoration of pressure causes first a slight expansion of the vessel followed by a gradual return down to the initial diameter. Modified from Johnson.[12]

Fig. 5.[12] Previously several researchers had suggested that as pressure increased for example, fluid would filter out of the blood through the capillary wall and increase hydrostatic pressure in the tissues, which would occlude the outflow vessels (the veins) to return flow to its previous level. This is the *tissue pressure hypothesis*[13] and is conceptually attractive since it does not require an active vascular response. Subsequent studies showed that this mechanism could only be present in encapsulated organs such as the brain and kidney in pathological situations where capillaries become abnormally leaky.

Studies of red cell velocity in the capillaries of this tissue revealed a periodic flow pattern (flomotion) that was very sensitive to arterial and venous pressures feeding and draining the organ as shown in Fig. 6. Stepwise reduction of arterial pressure in this study showed that red cell velocity actually increased in some capillaries at reduced arterial pressure.[14] Later studies showed an increase in volume flow with arterial pressure reduction in arterioles of skeletal muscle.[15] Since, according to the metabolic hypothesis, flow could not fully return to the previous level during arterial pressure reduction, these observations are not consistent with the metabolic hypothesis. Conversely, they are consistent with the myogenic hypothesis, which is not based on flow-dependency.

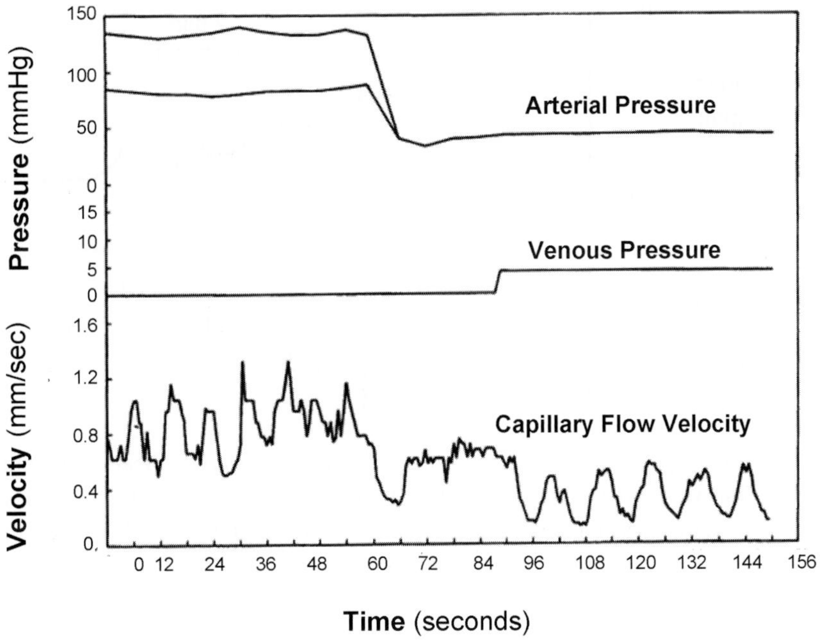

Fig. 6. Velocity of red cells in a capillary of mesentery during a step reduction in arterial pressure and a step increase in venous pressure. Periodic flow behavior in the control period is probably due to rhythmic constriction and dilation of the terminal arteriole immediately upstream (vasomotion). Note that when arterial pressure is reduced, flow initially falls and then returns toward the control level, but vasomotion is lost. A small increase in venous pressure causes a reduction in velocity and periodicity is restored. This likely is due to restoration of pressure and myogenic constriction in the terminal arteriole. Systolic and diastolic arterial pressures are plotted separately in the control period where values are substantially different. Modified from Johnson and Wayland.[14]

Other studies provided clear evidence for the myogenic hypothesis. As surgical techniques improved it became possible to isolate arterioles and small arteries (100–200 μm i.d.) and cannulate both ends of the vessel, thus allowing control of both flow and pressure in the vessel. The vessels were mounted in a chamber with physiological fluid surrounding the vessel and physiological fluid perfusing the vessel. When the internal pressure was increased in the absence of flow, these vessels initially dilated passively and then actively constricted to a smaller diameter that was maintained as long as the pressure was applied as shown in Fig. 7.[16] When the pressure was reduced in a single step there was an initial passive narrowing of the vessels followed by dilation.

Similar observations have been reported in isolated small arteries and arterioles from a variety of vascular beds indicating that the myogenic mechanism is of general importance. Since the myogenic response causes constriction of arterioles at normal arterial pressure, it is one of the principal mechanisms responsible for the partially constricted state of the arterioles under normal conditions. The other is the continuous

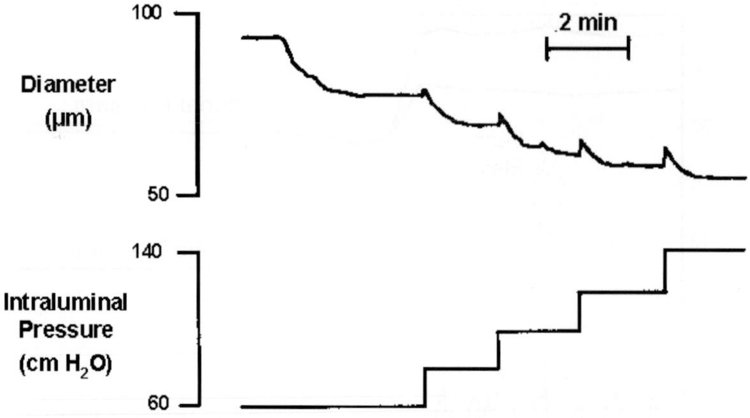

Fig. 7. Myogenic response of an arteriole surgically isolated from the heart and mounted in a chamber. Note that step increases in pressure lead first to a small dilation followed by a sustained constriction of the arteriole to a smaller diameter. Note the magnitude of the effect: the steady state diameter decreased by about 30% over this pressure range. Modified from Kuo *et al.*[16]

vasoconstrictor influence of the sympathetic nerves found in most blood vessels. Effective regulation of blood flow of course requires that there be robust mechanisms for both constriction and dilation of the vessels involved.

5.4 *Development of a theoretical basis for the myogenic hypothesis*

One of the features of the myogenic hypothesis that is not readily explained is the fact that the vessel actually becomes smaller and remains at reduced diameter when the pressure is elevated and maintained at an elevated level. There was no conceptual model to explain how this could happen. Smooth muscle from the intestine contracts when the muscle cells are stretched but the contraction can, at best, return the muscle to its original length. If the muscle returns to its original length there is no stimulus for contraction, assuming elongation of the muscle is the stimulus. Stretching the muscle provides the stimulus by deforming the membrane and opening stretch sensitive ion channels, leading to inflow of Ca^{2+} and excitation of the contractile machinery. To provide a solution to this dilemma, a German renal physiologist, Klaus Thurau[17] proposed that the vascular smooth muscle cells of the arteriole were sensitive to the circumferential wall tension or stress (hoop stress) of the vessel rather than muscle cell length. In a thin-walled cylinder, circumferential wall tension is a function of pressure and vessel radius according to the Law of Laplace: ($T = P \times r$) where T = tension, P = transmural pressure and r = radius. In Fig. 8 the hypothetical response of an arteriole to a step rise in internal pressure with regulated wall tension is shown.[18] When pressure increases the vessel initially dilates passively. Wall tension increases due to the rise in pressure and vessel radius. If the vessel is sensitive to

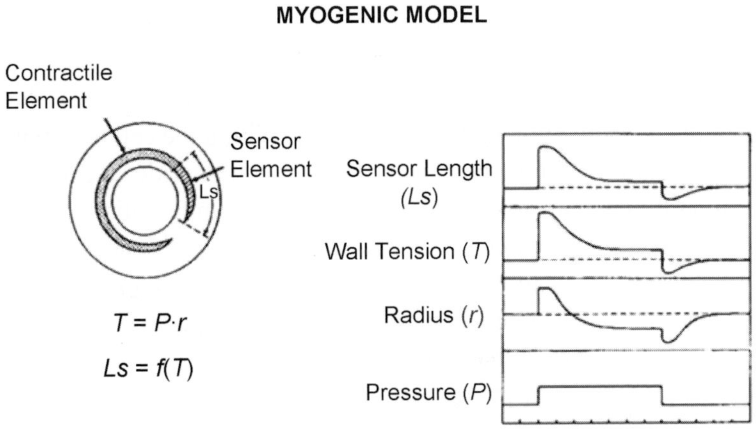

Fig. 8. Conceptual model of a vascular smooth muscle cell in an arteriole that reacts to changes in circumferential wall tension. The muscle cell model consists of a sensor element in series with the contractile machinery of the cell. It is hypothesized that passive elongation of the sensor with increased circumferential tension causes a signal to be sent to the contractile machinery to elicit contraction. The magnitude of the signal will be a function of internal pressure and vessel diameter according to the Law of Laplace. Modified from Johnson.[18]

wall tension, for example due to the presence of a passive sensor element in series with the contractile machinery, the vessel will contract. When the vessel returns to its initial diameter the wall tension will still be elevated and this error signal will allow the vessel to constrict to a smaller diameter. A steady state will be reached when the vessel diameter is less than control but wall tension is still somewhat elevated. Experimental studies in which the pressure in arterioles of the mesentery *in situ* was measured along with the arteriolar diameter showed that calculated wall tension did indeed follow the prediction shown in Fig. 8.

5.5 *Cellular basis of the myogenic response*

While the smooth muscle cell of the arteriole does not contain an anatomically separate sensor element as shown in the diagram there are specific regions (dense bodies) in the plasma membrane, where the force is transmitted between the contractile machinery and the extracellular matrix. In this region integrins are found at the junction between the cell and the matrix. It known from other studies that integrins function as mechanosensors and, in this case, an increase in force across integrins between the contractile machinery and the extracellular matrix could lead to contraction.[19] Also there is evidence (as noted above) for stretch sensitive channels in the plasma membrane. If such channels were located in the region of the dense body, they could lead to depolarization of the cell. It is known that an increase in intravascular pressure leads to a graded depolarization of the vascular smooth muscle cell depending on the magnitude of the pressure change. Depolarization of vascular smooth muscle leads to influx of Ca^{2+} from the extracellular fluid through voltage gated Ca^{2+} channels.[19] The Ca^{2+} interacts with calmodulin and causes activation of myosin light chain kinase and an increase in the interactions between actin and myosin filaments. There is also evidence[19] that the myogenic response may involve activation of protein kinase C within the cell to sensitize the contractile machinery to Ca^{2+}. Thus the current information provides a cellular basis for a myogenic response that tends to maintain a constant circumferential wall tension.

It should be noted that regulation of flow by the myogenic mechanism is incidental. It does tend to regulate capillary hydrostatic pressure, which is important in maintaining a normal fluid balance between blood and tissue. Models have been developed in which the tension sensor can have the effect of maintaining constant flow either by precise adjustment of the gain of the system or by a series of independent myogenic effectors along the arteriolar network.[6]

6. The Tubuloglomerular Hypothesis

Another mechanism that contributes to autoregulation is unique to the kidney. This is the *tubuloglomerular feedback* mechanism.[20] In an unusual anatomical arrangement called the juxtaglomerular apparatus, the ascending loop of Henle, which contains fluid filtered from the glomerular capillaries, is in close proximity to the afferent arteriole that feeds these capillaries. A thickened region of epithelium in the tubule called the macula densa senses the chloride level in the renal tubular fluid, which is flow rate dependent. The macular densa activates humoral mechanisms that tend to keep the tubular fluid chloride level constant by altering the diameter of the afferent arteriole. Note that this is a flow-dependent mechanism of autoregulation of blood flow.

The afferent arteriole and the upstream arterioles are also myogenically active. As a result, renal autoregulation is especially well regulated with flow being maintained at a very constant level over an arterial pressure range of 60 to 160 mmHg as shown in Fig. 2. If we put together the information we have gained on the myogenic response and tubuloglomerular feedback we can see that the myogenic response is a somewhat crude mechanism that tends to maintain a constant flow, the tubuloglomerular feedback provides fine tuning by sensing a variable that is sensitive to flow *per se*.

7. Clinical Significance of Autoregulation

7.1 *Autoregulatory adjustments with large artery disease*

Atherosclerosis is a disease of the large arteries that causes narrowing of the lumen, increased resistance in this region and reduction of the driving head of pressure to the downstream vessels. The autoregulatory response in the arterioles can compensate for the increased resistance caused by the atherosclerotic lesion and the person with the disease may be unaware of the problem. However in this circumstance the dilator reserve of the arterioles is reduced and the maximum blood flow that could be achieved during maximal exercise is decreased. If the disease occurs in a coronary artery the subject may experience angina (ischemic pain) or irregularities in the electrical conduction process in the heart during maximal exercise. A treadmill test therefore can reveal the presence of coronary artery disease that is normally masked by the autoregulatory response of the arterioles.

7.2 *Autoregulatory response in hemorrhage*

Another instance in which autoregulation is important clinically is during a drop in arterial pressure due to blood loss. In this condition, the pressure drop is detected by sensors in the carotid artery and aorta and a reflex response of the sympathetic nervous system causes constriction of the arterioles generally throughout the body. However, the arterioles in the brain and myocardium do not have significant sympathetic innervation so do not participate in the generalized vasoconstriction. The autoregulatory response in those vascular beds then helps to maintain adequate blood flow despite the reduced arterial pressure. The sympathetic nervous system overrides the autoregulatory mechanism in other, less essential, organs and flow in those organs decreases.

8. Summary and Conclusions

8.1 *The autoregulatory mechanism*

Autoregulation of blood flow can be attributed to several mechanisms. In some instances evidence supports a link between blood flow and metabolic requirements of the tissue. Other studies indicate that the myogenic response of the arterioles is important. In the kidney both the myogenic response and the tubuloglomerular feedback mechanism are involved. Autoregulation provides a mechanism for maintaining blood flow to vital organs when arterial pressure is reduced and in maintaining adequate flow to the tissues in the presence of vascular disease in large arteries.

Study of autoregulation has also been a vehicle elucidating the fundamental mechanisms of flow regulation. In putting together this information, we see that the vascular smooth muscle of the arteriole has a remarkable property that enables it to sense its internal pressure and certain chemical mediators in its immediate environment and respond accordingly to maintain adequate blood flow to the tissues under a variety of conditions.

8.2 *The process of scientific discovery*

Study of the development of our understanding of autoregulation also provides lessons in the process of scientific discovery. What are some of the lessons to be learned from this example? Several come to mind. First, if a mechanism seems intuitively correct, that should not prevent us from examining it critically. Second, if a proposed mechanism seems counter-intuitive, that should not discourage us from exploring it. Third, proof that one mechanism is involved in a phenomenon does preclude the possibility that there may be other mechanisms that produce similar results. Redundancy is common in biological systems, especially where important functions are involved. Fourth, it is important not to be limited by preconceived notions and presume the simplest explanation is more likely to be correct. The last point should provide encouragement to who are tempted to make bold predictions but feel constrained because of conventional wisdom.

Acknowledgments

The author's research is supported in part by NIH grants HL 66318 and HL 52684.

References

1. P. C. Johnson, Review of previous studies and current theories of autoregulation, *Circ. Res.* **15**(Suppl.): 2–9 (1964).
2. F. Crick, *What Mad Pursuit. A Personal View of Scientific Discovery* (Basic Books, 1988), p. 5.
3. T. S. Park, D. G. L. van Wylen, R. Rubio and R. M. Berne, Brain interstitial adenosine and autoregulation of cerebral blood flow in neonatal piglet, *Concepts Ped. Neurosurg.* **10**: 243–254 (1990).
4. U. Pohl and R. Busse, Hypoxia stimulates release of endothelium-derived relaxant factor, *Am. J. Physiol.* **256**: H1595–1600 (1989).
5. W. M. Bayliss, On the local reaction of the arteriolar wall to changes of internal pressure, *J. Physiol. (London)* **28**: 220–231 (1902).
6. P. Johnson, The myogenic response. In: *Physiology Handbook*, eds. B. F. David, A. P. Somlyo and H. V. Sparks, Jr.; exec. ed. S. R. Geiger (American Physiological Society, Bethesda, 1980), Chapter 15.
7. B. Folkow, Intravascular pressure as a factor regulating the tone of the small vessels, *Acta Physiol. Scand.* **17**: 289–310 (1949).
8. E. Landis, The capillary circulation. In: *Circulation of the Blood: Men and Ideas*, eds. A. P. Fishman and D. W. Richards (American Physiological Society, Bethesda, 1982), Chapter 6.
9. H. Wayland, Microcirculatory research methods, *Microvasc. Res.* **5**: 229–233 (1973).
10. M. Intaglietta, R. F. Pawula and W. R. Tompkins, Pressure measurements in the mammalian microvasculature, *Microvasc. Res.* **2**: 212–220 (1970).
11. M. Intaglietta, W. R. Tompkins and D. R. Richardson, Velocity measurements in the microvasculature of the cat omentum by on-line method, *Microvasc. Res.* **2**: 462–473 (1970).
12. P. C. Johnson, Autoregulatory responses of cat mesenteric arterioles measured *in vivo*, *Circ. Res.* **22**: 199–212 (1968).
13. L. B. Hinshaw, Mechanism of renal autoregulation: role of tissue pressure and description of a multifactor hypothesis, *Circ. Res.* **15**(Suppl.): 120–131 (1964).
14. P. C. Johnson and H. Wayland, Regulation of blood flow in single capillaries, *Am. J. Physiol.* **212**: 1405–1415 (1967).

15. P. Ping and P. C. Johnson, Arteriolar network response to pressure reduction during sympathetic nerve stimulation in cat skeletal muscle, *Am. J. Physiol.* **266**: H1251–1259 (1994).
16. L. Kuo, M. J. Davis and W. M. Chilian, Myogenic activity in isolated subepicardial and subendocardial coronary arterioles, *Am. J. Physiol.* **255**: H1558–1562 (1988).
17. K. W. Thurau, Autoregulation of renal blood flow and glomerular filtration rate, including data on tubular and peritubular capillary pressures and vessel wall tension, *Circ. Res.* **15**(Suppl.): 132–141 (1964).
18. P. C. Johnson, The microcirculation and local and humoral control of the circulation. In: *Cardiovascular Physiology*, *Int. Rev. Physiol. Ser.* Vol. 1., eds. A. C. Guyton and C. E. Jones (Baltimore University Park, 1974), pp. 163–195.
19. M. A. Hill, H. Zou, S. J. Potocnik, G. A. Meininger and M. J. Davis, Invited review: arteriolar smooth muscle mechanotransduction: Ca(2+) signaling pathways underlying myogenic reactivity, *J. Appl. Physiol.* **91**: 973–983 (2001).
20. J. Schnermann, Homer W. Smith Award lecture. The juxtaglomerular apparatus: from anatomical peculiarity to physiological relevance, *J. Am. Soc. Nephrol.* **14**: 1681–1694 (2003).

SECTION III

BLOOD CELL BIOENGINEERING

CHAPTER 8

MOLECULAR BASIS OF CELL AND MEMBRANE MECHANICS

Lanping Amy Sung

Abstract

To understand the molecular basis of cell and membrane mechanics, we cloned cDNAs that encode red blood cell membrane skeletal proteins. Characterizing genomic organization allowed us to disrupt a target gene in an embryonic stem cell and create knockout mouse models. Such models allowed us to establish the precise relationship between a molecular defect and mechanical changes of single cells by experiments. We expressed recombinant proteins and mapped their binding sites. From protein-protein interaction we constructed the first three-dimensional model for a junctional complex and explained how the junctional complex may, in turn, dictate the network topology of the membrane skeleton. The 3-D nano-mechanics of the red cell membrane skeleton may now be understood at the molecular level by simulation.

1. Introduction

Molecular biology is an important part of bioengineering. This chapter will show how molecular biology is being integrated into research in biomechanics. One of the systems my laboratory is working on is the cell membrane. In this research we address how individual components of the cell membrane, especially the protein molecules, may affect the mechanical properties of the cell and membrane.

The plasma membrane is a 5-nm thick film that serves as a barrier between the contents of the cell and its surrounding medium. This film is composed mainly of a lipid bilayer, which is penetrated by channel and pump proteins that transport specific substances into and out of the cell. The lipid bilayer is also associated with other proteins, including those acting as sensors to detect the changes in its environment, and those function as infrastructures to provide mechanical stability for the membrane when the cell is being stressed or deformed.

Some of the mechanical properties of the membrane (e.g., membrane fluidity) are dependent on the composition and the structure of phospholipids, sterols and glycolipids that form the lipid bilayer of the membrane. Yet, other mechanical properties of the membrane (e.g., the resistance to elastic deformation or fragmentation) are primarily attributable to the membrane proteins and their organization. Cell membrane proteins can be classified into three categories on the basis of their relationships with the lipid bilayer: (1) transmembrane proteins that span the lipid bilayer once or more; (2) acylated proteins (i.e. proteins having a fatty acid component) that are embedded, or partially embedded in the lipid bilayer; and (3) peripheral proteins that are non-covalently associated with the lipid bilayer or other proteins on either side of the membrane. Many of these membrane proteins interact with each other non-covalently. They form network structures that provide the stability for the lipid bilayer and mechanical properties for the cell membrane.

We investigate the functional role of specific proteins in the organization of the protein network, and how specific proteins affect the mechanical properties and stability of the cell membrane. We especially want to know the stress-strain relationship of the cell membrane when a specific protein is absent, over-expressed, under-expressed, or mutated. In other words, we want to know the molecular basis of cell membrane mechanics.

Our research helps us to understand how the cell membrane deforms in response to stress, how much the cell membrane deforms before it breaks, how does a membrane skeletal protein effect the mechanics of the membrane, and what molecular and mechanical defects may be present in certain hereditary diseases. Our research also involves creating cell membrane models in which one specific membrane network protein is mutated or depleted by genetic engineering. Generating and analyzing these disease models helps us to understand and appreciate the importance of biomechanics in living tissues. Furthermore, restoring normal mechanical stability and properties to these model membranes, and rescuing other disease phenotypes in hereditary diseases by gene therapy will be among the challenges facing the new generation of bioengineers.

1.1 *The membrane skeletal network*

The human erythrocyte membrane is the best characterized cell membrane system in living cells. It is the simplest in that the erythrocyte has no complicated cytoskeletons connecting the plasma membrane to other parts of the cell. It has only a thin (yet three-dimensional) layer of protein network underneath the lipid bilayer. Furthermore, it has no nucleus and other organelles in the cytoplasm to undergo cell division, protein synthesis, and many other functions of living cells.

The mechanical properties of the erythrocyte membrane have been well characterized. The molecular structures of the major proteins and lipids and their organization in the membrane have also been studied. Basically all major erythrocyte membrane skeletal proteins, and some minor ones, have been cloned and characterized in the past 20 odd years. Electron microscopy and biochemical analyses of human erythrocyte

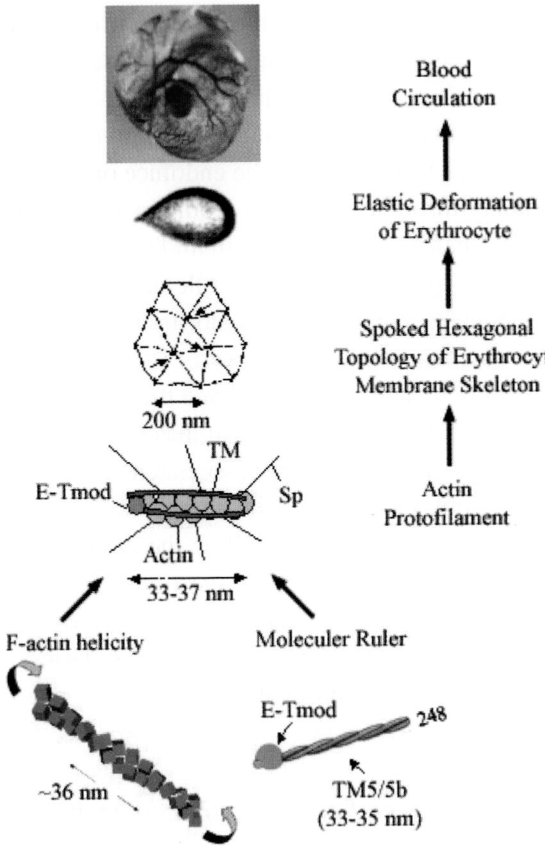

Fig. 1. Molecular basis of erythrocyte membrane mechanics. From bottom up: **(a)** Both the intrinsic properties of F-actin helix and the molecular ruler made out of E-Tmod and TM5/5b provide mechanisms to form actin protofilaments of 37 nm. In the molecular ruler, the E-Tmod-binding site positions the globular E-Tmod at the N-terminus of the rod-like TM5/5b molecule. **(b)** At a junctional complex the number of G actin in the protofilament (six per strand protected by the ruler) determines the number of spectrin dimers, six, associated with the filament. **(c)** The number of spectrin dimers dictated the "spoked" hexagonal topology of the membrane skeletal network. **(d)** The network provides the mechanical strength and elastic properties for the membrane, allowing repeated deformation of erythrocytes. **(e)** Normal circulation of *E-Tmod +/–* erythrocytes in the yolk sac and the embryo proper. The erythrocytes turn blue as the "knocked-in" *lacZ* reporter gene reports a high level expression of E-Tmod.

membranes have shown that these major membrane proteins form a relatively regular, thin protein network on the cytosolic side of the membrane (sketched in Fig. 1). This membrane skeletal network consisting of "spoked" hexagons supports the stability of the lipid bilayer and contributes to the elastic behaviors of the erythrocyte membrane. While the 200-nm long, spring-like spectrin tetramers are the thread of the hexagons, the short ~37-nm stubs of actin protofilaments are the key components of the knobs. The knob, called junctional complex (JC), is where the tail ends of spectrin dimers (usually six) meet. The head-to-head association of spectrin dimers occurs in the middle of the 200-nm long tetramers. The JC also contains several actin-associated proteins, such as tropomyosin isoforms 5 or 5b (TM5/5b) and erythrocyte tropomodulin (E-Tmod) that are involved in defining the uniform length of the actin protofilament.

This thin protein network is anchored to the endoface of the lipid bilayer in the erythrocyte membrane mainly through suspension complex (SC), which contains band 3, ankryin, and protein 4.2. A pair of SC is located in mid regions of the spectrin tetramers. Each of the proteins in a SC plays an important role in attaching the membrane skeletal network to the lipid bilayer: band 3 is the major transmembrane protein that serves as the anion exchanger (a channel protein); ankyrin is a peripheral membrane protein that binds strongly and non-covalently to band 3 and the β subunit of spectrin; and protein 4.2, which binds to both band 3 and ankyrin, has a fatty acid component, further enhances protein 4.2 and the entire SC tightly associated with the lipid bilayer. What is the advantage of such a network organization in providing the stability of the lipid bilayer? How does the network organization allow the elastic deformation of the erythrocytes? These are the important engineering questions because erythrocytes need to be strong in order to survive the flow dynamics of the cardiovascular system and yet highly deformable in order to negotiate through narrow capillaries, many of which have smaller diameters than that of the erythrocytes.

1.2 *The mechanical properties of erythrocyte membranes*

The mechanical properties of erythrocyte membranes have been characterized by several techniques using engineering principles and technologies. These include flow channel, micropipette aspiration, ektatocytometry, and microceiving techniques. While the first three techniques analyze the stress-strain relationship of the erythrocyte membrane by observing the degree of cell deformation, the microceiving technique measures the resistance of the flow in relation to the deformability of erythrocytes. In addition, the osmotic fragility is a measurement of the mechanical stability of the erythrocyte membrane when cells are subjected to low osmotic solutions, and the degrees of hemolysis of erythrocytes are quantified. An example of the elastic deformation of erythrocytes is shown in the middle of Fig. 1. The cell undergoes deformation in response to shear stresses when the cell is attached to the floor of a flow channel with a point attachment.

2. Defects of Membrane Mechanics in Knockout (KO) Mouse Models

To understand why the mechanical stability of the membrane or the deformabilty of diseased erythrocytes is altered in patients with hemolytic anemia or other disorders, and how the mechanical stability and properties may be restored, we must elucidate the molecular basis of membrane mechanics of both normal and diseased erythrocytes.

The general strategy we took was to develop and intertwine two parallel lines of researches: (1) from cDNA cloning to knockout mice and (2) from protein-protein interaction to 3-D model and nano-mechanics. Each of these key steps will be discussed, and one paper describing the creation of knockout mouse model from the embryonic stem cells[1] and another paper reporting the 3-D model of a JC[2] are cited at the end of this chapter.

In order to understand how membrane skeletal proteins interact with each other in forming the protein network and how the network organization may affect the mechanical properties of the erythrocyte membrane, the amino acid sequences or the primary structures of these membrane skeletal proteins are needed. The amino acid sequences are also essential to identify the molecular defects in patients and to study the effects of mutations on the mechanical properties of erythrocyte membranes.

Recombinant DNA technologies are the technologies that have been used in the past 20 odd years to reveal the primary structures of major membrane skeletal proteins. The recombinant DNA technologies include the use of restriction endonucleases, direct nucleotide sequencing, cDNA and genomic DNA cloning, site-specific mutagenesis, polymerase chain reaction (PCR), and generation of transgenic or knockout cells and animals.

2.1 *The cDNA cloning of human erythrocyte tropomodulin*

The cDNA cloning of human E-Tmod was done by screening a human fetal liver cDNA expression library with an E-Tmod specific antibody raised in rabbits. The additional 5′ sequences of the cDNA were obtained by PCR, a technique used to amplify a piece of DNA *in vitro* with a pair of sequence-specific primers. In this case the DNA *in vitro* was the cDNA derived from human reticulocytes (young erythrocytes newly released into circulation that no longer have nuclei, but still have some remaining mRNA). Nucleotide sequencing of these clones, and translating the open reading frame of the cDNA, predicted a protein of 359 amino acids. The authenticities of cDNA sequences were verified by several lines of evidence, including the match of several translated sequences with known peptide sequences.

The translated amino acid sequence also allowed the prediction of protein properties, such as molecular mass, phosphorylation sites, isoelectric pH, and secondary structure. The complete amino acid sequence also allowed us to search for proteins with similar amino acid sequences that have been published and collected in sequence databases. National Center for Biotechnology Information (NCBI, http://www.ncbi.nlm.gov), for example, provides such services. At the time of E-Tmod cloning, there was no protein with similar structure. Now there are at least 15 proteins that belong to this family, all were cloned after human E-Tmod. Homologous proteins often have related (not necessarily similar) functions.

Recombinant DNA technologies have been used to clone and characterize more than a dozen of major and minor human erythrocyte membrane proteins. In addition to

E-Tmod, we have also cloned the cDNA for human protein 4.2, a transglutaminase-like molecule located in the SC of the erythrocyte membrane network. Protein 4.2, however, has lost the enzymatic activity to crosslink substrates (e.g., coagulation factor XIIIa cross-links fibrin in forming clots). It may function to prevent other membrane protein substrates from being cross-linked by the real transglutamiase.

2.2 *The genomic cloning of human erythrocyte tropomodulin*

Recombinant DNA technologies have also been used to clone and characterize genomic DNA of humans and other species. We have cloned the genomic DNA encoding *E-Tmod* to understand how exons and introns are organized in the gene. Studies on the genomic organization allow us to understand how the expression of a specific gene may be regulated and how the splicing of the mRNA is achieved. For example, we discovered two alternative promoters to regulate the transcription of the *E-Tmod* gene, and that there are nine exons encoding the 359 residues in the protein. Genomic fragments are also needed to construct targeting vectors to knockout specific exons or genes in creating disease models.

Physical locations of genes on chromosomes may be mapped by FISH (fluorescent *in situ* hybridization). We have mapped the genes encoding E-Tmod and protein 4.2 to human chromosomes 9q22 and 15q15-20, respectively. Chromosomal mapping helps the localization and identification of disease genes in animal models and human patients.

2.3 *The E-Tmod knockout mouse model*

When there is no naturally occurring disease model, or the cause-effect relationship between a mutation and a disease is not clear, creating a knockout model deficient of a specific protein is a clean approach to demonstrate the role of a protein. This is accomplished by first constructing a targeting vector, which is a modified genomic fragment, and then introducing, by electroporation, the targeting vector into a cell line. The cell lines will be embryonic stem (ES) cell lines, if the intention is to generate knockout mouse models. Through the rare probability of homologous recombination between the targeting vector and the endogenous gene, the endogenous normal gene is replaced by the targeting vector that carries a portion of the gene that has been disrupted. As a result, the expression of this gene, or the synthesis of this protein is specifically blocked. Investigations of such cells will allow us to establish the functional role of this protein.

The embryonic stem cells, in which a targeted disruption of a specific gene has occurred, allow us to create knockout animal models. This is achieved by microinjecting the genetically altered embryonic stem cells into early embryos at the stage of tropoblasts, and then transplanting these embryos into pseudopregnant mice. The resulting chimeric mice may give rise to heterozygous knockout mice if some of their germ cells are derived from the genetically altered embryonic stem cells. Breeding between the heterozygous knockout mice may give rise to homozygous knockout mice if the homozygous null mutation is not a lethal mutation. The homozygous knockout mice are completely deficient of the protein encoded by that specifically disrupted gene. Several knockout mice or cell lines deficient of erythrocyte membrane skeletal proteins have been created. We have contributed to these studies by creating embryonic stem cells and mice in which one or both copies of the *E-Tmod* gene have been disrupted.[1]

To create E-Tmod KO mice, we first characterized the exon-intron organization of the mouse *E-Tmod* gene. We then constructed a targeting vector in which the exon containing the ATG initiation codon was replaced by a neomycin resistant (*Neo*^{r}) gene and a bacterial *lacZ* reporter gene. Such targeting vector was then used to create ES cell lines heterozygous or homozygous for the *E-Tmod* null mutation.[1] These ES cells can also be driven into erythroid and other cell lineage in culture to demonstrate the effect of E-Tmod deficiency on the mechanical properties of erythroid and non-erythroid cells.

The ES cells, whose one copy of the *E-Tmod* gene is disrupted, have been used to generate chimeric, heterozygous, and homozygous *E-Tmod* knockout mice. While the heterozygous knockout mice (*E-Tmod*$^{+/-}$) survive, homozygous knockout mice (*E-Tmod*$^{-/-}$) die during embryonic development. The results indicate that E-Tmod is essential for the embryogenesis, and that there are no other genes in the mouse genome to compensate for its function when both copies of the *E-Tmod* gene are disrupted. The disease phenotypes in *E-Tmod*$^{-/-}$ embryos include abnormalities in the heart development, remodeling of blood vessels, and mechanical properties of erythrocytes.[1] Therefore we demonstrated for the first time that the capping of the actin filament at the slow-growing end is essential to establish the embryonic circulatory system to provide much needed oxygen in the growing embryos.

Our mechanical analysis using micropipette aspiration revealed that primitive erythroid cells with *E-Tmod* null mutation (*E-Tmod*$^{-/-}$) are mechanically weakened in their membranes, exhibiting partial hemolysis, and more prone to fragmentation in response to mechanical stress (Fig. 2). Without a functional molecular ruler (a T-ruler missing the cap), actin protofilaments of 37 nm may not be generated. Without the "lumbers" of a uniform size to construct the regular junctional complexes with six spokes of Sp, a well-knitted network with "spoked" hexagons would not be formed to sustain the mechanical stress.

2.4 *Mutants and diseases*

Mutations or deficiencies of a protein best reveal its function. Many hemolytic anemic patients have deficiencies in major membrane skeletal proteins in their erythrocytes,

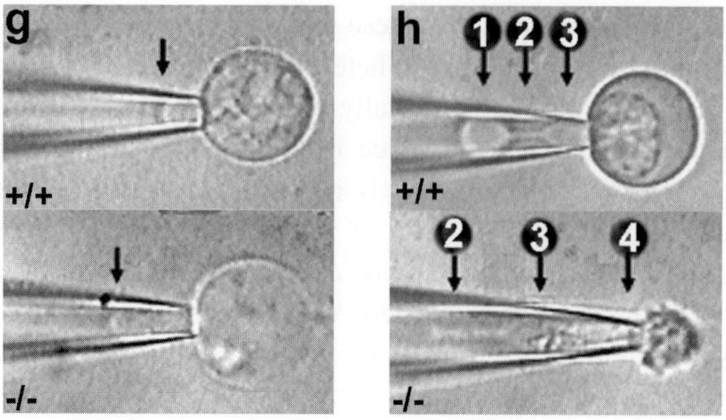

Fig. 2. Mechanical weakness of *E-Tmod* null erythroid cells tested by the micropipette aspiration technique. Deformation test on enucleated primitive erythrocytes (**g**) and fragmentation test on nucleated primitive erythroblasts (**h**) from E9.5 *E-Tmod* $^{+/+}$ (upper panel) and $^{-/-}$ (lower panel) yolk sacs were performed. Arrows in **g** indicate the length of cell membrane deformed into the pipette (radius of the pipette, Rp = 0.75 μm; aspiration pressure, ΔP = 30,000 dyne/cm^2; duration, t = 1300 msec). In the fragmentation test (**h**, Rp = 0.75 μm, aspiration force = 22.5 dyne/cm), the cell membrane showed progressive deformation with a nodule (arrow 1), a "neck" (narrowing membrane segment, arrow 2), cell body (arrow 3) and partially deformed nucleus membrane (arrow 4, not seen in wild type). Bar, 20 μm.

such as α and β spectrins, ankyrin, band 3, protein 4.1, and protein 4.2. In many cases the protein deficiencies are associated with point mutations or frame shift mutations of their nucleotide sequences. The erythrocytes of these patients exhibit sphereocytosis, ovalocytosis, or elliptocytosis and the membranes demonstrate osmotic frigidity. The existence of these patients demonstrates the importance of membrane skeletal proteins in maintaining the stability and elastic deformability of the erythrocyte membrane.

2.5 *Knock-in reporter gene*

A reporter gene, such as *lacZ* may be knocked-in downstream from a promoter. β-galactosidase, the gene product of *lacZ*, can be detected by its substrate, e.g., X-gal, or FDG, a fluorescent substrate. This would allow the identification of cells and tissues that normally express the gene. With the knock-in *lacZ* reporter gene, the expression pattern and function of E-Tmod in erythroid and non-erythroid cells and tissues may be studied *in vivo*. Furthermore, the phenotypes observed in knockout mice may serve as a guideline to screen human patients with similar blood, cardiovascular, or muscular diseases. The studies of knockout mice will, therefore, help us in identifying human diseases whose molecular defects are associated with mutations or abnormal expression of a specific gene.

3. Molecular Mechanics of a Junctional Complex

3.1 *Molecular ruler*

Although TM5 and TM5b are products of two different genes, they share several common features, including: (1) both are low molecular weight (LMW) isoforms, have 248 residues, and are ~33–35 nm long, (2) both have a high actin-binding affinity, and (3) both have a high E-Tmod-binding affinity. We discussed the actin affinity of TM and why it is important for the stability of protofilaments; the length of TM and how it defines the hexagonal topology of the membrane skeleton; and the E-Tmod affinity of TM and how that modulates the function of TM isoforms and the length of the actin filaments.[3]

First of all, high actin affinity allows TM5 and TM5b to form a more stable actin protofilament in JC to resist the pulling of spectrin in response to stresses. Secondly, LMW isoforms stabilize six G-actins per strand, allowing six Sp to bind to one protofilament, forming a hexagonal topology that allows a seamless continuation of the network. Lastly, their high E-Tmod affinity makes the TM/E-Tmod complex an effective measuring device, or a molecular ruler, capable of metering off long actin filaments to short protofilaments. Because such a molecular ruler is able to bind and block an actin filament at its pointed end, and not able to overlap with other TMs in a head-to-tail fashion along the actin filaments or at the barbed end, only the first six G-actin pairs located at the pointed end of the actin filaments are protected by TM. Given the stress and strain undergone in the erythrocyte membrane during circulation, segments of actin filaments not coated by TM are likely to be fragmented or depolymerized. Thus, only short segments of the actin filaments that are protected by 1 TM can survive.

3.2 *Mapping the E-Tmod binding site on TM5*

We characterized the E-Tmod-binding site by creating a series of deletion and missense mutations within the N-terminal region of TM5, followed by a solid-phase binding assay. I^7, V^{10}, and I^{14}, hydrophobic residues located at *a* and *d* positions of N-terminal heptad repeats involved in intertwine, are essential for E-Tmod-binding. R^{12}, a positively charged residue at the *f* position, is also involved in recognition. In contrast, A2R and G3Y mutations, each creating a bulky N-terminus, did not alter the binding. Results also suggest E-Tmod has a groove-type, rather than a cavity-type, binding site for hTM5.

3.3 *Mapping the TM5 binding site on E-Tmod*

We also mapped the TM5 binding site on human E-Tmod to include residues at L^{116}, E^{117} and/or E^{118} by identifying among 35 deletion clones and a series of point mutations that no longer bind to human TM5 and rat TM5b. Interestingly, upstream residues 71–104 contain an actin binding site. Futhermore, the N-terminal “KRK ring” of TM5 may participate in balancing electrostatic force with hydrophobic interaction in its dimerization and binding to E-Tmod.

Thus, site-specific mutagenesis allows us to change one or more specific residues in a molecule in order to understand whether these residues are involved in the protein interactions, and whether, in this case, important for the network organization and the mechanical properties.

3.4 *A 3-D model for a JC*

It is a long-standing mystery why erythrocyte actin filaments in the JC are uniformly ~37 nm and that the membrane skeleton consists of “spoked” hexagons. We have previously proposed that an external “molecular ruler” formed by E-Tmod and TM 5 or 5b functions to generate protofilaments of 12 G actin under mechanical stress. We have also illustrated that *intrinsic* properties of actin filaments, e.g., turns, chemical bonds, and dimensions of the helix, also favor fragmentation into protofilaments under mechanical stress (Fig. 1 bottom). Specifically, we constructed a 3-D model of a JC, as shown in Fig. 3a, in that a pair of G actin is wrapped around by a split α and β spectrin, which may spin to two potential positions, and stabilize to one when the tail end of Sp

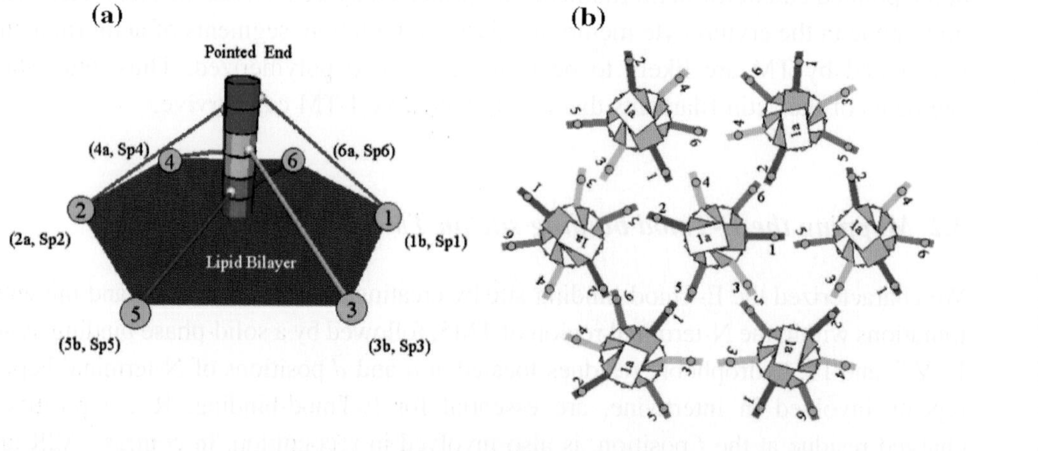

Fig. 3. A 3-D model of a junctional complex (JC) and a top view of six plus one JC in the membrane. **(a)** An actin protofilament that may function as the mechanical axis for three pairs of Sp. Each Sp may connect to a suspension complex (SC) in the lipid bilayer, forming a small hexagon, without physical edges. Both Sp pairs (top, middle, and bottom) and G actin pairs to which Sp pairs are attached are color-coded. The hexagon may be defined by the position of SC. In **(b)** six peripheral JC and a central one may rotate and connect to form a large “spoked” hexagon. All six edges and the spokes connecting the hub are made out of spectrin tetramers. All Sp heads are near SC (circle, in pink) and the plane of lipid bilayer, facilitating the collision among the heads. Note the order of SC.

is restricted (for details, see Ref. 2, which is attached). A reinforced protofilament may function as a mechanical axis to anchor three (top, middle and bottom) pairs of Sp. Each Sp pair may wrap around the protofilament with a wide dihedral angle (~166.2°) and a minimal axial distance (2.75 nm). Such three Sp pairs may spiral down (right-handed) the protofilament from the pointed end to the barb end with a dihedral angle of ~55.4° in between Sp pairs. This first 3-D model of JC may explain the "spoked" hexagonal topology of the erythrocyte membrane skeleton (Fig. 3b).

3.5 *3-D nano-mechanics of a JC*

We developed a mathematical model to compute the equilibrium states of a JC by dynamic relaxation.[4] We simulated deformations of a single unit in the network to predict the tension of each of the six Sp, and the attitude of the central actin protofilament [pitch (θ), yaw (Φ) and roll (ψ) angles] (Fig. 4). In equibiaxial deformation, six Sp would not begin their first round of "single domain unfolding in cluster" until the extension ratio (λ) reaches ~3.6, beyond the maximal sustainable λ of ~2.67. Before Sp unfolds, the protofilament would gradually raise its pointed end away from the membrane, while Φ and ψ remain almost unchanged. In anisotropic deformation, protofilaments would remain tangent but swing and roll at least once between $\lambda_i = 1.0$ and ~2.8, in a deformation angle- and λ_i-dependent fashion. Further study revealed that

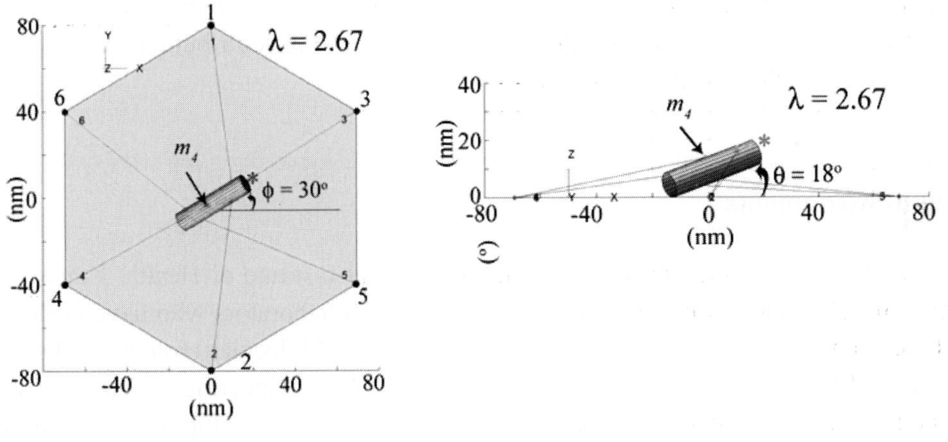

Fig. 4. The final equilibrium conditions [including the attitudes of an actin protofilament, i.e., pitch (θ), yaw (Φ) and roll (ψ) angles] at the equibiaxial deformation with λ = 2.67. The left panel is the top view; the right side views. The top view shows a protofilament connected through three pairs of Sp (i.e., Sp_1/Sp_2, Sp_3/Sp_4, and Sp_5/Sp_6) to six corresponding SC. The order of SC, numbered next to the hexagon, is according to the proposed sequence of the Sp attached to the protofilament[3] (also see Fig. 2) and numbered from the pointed end to the barbed end. The pointed end of the protofilament capped by E-Tmod is marked with *. Note that the line between each Sp-binding site on the protofilament and corresponding SC represents the distance (Sp "effective length") and not the Sp contour length.

wrapping spectrin around protofilament, rather than attaching directly onto it, would reduce the bifurcation of Φ angles. This structure-based prediction of nano-mechanics in response to deformations may reveal functional roles previously unseen for a JC during erythrocyte circulation.

4. Conclusions

This chapter demonstrates that the two seemingly unrelated fields, biomechanics and molecular biology, can be integrated. Together, they can address questions neither fields alone can address. One example given here is our research addressing the molecular basis of cell membrane mechanics. In this series of research we include cDNA and genomic cloning, creation of knockout mouse model, mechanical testing of genetically engineered erythrocytes, 3-D model of a JC, and mathematical simulation of the attitude of the actin protofilament and the tension of spectrin in defined network deformation.

Bioengineers, by integrating biomedical sciences and engineering, have improved our basic understanding of biological systems, have analyzed biomedical problems and have designed practical treatments of disease. Molecular biology and engineering are a powerful combination. The new generation of bioengineers who learn traditional engineering principles and techniques along with molecular biology, capable of performing recombinant DNA technologies, will clearly be at the frontier. As more bioengineers are able to integrate these two fields, more original and exciting contributions will be made at the interface of medicine and engineering.

Acknowledgments

The work was supported by grants from the National Institute of Health. I thank students, postdoctoral fellows, research associates, and collaborators who have made significant contributions in this series of research projects. Molecular biology works were conducted in the Biotech Core in the Department of Bioengineering which was established with the generous support from the Whitaker Development Award and Leadership Award.

References

1. X. Chu, J. Chen, M. C. Reedy, C. Vera, K.-L. P. Sung and L. A. Sung, E-Tmod capping of actin filaments at the slow growing end is required to establish mouse embryonic circulation, *Am. J. Physiol. Heart Circ. Physiol.* **284**: H1827–1838 (2003).

2. L. A. Sung and C. Vera, Protofilament and hexagon: a three-dimensional mechanical model for the junctional complex in the erythrocyte membrane skeleton, *Ann. Biomed. Eng.* **31**: 1314–1326 (2003).
3. L. A. Sung, K.-M. Gao, L. Y. Yee, C. J. Temm-Grove, D. M. Helfman, J. J.-C. Lin and M. Mehrpouryan, Tropomyosin isoform 5b is expressed in human erythrocytes: implication of tropomodulin-TM5 or tropomodulin-TM5b complexes in the protofilament and hexagonal organization of membrane skeletons, *Blood* **95**: 1473–1480 (2000).
4. C. Vera, R. Skelton, F. Bossens and L. A. Sung, 3-D nanomechanics of an erythrocyte junctional complex in equibiaxial and anisotropic deformations, *Ann. Biomed. Eng.* **33**: 1387–1404 (2005).

CHAPTER 9

CELL ACTIVATION IN THE CIRCULATION: THE AUTO-DIGESTION HYPOTHESIS

Geert W. Schmid-Schönbein

Abstract

In this chapter we present an idea for a 21st century biomedical engineer: get involved in the engineering analysis of an important human disease. There are many diseases that need attention and few have been subjected to a rigorous engineering analysis. The purpose of an engineering analysis is to find out what goes on and why an organ or a tissue fails. A good medical intervention is then based on such a rigorous engineering analysis. The time is right to start such an effort. I will illustrate the beginnings of an approach in one of the most lethal conditions, shock and multi-organ failure.

1. Introduction

As a broad but still young discipline, bioengineering offers a rich variety of opportunities to develop and improve techniques to treat disease and assist the disabled. You may wish to become involved in the design of new and better devices or tissue-engineered organs to treat persons with disease, or develop new diagnostics. There exist many opportunities. In this chapter I want to draw your attention to an opportunity for biomedical engineering to make another contribution to human disease and to medicine. While many medical disciplines are involved in the question on how to treat human disease, many times treatments are proposed that while being able to effectively alleviate symptoms, are not focused necessarily on the root cause of the disease. Before you treat a disease, you like to understand the actual cause and mechanisms that leads to disease in the first place. There are many cases in medicine where treatments are based on hard knowledge about the cause of the disease, like many infections. In such cases medicine has developed effective and relatively inexpensive preventions (like vaccinations) and treatments (like antibiotics). These are the great triumphs of medicine. But there are still many important diseases in which the root cause is still subject to uncertainty and speculation. Bioengineers are well equipped to focus on such questions, and there exists a great opportunity to contribute to some of the most important issues in life.

Our aspiration is that it may be possible in the future to carry out a detailed analysis of the failure of human tissues, perhaps with mathematical precision. This analysis should predict in a quantitative fashion the actual progression of the events that start with early organ dysfunction and eventually lead to catastrophic failure, such as a stroke or heart attack. At the moment we do not have such an analysis on hand. Yet the formulation of such an analysis will change the way we think about disease. It will give us a multitude of new suggestions how to reduce the risks and how to anticipate complications and perhaps give us new ideas how to treat.

At the moment this possibility still looks far in the future. But this book is about the future, your future. We should keep in mind that one does not understand a problem to the degree that we can control it in detail, until one has carried out an engineering analysis. In spite of the fact that the sinking of the oceanliner Titanic was observed by thousands of witnesses, only a recent stress analysis revealed the actual cause of the sinking of the ship after it rammed an iceberg. The medical literature is rich in detailed descriptions of the events that surround organ failure in patients, from modern imaging of blood flow and cell metabolism to detection of individual genes, yet we have only a limited ability to identify the origin of the problem or to predict the outcome, and then only in general terms and based on previous observations of similar cases.

Today's computers are powerful, thus we are in a position to solve complex sets of equations. All genes in man have been sequenced now and there is rapid progress in the identifications of the proteins they encode. Thus, you may argue, we should be in a position to develop a quantitative theory that will *predict* how living tissues work and fail. Bioengineering is the natural home for such an effort. Engineers like to build models, to test them and to improve them, so that they become truly useful. No model will ever be perfect.

What are the issues you need to focus your attention on and what are the elements that need to enter into such an analysis of organ failure and disease? There are many entry points. In fact you need to look at experimental observations and measurements on living tissues in order to start the analysis. The greatest challenge lies in the question how to think about and how to formulate the problem.

In this chapter we will look at cardiovascular diseases, which includes such conditions as a stroke, myocardial infarction (MI), and physiological shock, but also chronic conditions like a gradual degeneration of tissues (like joints, the brain, the eye, and others). These are truly important problems. On the surface, stroke, MI or shock look different, but they may have some common features. One of them that has been identified in recent time is the presence of a condition referred to as *inflammation*. Inflammation is a medical term, not an engineering term, and in a living tissue it is associated with a sequence of events that serves to initiate a process that in the end leads to repair of an injured tissue. Inflammation can be detected in a tissue in a number of ways. Clinicians can detect it by the presence of enhanced concentrations of certain proteins (like C-reactive proteins, fibrinogen, and others) in the plasma. Inflammation has now been detected in many different diseases, even cancer, and it seems inflammation can be triggered by different ways, some of them include foreign organisms (so called infectious inflammation) but others may not.

One of the events in inflammation is a dysfunction, if not complete failure, of the microcirculation. Thus in this chapter we will discuss the flow in microscopic vessels and examine mechanisms that may lead to their failure in man. If a blood vessel does not flow, we wish to identify the cause.

2. The Microcirculation

The microcirculation represents the region of the circulation that is made up of billions of capillaries designed to supply nutrients to the tissue cells and remove their metabolites.[1] All organs have a microcirculation. Interruption of the blood flow in the microcirculation leads to rapid cell death. Thus we will focus on the microcirculation. All the cells that make up a microcirculation have now been identified. They are the red cells (erythrocytes) (Fig. 1), platelets (thrombocytes) (Fig. 2), and the white cells (leukocytes) (Fig. 1) and the cells in the wall of the blood vessels, which includes the endothelial cells (Fig. 3), smooth muscle cells, pericytes and a few others, which for simplicity we will not mention further but may have to enter the analysis in the future.

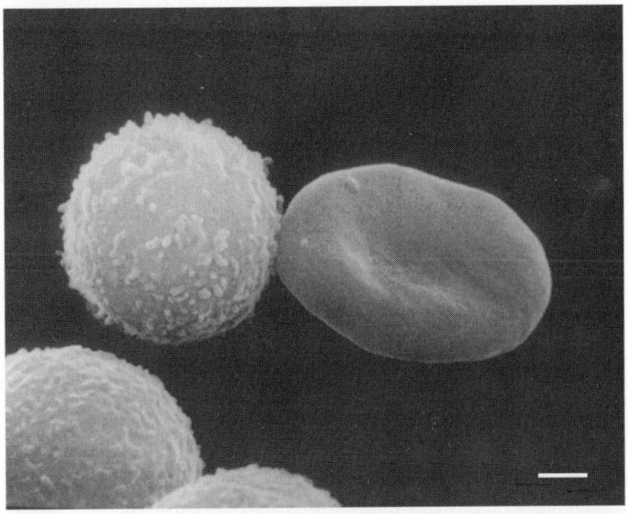

Fig. 1. Scanning electron micrograph of a typical human red cell (right) and leukocyte (left). Note that that such cell has to be fixed and dehydrated to be examined by scanning electron microscopy. Thus the cells are shrunk (about 30%–40%) but their shapes are preserved. Note that the white cell has many membrane folds, there are none on the red cell. The length of the crossbar is 1 μm.

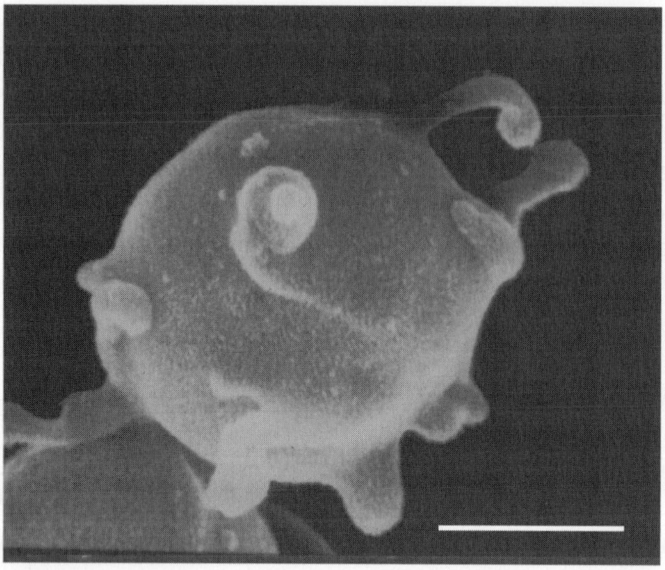

Fig. 2. Scanning electron micrograph of typical human platelets. Note they are much smaller than red or white cells.

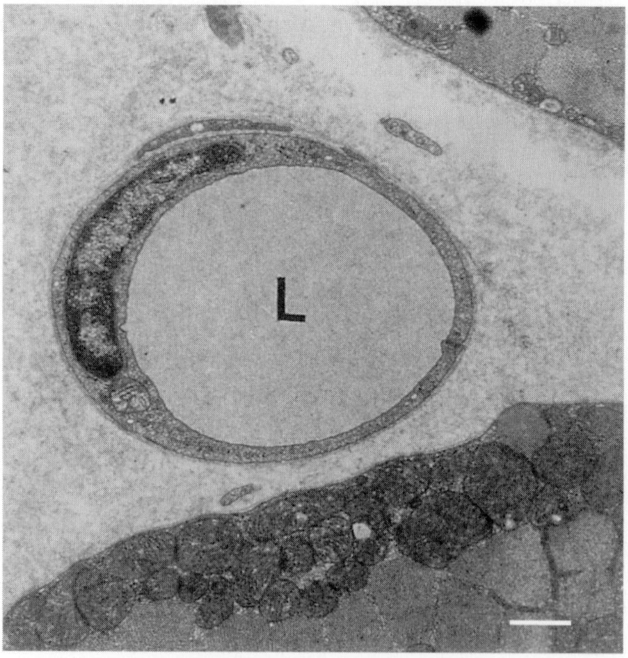

Fig. 3. Transmission electron micrograph of a capillary in rat skeletal muscle. The capillary endothelium is surrounded by three skeletal muscle cells. In transmission electron microscopy you are looking at the surface of a thin (about 0.1 μm) slice through the tissue. The length of the crossbar is 1 μm.

2.1 *The blood cells*

The red cells are carriers of hemoglobin, which serves to transport oxygen, nitric oxide and other gases, and the platelets serve to control bleeding out of blood vessels in case the wall of the blood vessels has been damaged. There are five classes of white cells, which in an adult are produced in the bone marrow: the lymphocytes (which make antibodies or also may kill tumor cells), the neutrophils and monocytes (which can clean up tissue debris by phagocytosis), the eosinophils and basophils which have more specific activities, like the release of physiological mediators (Fig. 4). The circulating cells are suspended in plasma which is an aqueous medium made up of water salts, proteins, lipids, sugars, i.e. it is a complex mixture of biological molecules, all at relatively low concentrations.

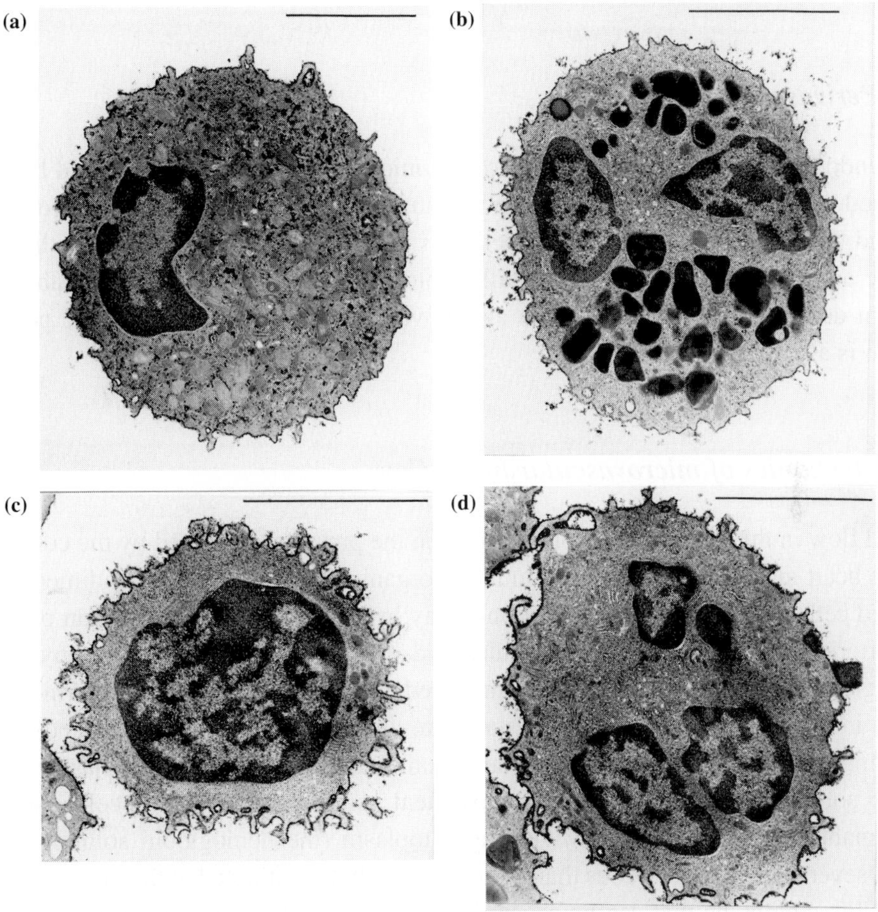

Fig. 4. Transmission electron micrograph of typical passive human white cells. Four types are shown: **(a)** neutrophil, **(b)** eosinophil, **(c)** lymphocyte, **(d)** monocyte. The length of the crossbar is 1 μm.

2.2 *The endothelial cell*

The walls of the blood vessels are lined by the endothelium. For that reason some people refer to the endothelium as the *container* for the blood. The endothelium is made up of individual cells that serve many purposes: they prevent the blood from clotting; they retain the plasma and its molecules inside the blood vessels; the endothelial cells permit some fluid to leak in small amounts out of the blood vessels into the tissue (the leakage is organ dependent); they permit circulating cells to adhere to their membrane and facilitate leukocytes to crawl across into the tissue;[2] endothelial cells metabolize hormones, synthesize and release cellular mediators into the blood stream or into adjacent smooth muscle cells, and they divide to form new blood vessels. Remarkably, the endothelial cells sense the magnitude of the fluid stresses on their membrane due to the blood flow[3] and they seem to adjust according to the blood stream (see Chapter 1 of this book for greater detail). Keep in mind, for none of these cells do we know all functions; the list is still quite incomplete.

2.3 *Perivascular cells*

The endothelial cells in the arterioles of the microcirculation are surrounded by a specialized muscle cell, the vascular smooth muscle. Vascular smooth muscle form coils around the arterioles and upon contraction is able to reduce the diameter of the lumen of the arterioles. Capillaries, our smallest blood vessels (their diameter is about 5 to 10 μm depending on the tissue) are also covered by a specialized cell, the pericytes, which is also present in venules.

2.4 *Mechanics of microvascular blood flow*

Blood flow in the microcirculation depends on the pressure generated by the contraction of the heart muscle. The capillaries in many organs are narrower than the dimensions of the red cells and the leukocytes. The only way by which circulating cells can pass from the arteries and arterioles to the venules and back into the large veins is by passing through the capillaries. In capillaries, most red and white cells have to deform in order to fit into the lumen. Thus the resistance to flow in the capillaries depends on the deformability of the blood cells and the endothelium. This problem that has been studied to some degree and you can find excellent summaries.[4,5] The red cell is a highly deformable particle, with a low viscous cytoplasm (the hemoglobin solution is only about seven times more viscous than water, although it is saturated with hemoglobin) and a flexible membrane that can be sheared, bend, twisted, but it resists an increase of its area. In contrast to the red cells, the white cells and the endothelial cells have a nucleus and their cytoplasm consists of a fiber network composed of actin filament[6] that are held together into a meshwork by "actin binding proteins" and "actin associated proteins."[7,8]

This actin is similar to the actin fibers found in muscle cells, but is used by the cell to form a meshwork that serves to build the actual cytoplasmic volume, just like the poles in circus tent. Our cells and organs are made up to a large extent by actin.

The cell cytoplasm is viscoelastic, i.e. it exhibits an elastic response and also creeping deformation. Compared to the red cells the white cells are much stiffer. Together with the fact that white cells have a larger volume than red cells this leads to a situation that the resistance imposed by a single white cell in capillaries is much larger than that of a single red cell.[9] Fortunately, we have much fewer white cells than red cells in the circulation. Platelets are smaller particles that usually have less influence on capillary blood flow. But platelets may aggregate into large clusters, in which case they may obstruct even larger blood vessels, especially if the endothelium has been damaged.[10]

3. Cardiovascular Cell Activation

3.1 *Cytoplasmic pseudopod formation*

While the majority of endothelial cells or leukocytes in a healthy microcirculation are in a relatively low state of activation, there is a minority of cells on which several signs of activation can be detected. The activation can be detected in the form of local cytoplasmic projections, so-called pseudopods (Fig. 5), which is a slow deformation of the cell cytoplasm due to actin polymerization.[11] Pseudopods are cytoplasmic regions that are mobile due to a dynamic actin polymerization and depolymerization (Fig. 6), but are also stiffer than the remaining regions of the cytoplasm away from the pseudopods.[12] Endothelial cells may also project pseudopods, which in capillaries block the blood flow to a significant degree (Fig. 7).

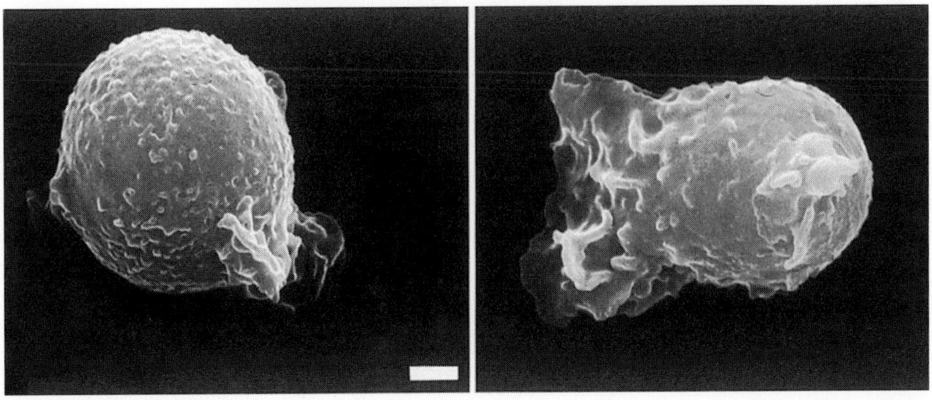

Fig. 5. Two examples of a spontaneously activated human white cell with characteristic pseudopods as seen by scanning electron microscopy. Compare this shape with passive leukocytes (Fig. 1) without such pseudopods. The length of the crossbar is 1 μm.

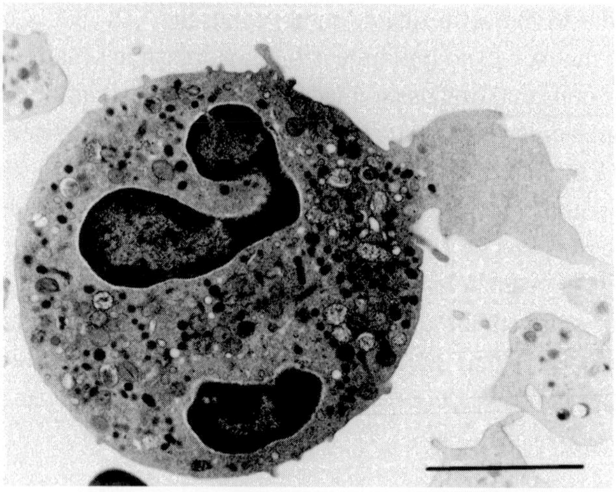

Fig. 6. Transmission electron micrograph of human neutrophil with pseudopod formation. The pseudopods contain actin and fewer cell organelles.

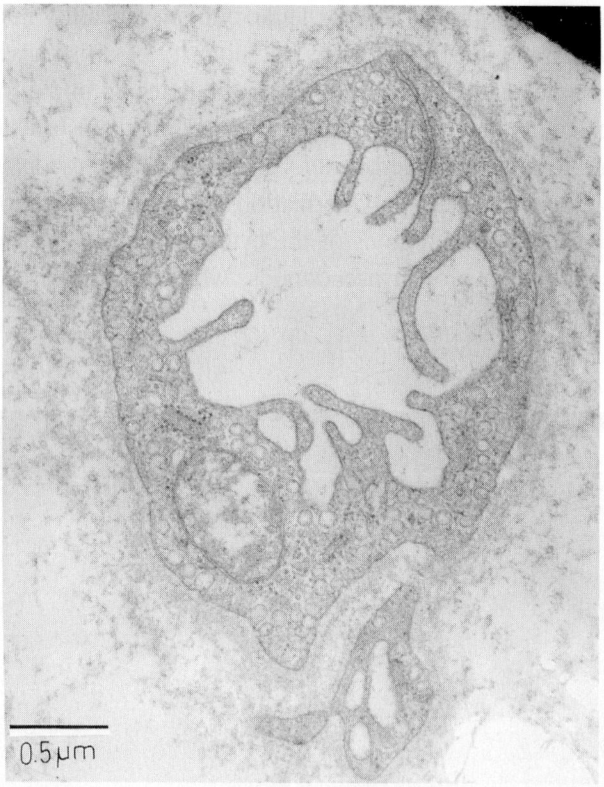

Fig. 7. Transmission electron micrograph of a capillary in rat skeletal muscle after stimulation (for details see Ref. 56).

3.2 *Oxygen free radical formation*

The cell activation may also be detected besides pseudopod formation in form of other cellular activities, such as the production of oxygen free radicals, oxygen molecules with one or more free electrons on their outer shell. Oxygen free radicals are comparatively reactive and interact with many different biological molecules. Radicals can peroxidize lipid membranes or break DNA.[13] In fact, neutrophils or monocytes utilize this process to peroxidize the membrane of bacteria as the first step of their destruction.

3.3 *Cell membrane adhesion*

Another form of cell activation is manifested in the form of the ability of the cell membrane to establish adhesive contact with other cells and surfaces. Adhesion in cells is a regulated process mediated by a specific set of glycoproteins located in the plasma membrane. In leukocytes L-selectin and several isoforms of integrins have been identified. Blockade of the selectins or integrins with monoclonal antibodies eliminates the ability of leukocytes to adhere to biological surfaces. There is also a form of selectin on endothelium, P-selectin, as well as several members of the immunoglobulin superfamily, such as the intercellular adhesion molecules (ICAM-1, ICAM-2) and the vascular adhesion molecules (VCAM-1, VCAM-2). P-selectin is stored in the cytoplasm of the endothelial cell in membrane bound vesicles, which upon stimulation of the endothelium leads to rapid discharge of the vesicles and incorporation of P-selectin into the endothelial membrane. In this way the endothelial cell becomes in a relatively short time period an adhesive surface for white cells, but less for red cells. Platelets may also adhere to endothelium and so may red cells, but only in selected diseases such as sickle cell anemia.

3.4 *Pro- and anti-inflammatory genes*

There are other forms of cell activation. White cells may release cytoplasmic granules which contain a spectrum of proteolytic enzymes and may lead to tissue destruction.[14] Endothelial cells, which under normal conditions express specific genes that serve to scavenge oxygen free radicals (superoxide dismutase, catalase, nitric oxide synthase, cyclooxygenase),[15] may start to express new genes, such as cellular growth factor (e.g. platelet derived growth factor, vascular endothelial derived growth factor), cytokines (e.g. tumor necrosis factor, interleukin-1), which in turn can activate white cells, as well as genes which promote coagulation of the blood.[16,17]

3.5 *Microvascular blood flow during cell activation*

The activation of cells in the microcirculation has many consequences, most of them cause a disturbance of the normal microcirculation, and thus may constitute the beginning of a disease process. Instead of normal passage of passive white cells through the capillary network, activation of white cells leads to an elevation of the hemodynamic resistance[18] and entrapment of white cells in the capillaries[19,20] with attachment to the endothelial surface (Fig. 8). In capillaries, the slower moving white cells disturb the otherwise well aligned motion of the red cells and raise the hemodynamic resistance even without attachment to the endothelium. Activated white cells become trapped in capillaries and may occlude them.[21–23] The activated leukocytes may also release humoral mediators which signal the arterioles to contract.

Activation of leukocytes by means of mediators that are released during oxygen free radical formation leads to production of oxygen free radical by both leukocytes and the endothelium.[24] Interestingly, activation of the microcirculation with inflammatory substances, like platelet activating factor (which in spite of its name is also an effective activator for endothelial cells or white cells), leads to migration of leukocytes across the wall of postcapillary venules into the tissue, but in a healthy tissue leads to cell injury mostly in the presence of a *secondary* stimulator.[25] Thus, a single stimulus is less effective in mediating cell activation and tissue cell injury compared with stimulation by a combination of stimulators. The exposure to multiple stimulations (e.g. from smoking, infections, dietary risk factors, arterial hypertension) may be an important issue also in human cardiovascular disease.

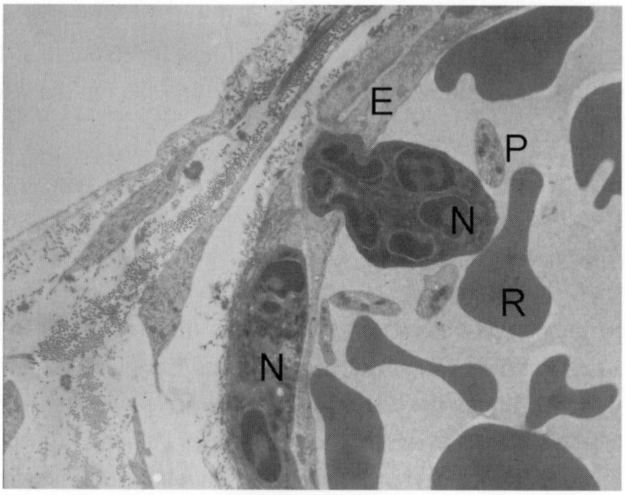

Fig. 8. Transmission electron micrograph of white cells crawling across the endothelium (E) in a small venule. Only part of the venular wall is shown with collagen (C) fiber in the tissue and red cells (R) in the lumen (L) of the venule. Note that one of the neutrophils has projected a major part of its cytoplasm *underneath* the endothelium.

3.6 *Cell activation in shock*

Perhaps one of the most direct ways to illustrate the importance of cell activation can be demonstrated in hemorrhagic shock. To induce hemorrhagic shock, the central blood pressure is reduced for a selected period of time by withdrawal of blood from a central artery or vein. Trauma victims with major blood loss may be subject to hemorrhagic shock. Such a reduction of the blood pressure may lead to significant organ dysfunction and lack of perfusion of the microcirculation in many organs. Depletion of white cells in the circulation serves to enhance the probability of survival following hemorrhagic shock. But in any group of animals, some survive and some will succumb to this type of challenge of the circulation by temporary pressure reduction. What could be the reason? While there are a number of differences between survivors and non-survivors, like heart rate, white cell counts and others, these parameters are only different as an average between the two groups, individual animals show large deviations from the average. But the level of cell activation prior to the pressure reduction exhibits a difference between the two groups: survivors have initially and during the course of the pressure reduction a low level of white cell activation, non-survivors have elevated levels[26] (Fig. 9).

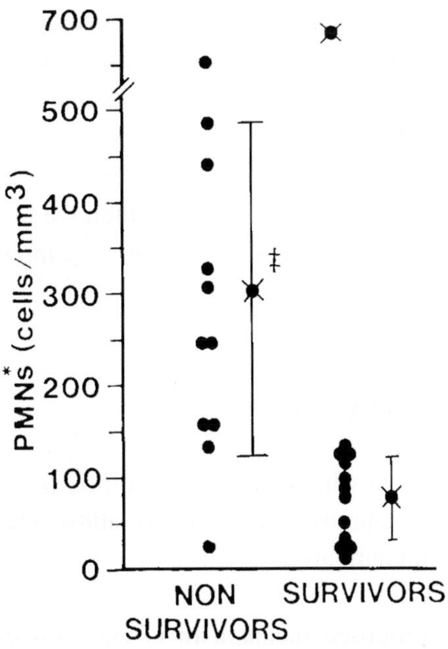

Fig. 9. The number of superoxide forming neutrophils (as determined by nitroblue-tetrazolium reduction) in the rat circulation *before* hemorrhagic shock by reduction of the blood pressure (to a mean arterial pressure of 40 mmHg for 90 minutes). The survival was determined by observation over a period of 24 hours after the hemorrhagic shock. Note the separate levels of cell activation between the two groups. There are animals which initially have low levels of cell activation (left), but in the course of hemorrhagic shock develop higher levels and die.

4. Cell Activation in Patients

There is evidence that patients with myocardial ischemia and stroke have elevated levels of cell activation in the circulation. The activation can be detected in the form of pseudopod formation on white cells,[27] oxygen free radical production,[28] adhesion,[29] reduced cell deformability or gene expression.[30] Recently a large number of clinical studies were published that have utilized measurement of the level of C-reactive protein in plasma as markers for inflammation.[31,32] Patients, who have a known risk factor for these conditions, also exhibit signs of activation in their white blood cells, such as diabetics, patients with arterial hypertension,[33] as well as smokers.[34] As we have seen above, activated cells may be trapped in capillaries. Thus the population of cells that we collect from the venous or in selected cases also arterial blood of patients may contain fewer of the activated cells. The activated cells may be trapped in the microcirculation of those organs that are subject to complications. Unfortunately there are currently no non-invasive techniques to study endothelial cell activation in patients. Thus there is a need that alternative techniques be developed to study the activation in patients. One such approach was developed by Pitzer *et al.*,[34] who collected fresh plasma and incubated it with naive donor cells from an individual without symptoms (typically a young college student who does not smoke or show other signs of activation). Incubation of the plasma of smokers with such *naive* leukocytes leads to significantly elevated levels of activation. This experiment illustrates that the plasma of smokers may contain a factor that serves to activate the white cells. Similar tests in hemorrhagic shock reveal that activated cells may not be detectable in blood samples from central blood vessels.[35] If the plasma contains an activator which can reach every organ and every capillary of the microcirculation, even innocent bystander organs, then we may be dealing with a form of activation which is less desirable and may lead to complications in microvascular perfusion and organ function.

5. Mechanisms of Cell Activation

The fundamental question with which we are confronted with in several cardiovascular diseases is: what are the mechanisms for cell activation. There are five general categories that could lead to cell activation:

(1) The most frequently proposed mechanism is via a *positive feedback* mechanism involving actual cell activators, i.e. a molecule or a set of molecules that directly stimulate leukocytes, endothelial cells or platelets. The list of molecules in this category is large and may span from bacterial (like endotoxin which are membrane fragments of bacteria) or virus derived products, to direct activators such as complement fragments (an important protein involved in immune reactions), lipid membrane derived products (e.g. platelet activating factor, lipoxygenase derived

products), small peptides (cytokines, lymphokines) and fragments derived from extracellular proteins, thrombotic products (e.g. thrombin, fibrin fragments), to peroxidized products (e.g. oxidized cholesterol).

(2) A mechanism via a *negative feedback* for cell activation. There are natural mediators that serve to downregulate cells *in vivo*, such as glucocorticoid (derived from the adrenal gland), adenosine (a metabolite), nitric oxide, and albumin (one of the main plasma proteins). Depletion of these mediators serves to upregulate the cells.

(3) There is evidence that adhesion molecules, like integrins, may serve not only as molecules for membrane attachment, but also as signaling molecules across the membrane to the cell cytoplasm. Thus it is possible that passive white cells may become activated upon attachment to activated endothelium, and vice versa activated white cells may activate passive endothelial cells upon membrane contact. This has been referred to as *junxtacrine activation.*

(4) *Fluid shear stress* serves to control the state of activation of cardiovascular cells. Perhaps this is the most surprising mechanism that controls cell activation.

(5) Activation by *physical transients*: Transients of gas (like oxygen, carbon dioxide, etc.) concentrations or also temperature transients have the ability to stimulate cell activation irrespective of the direction of the transient (up or down) but dependent on the magnitude of the transient.

6. Cell Activation by Fluid Shear Stress

In spite of the fact that fluid shear stress (the force tangential to the cell surface per unit area, typically about 10 dyn/cm^2) in the microcirculation is much lower that the fluid normal stress (the negative of the fluid pressure and about 10^3 to 10^5 dyn/cm^2), it has a powerful effect on endothelial cells,[36,37] red cells,[38] and platelets.[39] The mechanics in the microcirculation controls biology. Endothelial cells respond to shear stress by ion exchange, induction of membrane signaling pathways, by phosphorylation of G-proteins, rearrangement of the cytoplasmic actin, and expression of genes with shear stress specific promotors. Furthermore, endothelial cells are quite sensitive with respect to the details of the shear stress field; turbulent and unsteady stresses give a different response than steady fluid shear stress.[40,41]

The application of steady laminar shear stress serves to induce a specific set of genes in endothelial cells, which serve to protect against the detrimental action of the oxygen free radicals (cyclooxygenase, endothelial nitric oxide synthase, superoxide dismutase, and catalase) and may constitute the best of all situations in an adult circulation. Turbulence or the application of cytokines serves to disturb this unique pattern of gene expression[15,42] and lead to disturbance of the microcirculation. White cells also respond instantaneously to shear stress.[43–46] Physiological levels of fluid shear stress serves to downregulate white cells. Cessation of flow in the microcirculation at reduced fluid shear stress serves to upregulate the cells. Thus, just the reduction of fluid shear stress may serve to activate cells in the circulation.

7. Cell Activation in Shock

7.1 *Pancreatic digestive enzymes*

One of the strongest forms of inflammation and cell activation is encountered in shock; after all it is also one of the most lethal medical conditions. Evidence derived from the analysis of plasma suggests that in shock a plasma derived factor call be detected. The inflammatory mediators can be detected early in shock, sometimes in less than an hour. This suggests that the inflammatory mediator is already present rather than being newly synthesized. Basic analysis of which organ may mount an inflammatory reaction shows that while all organs may release some inflammatory mediators, there is one organ which stands out — the pancreas.[47] The pancreas serves to synthesize insulin, but it is also the source of digestive enzymes, the sole purpose of which is to be discharged into the lumen of the intestine and to digest food. Thus, digestive enzymes have the ability to break down just about all biological molecules as part of our nutrition, proteins (by proteases), lipids (by lipases), carbohydrates (by amylases) and nucleotides (by nucleases). If we introduce digestive enzymes (especially proteases and lipases) into other organs, they also become the source of powerful digestive enzymes.[48]

7.2 *The auto-digestion hypothesis*

Since pancreatic enzymes are a natural part of the digestion, the question of what protects us against self-digestion arises. There may be several mechanisms that protects us against our own enzymes in the lumen of the intestine. The most natural one is to compartmentalize the digestive enzymes in the lumen of the intestine, and not let them escape. The barrier that may serve this function is a specialized epithelial cell layer lining the lumen of the intestine — the mucosal epithelium. This barrier usually prevents digestive enzymes to escape from the lumen of the intestine. But under conditions that compromise the epithelial lining (e.g. reduction of oxygen supply to the intestine, injury by inflammatory mediators), the high concentrations of digestive enzymes in the lumen may escape into the wall of the intestine and initiate auto-digestion of the tissue that makes up the wall of the intestine. The current evidence suggests that in the process of such auto-digestion a new class of inflammatory mediators may be generated[49] that is carried away by the venous blood flow and the lymphatics into the central circulation.[50] As the inflammatory mediators enter into the circulation, they are carried to every organ, where they initiate cell activation and an inflammatory reaction.[51] The consequence is organ failure in short order, a process denoted as *multi-organ-failure*.

7.3 *Blockade of digestive enzymes*

One of the opportunities for intervention suggested by our current analysis is that it may be possible under limited circumstances to block the digestive enzymes in the lumen of the intestine, on one hand by intestinal lavage and rinsing of all intestinal enzymes out of the lumen of the intestine, and on the other hand by use of enzyme inhibitors to inactivate the pancreatic enzymes. Such intervention prevents to a large degree the formation of inflammatory mediators in experimental forms of shock, and consequently also serves to alleviate the symptoms of shock, such as cell activation and inflammation.[52–55]

Thus, self-digestion may be at the heart of inflammation. It is only an idea at the moment. Its clinical utility is untested. In the meantime it is an opportunity to explore a specific source for the origin of inflammation. The digestive system is present for our entire life, it is powerful, and it may be our downfall.

8. Synopsis

The study of cell activation and inflammation may serve as a key entry point to develop a systematic model of cardiovascular disease. This is the message of this chapter.

Acknowledgments

I am indebted to many colleagues and students in the Department of Bioengineering, who have inspired the ideas presented here: Drs. Jorge Barroso-Aranda, Brian Helmke, Michelle Mazzoni, Fariborz Moazzam, Scott Simon, Thomas C. Skalak, Allan Swei, Don Sutton, David Tung, Brian Helmke, Erik Kistler, Camille Wallwork, Alexander Penn, the students Jennifer J. Costa, Jo-Ellen Pitzer, Kevin L. Ohashi, Drs. Ricardo Chavez-Chavez, Shunichi Fukuda, Armin Grau, Anthony Harris, Fred Lacy, Jye Lee, Shou-Yan Lee, Phillip Pfeiffer, Kai Shen, Makoto Suematsu, Hidekazu Suzuki, Shinya Takase, Hiroshi Mitsuoka, Ayako Makino, Hainsworth Shin, Sheng Tong, and Yutaka Komai, my long term associate Frank A. Delano, and our friend and founder of Bioengineering at UCSD, the late Benjamin W. Zweifach. Special thanks to Drs. Robert L. Engler and Daniel T. O'Connor in the Department of Medicine at UCSD, Dr. John J. Bergan and David Hoyt in the Department of Surgery, and Dr. Gregory J. Del Zoppo at the Scripp Research Institute in La Jolla for their clinical perspective and enthusiastic support. Special thanks to Dr. Tony Hugli at the Torrey Pines Institute for Molecular Sciences for his long-term friendship and devotion to the analysis of the shock mediators. All of this work was only possible by the support from the National Institute of Health, grants HL 10881, HL 43024, and HL 67825.

References

1. B. W. Zweifach and H. H. Lipowsky, Pressure-flow relations in blood and lymph microcirculation. In: *Handbook of Physiology*, Section 2: The Cardiovascular System, eds. E. M. Renkin and C. C. Michel (American Physiological Society, Bethesda, MD, USA, 1984), pp. 251–307.
2. K. L. Ohashi, D. K. Tung, J. Wilson, B. W. Zweifach and G. W. Schmid-Schönbein, Transvascular and interstitial migration of neutrophils in rat mesentery, *Microcirculation* **3**: 199–210 (1996).
3. R. Skalak and N. Öskaya, Models of erythrocyte and leukocyte flow in capillaries. In: *Physiological Fluid Dynamics II*, eds. L. S. Srinath and M. Singh (McGraw-Hill, New Delhi, Tata, 1987), pp. 1–10.
4. M. J. Levesque and R. M. Nerem, The elongation and orientation of cultured endothelial cells in response to shear stress, *J. Biomech. Eng.* **107**: 341–347 (1985).
5. S. Chien, Red cell deformability and its relevance to blood flow, *Annu. Rev. Physiol.* **49**: 177–192 (1987).
6. R. Satcher, C. F. Dewey, Jr. and J. H. Hartwig, Mechanical remodeling of the endothelial surface and actin cytoskeleton induced by fluid flow, *Microcirculation* **4**: 439–453 (1997).
7. V. O. Paavilainen, E. Bertling, S. Falck and P. Lappalainen, Regulation of cytoskeletal dynamics by actin-monomer-binding proteins, *Trends. Cell. Biol.* **14**: 386–394 (2004).
8. C. Revenu, R. Athman, S. Robine and D. Louvard, The co-workers of actin filaments: from cell structures to signals, *Nat. Rev. Mol. Cell Biol.* **5**: 635–646 (2004).
9. G. W. Schmid-Schönbein, Rheology of leukocytes. In: *Handbook of Bioengineering*, eds. R. Skalak and S. Chien (Pergamon Press, New York, 1987), p. 13.
10. T. Palabrica, R. Lobb, B. C. Furie, M. Aronovitz, C. Benjamin, Y. M. Hsu, S. A. Sajer and B. Furie, Leukocyte accumulation promoting fibrin deposition is mediated *in vivo* by P-selectin on adherent platelets, *Nature* **359**: 848–851 (1992).
11. D. V. Zhelev, A. M. Alteraifi and R. M. Hochmuth, F-actin network formation in tethers and in pseudopods stimulated by chemoattractant, *Cell Motil. Cytoskeleton* **35**: 331–344 (1996).
12. G. W. Schmid-Schönbein, Leukocyte biophysics. An invited review, *Cell Biophys.* **17**: 107–135 (1990).
13. J. M. McCord, Oxygen-derived radicals: a link between reperfusion injury and inflammation, *Fed. Proc.* **46**: 2402–2406 (1987).
14. S. J. Weiss, Tissue destruction by neutrophils, *N. Engl. J. Med.* **320**: 365–376 (1989).
15. J. N. Topper, J. Cai, D. Falb and M. A. Gimbrone, Jr., Identification of vascular endothelial genes differentially responsive to fluid mechanical stimuli: cyclooxygenase-2, manganese superoxide dismutase, and endothelial cell nitric oxide synthase are selectively up-regulated by steady laminar shear stress, *Proc. Natl. Acad. Sci. USA* **93**: 10417–10422 (1996).
16. N. Resnick, H. Yahav, A. Shay-Salit, M. Shushy, S. Schubert, L. C. Zilberman and E. Wofovitz, Fluid shear stress and the vascular endothelium: for better and for worse, *Prog. Biophys. Mol. Biol.* **81**: 177–199 (2003).
17. Y. J. Shyy, H. J. Hsieh, S. Usami and S. Chien, Fluid shear stress induces a biphasic response of human monocyte chemotactic protein 1 gene expression in vascular endothelium, *Proc. Natl. Acad. Sci. USA* **91**: 4678–4682 (1994).
18. B. P. Helmke, S. N. Bremner, B. W. Zweifach, R. Skalak and G. W. Schmid-Schönbein, Mechanisms for increased blood flow resistance due to leukocytes, *Am. J. Physiol.* **273**: H2884–2890 (1997).
19. K. C. Warnke and T. C. Skalak, Leukocyte plugging *in vivo* in skeletal muscle arteriolar trees, *Am. J. Physiol.* **262**: H1149–1155 (1992).
20. L. S. Ritter, D. S. Wilson, S. K. Williams, J. G. Copeland and P. F. McDonagh, Early in reperfusion following myocardial ischemia, leukocyte activation is necessary for venular adhesion but not capillary retention, *Microcirculation* **2**: 315–327 (1995).

21. U. Bagge, B. Amundson, and C. Lauritzen, White blood cell deformability and plugging of skeletal muscle capillaries in hemorrhagic shock, *Acta Physiol. Scand.* **108**: 159–163 (1980).
22. R. L. Engler, G. W. Schmid-Schönbein and R. S. Pavelec, Leukocyte capillary plugging in myocardial ischemia and reperfusion in the dog, *Am. J. Pathol.* **111**: 98–111 (1983).
23. G. J. del Zoppo, G. W. Schmid-Schönbein, E. Mori, B. R. Copeland and C. M. Chang, Polymorphonuclear leukocytes occlude capillaries following middle cerebral artery occlusion and reperfusion in baboons, *Stroke* **22**: 1276–1283 (1991).
24. M. Suematsu, H. Suzuki, H. Ishii, S. Kato, T. Yanagisawa, H. Asako, M. Suzuki and M. Tsuchiya, Early midzonal oxidative stress preceding cell death in hypoperfused rat liver, *Gastroenterology* **103**: 994–1001 (1992).
25. D. K. Tung, L. M. Bjursten, B. W. Zweifach and G. W. Schmid-Schönbein, Leukocyte contribution to parenchymal cell death in an experimental model of inflammation, *J. Leukoc. Biol.* **62**: 163–175 (1997).
26. J. Barroso-Aranda and G. W. Schmid-Schönbein, Transformation of neutrophils as indicator of irreversibility in hemorrhagic shock, *Am. J. Physiol.* **257**: H846–852 (1989).
27. R. R. Chang, N. T. Chien, C. H. Chen, K. M. Jan, G. W. Schmid-Schönbein and S. Chien, Spontaneous activation of circulating granulocytes in patients with acute myocardial and cerebral diseases, *Biorheology* **29**: 549–561 (1992).
28. I. Ott, F. J. Neumann, M. Gawaz, M. Schmitt and A. Schomig, Increased neutrophil-platelet adhesion in patients with unstable angina, *Circulation* **94**: 1239–1246 (1996).
29. J. Grau, E. Berger, K. L. Sung and G. W. Schmid-Schönbein, Granulocyte adhesion, deformability, and superoxide formation in acute stroke, *Stroke* **23**: 33–39 (1992).
30. N. Marx, F. J. Neumann, I. Ott, M. Gawaz, W. Koch, T. Pinkau and A. Schomig, Induction of cytokine expression in leukocytes in acute myocardial infarction, *J. Am. Coll. Cardiol.* **30**: 165–170 (1997).
31. P. M. Ridker, Inflammation in atherothrombosis: how to use high-sensitivity C-reactive protein (hsCRP) in clinical practice, *Am. Heart Hosp. J.* **2**: 4–9 (2004).
32. P. Libby and P. M. Ridker, Inflammation and atherosclerosis: role of C-reactive protein in risk assessment, *Am. J. Med.* **116**(Suppl. 6A): 9S–16S (2004).
33. F. Lacy, M. T. Kailasam, D. T. O'Connor, G. W. Schmid-Schönbein and R. J. Parmer, Plasma hydrogen peroxide production in human essential hypertension: role of heredity, gender, and ethnicity, *Hypertension* **36**: 878–884 (2000).
34. J. E. Pitzer, G. J. Del Zoppo and G. W. Schmid-Schönbein, Neutrophil activation in smokers, *Biorheology* **33**: 45–58 (1996).
35. K. Shen, R. Chavez-Chavez, A. K. L. Loo, B. W. Zweifach and G. W. JS-S Barroso-Aranda, Interpretation of leukocyte activation measurements from systemic blood vessels. In: *Cerebrovascular Diseases*, 19th Princeton Stroke Conference, eds. M. A. Moskowitz and L. R. Caplan (Butterworth-Heinemann, Boston, Massachusetts, 1995), Chap. 6, pp. 59–73.
36. C. F. Dewey, Jr., S. R. Bussolari, M. A. Gimbrone, Jr. and P. F. Davies, The dynamic response of vascular endothelial cells to fluid shear stress, *J. Biomech. Eng.* **103**: 177–185 (1981).
37. P. F. Davies, Flow-mediated endothelial mechanotransduction, *Physiol. Rev.* **75**: 519–560 (1995).
38. R. M. Johnson, Membrane stress increases cation permeability in red cells, *Biophys. J.* **67**: 1876–1881 (1994).
39. K. Konstantopoulos, K. K. Wu, M. M. Udden, E. I. Banez, S. J. Shattil and J. D. Hellums, Flow cytometric studies of platelet responses to shear stress in whole blood, *Biorheology* **32**: 73–93 (1995).
40. N. DePaola, M. A. Gimbrone, Jr., P. F. Davies and C. F. Dewey, Jr., Vascular endothelium responds to fluid shear stress gradients, *Arterioscler. Thromb.* **12**: 1254–1257 (1992).
41. M. Andersson, L. Karlsson, P. A. Svensson, E. Ulfhammer, M. Ekman, M. Jernas, L. M. Carlsson and S. Jern, Differential global gene expression response patterns of human endothelium exposed to shear stress and intraluminal pressure, *J. Vasc. Res.* **42**: 441–452 (2005).

42. J. A. Frangos, T. Y. Huang and C. B. Clark, Steady shear and step changes in shear stimulate endothelium via independent mechanisms — superposition of transient and sustained nitric oxide production, *Biochem. Biophys. Res. Commun.* **224**: 660–665 (1996).
43. F. Moazzam, F. A. DeLano, B. W. Zweifach and G. W. Schmid-Schönbein, The leukocyte response to fluid stress [see comments]. *Proc. Nat. Acad. Sci. USA* **94**: 5338–5343 (1997).
44. M. F. Coughlin and G. W. Schmid-Schönbein, Pseudopod projection and cell spreading of passive leukocytes in response to fluid shear stress, *Biophys. J.* **87**: 2035–2042 (2004).
45. S. Fukuda, T. Yasu, N. Kobayashi, N. Ikeda and G. W. Schmid-Schönbein, Contribution of fluid shear response in leukocytes to hemodynamic resistance in the spontaneously hypertensive rat, *Circ. Res.* **95**: 100–108 (2004).
46. S. Fukuda, T. Yasu, D. N. Predescu and G. W. Schmid-Schönbein, Mechanisms for regulation of fluid shear stress response in circulating leukocytes, *Circ. Res.* **86**: E13–18 (2000).
47. E. B. Kistler, T. E. Hugli and G. W. Schmid-Schönbein, The pancreas as a source of cardiovascular cell activating factors, *Microcirculation* **7**: 183–192 (2000).
48. S. W. Waldo, H. S. Rosario, A. H. Penn and G. W. Schmid-Schönbein, Pancreatic digestive enzymes are potent generators of mediators for leukocyte activation and mortality, *Shock* **20**: 138–143 (2003).
49. W. J. Kramp, S. Waldo, G. W. Schmid-Schönbein, D. Hoyt, R. Coimbra and T. E. Hugli, Characterization of two classes of pancreatic shock factors: functional differences exhibited by hydrophilic and hydrophobic shock factors, *Shock* **20**: 356–362 (2003).
50. G. W. Schmid-Schönbein and T. E. Hugli, A new hypothesis for microvascular inflammation in shock and multiorgan failure: self-digestion by pancreatic enzymes, *Microcirculation* **12**: 71–82 (2005).
51. F. Fitzal, F. A. DeLano, C. Young, H. S. Rosario and G. W. Schmid-Schönbein, Pancreatic protease inhibition during shock attenuates cell activation and peripheral inflammation, *J. Vasc. Res.* **39**: 320–329 (2002).
52. H. Mitsuoka, E. B. Kistler and G. W. Schmid-Schönbein, Generation of *in vivo* activating factors in the ischemic intestine by pancreatic enzymes, *Proc. Natl. Acad. Sci. USA* **97**: 1772–1777 (2000).
53. H. Mitsuoka, E. B. Kistler and G. W. Schmid-Schönbein, Protease inhibition in the intestinal lumen: attenuation of systemic inflammation and early indicators of multiple organ failure in shock, *Shock* **17**: 205–209 (2002).
54. E. A. Deitch, H. P. Shi, Q. Lu, E. Feketeova and D. Z. Xu, Serine proteases are involved in the pathogenesis of trauma-hemorrhagic shock-induced gut and lung injury, *Shock* **19**: 452–456 (2003).
55. J. J. Doucet, D. B. Hoyt, R. Coimbra, G. W. Schmid-Schönbein, W. G. Junger, L. W. Paul, W. H. Loomis and T. E. Hugli, Inhibition of enteral enzymes by enteroclysis with nafamostat mesilate reduces neutrophil activation and transfusion requirements after hemorrhagic shock, *J. Trauma* **56**: 501–510; discussion 510–511 (2004).
56. J. Lee and G. W. Schmid-Schönbein, Biomechanics of skeletal muscle capillaries: hemodynamic resistance, endothelial distensibility, and pseudopod formation, *Ann. Biomed. Eng.* **23**: 226–246 (1995).

CHAPTER 10

BLOOD SUBSTITUTES AND THE DESIGN OF OXYGEN NON-CARRYING AND CARRYING FLUIDS

Marcos Intaglietta

Abstract

Analysis of oxygen transport at the level of microscopic blood vessels has resulted in the development of blood substitutes with novel properties that are different from blood. These include a high oxygen affinity, which targets oxygen delivery to tissue regions with low pO_2, and a high viscosity, which insures the maintenance of functional capillary density in conditions of extreme anemia. These materials are based on the conjugation of polyethylene glycol with molecular hemoglobin, and are effective in treating hemorrhage and extreme hemodilution at low concentration, providing a realistic re-deployment of existing resources of human blood.

1. Introduction

Traumatic injury usually involves the loss of circulating blood, an event for which the organism has a narrow tolerance since the rapid loss of as little as 10% blood volume leads to severe cardiovascular perturbations. A 30% deficit in blood volume results in irreversible shock if not rapidly corrected. As a consequence, maintenance of normovolemia is the objective in all forms of resuscitation. Losses in blood volume imply the decrease of oxygen carrying capacity, or the amount of molecular oxygen carried by a volume of blood. However, these have a significantly greater margin of safety, whereby losses of up to 50% of the red blood cells (RBCs) mass do not affect survival of the organism at rest. Consequently, correction of blood losses comprises the intertwined and sequential processes of restitution of blood volume and circulatory oxygen carrying capacity.

Initially blood losses are effectively corrected by the introduction of volume restitution fluids, or plasma expanders, in order to maintain blood pressure and perfusion of the tissues, thus ensuring that the remaining oxygen carrying capacity effectively delivers oxygen to the tissue and the continuation of the extraction of the products of metabolism from the tissues. As the loss of blood extends, clinical indications and medical experience indicate the need to restore oxygen carrying capacity by the transfusion of blood. These conditions are referred to as the "transfusion trigger," and are summarized by the blood hemoglobin content, or hematocrit.

Most vertebrates respond to blood losses by autotransfusion, which attempts to remedy the loss of circulating volume by transferring fluid from the tissue to the circulation by means of osmotic effects. Blood and the circulation are the result from the evolution from single cell organisms into multicellular life forms, however blood transfusion was not a part of the evolutionary process, and the organism is not necessarily adapted to this form of blood volume/oxygen carrying capacity restoration.

Since we are not specifically adapted to the transfusion of blood we can assume that fluids may be devised that are better than blood for resuscitating an organism and since restoration of circulating volume is a priority in correcting blood losses, the development of an effective blood substitute must also consider the optimal procedures and fluids for accomplishing this purpose, independently of the restoration of oxygen carrying capacity.

2. The Microcirculation

Blood exerts its functions in the microcirculation system of microscopic vessels that brings blood in contact with virtually every cell of the organism. This system is comprised of the distributing vessels, or arterioles and the collecting vessels, or venules, which are connected by the capillary network (Fig. 1). The role of capillaries as suppliers of oxygen followed observations in the lung, where it was found that these conduits delivered atmospheric oxygen to blood. The presence of a vast network of tissue capillaries in the tissue suggests that their function would be a reversal of that in the lung, unloading blood oxygen to tissue. However, the lung is the only organ whose tissue is mostly capillaries with very large oxygen gradients between blood and atmospheric oxygen, while tissue capillaries present very small oxygen gradients. Furthermore capillaries are not the sole source of oxygen in the microcirculation of other tissues as shown by Duling and Berne,[1] who demonstrated that significant amounts of oxygen are delivered from the arterioles.

The parameter that characterizes the role of capillaries as a system for the exchange of materials between blood and tissue is functional capillary density (FCD) defined as the number of capillaries that possess red blood cell transit. Tissue survival is critically dependent on functional capillary density, as shown by the study of Kerger *et al.*,[2] who found that this is the principal functional anatomical event that separates surviving from non-surviving animals subjected to hemorrhagic shock, and not tissue oxygenation which was not different between groups. Consequently the clinical evidence calling for the transfusion of blood could be related to the onset of capillary collapse, leading to the accumulation of products of metabolism in the tissue.

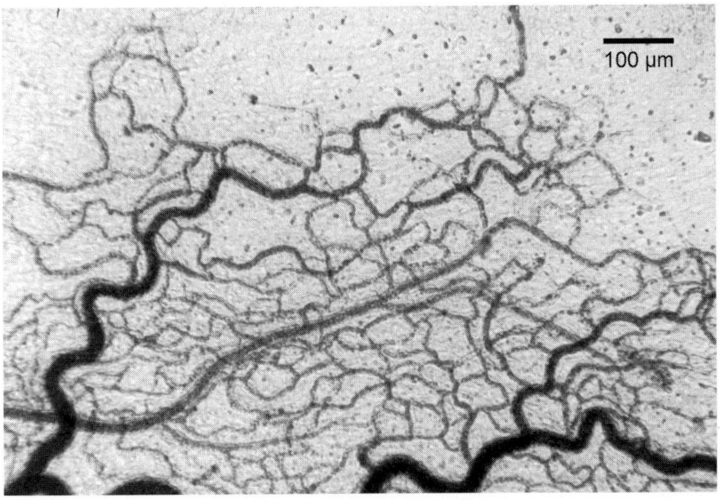

Fig. 1. Photo of the microcirculation of the rabbit omentum. The microcirculation is the smallest system of blood vessels of the organism, and the location of oxygen exchange. It is a very large system comprising about 1000 km of tubes per kg of tissue. The system of tubes connects arteries and veins and insures that each cell of the organism is within the diffusion field for oxygen delivered by RBCs.

3. Plasma Expanders

Blood transfusions were conceived as a means for replacing lost volume in severe blood losses since the discovery of the circulation (Fig. 2). The earliest successful volume replacement fluid was physiologic saline, a so-called plasma expanders, which was introduced around 1875. Blood transfusion remained the major goal in restoring tissue oxygenation, and the use of human blood became possible when substances responsible for incompatibility reactions and hemolysis of blood were discovered in 1900. Blood typing and the use of sodium citrate and glucose to prevent the coagulation of blood was the final step in the introduction and relatively safe use of blood transfusions. Blood banking was developed during World War I and the American Red Cross begun establishing blood banks in 1947.

The addition of potassium and calcium ions significantly improves the effectiveness of saline in maintaining the viability of tissue, and lactate is added to this composite solution, which is gradually converted into sodium bicarbonate, which prevents alkalosis. This constitutes "lactate-Ringers"which is the most widely used fluid for blood volume restitution. This fluid is given in volumes that are as much as three times the treated blood loss, since the dilution of plasma proteins lowers the plasma osmotic pressure causing an imbalance of the fluid exchange favoring microvascular fluid loss and edema. The advantage of crystalloid solutions is that large volumes can be given over short periods, and excess volumes are rapidly cleared from the circulation by diuresis, which is beneficial in the treatment of trauma.

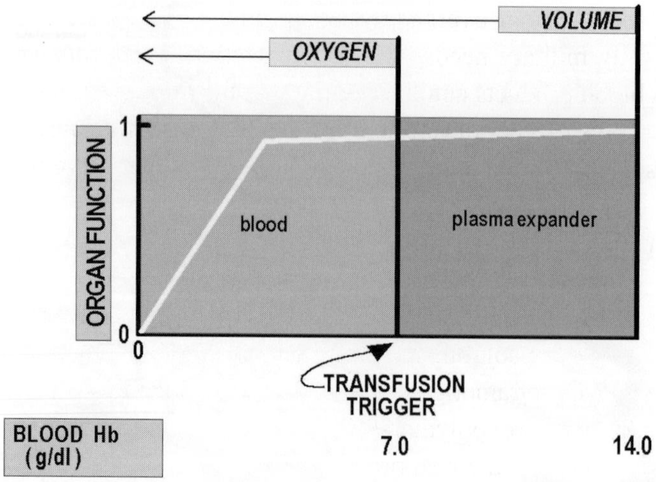

Fig. 2. The transfusion trigger is the point at which restitution of oxygen carrying capacity by blood transfusions or an artificial blood substitute is required. This number is set by medical practice and it represents the point at which the patient is believed to be at risk of ischemic episodes. The transfusion trigger is often set at a higher value in order to insure safety, while experimental studies show that the organism will tolerate and adapt to very low hemoglobin values. Blood, or an artificial blood should be transfused prior to reaching the transfusion trigger. Crystalloids such as Ringer's lactate, or colloids, mainly albumin, dextran or starch are used prior to reaching the transfusion trigger.

Blood plasma (the fluid portion of blood) and serum (blood plasma with the coagulation factors and platelets removed) are superior to crystalloid solutions in maintaining tissue viability due to the osmotic pressure exerted by albumin (molecular weight 69 kDA). High molecular weight polysaccharides consisting of linked glucose molecules with molecular weights 40–70 kDA (Dextrans) have similar osmotic pressures as albumin and were used for volume replacement and shock treatment until recently in Europe. They are being replaced worldwide by hydroxyethyl starch, a chemically modified starch (molecular weight 40–450 kDA). Gelatins (molecular weight 30–35 kDA) were in use in the United States until 1978 when they were declared unsafe by the US Food and Drug Administration, however these compounds are still in use in many countries.

4. Oxygen Carrying Fluids

Oxygen carrying plasma expanders (OCPE) were first developed using solutions of cell free human hemoglobin in saline, an approach that did not succeed because the hemoglobin molecule in plasma rapidly broke up into dimers and monomers, overwhelming the ability of the clearing capacity of the liver, leading eventually to kidney failure. Hemoglobin was also found to be vasoactive and to cause hypertension. Chemical modification of hemoglobin overcame these problems as the demand for an OCPE increased driven by military needs, and the realization of the possibility of infection from blood and blood products and the finding that the human immunodeficiency virus (HIV) and other virus can be transmitted through blood transfusions.

Three classes of oxygen carrier compounds are under development: modified molecular Hb solutions (HBOCs), fluorocarbon-based oxygen carriers (FBOCs), and vesicle encapsulated, i.e. non-molecular, cell-like suspension of oxygen carrying compounds. The oxygen carrying material to date is either hemoglobin that carries oxygen through a reversible pO_2 dependent chemical reaction, or fluorocarbons, which carry oxygen by virtue of their solubility. The molecular modifications of hemoglobin are aimed at stabilizing its tetrameric structure using glutaraldehyde to crosslink two or more Hb molecules and form polymers. Several of these products advanced up to stage III clinical trials. However, their embodied theoretical shortcomings ultimately determined that in an organism at risk due to trauma or disease, their use could aggravate the problem to be remedied. The outcome was that most candidates for market entry met with an unfortunate but predictable failure in advanced stages of clinical trials.

The principal problem was vasoactivity, a phenomenon that was linked to the scavenging of nitric oxide (NO) by hemoglobin, a process that appeared to be in part related to the extravasation of the hemoglobin molecule. Since acellular modified molecular hemoglobin became the oxygen carrier of choice, efforts were directed at increasing molecular size, in order to impede extravasation, and lowering the affinity for NO and thus impede vasoconstriction by reducing NO scavenging. Molecular size and therefore extravasation was found to be inversely related to vasoactivity, the smaller the hemoglobin molecule the greater the hypertensive effect. Notably both vasoactive and vasoinactive hemoglobin molecules have the same NO binding constant.[3]

In general, the failed attempts to produce substitutes were due to the assumption that a blood substitute should restore the oxygen carrying capacity of the shed blood, that it is beneficial the resulting mixture of remaining blood and resuscitation fluid has a viscosity lower than natural blood, and that oxygen affinity should be like blood or lower, to facilitate oxygen release from the circulation into the tissue. Developments in the understanding of how microvascular function is affected by blood substitutes, many made at University of California San Diego (UCSD), showed that practical blood substitutes required properties that were significantly different from the initial conceptualization (Fig. 3).

5. Blood Viscosity

An important characteristic of blood is its viscosity, which is determined by the concentration of RBCs or hematocrit (% of RBCs by volume versus the volume of whole blood). Normal blood has a viscosity of about 4 centipoise (cp) while plasma is approximately 1 cp. Blood viscosity is approximately proportional to the hematocrit squared; therefore small changes in RBC concentration have important effects on blood viscosity

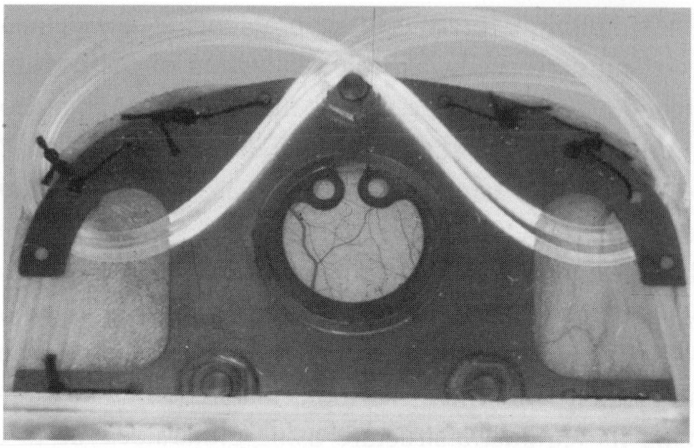

Fig. 3. Experimental preparation for the study of the microcirculation in the hamster. This animal model consisting of window chamber that contains a skin fold permits to carry out studies for several days, and in the absence of anesthesia. The effect of blood substitutes can be evaluated by measuring blood flow, the reaction of arterioles and venules, oxygen delivery from the microvessels, and tissue oxygenation at the microscopic level.

when hematocrit is normal. Colloidal solutions at about 10% concentration have viscosities similar to that of plasma; therefore, hemodilution with most plasma expanders and hemoglobin solutions significantly reduces blood viscosity.

The lowered blood viscosity due to hemodilution allows the heart to pump blood at higher volumetric rate without increased expenditure of energy, since the power (or metabolic) requirement of the heart, when blood pressure is maintained, is equal to the product of volumetric output (cardiac output) and blood viscosity. Physiologically, lowered blood viscosity increases central venous pressure and the return of venous blood to the heart, which improves cardiac performance and increases cardiac output. This causes increased blood flow velocity and shear rate at the vessel walls.

The decrease in hematocrit lowers the intrinsic oxygen carrying capacity of blood, but this effect is compensated by the increased blood flow velocity due to the lowered blood viscosity, which increases the rate at which oxygen carrying RBCs are delivered to the microcirculation. As a consequence, both systemic and capillary oxygen carrying capacities remain approximately normal down to an arterial hematocrit of about 25% (normal is 45%). This explains why the organism can easily sustain blood losses that lower the number of RBCs to half of its original value.

6. Hemodilution with Hemoglobin Base Oxygen Carrying Blood Substitutes

During hemodilution with non-oxygen carrying colloids, the organism compensates for the lost oxygen carrying capacity and there is no advantage in restituting blood oxygen carrying capacity until about half of the red blood cells are lost. This point identifies the "transfusion trigger"currently set at a hemoglobin concentration of about 7% (Fig. 2). The transfusion trigger determines the point at which the organism may be in danger of becoming ischemic, a condition that must be corrected by the transfusion of blood or a blood substitute. Therefore in most cases oxygen carrying capacity restitution is made when blood is hemodiluted, and the effects of the newly added oxygen carrying capacity are superposed to be those due to the hemodiluted blood.

The formulation fluid that restores oxygen delivery capacity requires understanding of the underlying physical principles involved. The principal barrier for the exit of oxygen from the blood vessels is the resistance to diffusion through the tissue, which is uniform and similar to that of oxygen through water. This is valid for most soft tissues including the blood vessel wall. As a consequence, oxygen leaks out continuously from the blood column and upon arrival to the microcirculation virtually half of the oxygen gathered in the lung has been lost. When this process is analyzed mathematically it is evident that the most important parameter in insuring the proper function of an oxygen carrying molecule in the circulation is the viscosity resulting from the mixture of blood and the blood substitute.[4]

Paradoxically lowering blood viscosity while maintaining oxygen carrying capacity, as it occurs with most of the presently developed artificial bloods is counter-productive, because as a consequence of the increased flow velocity due to the lower number of RBCs flow velocity is increased and too much oxygen arrives to the microcirculation. This overabundance of oxygen delivery is sensed by the mechanism of "autoregulation"which is designed to maintain oxygen delivery constant, which causes constriction of the blood vessels in order to limit the oxygen supply, and hypertension. An additional adverse effect of these hemoglobin solutions is due the shear stress at the vessel wall (i.e. the frictional force caused by the passage of blood over the blood vessel wall), which causes the release of vasodilators (EDRF). Shear stress is proportional to the product of blood flow velocity and viscosity, and lowering both factors significantly reduces shear stress and the release of EDRF, aggravating the vasoconstrictor effect. In fact most of "artificial blood"developed so far cause hypertension.

There is a critical limit to which blood viscosity can be lowered, and when this is reached blood viscosity can be restored also by increasing plasma viscosity, obtaining beneficial effect in maintaining functional capillary density and normal levels of NO at the microvascular wall without introducing red blood cells as shown by experimental studies in the hamster window preparation (Fig. 3).[5,6]

7. Optimal Oxygen Affinity

Oxygen affinity measures the strength of the bond between the oxygen carrier and oxygen. The higher the affinity the stronger the bond, and the greater the gradient in oxygen concentration is needed for the carrier to release its oxygen. In cardiovascular

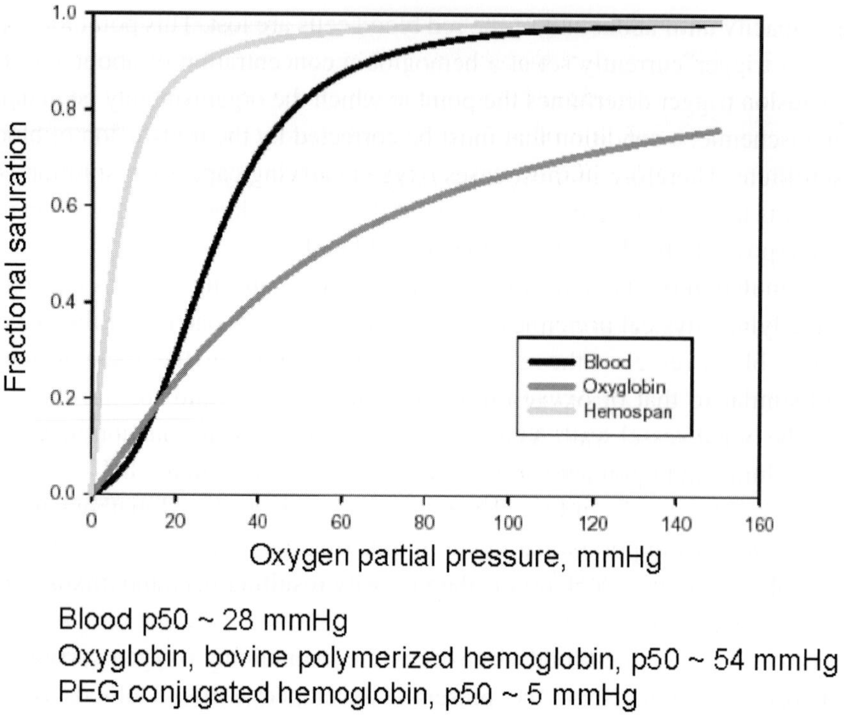

Fig. 4. Oxygen dissociation curves for blood, the high p50 oxygen carrier oxyglobin, manufactured by Biopure Inc., Boston, MA, and polyethylene glycol modified hemoglobin, MP4, manufactured by Sangart Inc., San Diego, CA.

physiology, oxygen affinity of hemoglobin is measured by the value of p50 or the partial pressure of oxygen at which hemoglobin is 50% saturated. Molecular modifications of hemoglobin with a high p50 readily release their oxygen, and *vice versa*. Human blood and many mammalian species have a p50 of about 28 mmHg, and since capillary pO_2 is about this value, blood releases half of its oxygen prior to the capillary circulation. Molecular hemoglobin based oxygen carriers tended to have a p50 similar or higher than blood, under the assumption that the oxygen dissociation curve should be right shifted, thus facilitating the release of oxygen into the tissue. High p50 oxygen carriers may have the opposite effect in terms of delivering oxygen to the tissue, since a right shifted dissociation curve (Fig. 4) favors oxygen unloading from small arteries and arterioles, producing a signal of increased oxygen delivery that leads to vasoconstriction and decreased oxygen supply to the capillaries.

When average tissue pO_2 is low, there is a distribution value with regions of high and low pO_2. Introducing a small quantity of a low p50 hemoglobin oxygen carrier into the circulation delivers oxygen only to portions of the tissue with low pO_2 and particularly those regions that are anoxic. Conversely, introducing even significant amounts of right shifted hemoglobin would have no effect since the bound oxygen would be unloaded in oxygenated regions. Therefore, low p50 prevents tissues from developing local anoxia due to the inherent variability of tissue oxygenation, and "targets" oxygen delivery to those regions where tissue oxygen is low.

Cross-linked or polymerized hemoglobins developed so far as the oxygen carrier for OCPEs have a high value of p50. Conversely, surface decoration of hemoglobin with polyethylene glycol (PEG) lowers the p50, and hemoglobin vesicles can incorporate materials that shift the p50.[7]

8. Colloid Osmotic Pressure

There is no well-defined consensus on the most desirable value of the colloid osmotic pressure for plasma expanders and OCPEs, although it has usually been assumed that the colloid osmotic pressure should be similar to that of blood and in the range of 20–25 mmHg. Saline and Ringer's lactate have zero colloid osmotic pressure. There may be some advantages in using fluids with high colloidal and osmotic pressures, since they cause tissue fluid to come into the vascular compartment. In conditions of hemorrhage there is endothelial edema and the decrease of functional capillary density, which has been demonstrated to be rapidly reversed upon the introduction of hyperosmotic and hyperoncotic fluids.[8] This effect is especially relevant in view of the importance of restoring functional capillary density as a part of the process of recovery of the injured organism.

High oncotic pressures move fluid from the tissue into the circulation, increasing circulating volume, which in the absence of vasoconstriction may cause a mild increase in systemic pressure and increased cardiac output. However, these two effects do not change peripheral vascular resistance and therefore high oncotic pressure does not cause vasoconstriction.

9. Definition of an Effective Oxygen Carrying Blood Substitute

It should be apparent that OCPEs that have transport properties similar to blood do not result in optimal blood replacement fluids. The two salient features that set an optimal OCPE apart from blood are the low p50 and high viscosity. The most effective oxygen carrying resuscitation fluids are found to have a p50 is in the neighborhood of 5 mmHg, and viscosity in the range of 5–8 cP, so that when the fluid is introduced into the circulation plasma viscosity rises to about 3 cP. These properties can be obtained by conjugating hemoglobin with PEG and various formulations have been tested in both animal experiments and human trials with excellent results.[9] Furthermore, the NO scavenging characteristics do not appear to be relevant or significant since these fluids have the same NO binding constant as other formulations that are vasoactive.[3]

In shock and extreme hemodilution, these fluids appear to be more effective than blood because they specifically maintain functional capillary density, which is a critical component of the resuscitation process, and an outcome that cannot be achieved with a blood transfusion because blood does not increase plasma viscosity, which is necessary for re-pressurizing the capillaries.[10] PEG-hemoglobin also appears to be free of most of the toxicities seen with other proposed oxygen carriers.

An OCPE designed according to the premises presented require comparatively small amounts of hemoglobin to be effective, which is important since to the present human hemoglobin appears to be the oxygen carrier of choice.[11] Experimental results show that concentrations of hemoglobin of 4% are as effective as blood; therefore, about three equivalent units of product can be obtained from a unit of blood. A key factor is that these materials restore microvascular function and particularly functional capillary density, therefore fluids based on PEG may also find therapeutic applications in disease conditions that require treatment aimed at improving the microcirculatory function.

Acknowledgments

This work was supported by the Bioengineering Research Partnership grant R24-HL64395 and grants R01-HL62354 and R01-HL62318 to MI.

References

1. B. R. Duling and R. M. Berne, Longitudinal gradients in periarteriolar oxygen tension. A possible mechanism for the participation of oxygen in the local regulation of blood flow, *Circ. Res.* **27**: 669–678 (1970).
2. H. Kerger, D. J. Saltzman, M. D. Menger, K. Messmer and M. Intaglietta, Systemic and subcutaneous microvascular pO_2 dissociation during 4-h hemorrhagic shock in conscious hamsters, *Am. J. Physiol.* **270**: H827–836 (1996).
3. R. J. Rohlfs, E. Brunner, A. Chiu, A. Gonzales, M. L. Gonzales, D. Magde, M. D. Magde Jr., K. D. Vandegriff and R. M. Winslow, Arterial blood pressure responses to cell-free hemoglobin solutions and the reaction with nitric oxide, *J. Biol. Chem.* **273**: 12128–12134 (1998).
4. M. Intaglietta, Whitaker lecture 1996: microcirculation, biomedical engineering and artificial blood, *Ann. Biomed. Eng.* **25**: 593–603 (1997).
5. A. G. Tsai, C. Acero, P. R. Nance, P. Cabrales, J. A. Frangos, D. G. Buerk and M. Intaglietta, Elevated plasma viscosity in extreme hemodilution increases perivascular nitric oxide concentration and microvascular perfusion, *Am. J. Physiol. Heart Circ. Physiol.* **288**: H1730–1739 (2005).
6. A. G. Tsai, B. Friesenecker, M. McCarthy, H. Sakai and M. Intaglietta, Plasma viscosity regulates capillary perfusion during extreme hemodilution in hamster skin fold model, *Am. J. Physiol.* **275**: H2170–2180 (1998).
7. P. Cabrales, H. Sakai, A. G. Tsai, S. Takeoka, E. Tsuchida and M. Intaglietta, Oxygen transport by low and normal oxygen affinity hemoglobin vesicles in extreme hemodilution, *Am. J. Physiol. Heart Circ. Physiol.* **288**: H1885–1892 (2005).
8. M. C. Mazzoni, P. Borgström, M. Intaglietta and K. E. Arfors, Capillary narrowing in hemorrhagic shock is rectified by hyperosmotic saline-dextran reinfusion, *Circ. Shock* **31**: 407–418 (1990).

9. R. M. Winslow, A. Gonzales, M. L. Gonzales, M. Magde, M. McCarthy, R. J. Rohlfs and K. Vandegriff, Vascular resistance and the efficacy of red cell substitutes in a rat hemorrhage model, *J. Appl. Physiol.* **85**: 993–1003 (1998).
10. R. Wettstein, A. G. Tsai, D. Erni, R. M. Winslow and M. Intaglietta, Resuscitation with MalPEG-Hemoglobin improves microcirculatory blood flow and tissue oxygenation after hemorrhagic shock in awake hamsters, *Crit. Care Med.* **31**: 1824–1830 (2003).
11. R. M. Winslow, MP4, a new nonvasoactive polyethylene glycol-hemoglobin conjugate, *Artif. Organs* **28**: 800–806 (2004).

Additional Reading

Blood Substitutes, ed. R. M. Winslow (Elsevier, London, 2005).

SECTION IV

RESPIRATORY-RENAL BIOENGINEERING

CHAPTER 11

ANALYSIS OF HUMAN PULMONARY CIRCULATION: A BIOENGINEERING APPROACH

Wei Huang, Michael R. T. Yen and Qinlian Zhou

Abstract

This chapter presents a bioengineering approach to study human pulmonary circulation based on the principles of continuum mechanics in conjunction with detailed measurements of pulmonary vascular geometry, vascular elasticity, and blood rheology. Experimental data are used to construct a mathematical model of pulsatile flow in the human lung. Input impedance of every order of pulmonary blood vessels is calculated under physiological condition, and pressure-flow relation of the whole lung is predicted theoretically. The influence of variations in vessel geometry and elasticity on impedance spectra is analyzed. The goal is to understand the detailed pulmonary blood pressure-flow relationship in the human lung for clinical application.

1. Introduction

Bioengineering is engineering of living organisms. Like all other living organisms, pulmonary circulation is a marvelous engineering system. An engineering analysis on pulmonary circulation can help us better understand its operating mechanism.

The human pulmonary circulation system carries blood from the right ventricle to the left atrium of the heart. Blood flows from the right ventricle to the pulmonary arterial trunk which is divided into the main right and left pulmonary arteries. The pulmonary arteries enter the lungs, and divide again and again into smaller and smaller branches like a tree in the lungs. The branching process continues to the pre-capillary arterioles, and ending in the pulmonary capillary sheets. In the capillary sheets the blood acquires oxygen and releases carbon dioxide into the air sacs called pulmonary alveoli. Then the blood flows from the capillary sheets into a venous tree, terminating in the left atrium of the heart. The blood flow is pulsatile, the system is complex, and the *in vivo* measurement is difficult. The chances to do invasive measurement in human subjects are limited. However, the pressure-flow relationship in pulmonary circulation is extremely important to the understanding of diseases in the lung and the heart, such as pulmonary hypertension, atherosclerosis, edema, thromboembolism, various respiratory disorder, and diabetes. In this regard a thorough theoretical understanding with full experimental verification is essential.

The pulsatile pressure-flow relations in the lungs are often expressed in electric engineering terminology. For example, in clinical applications it is important to estimate the pulmonary arterial input impedance, which is the ratio of the amplitude of oscillatory arterial pressure to the oscillatory inflow rate at a given frequency. The impedance as a function of frequency is referred to as a spectrum. Two general theoretical approaches are used to study pulsatile hemodynamics. One is the electric circuit analog, or the "lumped parameter" model, in which the anatomic details are ignored. The other is the bioengineering approach, or the "continuum" model. Both approaches have been used in the studies of pulmonary pulsatile flow.[1] Whereas the lumped parameter approach is often convenient and effective, the continuum approach is more fundamental. To evaluate the values of the lumped parameters theoretically, for example, one has to use the continuum model.

The bioengineering approach is based on the principles of continuum mechanics in conjunction with detailed measurement of vascular geometry, vascular elasticity and blood rheology. The method is uniquely suitable to the study of pulmonary blood flow. It provides an analytical tool to analyze the physiological problems, and to synthesize the many components of a complex problem so that quantitative predictions are made possible. Zhuang *et al.* used this analytical approach and successfully developed a hemodynamic model for steady blood flow in cat lung.[2] In their calculations, the sheet flow theory[3] was used for pulmonary capillary blood flow and an analogous "fifth power" was used for flow in the arteries and veins. Later on, experiments were performed to validate the theoretical predictions.[4] So far, the bioengineering approach has been used to compute the impedance for pulsatile flow in dog lung,[5] cat lung,[6] and human lung.[7] The agreement between theory and experiment in dog[5] and cat[6] was satisfactory. We shall describe this bioengineering approach below.

2. The Bioengineering Approach Defines What Experimental Data are Needed

Knowledge of the various factors controlling pulmonary blood flow is important in human health and disease, and today our knowledge and understanding of these factors is incomplete. For an experimenter, one should ask: what precise experimental data are required? To answer this question, one should seek theoretical guidance. For the bioengineering approach to pulmonary circulation, the answer is that we require (1) detailed description of vascular geometry, (2) measured elasticity of blood vessels, and (3) rheology of blood in blood vessels. With the information listed above, and together with the basic physical laws and appropriate boundary conditions, differential equations can be written and the problems solved either analytically or numerically.

2.1 *Morphometry of pulmonary vasculature in human lung*

Quantitative studies of pulmonary vasculature were started by Malpighi in 1661. The early history of the pulmonary vascular morphometry has been summarized by Miller.[1] Figure 1a shows a typical cast of a small segment of an arterial tree in human lung. The arterial cast was prepared by the silicone elastomer casting method.[8] The complex structure of the pulmonary structure has been described by three schemes: the Weibel model (Fig. 1b), the Strahler model (Fig. 1c), and the diameter-defined Strahler system (Fig. 1d). Weibel published his work in 1962. Cumming, Horsfield and their co-workers used Strahler model in 1970–1981. The concept of the diameter-defined Strahler system was used by Fung and Yen for the pulmonary arteries of the dog in 1987 (unpublished). Later on, the diameter-defined Strahler system was developed and used in the studies of the coronary blood vessels in the pig,[9] the rat pulmonary arterial tree,[10] the dog pulmonary venous tree[11] and arterial tree,[12] and the human pulmonary vasculature.[8]

The Weibel model assumes a symmetrical dichotomy as shown in Fig. 1b. The largest vessel is designated as a vessel of generation 1. After each bifurcation, the generation number of the offspring is increased by 1. The two offspring at each bifurcation are assumed to be equal in diameter and length. The average of the two offspring is used as the morphological data. Weibel and Gomez reported a total of 28 generations in the human pulmonary arterial tree, and 23 generations in the bronchial tree.[13]

The Strahler model avoids the symmetric dichotomy assumption of the Weibel model. In the Strahler model (Fig. 1c), the smallest non-capillary blood vessel is defined as vessels of order 1. When two vessels of the same order meet, the order number of the confluent vessel is increased by 1, and so on. Investigators have used Horsfield's adoption of the Strahler system to study the human and cat lungs.[14–19] Their data show very large overlaps of the diameters in the successive orders of vessels. Calculations of vascular resistance that is proportioned to the fourth power of the diameters became horribly inaccurate based on such data. The diameter-defined Strahler's system solves this problem.[10]

In the diameter-defined Strahler's system (Fig. 1d), a new rule is added: when a vessel of order n with diameter D_n meets another vessel of order n, the confluent vessel is called a vessel of order $n + 1$ *if and only if* its diameter is larger than $D_n + (S_n + S_{n+1})/2$, where S_n and S_{n+1} are the standard deviations of the diameters of orders n and $n + 1$. The right hand schematic sketch shown in Fig. 1d shows the scheme of the diameter-defined Strahler's ordering system.

We used the diameter-defined Strahler's system to study the human pulmonary vasculature.[8] Data on pulmonary arteries and veins are summarized in Tables 1 and 2, respectively. The silicone elastomer casts of pulmonary blood vessels were prepared from two postmortem human lungs. The branching pattern and vascular geometry of the human pulmonary arterial and venous trees were measured. There were 15 orders of pulmonary arteries between the main pulmonary artery and the capillaries, and 15 orders of pulmonary veins between the capillaries and the left atrium.[8] In order to

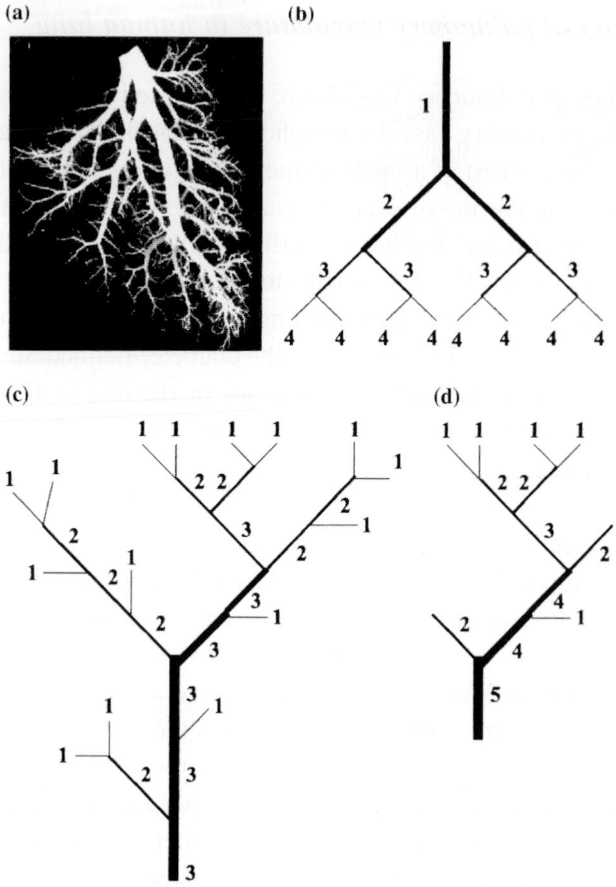

Fig. 1. **(a)** Typical cast of a small segment of arterial tree in human lung shows the complex structure of the vascular tree. The proposed three schemes describing this complex structure are illustrated by the "Weibel model" **(b)**, the "Strahler model" **(c)**, and the "diameter-defined Strahler system" **(d)**. In the Weibel model the largest vessel is designated as a vessel of generation 1. After each bifurcation, the generation number of the offsprings is increased by one. The exact opposite is true in the "Strahler model" and the "diameter-defined Strahler system," in which the smallest non-capillary blood vessel is defined as vessels of order 1. In the "Strahler model" when two vessels of the same order meet, the order number of the confluent vessel is increased by 1 while when two vessels of different order meet, the order number of the confluent vessels remains the same as the larger of the two. In the "diameter-defined Strahler system" when two vessels of different order and diameter meet one another, the order number of the confluent vessel is increased by 1 if and only if its diameter is larger than either of the two segments by a certain amount. Otherwise, the order number of the confluent segment is not increased. From Mandegar *et al.*,[42] by permission.

distinguish the series and parallel vessels of the same order, the concepts of segments and elements were defined: a segment is the blood vessel between two successive bifurcation points, and an element is all segments of the same order connected in series. In our model, the pulmonary blood vessels of the same order are considered to

Table 1. Morphometric and elastic data of the pulmonary arteries in human lung. Data are from Refs. 4, 8 and 26.

Order	Number of branches N_n	Vessel diameter D_n (cm)	Vessel length L_n (cm)	Apparent viscosity μ (cP)	Compliance β (10^{-4}Pa^{-1})
1	102411624	0.0020	0.0220	2.5	0.698
2	28114394	0.0036	0.0260	3.0	0.482
3	10203806	0.0056	0.0360	3.5	0.361
4	4513692	0.0097	0.0450	4.0	0.361*
5	1348338	0.0150	0.0680	4.0	0.361*
6	571544	0.022	0.108	4.0	1.135
7	172040	0.034	0.192	4.0	0.932
8	44008	0.051	0.281	4.0	0.877
9	12450	0.077	0.373	4.0	1.238
10	3448	0.116	0.658	4.0	1.281
11	900	0.175	1.235	4.0	1.281*
12	254	0.271	1.807	4.0	1.281*
13	86	0.416	2.597	4.0	1.281*
14	14	0.734	3.569	4.0	1.281*
15	4	1.480	2.53	4.0	1.281*
16	1	3.0†	9.05†	4.0	1.281*

Compliance data were obtained with $P_A = 0$ and $P_{PL} = -10$ cmH$_2$O. *Unknown compliance coefficients obtained by extrapolating from the nearest order of vessels. † Data from Singhal *et al.*[17]

be parallel, and the vessels of different orders are connected in series. Hence, the diameter and length of elements, not segments, of arteries and veins are used. For the diameter and length of the main pulmonary artery, the data reported in Singhal *et al.*[17] is used.

2.2 *Elasticity of human pulmonary blood vessels*

Vascular elasticity, or distensibility, is a mechanical property of blood vessels which determines a change in the diameter of blood vessels as a result of the change in blood pressure. The mechanical properties of pulmonary blood vessels are essential factors influencing the distribution of pulmonary blood pressure, regional distribution of pulmonary blood volume, transit time distribution of blood in the lung, and pulse-wave attenuation through the lung. The mechanical properties also affect the pressure-flow relationship and the change in total blood volume in response to an alteration of blood pressure.

In most cases, data on elasticity of mammalian blood vessels have been obtained from postmortem tissues. Three methods have been used to obtain the elasticity data on

Table 2. Morphometric and elastic data of the pulmonary veins in human lung. Data are from Refs. 4, 8 and 26.

Order	Number of branches N_n	Vessel diameter D_n (cm)	Vessel length L_n (cm)	Apparent viscosity μ (cP)	Compliance β ($10^{-4}Pa^{-1}$)
15	4	1.2970	3.5680	4.0	0.708*
14	8	0.8650	3.4990	4.0	0.708*
13	22	0.5860	1.9490	4.0	0.708*
12	62	0.4000	2.6490	4.0	0.708*
11	126	0.2880	1.7900	4.0	0.708*
10	286	0.1990	1.4780	4.0	0.708*
9	870	0.1420	1.1240	4.0	0.708*
8	2134	0.0900	0.6780	4.0	0.708
7	7442	0.0620	0.4790	4.0	0.887
6	31134	0.0380	0.2920	4.0	1.260
5	146050	0.0230	0.1500	4.0	1.260
4	906102	0.0130	0.1060	4.0	1.798
3	4332666	0.0067	0.0380	3.5	0.433
2	16988490	0.0031	0.0210	3.0	0.625
1	79647106	0.0018	0.0130	2.5	1.010

Compliance data were obtained with $P_A = 0$ and $P_{PL} = -10$ cmH_2O. *Unknown compliance coefficients obtained by extrapolating from the nearest order of vessels.

pulmonary arteries and veins: the biaxial testing machine for larger specimens, the X-ray photography for smaller blood vessels with diameters larger than 100 μm, and the silicon elastomer method for vessels smaller than 100 μm in diameter. The X-ray image technique of an isolated lung is commonly used to determine the distensibility of pulmonary arteries and veins in different species. The earlier work began with Patel's direct measurement of the canine main pulmonary artery with an electronic caliper in 1960.[20] The same vessel was measured by Frasher and Sobin in 1965 by making silicone polymer casts under various controlled transmural pressures.[21] Later Caro and Saffman used X-ray photography to record elasticity of pulmonary blood vessels of the rabbit in the range of 1~4 mm diameter.[22] Maloney *et al.* determined the elasticity of canine pulmonary blood vessels in the diameter range of 0.8~3.6 mm.[23] The elasticity of the cat pulmonary arteries in the diameter range of 100~1600 μm was measured by Yen *et al.* using X-ray photographic method.[24] By the same method, the elasticity of pulmonary veins in the diameter range of 100~1200 μm was determined.[25] The elasticity data of dog pulmonary blood vessels were obtained from the X-ray radiographs of the dog's isolated right lung by Gan and Yen.[5] They measured the elasticity of blood vessels in the diameter range of 100~4500 μm.

In the computational model, the compliance, β, of a blood vessel is used to represent the elasticity of the blood vessel. According to the Laplace formula, the tensile stress σ in the wall of a pulmonary blood vessel of diameter D is

$$\sigma = (P - P_{PL})\left(\frac{D}{2h_w}\right) \quad (1)$$

where h_w is the wall thickness of a blood vessel, P is the local blood pressure, P_{PL} is the pleural pressure. The strain is σ/E, a ratio of the tensile stress σ in the vessel wall versus the Young's modulus E of the vessel wall material, which is also the change in diameter ΔD divided by the diameter at zero transmural pressure D_0. Hence, the compliance is

$$\beta = \frac{\Delta D / D_0}{P - P_{PL}} = \frac{D}{2h_w}\frac{1}{E}. \quad (2)$$

The compliance of a blood vessel is expressed by the percentage changes of diameter versus change in transmural pressure. It depends partly on the ratio of the vessel wall thickness h_w to vessel diameter D, and partly on the Young's modulus E of the vessel wall material.

The data on compliance β of human pulmonary arteries in different orders are summarized in Table 1, and the data for pulmonary veins are listed in Table 2. The elasticity of the human pulmonary arterioles and venules for vessel diameter less than 100 μm (including the pulmonary blood vessels of orders 1–3) was determined by Yen and Sobin applying the silicone elastomer method.[4] By using the X-ray photographic method, Yen *et al.* measured the elasticity of the human pulmonary arteries in the diameter range of 200~1600 μm (orders 6–10 in branching pattern), and the veins in the diameter range of 100~1200 μm (orders 4–8 in branching pattern).[26] The elasticity data of the human pulmonary arteries of orders 4 and 5, and orders 11–16, and the pulmonary veins of orders 9–15 are not available in literature. The data we used in the model are obtained by extrapolating from the nearest order of the pulmonary blood vessels.

2.3 *Human pulmonary capillaries*

The structure of pulmonary capillaries is very different from that of pulmonary arteries and veins. The pulmonary capillaries form a continuous network in a flat sheet which comprises the alveolar wall.[27,28] The alveolus in the lung is the smallest unit of the air space. The dense network of the pulmonary capillary blood vessels of the cat's lung is shown in Figs. 1 and 2 of Sobin *et al.*[27]

Fung and Sobin described the capillary flow as a "sheet flow" taking place between two relatively flat sheets which are connected by a large number of closely spaced posts.[3,29] By the sheet-flow model, the blood is not channeled in tubes, but has freedom to move in any way between the posts. This model greatly simplifies the hydrodynamic analysis of flow in the alveolar wall and the structural analysis of the elastic deformation of the sheet. The morphometric basis of the sheet-flow theory was investigated by Sobin *et al.* in 1970.[27] They measured the sheet dimensions by optical microscopy on the specimens prepared by solidifying a silicone elastomer perfused into the capillaries.[27] Additionally, Sobin *et al.* used a microvascular casting method to measure the elasticity of the pulmonary capillary sheet in the cat.[30] The variation of alveolar sheet thickness in cat lung has been shown with respect to change in transmural pressure.[30,31] When the transmural pressure is positive and less than 35 cmH_2O, the capillary sheet thickness h increased linearly with increasing pressure as follows[30]

$$h = h_0 + \beta^*(P - P_A) = h_0 + \beta^* \Delta P \tag{3}$$

where h is the thickness of the capillary sheet, h_0 is the thickness at zero transmural pressure, β^* is the compliance of the sheet, P is the local blood pressure, and P_A denotes the alveolar gas pressure. At higher values of ΔP, the compliance becomes very small. For negative values of ΔP, the vessel collapses and h tends to zero.

Table 3 summarizes the set of parameters for the human pulmonary capillary sheet used in our computation and the references where those data were reported. The data for the average length of capillaries, $\bar{L}$, in human lung is not available in literature. We made an estimate based on the ratio of the alveolar diameter D in human, dog, and cat: 250: 94: 117 reported by Gan and Yen.[5] Sobin *et al.* have found that in cat lungs $\bar{L} = 556 \pm 286$ μm.[32] Gan and Yen estimated that in dog lungs $\bar{L} = (94 \times 556)/117 = 447$ μm[5]. In human lung, we estimate $\bar{L} = (250 \times 556)/117 = 1188$ μm.

Table 3. Morphometry, elasticity and blood viscosity data of the capillary sheet in human lung.

Symbol	Value	Reference no.
h_0	3.5 μm	44
α	0.127 μm/cmH_2O	44
$\bar{L}$	1188 μm	Estimated (see text)
S	0.88	44
A	126 m^2	45
μ	1.92 cP	34
K	12	38
F	1.8	46

2.4 *Blood viscosity in human pulmonary blood vessels*

When the shear strain rate of blood is smaller than 100 sec^{-1}, the coefficient of blood viscosity is inversely proportional to strain rate. When the strain rate approaches zero, the blood behaves like a plastic solid that has a small yield stress and an infinite coefficient of viscosity. However, when the shear strain rate exceeds 100 sec^{-1}, viscosity coefficient of blood is approximately a constant, which is dependent upon the hematocrit. That is, at high shear rate, whole blood behaves like a Newtonian fluid. In other words, as the shear rate increases from zero to a high value, the stress-strain relation of blood changes from non-Newtonian to Newtonian behavior. The transition to the Newtonian region depends on the hematocrit. The effect of the non-Newtonian rheological behavior of blood in pulmonary arteries and veins is quite small.[33]

Yen and Fung studied the blood viscosity on a scale model of the pulmonary alveolar sheet, with the red blood cells simulated by soft gelatin pellets and the plasma simulated by a silicone fluid.[34] They demonstrated that the relative viscosity of blood μ_r with respect to plasma depends on hematocrit H in the following manner

$$\mu_r = 1 + aH + bH^2 \tag{4}$$

where a and b can be determined from the experiment. Applying this result to blood, they found that the viscosity of blood in the alveoli is related to the viscosity of the plasma and the hematocrit in a quadratic relationship:

$$\mu_{\text{blood in alveoli}} = \mu_{\text{plasma}}(1 + aH + bH^2). \tag{5}$$

The variation of the apparent viscosity of blood in the pulmonary capillary sheet with hematocrit was determined in model experiments.[34] The apparent viscosity is approximately 1.92 cp when the hematocrit is 30%. The apparent viscosity is assumed to be 4.0 cp in larger vessels. Then the apparent viscosity of blood in small vessels in the orders of 1–3 is obtained by linear interpolation as 2.5, 3.0, and 3.5, respectively.

3. How to Use the Experimental Data in the Computational Model

After collecting the experimental data in morphometry, elasticity, and blood viscosity, one may ask how to use these data. With specifying boundary conditions, we can use these experimental data to compute the pressure-flow relationship, the transit time, the wave propagation, and the stress-strain relationship in the whole lung, or anywhere in the lung. The basic laws are the conservation of mass, momentum, and energy. A great deal of examples has been shown in Fung.[35,36] In the following, we present how to carry out theoretical analysis of the pulsatile blood flow in the human pulmonary circulation based on the experimental data.

3.1 *Impedance of pulmonary arteries and veins*

In the model studies, the pulmonary vasculature is separated into two basic components, the arterial and venous vessels versus the capillary network. The arteries and veins are treated as elastic tubes and the capillaries as two-dimensional sheets. The macro- and microcirculation is transformed into an electrical circuit analog. The pulmonary vascular input impedance and the characteristic impedance of each order in the pulmonary arterial tree are calculated under normal physiological conditions. In the model, a blood vessel is assumed as a thin-walled circular cylindrical elastic tube of uniform material and mechanical properties, and blood is a homogeneous, incompressible, Newtonian fluid. The characteristic impedance, Z_C, is given by Womersley[37]

$$Z_c = \frac{4}{\pi D^2}\sqrt{\frac{\rho}{2\beta(1-\upsilon^2)(M_{10}\exp(i\varepsilon_{10}))}} \tag{6}$$

where D is the diameter of a blood vessel, ρ is the density of blood, M_{10} and ε_{10} are functions of Bessel functions of the Womersley's non-dimensional parameter α. β is the compliance of a blood vessel expressed in equation (2), and υ is the Poisson's ratio. The complex wave velocity, c, is given by

$$c = \sqrt{\frac{M_{10}\exp(i\varepsilon_{10})}{2\rho\beta(1-\upsilon^2)}}. \tag{7}$$

The relationship between input impedance Z_{in} and characteristic impedance Z_C depends on the structure of the vascular tree, and is put in the following form by Gan and Yen[5]

$$Z_{in} = \frac{Z_C}{Z_T}\frac{1-i\lambda\tan(\kappa l)}{\lambda - i\tan(\kappa l)}Z_C, \tag{8}$$

where l is the distance from the point in question to the distal end of the tube where reflection is generated, and κ is the complex propagation coefficient. The discontinuity coefficient λ is a function of the terminal impedance Z_T, which is the impedance faced by the end of the vessel, and the characteristic impedance Z_C,

$$\lambda = \frac{Z_C}{Z_T}. \tag{9}$$

In applying the above formulas to the lung, the variation of the physical properties along the vascular tree is taken into account by assigning different values for the geometrical and deformational properties of the vessels in each successive order of branch.

The continuity of pressure and flow at each branch point is imposed as the boundary conditions of each vessel. The pulmonary microvascular impedance model described in the following section is used to connect the pulmonary arterial and venous trees.

3.2 *Microcirculation impedance*

A one-dimensional solution of transient flow through the alveolar sheet that contains the basic feature of the more complex two-dimensional analysis was presented by Fung.[38,39] In our two-dimensional case, each alveolar sheet is supposed to stretch between parallel arterioles and venules. If the permeability of water across the endothelium can be ignored and the amplitude of the thickness fluctuations is small compared with the mean pulmonary alveolar sheet thickness, the basic equation for a transient flow can be expressed as:[38]

$$\frac{d^2h^4}{dx^2} = 4\mu kf\beta\frac{dh}{dt} \tag{10}$$

where h is the thickness of capillary sheet; μ is the apparent viscosity of the blood in the capillary sheet; k is a function of the ratio of the thickness to width of the capillary sheet and has a numerical value of about 12; f is a friction factor, which is a function of the ratio of the post-diameter to the sheet thickness and other flow parameters, and has a value of about 1.6; and β is the compliance constant of the capillary sheet.

Equation (10) is a nonlinear differential equation and does not have a harmonic solution with respect to time. Only in small perturbations the basic equations can be linearized and the concept of impedance can be useful. Linearization can be justified if the amplitude of the pressure oscillation, $H(x)$, is small compared with the mean pulmonary arterial pressure. Under this condition, the solution of the governing equation is set in the following form:

$$h(x,\ t) = h_{SI}(x) + e^{i\omega t}H(x). \tag{11}$$

Here the amplitude of the pressure oscillation $H(x)$ is assumed to be much smaller than the steady-impervious thickness of the capillary sheet, $h_{SI}(x)$. $h_{SI}(x)$ is the solution of equation (10) for steady flow and is equal to:

$$h_{SI}(x) = \left[h_a^4 - (h_a^4 - h_v^4)\frac{x}{\bar{L}}\right]^{1/4}, \tag{12}$$

where x is the distance measured from the inlet, $\bar{L}$ is the average length of blood pathway between the inlet and outlet, h_a is the steady-state sheet thickness at the arteriole inlet, and hv is that at the venule outlet. The pressure per unit width, $p(x, t)$, can be represented as the sum of the steady-impervious term p_{SI} and the oscillatory term of blood pressure P:

$$p(x, t) = p_{SI}(x) + e^{i\omega t} P(x). \tag{13}$$

Similarly, the flow per unit width, $q(x, t)$, can also be represented as the sum of the steady-impervious term q_{SI} and the oscillatory term of blood flow $\dot{Q}$:

$$q(x, t) = q_{SI}(x) + e^{i\omega t} \dot{Q}(x). \tag{14}$$

The variables are written in dimensionless form. The non-dimensional variables are related to their dimensional counterparts, denoted by circumflex (˜). Then the governing equation for pulsatile flow in the capillary sheet becomes

$$\left(\frac{d^2}{d\tilde{x}^2}\right)(\tilde{h}_{SI}^2 \tilde{H}) = i\Omega\tilde{H} \tag{15}$$

where Ω is the dimensionless frequency parameter and is given by

$$\Omega = \mu kf\, \beta^* \omega \overline{L}^2 / h_o^3 \tag{16}$$

The relationship between the blood pressure and the thickness of capillary sheet is

$$\tilde{p} = \tilde{p}_{SI} + e^{i\omega t}\tilde{H}, \tag{17}$$

and the relationship between the blood flow rate and the thickness of capillary sheet is

$$\tilde{\dot{Q}} = -\frac{d}{d\tilde{x}}(\tilde{h}_{SI}^3 \tilde{H}). \tag{18}$$

Let the oscillatory pressure and flow at the arteriole and venule edges of the sheet be designated by $\tilde{P}_a$, $\tilde{\dot{Q}}_a$, $\tilde{P}_v$, $\tilde{\dot{Q}}_v$, respectively. The relationships between these non-dimensional quantities can be expressed in the following matrix form:

$$\begin{Bmatrix} \tilde{P}_a \\ \tilde{\dot{Q}}_a \end{Bmatrix} = \begin{Bmatrix} A_{11} & A_{12} \\ A_{21} & A_{22} \end{Bmatrix} \begin{Bmatrix} \tilde{P}_v \\ \tilde{\dot{Q}}_v \end{Bmatrix}. \tag{19}$$

Hence, the capillary network can be inserted between the arteries and veins to complete the circuit.

4. Human Pulmonary Arterial Impedance

Based on the equations outlined in Sec. 3 above and the available experimental data summarized in Tables 1–3, we computed the pulmonary arterial impedance in normal human lung and presented a pulmonary arterial impedance spectrum[7] in Fig. 2. Figure 2a shows modulus plotted against frequency, and Fig. 2b shows phase plotted against frequency. In computation, the following conditions were applied: airway pressure $P_A = 0$ cmH$_2$O, pleural pressure $P_{PL} = -7$ cmH$_2$O, venous pressure $P_v = 2$ cmH$_2$O, and flow rate $\dot{Q} = 5$ l/min.

The solid line in Fig. 2a represents the input impedance modulus with a unit of dyn-sec/cm^5 at the main pulmonary artery in human. There are four peaks between 0 and 15 Hz. The highest peak is the first peak with a value of 169.4 dyn-sec/cm^5 at 0 Hz, and then the modulus decreases rapidly to a valley with a minimum value of 8 dyn-sec/cm^5 around 1.5 Hz. The second peak is 70 dyn-sec/cm^5 at 4 Hz, and the third peak is 50 dyn-sec/cm^5 at 9.5 Hz. The lowest peak is the fourth peak with a value of 28 dyn-sec/cm^5 at 13 Hz. The dashed line in Fig. 2a represents the characteristic impedance at the main pulmonary artery. The highest characteristic impedance is 21.9 dyn-sec/cm^5 when the

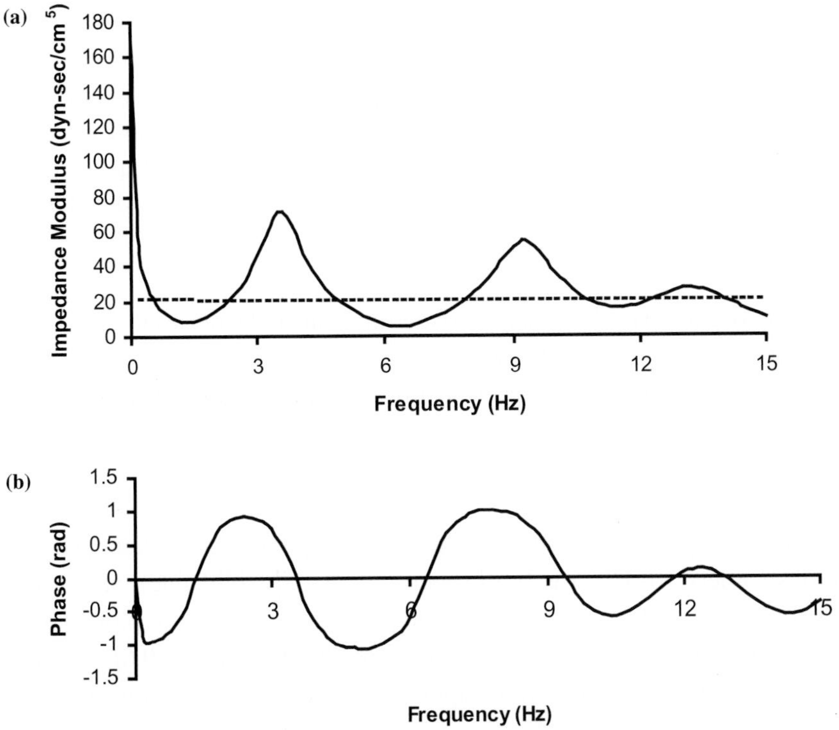

Fig. 2. Input impedance at the main pulmonary artery in normal human lung when airway pressure $P_A = 0$ cmH$_2$O, pleural pressure $P_{PL} = -7$ cmH$_2$O, venous pressure $P_v = 2$ cmH$_2$O, and flow rate = 5 l/min. (**a**) Modulus. (**b**) Phase. Zero-frequency input impedance modulus is 169.4 dyn-sec/cm^5. From Zhou *et al.*,[7] by permission.

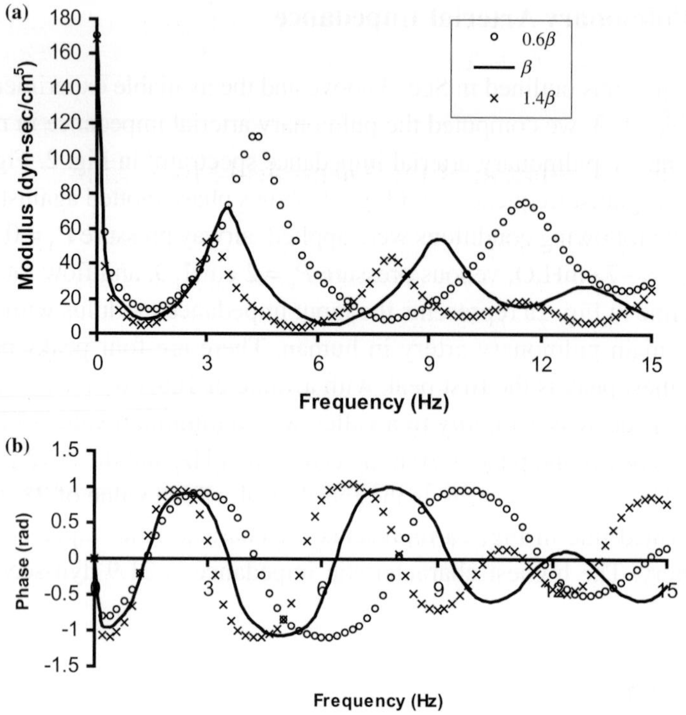

Fig. 3. The input impedance spectrums when the compliance coefficient β of the proximal arteries (orders 10–16) was 0.6β (open circles), β (solid line), and 1.4β (cross), respectively. The following conditions were applied: airway pressure $P_A = 0$ cmH$_2$O, pleural pressure $P_{PL} = -7$ cmH$_2$O, venous pressure $P_v = 2$ cmH$_2$O, and flow rate = 5 l/min. **(a)** Modulus. **(b)** Phase. From Zhou *et al.*,[7] by permission.

frequency is close to 0 Hz. The characteristic impedance decreases to 20.7 dyn-sec/cm^5 at 9 Hz, and remains stable after 9 Hz. Figure 2b shows phase plotted against frequency. The phase at the main pulmonary artery features a valley at about 0.2 Hz before climbing up the first peak around 2.5 Hz. The second peak is at 7.5 Hz, and the last peak is at 12.3 Hz.

The effects of compliance coefficient β on the impedance were studied. Figure 3 shows the input impedance spectrums when the compliance coefficient β of the proximal arteries (orders 10–16) was 0.6β (open circles), β (solid line), and 1.4β (cross), respectively. The modulus in Fig. 3a and the phase in Fig. 3b were plotted against frequency. The modulus and phase spectrums are different when 0.6β (open circles), β (solid line), and 1.4β (cross) were applied. When the pulmonary arteries become stiffer, the modulus rises dramatically at higher frequencies. Besides the pulmonary input impedance presented in Fig. 3, we calculated the pulmonary characteristic impedance. In experimental studies,[40] the pulmonary characteristic impedance was defined as the average of characteristic impedance moduli at frequencies between 2 and 12 Hz. We determined the characteristic impedance according to this definition. The characteristic impedance was $Z_C = 44.1$ dyn-sec/cm^5 when 0.6β was applied, $Z_C = 27.9$ dyn-sec/cm^5 when β was applied, and $Z_C = 20.1$ dyn-sec/cm^5 when 1.4β was applied.

5. Experimental Validation of the Theoretical Model Study

Through the bioengineering approach, a theoretical model of pulsatile flow in the pulmonary circulation can be developed based on the experimental data of vascular morphometry, elasticity, and blood viscosity. The theoretical pulsatile pressure-flow relations can be predicted. However, the predictions need to be validated in experimental studies.

To validate the theoretical predictions in the dog model, the experimental studies of pulmonary arterial pressure-flow relations in the normal and chronic pulmonary thromboembolic dogs were completed.[41] A fluid-filled pulmonary arterial catheter was placed percutaneously for measurement of pulmonary arterial pressure. A radionuclide perfusion scan was performed for measurement of mean pulmonary blood flow. The pressure and flow of pulmonary arteries were measured in days 0 and 30. The large pulmonary arteries were chronically obstructed with lysis-resistant thrombi since day 0. Comparison of experimentally measured and model-derived pulmonary arterial pressure-flow is shown in Fig. 2 of Olman *et al.*[41] The model-derived pulmonary arterial pressure was within 1 mmHg of the baseline (day 0) measured pulmonary arterial pressure.

Experiments on isolated perfused cat lungs were carried out to validate the theoretical model.[6] The pulsatile blood pressure in the pulmonary arterial trunk was measured with a fluid-filled Teflon tubing. The flow in the pulmonary arterial trunk was measured with a flow probe around the vessel. The resulting data of pressure and flow were subjected to a frequency analysis. The modulus and phase of pressure and flow waves were calculated from the Fourier coefficients.[6] Comparison of the experimental and model-derived pulmonary vascular input impedance in cat lung is shown in Fig. 2 of Huang *et al.*[6] In general, the model-derived input impedance fits the animal data well.

6. Challenges for the Future

As shown in Fig. 3, the model-derived input impedance is very sensitive to geometrical and mechanical properties of large arteries. To better simulate the pulmonary vascular impedance theoretically, we need to include the information of pulmonary trunk (i.e. curvature, length, diameter), and the angles between the big branches. So far, these data are not available. Additionally, the influence from heartbeat should be considered in the theoretical model.

Basic researches in biomedical science are aimed to improve the diagnosis and treatment of human diseases. It is our objective to understand the detailed pulmonary blood pressure-flow relationship in human lung for clinical application. After the establishment of a theoretical model on the pulsatile flow in human lung,[7] experimental studies are necessary to evaluate the theoretical model. Refinement of the theoretical model is possible based on the experimental results.

Studies on the molecular mechanisms of pulmonary vascular diseases have attracted a greal deal of attention in recent years.[42] The changes in molecular and cellular levels will lead to the changes of the vascular geometry, the viscoelastic properties of the vessel walls, and the blood viscosity in pulmonary vascular diseases, such as pulmonary hypertension, atherosclerosis, edema, thromboembolism, various respiratory disorder, and diabetes. As we discussed above, the blood pressure-flow relationship is determined by the vascular morphometry, the vascular mechanical properties, and the rheology of the blood. Therefore, the pulmonary vascular impedances are changed in the disease lungs.[43] The theoretical models of pulmonary vascular impedances in human diseases will be useful in understanding the mechanism of the diseases, and eventually helpful in the diagnosis and treatment of the human lung diseases.

References

1. W. S. Miller, *The Lung*, 2nd ed. (Thomas, Springfield, 1947).
2. F. Y. Zhuang, Y. C. Fung and R. T. Yen, Analysis of blood flow in cat's lung with detailed anatomical and elasticity data, *J. Appl. Physiol.* **55**: 1341–1348 (1983).
3. Y. C. Fung and S. S. Sobin, Theory of sheet flow in lung alveoli, *J. Appl. Physiol.* **26**: 472–488 (1969).
4. R. T. Yen and S. S. Sobin, Elasticity of arterioles and venules in postmortem human lungs, *J. Appl. Physiol.* **64**: 611–619 (1988).
5. R. Z. Gan and R. T. Yen, Vascular impedance analysis in dog lung with detailed morphometric and elasticity data, *J. Appl. Physiol.* **77**: 706–717 (1994).
6. W. Huang, Y. Tian, J. Gao and R. T. Yen, Comparison of theory and experiment in pulsatile flow in cat lung, *Ann. Biomed. Eng.* **26**: 812–820 (1998).
7. Q. Zhou, J. Gao, W. Huang and M. Yen, Vascular impedance analysis in human pulmonary circulation, *Biomed. Sci. Instrum.* **42**: 470–475 (2006).
8. W. Huang, R. T. Yen, M. McLaurine and G. Bledsoe, Morphometry of the human pulmonary vasculature, *J. Appl. Physiol.* **81**: 2123–2133 (1996).
9. G. S. Kassab, C. A. Rider, N. J. Tang and Y. C. Fung, Morphometry of pig coronary arterial trees, *Am. J. Physiol.* **265**: H350–365 (1993).
10. Z. L. Jiang, G. S. Kassab and Y. C. Fung, Diameter-defined Strahler system and connectivity matrix of the pulmonary arterial tree, *J. Appl. Physiol.* **76**: 882–892 (1994).
11. R. Z. Gan, Y. Tian, R. T. Yen and G. S. Kassab, Morphometry of the dog pulmonary venous tree, *J. Appl. Physiol.* **75**: 432–440 (1993).
12. Y. Tian, Dissertation, University of Memphis, Memphis, TN, 1993 (unpublished work).
13. E. R. Weibel and D. M. Gomez, Architecture of the human lung. Use of quantitative methods establishes fundamental relations between size and number of lung structures, *Science* **137**: 577–585 (1962).

14. G. Cumming, L. K. Harding, K. Horsfield, K. Prowse, S. S. Singhal and M. J. Woldenberg, Morphological aspects of the pulmonary circulation and of the airway, *Fluid Dyn. Blood Circ. Respir. Flow* **65**: 23-0–23-6 (1970).
15. K. Horsfield, Morphometry of the small pulmonary arteries in man, *Circ. Res.* **42**: 593–597 (1978).
16. K. Horsfield and I. Gorden, Morphometry of pulmonary veins in man, *Lung* **159**: 211–218, (1981).
17. S. Singhal, R. Henderson, K. Horsfield, K. Harding and G. Cumming, Morphometry of the human pulmonary arterial tree, *Circ. Res.* **33**: 190–197 (1973).
18. R. T. Yen, F. Y. Zhuang, Y. C. Fung, H. H. Ho, H. Tremer and S. S. Sobin, Morphometry of cat pulmonary venous tree, *J. Appl. Physiol.* **55**: 236–242 (1983).
19. R. T. Yen, F. Y. Zhuang, Y. C. Fung, H. H. Ho, H. Tremer and S. S. Sobin, Morphometry of cat's pulmonary arterial tree, *J. Biomech. Eng.* **106**: 131–136 (1984).
20. D. J. Patel, D. P. Schilder and A. J. Mallos, Mechanical properties and dimensions of the major pulmonary arteries, *J. Appl. Physiol.* **15**: 92–96 (1960).
21. W. G. Frasher, Jr. and S. S. Sobin, Pressure-volume response of isolated living main pulmonary artery in dogs, *J. Appl. Physiol.* **20**: 675–682 (1965).
22. C. G. Caro and P. G. Saffman, Extensibility of blood vessels in isolated rabbit lungs, *J. Physiol.* **178**: 193–210 (1965).
23. J. E. Maloney, S. A. Rooholamini and L. Wexler, Pressure-diameter relations of small blood vessels in isolated dog lung, *Microvasc. Res.* **2**: 1–12 (1970).
24. R. T. Yen, Y. C. Fung and N. Bingham, Elasticity of small pulmonary arteries in the cat, *J. Biomech. Eng.* **102**: 170–177 (1980).
25. R. T. Yen and L. Foppiano, Elasticity of small pulmonary veins in the cat, *J. Biomech. Eng.* **103**: 38–42 (1981).
26. R. T. Yen, D. Tai, Z. Rong and B. Zhang, Elasticity of pulmonary blood vessels in human lungs. In: *Respiratory Biomechanics-Engineering Analysis of Structure and Function*, eds. M. A. F. Epstein and J. R. Ligas (Springer-Verlag, New York, 1990), pp 109–116.
27. S. S. Sobin, H. M. Tremer and Y. C. Fung, Morphometric basis of the sheet-flow concept of the pulmonary alveolar microcirculation in the cat, *Circ. Res.* **26**: 397–414 (1970).
28. S. S. Sobin and Y. C. Fung, Response to challenge to the Sobin-Fung approach to the study of pulmonary microcirculation, *Chest* **101**: 1135–1143 (1992).
29. Y. C. Fung and S. S. Sobin, Elasticity of the pulmonary alveolar sheet, *Circ. Res.* **30**: 451–469 (1972).
30. S. S. Sobin, Y. C. Fung, H. M. Tremer and T. H. Rosenquist, Elasticity of the pulmonary alveolar microvascular sheet in the cat, *Circ. Res.* **30**: 440–450 (1972).
31. S. S. Sobin, R. G. Lindal, Y. C. Fung and H. M. Tremer, Elasticity of the smallest noncapillary pulmonary blood vessels in the cat, *Microvasc. Res.* **15**: 57–68 (1978).
32. S. S. Sobin, Y. C. Fung, R. G. Lindal, H. M. Tremer and L. Clark, Topology of pulmonary arterioles, capillaries, and venules in the cat, *Microvasc. Res.* **19**: 217–233 (1980).
33. Y. C. Fung, Dynamics of blood flow and pressure-flow relationship. In: *The Lung: Scientific Foundations*, eds. R. G. Crystal and J. B. West (Raven Press, New York, 1991), pp. 1121–1134.
34. R. T. Yen and Y. C. Fung, Model experiments on apparent blood viscosity and hematocrit in pulmonary alveoli, *J. Appl. Physiol.* **35**: 510–517 (1973).

35. Y. C. Fung, *Biomechanics: Circulation*, 2nd ed. (Springer-Verlag, New York, 1996).
36. Y. C. Fung, *Selected Works on Biomechanics and Aeroelasticity* (World Scientific, New Jersey, 1997).
37. J. R. Womersley, Oscillatory flow in arteries: the constrained elastic tube as a model of arterial flow and pulse transmission, *Phys. Med. Biol.* **2**: 178–187 (1957).
38. Y. C. Fung, Theoretical pulmonary microvascular impedance, *Ann. Biomed. Eng.* **1**: 221–245 (1972).
39. Y. C. Fung, Fluid in the interstitial space of the pulmonary alveolar sheet, *Microvasc. Res.* **7**: 89–113 (1974).
40. J. P. Murgo and N. Westerhof, Input impedance of the pulmonary arterial system in normal man. Effects of respiration and comparison to systemic impedance, *Circ. Res.* **54**: 666–673 (1984).
41. M. A. Olman, R. Z. Gan, R. T. Yen, I. Villespin, R. Maxwell, C. Pedersen, R. Konopka, J. Debes and K. M. Moser, Effect of chronic thromboembolism on the pulmonary artery pressure-flow relationship in dogs, *J. Appl. Physiol.* **76**: 875–881 (1994).
42. M. Mandegar, Y. C. Fung, W. Huang, C. V. Remillard, L. J. Rubin and J. X. Yuan, Cellular and molecular mechanisms of pulmonary vascular remodeling: role in the development of pulmonary hypertension, *Microvasc. Res.* **68**: 75–103 (2004).
43. W. W. Nichols and M. F. O'Rourke, *McDonald's Blood Flow in Arteries: Theoretic, Experimental, and Clinical Principles*, 4th ed. (Oxford University Press, New York, 1998).
44. S. S. Sobin, Y. C. Fung, H. M. Tremer and R. G. Lindal, Distensibility of human pulmonary capillary blood vessels in the interalveolar septa, *Fed. Proc.* **38**: 990 (1979).
45. P. Gehr, M. Bachofen and E. R. Weibel, The normal human lung: ultrastructure and morphometric estimation of diffusion capacity, *Resp. Physiol.* **32**: 121–140 (1978).
46. J. S. Lee, Slow viscous flow in a lung alveoli model, *J. Biomech.* **2**: 187–198 (1969).

CHAPTER 12

PULMONARY GAS EXCHANGE

Peter D. Wagner

Abstract

This chapter describes how the lungs perform their principal task — the uptake of O_2 from the air into the blood, and the simultaneous movement of CO_2 from the blood to the air. The unique structure of the lungs as 300 million separate alveoli (300 μm diameter gas-filled spaces whose walls are essentially made of capillary networks) facilitates this exchange through a linked series of transport functions that employ both convective and diffusive movements of gas. The alveoli lie at the distal ends of a complex branching network of hollow airways much as grapes connect to a stalk. The alveolar wall capillaries are fed by a similar, branching pulmonary arterial tree, and are drained by corresponding pulmonary veins. Ventilation (a convective process) brings O_2 from the air to the alveoli during inspiration, while during expiration the gas in the alveoli is moved back to the outside air by flow reversal through the same airways to eliminate CO_2. The two gases exchange between blood and air in the alveoli by diffusion. This avoids energy expenditure and accounts for the very large number of very small exchange units, a strategy that greatly increases surface area without also requiring a large lung volume. The third process is blood flow, again convective, that moves the blood out of the alveolar capillaries and back to the left heart for distribution to the tissues. These three transport functions are well-understood and can be modeled mathematically with remarkable accuracy using simple mass conservation principles. The lung is the only organ whose major function can be described and thus understood adequately in terms of such simple transport equations, as this chapter shows. More complete descriptions can be found in the references,[1–6] which are intended for further reading for those interested in additional details.

1. The Lung is for Gas Exchange

More than any other organ, the lung should appeal to bioengineers. It is the only organ whose principal function, exchange of gases, can be accurately described by simple equations. This chapter is intended to explain how the structure of the lungs dictates its transport functions, leading to such a precise quantitative description of whole organ gas exchange function. Thus, the structure is first described, followed by the main transport functions, and then these functions are brought together into an equation system whose solution describes gas exchange in surprisingly simple yet biologically accurate terms.

Even more remarkable, these equations apply not just in health but also in disease and allow gas exchange in the diseased lung to be understood in the same manner as in health.

1.1 *Structure of the lungs*

As the chapter title says, the lung is for gas exchange. That means the following: the lung supports basic cell metabolism in which O_2 is consumed, producing energy for cell function. This process generates CO_2 and other by-products. How do the cells get their O_2? And how do they get rid of their CO_2? The answer for both gases is via the lungs: that is what the lungs are there for.

The lungs are interposed between the circulating blood and the air we breathe. Breathing delivers the O_2 we inhale to the pulmonary blood capillary surfaces by a process we call ventilation. The unique structure of the lung allows passive transfer of O_2 from that air into (and CO_2 out of) the blood by a **diffusion** process. The evolved CO_2 is then eliminated into the air by the same process of ventilation that brought O_2 in. The O_2-enriched blood is next pumped to the tissues by the process of **perfusion**. When it reaches the tissues, the blood gives up its O_2 to the cells to enable cell metabolism. At the same time, it picks up the CO_2 previously mentioned as produced from these metabolic processes. The blood that leaves the tissues is therefore low in O_2 and high in CO_2, and is pumped back to the lung in that state. When it reaches the lung, the process begins all over again — O_2 is transferred into the blood, CO_2 transferred out of the blood, and the blood is pumped again around the body as before.

To understand this set of processes one must know something about the structure of the lungs. A good, even if overly simple, starting point is to think of the lungs as a giant bunch of grapes still attached to their branched stalk. The good news is that if we can understand how one such "grape" looks and functions, we essentially know how the whole lung looks and functions, because all "grapes" look and function similarly. In the lung, the word we use instead of "grape" is "functional lung unit." While real grapes have a skin, a semi-solid interior pulp, and a solid stalk, the functional lung unit has a "skin" or wall that is made up of a network of short interconnected capillary segments through which blood flows. In fact, the great majority of the "skin" is just these capillary networks — there is very little other tissue in the wall of the lung unit. There is just enough collagen and elastin to give it support and strength, but not so much that it gets in the way of diffusion. And instead of a pulpy interior, the lung unit's interior

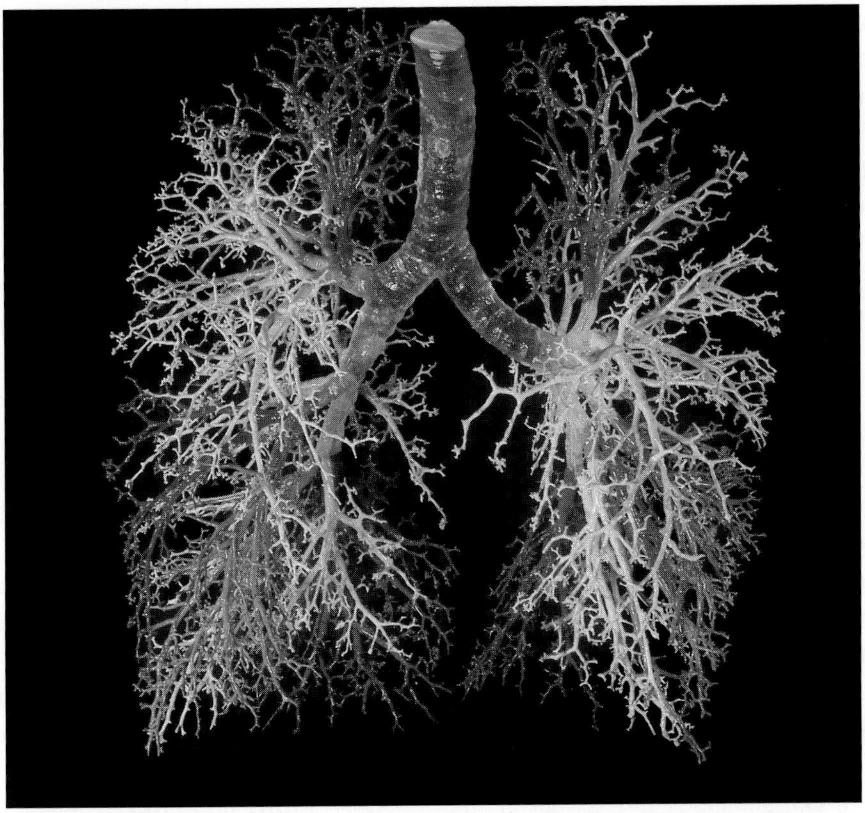

Fig. 1. Plastic cast of the airways of a canine lung showing the single large trachea at the top branching progressively into ever smaller daughter branches. There are some 17 orders of branching before these branches give way to alveoli where gas exchange occurs. The alveoli are not present, having been destroyed in the process of making the cast. The smallest branches seen are probably terminal bronchioles.

is simply air, just like a soap bubble. Finally, the stalk is actually a branched series of hollow tubes we call the airways that connect the mouth to the air-filled interior of every "grape." They conduct the O_2-rich air we breathe down to the "grapes" during inhalation, and again conduct the CO_2-rich air back out to the environment during exhalation.

The airways branch essentially dichotomously, like branches of a tree (Fig. 1). There are about 16–17 orders of branching, and as a result, about 2^{16} or 2^{17} "grapes," or functional lung units. In round numbers, there are about 100,000 functional lung units. Remember, just like real grapes, each lung unit is separately connected in parallel with all the others via its own airway ("stalk") through which O_2 enters and CO_2 leaves with every breath. But the branching airway pattern means that, like streams flowing into a large river, eventually the airways join together and form a single large air passage we call the trachea (in common language, the windpipe). It is the counterpart of the single major stalk that connects the real bunch of grapes directly to the grape vine. The trachea runs vertically in the neck, and is joined to the mouth via the larynx and pharynx.

The analogy to a bunch of grapes is visually and conceptually useful, but is limited. One limitation is that a bunch of grapes has no vascular tree. In the real lung, in addition to airways (stalks) feeding into functional lung units (grapes), there is a complex **arterial** vascular tree that carries blood, returning from the tissues low in O_2 and high in CO_2, into the capillary networks making up the walls of the functional units, and an equally complex **venous** vascular tree carrying blood now high in O_2 and low in CO_2 (after gas exchange has occurred in the functional lung units) back to the left side of the heart. This blood will be pumped around the body to supply cells with their O_2 and pick up their CO_2, which will be returned once again to the lungs.

The arterial and venous vascular trees are also dichotomously branched sets of conducting hollow tubes that have a similar branching architecture as do the airways. The difference is mostly in what fluid the various vessels transport. Clearly, air flows through the airways, and blood through the arterial, capillary, and venous networks. We call the entire blood vessel system the pulmonary (or lung) vasculature. Remember that the capillaries basically constitute the entire wall of the lung units.

Figure 2 shows a drawing of the functional lung unit from the work of Netter. You can see the lung unit ("grape") as all the alveoli distal to each terminal bronchiole

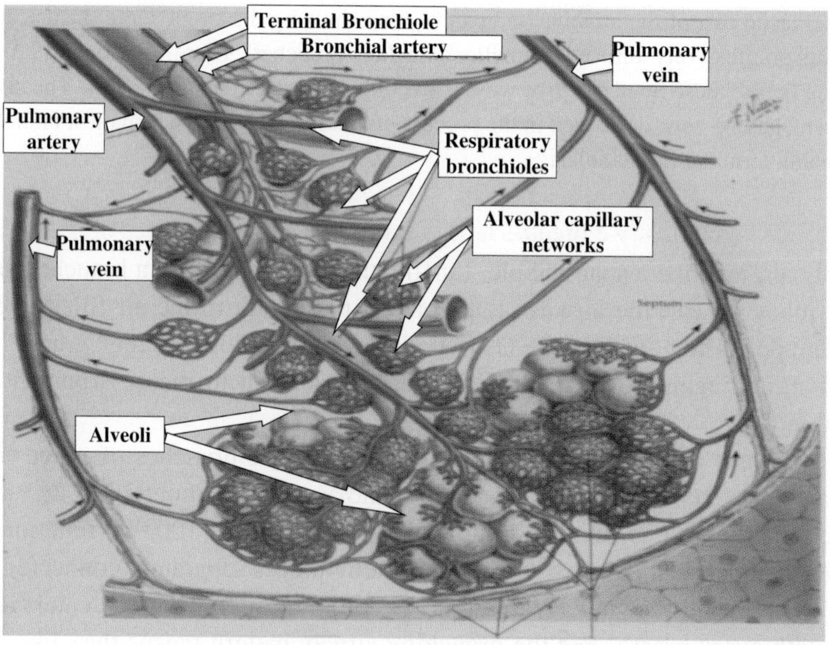

Fig. 2. Drawing of the airways, alveoli, and pulmonary vasculature constituting part of one acinus. Each alveolus is wrapped in a dense network of capillaries where gas exchange occurs.

("stalk") with the "skin" consisting of a network of capillaries fed by a pulmonary artery and drained by a pulmonary vein. Some key observations are in order:

(1) Each functional unit does not look like just a single grape, but rather like a cluster of smaller grapes that somehow share walls and communicate with a single airway. In fact, there are so many smaller "grapes" that cluster together make up a single functional lung unit that we have another name for each of them — we call them alveoli. It turns out there are about 3000 of these smaller grapes that together make up a single functional lung unit, making some 300 million smaller grapes (alveoli) in all (because there are about 100,000 functional lung units as mentioned).
(2) Why then have we defined the functional unit as such a collection of several thousand "subunits"? Why is the individual, smaller alveolus not the functional unit of the lung? The answer is that because these subunit grapes are so small, gas concentrations are uniform throughout the 3000 or so that make up each functional lung unit. The capillary networks are so rich that any potential vascular obstructions are easily compensated by collateral blood flow around them. Thus, the collection of 3000 contiguous alveoli and their airways and blood vessels described above as the "functional lung unit" have been found to be the physically smallest unit of lung that can act as a homogeneous gas exchange unit. That means that all throughout the unit, concentrations of any gas will be uniform. The implication is that neighboring functional lung units, while also themselves uniform, may show a different level of O_2 or CO_2. Why this may happen is a key issue, but its explanation will be kept to later in this chapter. A shorter name for the functional lung unit is the acinus.
(3) Ventilation, or breathing, is an in-and-out process with gas flow reversing direction between inhalation and exhalation but using exactly the same airways in each case. Thus inhalation is followed by exhalation, which is followed by the next inhalation and so on, alternately bringing fresh air into the alveoli and taking air low in O_2 and high in CO_2 back out to the air around us. Each inhalation and exhalation is called a breath, and at rest we breathe at a frequency of about 15 times/minute. This rate may rise to 60/minute during very heavy exercise. Each breath moves a certain volume of air into, and then essentially the same volume back out of, the lungs. We call this the tidal volume. At rest, tidal volume is about 500 ml, but during very heavy exercise tidal volume can reach as much as three liters. The whole breathing process is called ventilation, and the amount of air inhaled or exhaled each minute is called the minute ventilation. It is the product of tidal volume and frequency. At rest, you can see that this would be about 7.5 liters/min, while during very heavy exercise this could reach 180 liters/min.
(4) Blood flow through the lungs is fundamentally different from ventilation. Rather than being an in-and-out process with flow reversal through a single set of conducting vessels, blood flow is a unidirectional or "through-and-through" process with blood pumped from the right heart through the pulmonary arteries, into the alveolar capillary network, and then onwards into the pulmonary veins and finally

back to the left heart. But just as with breathing, blood flow is the result of discrete pumping events carried out a number of times per minute. Thus, the heart, which of course is the blood pump, beats about 70 times/min at rest, and up to 180/min during very heavy exercise. Each beat of the heart pumps about 80 ml of blood at rest, and somewhat more, perhaps 120 ml, during very heavy exercise. The product of volume per beat (known as stroke volume) and heart rate is called cardiac output, and this ranges therefore from about 6 liters/min at rest to about 20 liters/min (or more) during very heavy exercise.

(5) Ventilation gets the O_2-rich air down to the alveoli and blood flow gets the O_2-rich blood from the pulmonary capillaries to the body tissues via the left heart. But how does the O_2 get from the alveoli to the blood? By simple, passive diffusion, O_2 molecules simply diffuse from the alveolar gas across the walls of the pulmonary capillaries and into the red blood cells where they combine with hemoglobin. The energy for this diffusive transport of O_2 molecules is their thermal kinetic energy, which produces random Brownian motion. No cellular energy is expended in this transport process — there are no ion pumps or other energy-consuming processes involved.

(6) It is this simplicity of transport — ventilation, diffusion and blood flow — that basically allows the lungs to be analyzed in simple yet accurate mathematical terms when considering both O_2 and CO_2 exchange.

1.2 *Transport functions of the lungs*

1.2.1 *Ventilation*

How ventilation happens: the chest wall, the respiratory muscles and the lungs

The lungs are contained within the chest cavity. If you remove the lungs from the chest and place them on a table, will they "breathe" by themselves? The answer is no. The lungs are an elastic structure much like a simple elastic party balloon. A balloon will not inflate and deflate by itself. The balloon (and the lungs alike) will simply collapse if laid on a table. Inflation of a balloon requires a positive pressure difference between its interior and exterior. This can be accomplished by: (a) positive pressure inflation by blowing into the neck of the balloon, or (b) negative pressure inflation by reducing the pressure around the balloon. This can be done by enclosing the balloon in an airtight box except for the neck of the balloon that protrudes into the air, and then vacuuming out the air in the box outside the balloon. The balloon will then inflate. Either way, inflation requires expenditure of energy to overcome the elastic recoil (Hooke's Law). However, because of this same recoil, deflation can be passive, recovering the stored elastic potential energy.

In normal life, the lungs operate in mode (b) above. Sitting inside the chest wall, which is sealed completely from the air around us, the lungs have elastic recoil that tends to make them collapse away from the chest wall. However, collapse does not occur because the recoil is balanced by outward springiness of the chest wall. As a result, the lungs stay inflated but to do this there must be a negative pressure between the lungs and chest wall. The act of inhaling air is the result of neural signals from the brain that instruct the muscles of respiration to contract. When these muscles (diaphragm and intercostal muscles) contract, they expand the volume of the chest cavity, thereby further reducing the already negative pressure around the lungs. That causes the lungs to inflate just as would a balloon in a box when the pressure is reduced around the balloon. Exhalation on the other hand is passive — the neural signals to contract the muscles are turned off, allowing passive deflation fueled by the elastic recoil of the lungs. It should be noted that, when higher levels of ventilation are required as during exercise, expiratory muscles can be activated to accelerate the speed of exhalation.

This mode of breathing (external negative pressure) is identical in principle to how the iron lungs of the last century worked. Respiratory muscle contraction was replaced by a mechanical pump that produced similar pressure changes around the chest of a patient enclosed in an iron cylinder in an airtight manner with just the head protruding into the room. This was needed when the respiratory muscles had been paralyzed, most often by polio.

While the lungs normally inflate by negative external rather than by positive internal pressure, it should be clear that either possibility will work. This realization led to positive pressure mechanical ventilation, commonly used during anesthesia and in very ill patients needing assisted ventilation in the intensive care unit. The problem here of course is how to connect the patient to the mechanical ventilator. A tightly fitting mask or a plastic tube inserted into the trachea and sealed tightly are two common ways to do this, but both are invasive and difficult to tolerate in conscious patients.

Finally, it should be noted that ventilation is a convective process much like water flowing down a garden hose from a region of high to low pressure.

Intrinsic design faults in the structure of the lungs and chest wall

From the above descriptions of the structure and ventilatory functions of the lungs, several fundamental design problems should be evident. That they have been largely overcome (in health) is testimony to evolution; however they become clinically very significant in many lung diseases. Evolution has not yet coped with the added insult of lung disease very well.

(1) Pneumothorax

Recall that the lungs are elastic and want to collapse away from the chest wall around them, an event that would be disastrous because breathing depends on lung inflation. It should come as no surprise that if the chest wall is injured enough that a hole allowing air into the chest cavity develops, the lungs will deflate immediately as the negative pressure around the lungs is lost. The same end result happens if alveoli at the lung surface tear. In the balloon-in-a-closed-box analogy, the balloon will collapse: (a) if a hole is drilled into the box, or (b) if the balloon tears. Either alone will suffice. This event is called a pneumothorax and can be rapidly fatal. Luckily, in humans, the right and left lungs are contained in separate chest cavity regions — they are each sealed individually. So if collapse occurs on, say, the right side from a knife wound to the chest, the left lung can continue to breathe and keep the patient alive.

(2) Deadspace

A far less ominous, but ever-present, defect in design results in what we call deadspace. It is simply the consequence of an in-and-out system of ventilation through the hollow airways that feed each functional lung unit. The airways are just that — plumbing — and have no ability to exchange O_2 or CO_2 themselves. Recall that there are some 16–17 generations of these airways. In a normal human, the total, summed volume of air contained in all members of all generations of airways is about 150 ml. Thus, if tidal volume is 500 ml (see above), the last 150 ml of each breath taken in simply stays in these airways and does nothing for gas exchange (hence the name "dead" space). Put another way, deadspace is 30% of tidal volume (for these particular numbers). That is a heavy tax on the energy-requiring process of muscle contraction needed to effect inhalation. We have to breathe 30% more, and expend 30% more energy, on breathing than if ventilation was, like blood flow, a through-and-through process (where the branching system would not constitute deadspace). In health this is a negligible energy burden, but in many diseases of the lungs, this becomes a clinically significant problem.

(3) Maldistribution

When there are 16–17 generations of dichotomously branching airways resulting in about 100,000 functional lung units as mentioned earlier, it is impossible to guarantee equal distribution of inhaled gas to all. No class of 100,000 students will have identical height, weight or blood pressure over all members. Thus the distribution of ventilation must be to some extent uneven. Suppose that at any airway branch point the air is not divided exactly 50/50, but that one branch gets $(50 + X)\%$, while the other therefore gets $(50 - X)\%$ of the air. If for illustration purposes, X is the same number systematically at each and every one of the, say, 17 branch points from the mouth down to the function lung units, the ratio of volume of air delivered to the extreme unit receiving $(50 + X)\%$ at every branch point to that in the other extreme unit receiving $(50 - X)\%$ at each branch point must be: ventilation ratio = $[(50 + X)/(50 - X)]^{17}$. If X is only 1.0 (a 2% failure of perfectly even distribution), the ventilation ratio of the best to worst ventilated units will be $(51/49)^{17}$, or 1.974. If X is 2.0, the ratio increases to 3.899. Clearly, minor inequalities perpetuated along all branch points have the potential to make huge disparities in the amount of air delivered to the various functional lung units. It continues to amaze many authors that in the face of this potential design disaster, there is actually very little ventilatory unevenness throughout the lungs in normal people. However, in many lung diseases, especially asthma and emphysema, this becomes a huge problem, and can be so severe as to be fatal.

(4) Particle deposition

The next intrinsic flaw relates to particle deposition in the lungs. The air we breathe is laden with small particles and gases many of which can injure the tissues of the airways and lungs when inhaled if they come in contact with the airway lining cells. An intrinsic outcome of the 17 orders of airway branching is that with each successive branch, each airway segment becomes not only shorter but also narrower. Despite this, the forward velocity of air during inhalation slows progressively as air moves down the airway tree from mouth to functional lung unit. This is because the total, summed cross-sectional area of each of the 17 generations increases exponentially with generation number. This in turn reflects the binary increase in airways at each generation: 1, 2, 4, 8, etc. for 17 branchings. This increase in number far outweighs the reduction in diameter mentioned above and explains the increase in cross-sectional area. Because the actual gas volume passing through each generation must be constant as cross-sectional area increases (the gas has nowhere else to go), velocity must decrease.

Thus, we have a situation of ever decreasing airway diameter combined with ever decreasing airflow velocity as air gets deeper into the lungs. Both factors separately will favor increasing particle deposition on the wet airway walls, leading to tissue injury deep in the lungs, if those particles are injurious in nature. A classic example of this is coalminers' "black lung" disease.

This design fault has led to development of the muco-ciliary transport system in the airways. The 17 generations of airways contain mucous glands and cells in their walls. They secrete mucus onto the cell surfaces lining the airways, and then a highly integrated ciliary beating mechanism transports this mucus proximally to the pharynx where it is either swallowed or expectorated as sputum (phlegm). The mucus acts like flypaper in attracting and holding small particles that might otherwise prove injurious to the cells in the lung. While this process is beneficial in clearing inhaled matter from the lungs, when too much mucus is produced it can overwhelm the ciliary transport system and obstruct the airways. This is often seen in both asthma and chronic bronchitis.

(5) Airflow resistance

For exactly the same anatomical reason as explained in (4), the summed resistance to airflow provided by all members of a given airway generation *decreases* exponentially as airways branch into smaller and smaller segments. Do not be confused by the fact that each segment, in becoming narrower than its parent, has a higher *individual* airflow resistance than its parent. It is the binary expansion of airway numbers with each generation that determines the outcome under discussion. Put another way, if one were to measure the intraluminal pressure in each of the 17 airway generations during constant flow as during inhalation, most of the pressure drop occurs in the first few generations from the mouth on down, even though these are the largest airways individually. This is of course a good thing in health — not having to provide large pressures to force air through narrow peripheral airways. But in disease, where one needs to measure airway resistance, the fact that most of the pressure drop occurs in the large, proximal airways means that disease of the small peripheral airways is very hard to assess.

(6) Alveolar instability

A final and potentially life-threatening design flaw is in the alveolar structure itself. You may have asked earlier why the lung has been divided up into so many (about 300 million) very small (about 300 micron diameter) alveoli. The answer is that for passive diffusion of O_2 and CO_2 to allow a high enough flow of gas between alveoli and blood, there must be a very large surface area through which the gas flows (Fick's law of diffusion). Because surface area of a sphere is proportional to radius squared, while its volume is proportional to radius cubed, one can pack a large total surface area (80 square meters) into a small total volume (4 liters) by having 300 million small alveoli rather than one very large one. However, like a collection of soap bubbles, this kind of structure is inherently unstable, and is subject to collapse. There is an air/liquid interface over the inner surface of every (roughly spherical) alveolus. This curved surface generates surface tension that works in a direction to shrink the surface and thereby promote collapse of the alveoli. Classic arguments based on the laws of surface tension have small bubbles blowing up bigger bubbles and thus themselves collapsing. A collapsed alveolus is an alveolus that cannot engage in gas exchange.

Why do we not all have collapsed lung areas then? There are two principal reasons. One is that unlike grapes on a stalk, the entire set of 300 million alveoli do not just hang in space but are attached to one another and in fact share common walls just as do two neighboring bedrooms inside a house. What that in turn does for stability is to enhance it. As soon as a given alveolus might start to collapse, its physical attachment to its neighbors helps splint or stabilize it, opposing collapse by direct connection. This is called mechanical interdependence. The second reason is the presence of a chemical compound lining the alveolar surface. This is surfactant, and is a lipoprotein that is secreted by alveolar epithelial cells. It lowers surface tension an order of magnitude below that of water, and thus reduces the collapsing tendency. It also makes the effort of inhaling much less than it would otherwise be because so much less work needs to be done when surface tension is low. The breathing disease of premature babies (respiratory distress of the newborn) is in fact caused specifically by absence of surfactant at the time of birth. It is characterized by collapsed alveoli that fail to provide gas exchange, and was fatal until surfactant replacement therapy was developed.

1.2.2 *Blood flow*

Since the entire cardiac output must flow through a single organ, i.e. the lungs, before being distributed to the rest of the body, one might at first think that the pulmonary circulation must operate under a very high pressure, and that the systemic circulation, due to its multiple parallel circuits that deliver blood to the many tissues and organs, can operate at much lower pressures. In fact the converse is true. The systemic circulation is the high pressure circuit and the pulmonary circulation operates at low pressure.

The likely reasons for the high pressures in the systemic circulation are: (a) the roughly two meter height span of normal humans requiring pressures that are high enough to perfuse the brain in particular, and (b) the need for tight control of blood flow distribution among organs according to need at any given moment — to the skin when hot, to the muscles when active, to the gut when digesting and so on. When all tissues in parallel contain arteries that have muscular control over arterial diameter and thus resistance, regional flow can be regulated according to need.

The likely reason why the pulmonary circuit is at low pressure is that the delicate alveolar structure would be destroyed by high vascular pressures that might rupture poorly supported blood vessels. Support is poor because unlike most other (solid) tissues, the blood vessels are contained in the very delicate alveolar wall. There is no biological need to have precise control over distribution of flow as the entire cardiac output needs to flow through the lungs. And, finally, there already is a mechanism for at least partial blood flow distribution regulation in the lungs, known as hypoxic pulmonary vasoconstriction. This process causes vasoconstriction in a region of lung where the O_2 concentration is low. This in turn helps to even out flow distribution anomalies between regions, and improve gas exchange (see below).

Because the pulmonary vascular tree seems anatomically so similar to the airways in numbers and sizes of branchpoints and segments, one might think that the airways design issues discussed above can easily be transposed to the blood vessels. However, this is not the case. The vascular tree mentioned earlier is a flow-through system such that deadspace does not exist. In addition, it is not directly exposed to the environment as are the airways. There is also no air/liquid interface, and flow occurs not through a system dependent on external negative pressure but rather on cardiac function. The major point of similarity is in the potential for maldistribution of blood flow among the roughly 100,000 functional lung units. Even in health, there is significant blood flow heterogeneity. This is in two forms. First, because of the weight of blood, there is a gravitational basis for uneven distribution, with more blood flowing to dependent than non-dependent regions. Second, as with the airways, there are intrinsic structural differences throughout the lung causing further unevenness of flow distribution.

Thus, the distributions of both blood flow and of ventilation are predictably non-uniform even in the normal lungs. Some regions of the lungs will be over-ventilated and under-perfused. Others will be under-ventilated and over-perfused. Overall, there will be a range of values for the *ratio* of ventilation to blood flow. In health, this range has been found to cover about one decade from about 0.3 to about 3, with a mean close to 1. (The mean of 1 reflects the fact that total ventilation is about 6 liters/min and total pulmonary blood flow is also about 6 liters/min. This gives an overall ratio of about 1.) Why the ratio of ventilation to blood flow, called the $\dot{V}/\dot{Q}$ ratio, is important will become apparent below when gas exchange is discussed.

There is one other design flaw in the pulmonary circulation. It is the target capillary network for venous blood coming from all the body's tissues, via their veins. All too commonly, venous thrombosis occurs in leg and pelvic veins in susceptible patients. These blood clots often break apart in the leg or pelvic vein, and then are swept into the venous system and transported to the lungs like a branch of wood floating down a river. When such a clot enters the pulmonary circulation it obstructs an artery of about its own size. This causes two adverse events — an increase in pulmonary vascular resistance, making the right heart work harder to pump blood into the lungs, and maldistribution of blood flow causing the range of $\dot{V}/\dot{Q}$ ratios to broaden. This will interfere with O_2 and CO_2 exchange and if severe, can be fatal.

1.2.3 *Diffusion*

To this point we have discussed some of the basic principles underling the provision of ventilation and blood flow, and related them to the innate structure of the lungs. While the structure does allow the lungs to perform their allotted duty of exchanging the O_2 and CO_2 necessary for cell metabolism around the body, it has been pointed out that the design of the lungs, while adequate in health, is far from optimal in the presence of lung disease. To finish this part of the story we must discuss the process of diffusion. Recall that ventilation, largely a convective process fueled by chest wall muscle contraction, brings fresh air down to the alveolar wall surfaces and that blood flow, also a convective process and fueled by cardiac function, distributes O_2 to all of the tissues of the body, starting from the lung capillaries.

The question is how O_2 and CO_2 move between the alveolar gas and the lung capillaries, and it was previously stated that this was accomplished by simple diffusion. Thus, O_2 moves by diffusion from alveolar gas through the alveolar wall and into the capillary ending up attached to hemoglobin inside the red cells (Fig. 3). In turn, O_2 traverses: (a) the surfactant fluid lining layer (not seen in Fig. 3), (b) the thin cytoplasm of one layer of alveolar epithelial cells, (c) the alveolar interstitial region where the collagen and elastin matrix sits, (d) the thin cytoplasm of one layer of capillary endothelial cells, (e) the plasma inside the capillary, and (f) the red blood cell interior where the hemoglobin is carried. CO_2 traverses the same path, but in the opposite direction because it is coming out of the blood. And of course, CO_2 is carried in the blood in forms additional to direct combination with hemoglobin. Thus for O_2, about 98% is carried bound to hemoglobin, the rest being physically dissolved in the water in the blood. For CO_2, about 90% is carried converted to bicarbonate ion. The remaining 10% is

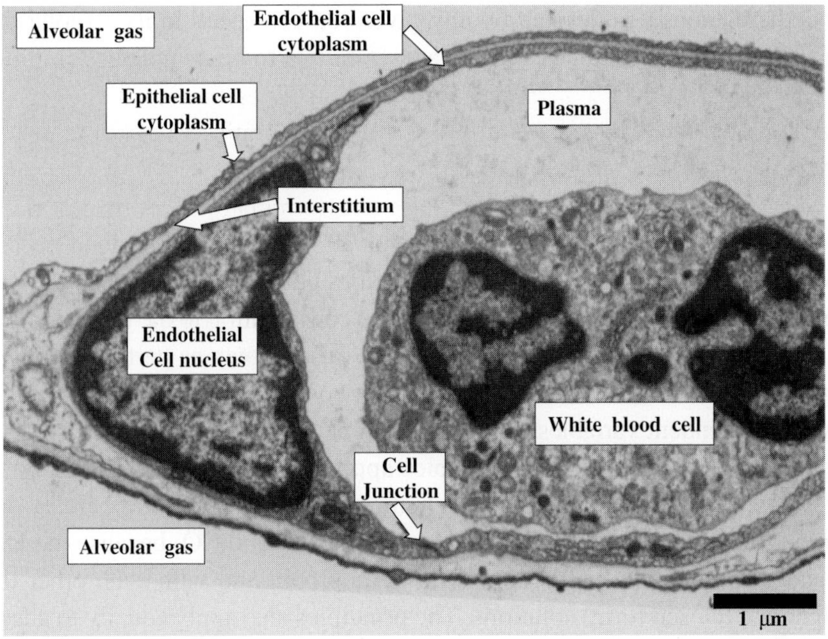

Fig. 3. Transmission electron micrograph showing the alveolar wall separating alveolar gas from capillary blood. The main point to notice is how thin the blood gas barrier is — perhaps one-third of one micron at its thinnest. This is necessary to enable adequate diffusive gas exchange.

carried split roughly equally between two forms — hemoglobin-bound (carbamino hemoglobin) and physically dissolved in the water of the blood.

While the successive structures through which these gases must diffuse are biologically and chemically different and will have different gas diffusivities, it suffices for teaching purposes to consider this pathway as simply reflecting a single resistance to (or conductance supporting) diffusion. We can call the conductance of this compound "membrane" (D). Fick's law of diffusion can now be applied to this system in its simplest fashion:

The net flux by diffusion of a gas from the alveolar gas into the capillary blood per unit time will depend on the conductance D. D in turn varies directly with the total lung capillary surface area available for diffusion, A, and inversely with the thickness (d) of the "membrane".

The net flux is also proportional to the difference in partial pressures between alveolar gas and capillary blood, ($P_{\text{alveolar}} - P_{\text{capillary}}$). In fact, the overall net flux is the product of D and ($P_{\text{alveolar}} - P_{\text{capillary}}$) — that is Fick's law.

Both surface area and membrane thickness play an intuitively reasonable role in diffusion and have been long substantiated as important. Fick's law allows us to develop a simple first order linear differential equation describing the time course of gas flow across the alveolar wall. We can answer, with adequate input data, the questions of: (a) what is the time course of diffusive movement of any gas across the alveolar wall and (b) is there enough time spent by any given red blood cell in its alveolar capillary transit to allow diffusive transfer of the gases it carries to be completed?

1.2.4 *Gas Exchange Equations*

To this stage, we have covered all the anatomical and functional ground necessary to begin building the quantitative model that describes pulmonary gas exchange. Basically, one needs only to apply simple equations of mass conservation to the three domains considered — ventilation, diffusion, and blood flow (also known as perfusion). In fact, for any single gas, these considerations result in just two equations and it turns out that there are only two dependent variables, assuring a unique solution to the system.

Importantly, the concepts and principles apply not only to O_2 and CO_2 but also to all gases. Gases that do not combine with hemoglobin (which we will here by define as inert gases) form an easier model to develop than do O_2 and CO_2 because the latter two gases engage in complex interactions with hemoglobin and with each other that preclude algebraic (closed form) solutions. The principles also apply equally to gases being taken in from the air as to those being eliminated from the blood (such as CO_2, or anesthetic gases washing out after an operation, for example).

The two equations mentioned describe: (1) the diffusive transport of the gas across the alveolar wall and (2) the convective transport of the gas between blood and external air as effected by the processes of ventilation and blood flow. It should be apparent that the diffusive and convective parts of gas transport must "talk to each other" — in the end, a given gas undergoes ventilation, diffusion and perfusion in interdependent sequence that results in a transport rate that is the same across all three phases.

Figure 4 is a simple model of the alveolus showing the alveolar gas, the capillary blood, and the alveolar wall (across which diffusion occurs) separating them. To set the stage for developing the diffusion equation, the question is to write an equation whose solution indicates the time course of change in gas partial pressure along the capillary

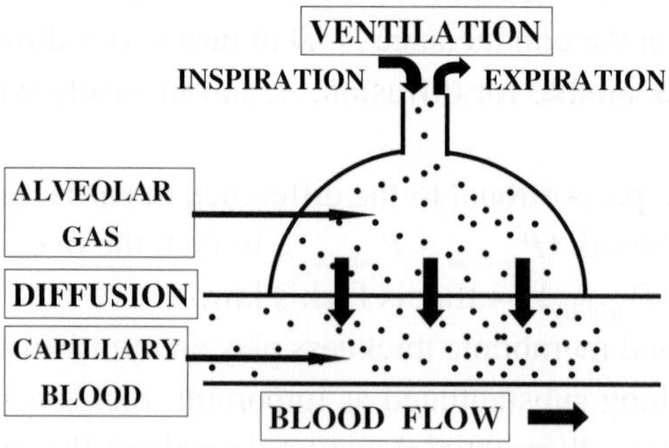

Fig. 4. Simple model of an alveolus showing all of the principal transport pathways — ventilation, diffusion, and perfusion (blood flow).

from the input (pulmonary arterial) end (left side) to the output (pulmonary venous) end (right side). In this construct, the alveolar gas is considered well-mixed spatially because diffusion in the gas phase is some 10^5 times faster than in tissue. In addition, it is taken to be constant in time because ventilation rapidly replenishes gas absorbed into the blood in a steady state. Independent variables (also taken to be both constant and known) are the diffusive conductance, D, incorporating structural dimensions (area A and thickness d of the membrane, as above); alveolar and pulmonary arterial partial pressures of the gas; solubility of the gas in tissue (or blood, generally considered to be the same). Dependent variables are the capillary partial pressure along the capillary, and the resulting net amount of gas transported across the membrane per unit time by the diffusive process, which can be calculated once the capillary partial pressure has been determined.

Diffusion equation

As detailed above, the instantaneous net rate of transport, $\dot{V}(t)$, of a gas by diffusion across the alveolar membrane, depends on (1) the diffusive conductance, D, which in turn depends on the total lung capillary surface area available for diffusion, A, and the thickness, d, of the "membrane" separating alveolar gas from capillary blood and (2) the difference in partial pressures between alveolar gas and capillary blood, $(P_{\text{alveolar}} - P_{\text{capillary}})$. Therefore,

$$\dot{V}(t) = D \times [PA - Pc(t)] = K \times A \times [PA - Pc(t)]/d. \tag{1}$$

K is a constant such that $D = K \times A/d$. PA is alveolar and $Pc(t)$ is capillary gas partial pressure at time t along the capillary, starting at $t = 0$ at the pulmonary arterial input end of the capillary.

$\dot{V}(t)$ can be expanded because it represents the instantaneous rate of change of concentration of the diffusing gas in the flowing capillary blood, $Cc(t)$. Thus,

$$\dot{V}(t) = (Vc/100) \times dCc(t)/dt. \tag{2}$$

Here, since $Cc(t)$ is expressed per unit volume of blood (traditionally in dl), yet A and d reflect whole lung values, we need to express $\dot{V}(t)$ for the whole lung. Vc is total capillary blood volume, in ml. Accordingly,

$$dCc(t)/dt = 100 \times K \times A \times [PA - Pc(t)]/(d \times Vc). \tag{3}$$

Now, if we limit the equation to inert gases, defined as gases whose concentration in blood obeys Henry's Law (concentration = solubility × partial pressure), we have:

$$\beta \times dPc(t)/dt = 100 \times K \times A \times [PA - Pc(t)]/(d \times Vc) \tag{4}$$

where β is solubility in blood.

Finally, K is (1) directly proportional to the solubility of the gas in the alveolar wall membrane, generally the same as β. However, because of the absence of hemoglobin in that membrane, yet the presence of hemoglobin in the blood, this may not always be the case, so we shall retain a different symbol, α, for the solubility of the gas in the membrane; (2) inversely proportional to the square root of the molecular weight of the gas. This comes from the kinetic theory of gases which says that the kinetic energy of all gases, $\frac{1}{2}mv^2$, is the same at any given temperature. Thus, the rate of diffusion, v, is inversely proportional to the square root of molecular mass, m. Suppose $K = k \times \alpha/m^{0.5}$. Incorporating all of these points results in:

$$dPc(t)/dt = 100 \times k \times \alpha \times A \times [PA - Pc(t)]/(d \times Vc \times \beta \times m^{0.5}) \tag{5}$$

collecting terms,

$$dPc(t)/dt = [100 \times k \times A/(d \times Vc)] \times [\alpha/(\beta \times m^{0.5})] \times [PA - Pc(t)]. \tag{6}$$

This equation describes the time course of capillary partial pressure change along the capillary in terms of previously defined *lung-related structural variables* (k, A, d, Vc) and *gas-related transport variables* (α, β, m). It is a first order linear differential equation that is easily integrated. Boundary conditions are (1) the capillary partial pressure at the pulmonary arterial input end at time $t = 0$, usually written as Pv and (2) the alveolar partial pressure PA. Both of these partial pressures are considered constant in time and known. Then, the transit time in the capillary, T, will be Vc/Q where Q is total blood flow through the capillary vasculature. At the whole lung level, Q is cardiac output and corresponds to A, d, Vc and k (which all also reflect whole lung values). Integrating Eq. (6) and using these boundary conditions yields:

$$[PA - Pc(t)]/[PA - Pv] = \exp(-W \times t) \tag{7}$$

where

$$W = [100 \times k \times A/(d \times Vc)] \times [\alpha/(\beta \times m^{0.5})]$$

Note that if $PA > Pv$, gas will diffuse from alveolar gas into the capillary blood, causing uptake of gas by the blood, but that if $Pv > PA$, gas will diffuse from capillary blood into alveolar gas, causing gas elimination from the blood. The equation holds for diffusion in either direction.

Finally, the value of Pc at the end of the capillary, when t is $T = Vc/Q$, is:

$$[PA - Pc(T)]/[PA - Pv] = \exp(-DL/(\beta \times \dot{Q})) \tag{8}$$

where

$$DL = [100 \times k \times A/d] \times [\alpha/(m^{0.5})].$$

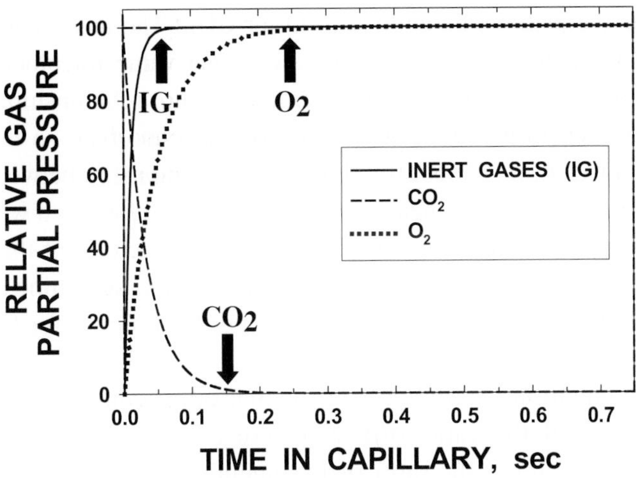

Fig. 5. Time course of change of gas partial pressure in a red cell as it travels along a pulmonary capillary from left to right, a journey of normally 0.75 seconds. Curves are shown for inert gases, CO_2 and O_2 under normal, resting conditions. The ordinate is scaled such that 0 represents incoming pulmonary arterial blood for inert gases and O_2, while 100 represents alveolar inert gas and O_2 levels. For CO_2, the scale is reversed (100 represents pulmonary arterial blood and 0 alveolar gas). Note that inert gases equilibrate (i.e. capillary blood reaches alveolar levels) within about 0.05 seconds, CO_2 equilibrates within about 0.15 seconds, and O_2 by about 0.25 seconds. Time courses are roughly exponential, and all three gases equilibrate well before the end of the capillary transit.

Figure 5 shows the time course for uptake of an inert gas using appropriate values for the independent variables, and assuming that membrane and blood solubilities are equal, that is, $\alpha = \beta$. Note that T is normally about 0.75 seconds because Vc is usually about 75 ml while Q is about 6 l/min. *Thus, inert gases complete their diffusive exchange in the first 50 milliseconds of the capillary transit, well within the available time. This makes their gas exchange perfusion-limited rather than diffusion-limited. Moreover, the time course is independent of absolute gas solubility, because* α *is in the numerator,* β *is in the denominator, and, as long as* $\alpha = \beta$*, they cancel.*

The reason for first discussing inert gases is not simply the theoretical advantage of describing diffusive gas exchange by a single exponential equation, allowing the defining principles to be laid out. There are practical reasons as well, because inert gases are used in human research and clinical care. Gases such as acetylene and nitrous oxide are used in trace concentrations to measure cardiac output because they are flow-limited. If they were diffusion limited, not flow limited, this would be very difficult to do. In addition, several anesthetic gases in current usage function as inert gases in how they are taken up and eliminated, and understanding how they are exchanged is important in determining the correct doses so that patients are neither over- nor under-anesthetized during surgery.

Can the above equations be applied to O_2 or CO_2? The answer is "yes" up to the point of assuming Henry's law (specifying a constant value for blood solubility, β in equation (4)). For CO_2, its concentration/partial pressure relationship is nearly linear, and the error induced in assuming a constant β is unimportant, so the equations can really be used. For CO_2, β is about 0.8. For O_2, the relationship between concentration and partial pressure is quite nonlinear, so no single value of β (which would be the slope of that relationship) will be appropriate. That said, the principles of the above equation system remain correct for O_2, and even using the equations calculating a mean value for β is useful for illustrating the differences between diffusion rates for O_2 (and CO_2) and inert gases.

Referring to equation (7), the only terms that differ compared to inert gases are molecular weight and the ratio α/β. Neglecting small differences in molecular weight, the critical factor is the ratio α/β. Remember that for inert gases, $\alpha = \beta$, so that $\alpha/\beta = 1$. For O_2, $\alpha = 0.003$ ml/(dl-mmHg). This number reflects O_2 solubility in water, the major component of cells and tissues. However, for β, we must look at the O_2Hb dissociation curve and determine the average slope between the pulmonary arterial and pulmonary venous points. This slope comes to about 0.083 ml/(dl-mmHg), because of the 5 ml O_2/dl difference in concentration and 60 mmHg difference in partial pressure between those two points. As a result, α/β for O_2 averages 0.003/0.083, or roughly 1/30.

This approximation says that the diffusion process for O_2 runs some 30-fold more slowly than for all inert gases, as shown in Fig. 5. It means that it takes much longer for the diffusive process to be complete for O_2 compared to inert gases. *It turns out that from numerical analytical calculations, the time to complete O_2 exchange is found to be about 0.25 seconds. While an order of magnitude longer than for inert gases, this is still well within the 0.75 seconds available (see above). Thus, normally, O_2 is also perfusion- rather than diffusion-limited. However, should capillary transit time fall (as it does during exercise), should alveolar surface area fall (as it does in some lung diseases), should the alveolar wall thickness increase (as it does in some lung diseases), or should one ascend to altitude (where β increases because pulmonary arterial and venous partial pressures fall onto the much steeper part of the O_2 dissociation curve because of hypoxia), O_2 can become diffusion-limited.*

What about CO_2? For this gas, $\alpha = 0.067$ ml/(dl-mmHg) and $\beta = 0.8$ ml/(dl-mmHg). The ratio α/β is thus about 1/12. As a result, CO_2 exchanges faster than O_2 but considerably slower than inert gases. It turns out that time to complete exchange is about half that of O_2, and CO_2 is rarely if ever found to be diffusion limited (Fig. 5).

An interesting case is the highly poisonous gas carbon monoxide, CO. For this gas, $\alpha = 0.0024$ ml/(dl-mmHg), close to that for O_2 above. However, β is very different. CO binds to Hb at the same sites as does O_2, but about 240 times more avidly (i.e. same concentration at a partial pressure 240 times lower). Thus, α/β comes to about 1/8300. As this calculation suggests, CO is highly diffusion limited even when transit time is normal at 0.75 seconds. In fact, after 0.75 seconds, exchange is less than 5% complete.

This result has important implications. First, as CO is inhaled from a source such as car exhaust, it takes longer to reach high, toxic levels than if it exchanged at the rate of O_2. That is of course good. But, when CO is being eliminated during recovery from poisoning, it also takes longer to get rid of because of slow diffusional transport, delaying recovery and possibly increasing hypoxic tissue damage. Second, the very high degree of diffusion limitation of CO makes it the best gas to use to measure the diffusing properties of the lungs. Of course, in so doing, we use only trace concentrations of CO and for only a few seconds, to avoid CO poisoning. Indeed, for about a century, CO has been a widely used gas for just this measurement, and has become very useful in both diagnosis and monitoring treatment of many lung diseases, reflecting their severity.

Convection equation

We need to put aside considerations of diffusion for now, and assume that all gases do complete diffusive exchange within the 0.75 seconds available transit time. Thus, in the above terminology, $PA = Pc(T)$. In words, this means that the partial pressure in the capillary at the end of the 0.75 second exposure to alveolar gas equals that in the alveolar gas. We also assume a steady state of gas exchange and constant values in time for all variables (described below).

We can then write two simultaneous mass-conservation equations for overall transfer rate ($\dot{V}$) of a gas: one for exchange between the air around us and the alveolar gas, and a second for exchange between the alveolar air and the capillary, as follows:

$$\dot{V} = \dot{V}_A \times FI - \dot{V}_A \times FA \tag{9}$$

and

$$\dot{V} = \dot{Q} \times Cc - \dot{Q} \times Cv. \tag{10}$$

In these equations, $\dot{V}_A$ is total alveolar ventilation (total ventilation minus that left in the conducting airway deadspace discussed earlier). $\dot{Q}$ is again cardiac output, as used before. *FI* is the fractional concentration of the gas in the inhaled air, and *FA* the fractional concentration of the same gas in the alveolar gas region. Note that this will be the concentration of the gas in the exhaled breath. For O_2, present in the air at 21%, *FI* would simply be 0.21. *Cc* and *Cv* are the gas concentrations in the capillary blood at the outflowing and inflowing ends of the pulmonary capillary, respectively.

These two equations are expressions of mass conservation. equation (9) says that the amount of gas taken up by the blood from the surrounding air is the difference between the amount inhaled and the amount exhaled, all in any given time period. equation (10) says that the amount of gas taken up can also be expressed as the difference between the amount leaving the lungs (in the outflowing blood) and that entering the lungs (in the inflowing blood). In a steady state, equations (9) and (10) are simultaneous and $\dot{V}$ is the same in both. Equating them, we get:

$$\dot{V}A \times [FI - FA] = \dot{Q} \times [Cc - Cv]. \quad (11)$$

Note that for gas being taken up from the air (such as O_2), $FI > FA$ and $Cc > Cv$. For a gas being eliminated (such as CO_2), $FI < FA$ but $Cc < Cv$. Thus the equations work in either direction with the negative signs canceling for elimination. Equation (11) is rearranged:

$$\dot{V}A/\dot{Q} = [Cc - Cv]/[FI - FA]$$

or

$$\dot{V}A/\dot{Q} = k \times [Cc - Cv]/[PI - PA] \quad (12)$$

where k is a constant relating partial pressure P to fractional concentration F (Dalton's law of partial pressures). In this equation, we take as independent and known quantities $\dot{V}A$, $\dot{Q}$, k, Cv, PI and the relationship between partial pressure and concentration in blood for the gas in question (for an inert gas this means the solubility, β, as discussed above; for O_2 or CO_2 or CO, their blood dissociation curves). Thus, PA becomes the only unknown and can be calculated easily.

This critically important equation says that the alveolar partial pressure PA (and, as a result, $\dot{V}$ from equation (9)) is determined by three interacting factors:

(1) the ratio of alveolar ventilation to blood flow, $\dot{V}A/\dot{Q}$;
(2) the boundary conditions (i.e. values for inspired and pulmonary arterial gas levels as indicated above by PI and Cv); and
(3) the binding characteristics of the gas in blood (β or the dissociation curve).

The ratio $\dot{V}A/\dot{Q}$ is called the ventilation/perfusion ratio and is of major importance. For a given gas and set of boundary conditions, variation in regional $\dot{V}A/\dot{Q}$ ratio directly causes changes in regional gas levels. For O_2, when $\dot{V}/\dot{Q}$ is low so too is alveolar PO_2; when $\dot{V}A/\dot{Q}$ is high, so too is alveolar PO_2. The separate solutions to equation (12) for O_2 and for CO_2 are given in Fig. 6 as the solid curved lines, one for each gas. Here the vertical dashed line is positioned at a $\dot{V}A/\dot{Q}$ ratio of 1.0, the normal average value in health. Where it intersects the O_2 and CO_2 curves, one may read off the corresponding values of PO_2 and PCO_2, shown by the horizontal dashed lines. Thus, normal alveolar PO_2 is about 100 mmHg, and normal alveolar PCO_2 about 40 mmHg.

When $\dot{V}A/\dot{Q}$ ratios vary throughout the lungs, $\dot{V}A/\dot{Q}$ inequality is said to exist, and gas exchange is impaired. As mentioned earlier, even normal lungs show some variation

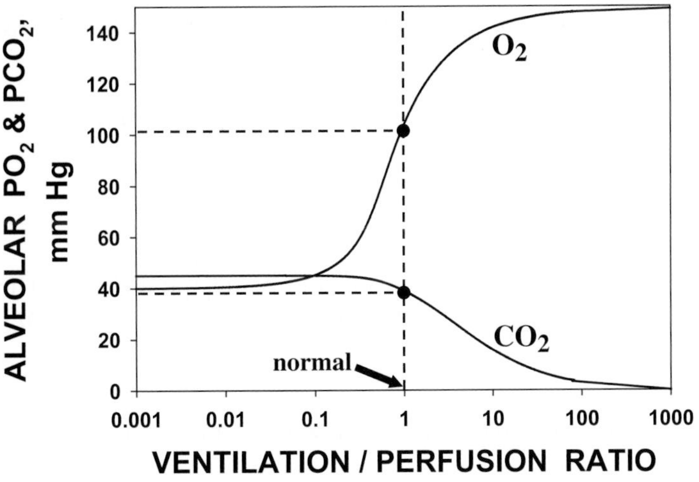

Fig. 6. Dependence of alveolar Po_2 and Pco_2 on the ventilation/perfusion ratio. The normal $\dot{V}A/\dot{Q}$ ratio of about 1 results in alveolar Po_2 of 100 and Pco_2 of 40 mmHg. When $\dot{V}A/\dot{Q}$ ratios fall, Po_2 falls and Pco_2 rises. The opposite is seen when $\dot{V}A/\dot{Q}$ ratio rises. When $\dot{V}A/\dot{Q}$ ratio approaches zero, gas partial pressures approach those of pulmonary arterial blood; when the ratio approaches infinity, gas partial pressures approach those of inspired gas.

in regional $\dot{V}A/\dot{Q}$ ratios, but the spread is insufficient to significantly affect overall gas exchange. This is not the case in many lung diseases, where $\dot{V}A/\dot{Q}$ inequality disrupts gas exchange in a major way, sometimes sufficiently to cause death from hypoxia.

Equation (12) can be solved for O_2 and CO_2 only using numerical analysis, but for inert gases, a closed form solution can be derived, as was also the case for diffusion. Thus, if we use the fact that from Henry's law, $Cv = \beta \times Pv$ and $Cc = \beta \times PA$, where β is inert gas blood solubility, equation (12) becomes:

$$\dot{V}A/\dot{Q} = k\beta \times [PA - Pv]/[PI - PA]. \quad (13)$$

We often use the symbol λ for the product $k\beta$; λ is known as the blood : gas partition coefficient. When terms are rearranged to make PA the subject, we have:

$$PA = [PI \times \dot{V}A/\dot{Q} + Pv \times \lambda]/[\dot{V}A/\dot{Q} + \lambda]. \quad (14)$$

This algebraic solution powerfully demonstrates how PA depends on the three factors listed above (namely, $\dot{V}A/\dot{Q}$ ratio; partition coefficient (λ) and boundary conditions (PI and Pv)).

Diffusion and convection together

Equation 8 (diffusion) and equation (13) (convection) can be combined when diffusion limitation is present, meaning that the assumption that alveolar (PA) and end-capillary

($Pc(T)$) partial pressures are the same would not be correct. As derived above, we had:

$$[PA - Pc(T)]/[PA - Pv] = \exp(-DL/(\beta \times \dot{Q})) \tag{8}$$

$$\dot{V}A/\dot{Q} = \lambda \times [Pc(T) - Pv]/[PI - PA]. \tag{13}$$

This is a set of two equations in two unknowns (PA and $Pc(T)$). The known variables are $\dot{V}A$, $\dot{Q}$, k, β (hence λ), Pv, PI and DL. If we set $K = \exp(-DL/(\beta \times \dot{Q}))$, and solve, we get:

$$PA = [PI \times \dot{V}A/(\dot{Q}(1-K)) + Pv \times \lambda]/[\lambda + \dot{V}A/(\dot{Q}(1-K))] \tag{15}$$

and

$$Pc(T) = [PI \times \dot{V}A/\dot{Q} + Pv \times (\lambda + K \times \dot{V}A/(\dot{Q}(1-K)))]/[\lambda + \dot{V}A/(\dot{Q}(1-K))]. \tag{16}$$

These two equations show the complex interactions among diffusion and convection that are possible when both diffusion limitation ($K > 0$) and $\dot{V}A/\dot{Q}$ ratio affect gas exchange at the same time. Note that when diffusion limitation is not present, $K = 0$ and equations (15) and (16) reduce to the same relationship, i.e. equation (14). When diffusion limitation is complete such that there is no gas exchange, $K = 1$ and PA equals PI while $Pc(T)$ equals Pv. Figure 7 shows the solutions to equations (15), for alveolar gas (solid lines), and (16), for endcapillary blood (dashed lines) for inert gases being taken up from the atmosphere into the blood. In each panel, calculations are shown for no diffusion limitation ($K = \exp(-D/(\beta Q)) = 0$); mild diffusion limitation ($K = 0.25$); moderate diffusion limitation ($K = 0.50$); severe diffusion limitation ($K = 0.75$); and almost complete diffusion limitation ($K = 0.99$ and 0.999). The top two panels reflect a gas of low solubility (partition coefficient = 0.1). The middle two panels reflect a gas of medium solubility (partition coefficient = 1.0), and the lower two panels a gas of high solubility (partition coefficient = 10.0). The left hand panels show alveolar and capillary partial pressures as a function of ventilation/perfusion ratio, while the right hand panels show total gas volume flux through the lungs also as a function of ventilation/perfusion ratio. Note that this total flux is pictured as a percentage of that occurring in the complete absence of diffusion limitation.

These complex figures show in particular that:

(1) For any given gas, no matter its solubility, a given degree of diffusion limitation impairs gas flux more in alveoli with high than low $\dot{V}A/\dot{Q}$ ratio.
(2) A poorly soluble gas is impaired more than a soluble gas at a given $\dot{V}A/\dot{Q}$ ratio (for any given degree of diffusion limitation).

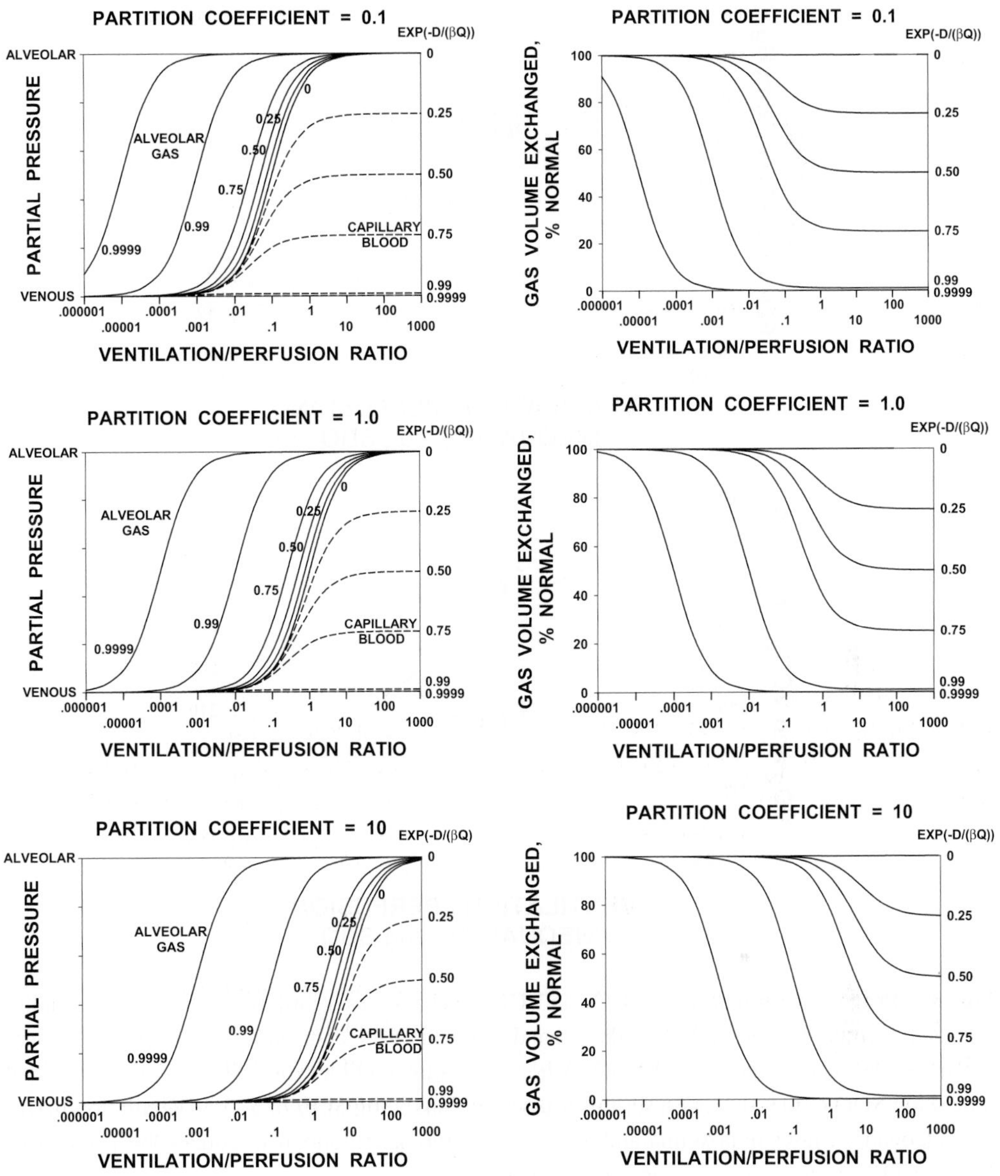

Fig. 7. Graphical solutions to equations (15) and (16), for alveolar and endcapillary partial pressures respectively for gases of low, medium and high solubilities, when diffusion limitation exists in lung units of varying ventilation/perfusion ratio. See text for detailed description.

1.2.5 *Consequences of $\dot{V}A/\dot{Q}$ inequality for gas exchange*

As mentioned above, when the $\dot{V}A/\dot{Q}$ ratio is not everywhere the same throughout the lungs, $\dot{V}A/\dot{Q}$ inequality is said to exist. When this happens, exchange of all gases — O_2, CO_2, and inert — is impaired. This impairment is manifested in several ways: by a

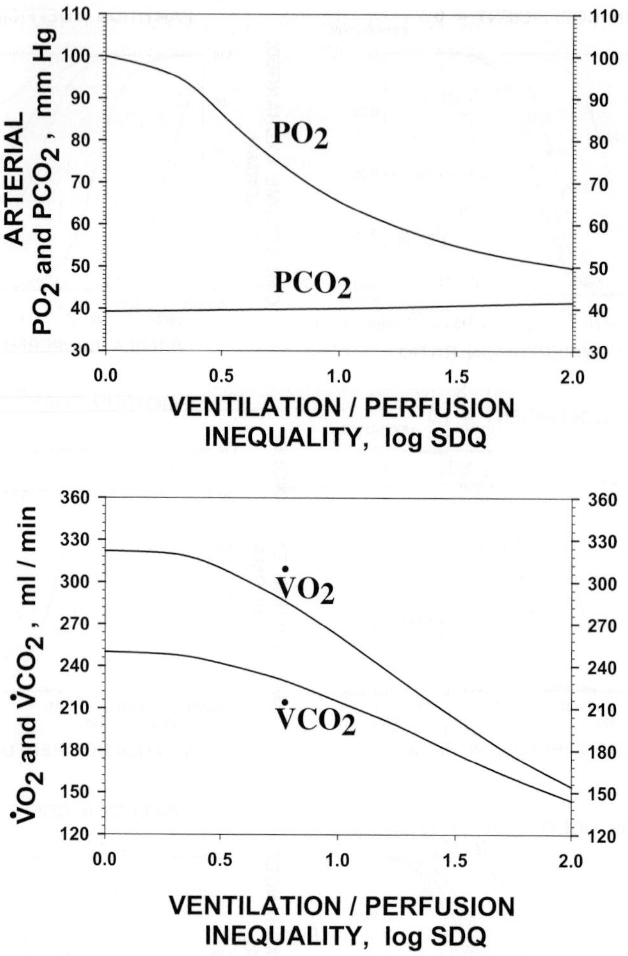

Fig. 8. Progressive impairment of O_2 and CO_2 exchange as ventilation perfusion inequality worsens. Inequality is described by the second moment of the distribution of $\dot{V}_A/\dot{Q}$ ratios (logarithmic scale), abbreviated to log SDQ. A homogeneous lung has a log SDQ of zero, a normal lung a log SDQ of 0.3–0.4 and a seriously diseased lung a log SDQ of 2–2.5. Ventilation/perfusion inequality causes hypoxemia and hypercapnia (top panel) and also reduces the volumes of O_2 and CO_2 the lungs can exchange (lower panel).

reduction in arterial P_{O_2} (hypoxemia), increase in arterial P_{CO_2} (hypercapnia) and reduction in the volumes of these two gases that the lungs can take up/eliminate.

Figure 8 shows how progressively more severe $\dot{V}_A/\dot{Q}$ inequality disturbs gas exchange, causing the arterial P_{O_2} to fall, the arterial P_{CO_2} to rise, and the ability of the lungs to exchange both O_2 and CO_2 (that is, $\dot{V}_{O_2}$ and $\dot{V}_{CO_2}$) to fall. In each panel, the abscissa is a parameter of the extent of ventilation/perfusion inequality called log SDQ This is the second moment, or dispersion, of the distribution of blood flow ($\dot{Q}$) on a logarithmic scale. When the distribution is normal, this parameter is the same as the standard deviation (SD). A homogeneous lung in which every gas exchange unit has the exact same $\dot{V}A/\dot{Q}$ ratio would have log SDQ = zero. For reference, a normal human would have a value of about 0.3–0.4. A patient with moderate lung disease would have a log SDQ ~ 1, while in severe lung disease, log SDQ is often 2 to 2.5. What Fig. 8 shows (top panel) is how arterial PO_2 falls and PCO_2 rises as $\dot{V}A/\dot{Q}$ inequality becomes more severe. The fall in PO_2 is substantial, while the rise in PCO_2 is much less. The difference is explained by the much steeper and more linear CO_2 dissociation curve compared to that of O_2. Note however that the relative reduction in gas volumes exchanged (lower panel) is about the same for the two gases, again explained by their dissociation curves. Thus, because that of CO_2 is steep, a small rise in PCO_2 equates to substantial interference with CO_2 elimination.

The calculations of Fig. 8 are done on the presumption that the body would not compensate for these changes as more and more inequality develops. In reality, there are three main ways in which the body will compensate: (1) reducing venous PO_2 and increasing venous PCO_2 to re-establish the normal arterio-venous gas concentration differences, (2) increasing ventilation to raise alveolar PO_2 and lower alveolar PCO_2, and (3) increasing cardiac output to raise mixed venous PO_2 and reduce mixed venous PCO_2. By a combination of these three compensating mechanisms, $\dot{V}O_2$ and $\dot{V}CO_2$ can be restored to normal, and arterial PCO_2 can also be normalized. However, arterial PO_2 generally fails to normalize. Again, the ability of CO_2 to normalize, and the failure of O_2, are both explained by the shapes and slopes of their dissociation curves.

1.2.6 *Implications for disease: causes of arterial hypoxemia and hypercapnia*

The entire preceding analysis can be summarized in terms of the causes of arterial hypoxemia and hypercapnia. There are four such causes emerging from this analysis:

(1) *Hypoventilation.* Equation (9) shows that for a given metabolic rate ($\dot{V}O_2$ and $\dot{V}CO_2$), the level of alveolar ventilation determines alveolar gas concentrations (and thus partial pressures). Applied to O_2 uptake and CO_2 elimination, a reduction in $\dot{V}A$ reduces alveolar and thus arterial PO_2, and increases alveolar and thus arterial PCO_2. This is typically seen in patients overdosing with narcotic drugs that depress the drive to breathe.

(2) *Diffusion limitation.* As equation (8) shows, if diffusional conductance is sufficiently reduced, the PO_2 in the capillary may never reach that of alveolar gas, resulting in hypoxemia. However, as suggested by Fig. 5, CO_2 is much less vulnerable than O_2 to this and hypercapnia is not seen. Diffusional limitation is seen in patients with lung fibrosis both because there are fewer capillaries and greater distances for O_2 to diffuse due to collagen deposition in the alveolar walls.

(3) *Ventilation/perfusion inequality.* When all alveoli enjoy the exact same $\dot{V}A/\dot{Q}$ ratio, there is no $\dot{V}A/\dot{Q}$ inequality and gas exchange is maximally efficient. However,

when areas of below normal and above normal $\dot{V}A/\dot{Q}$ ratio develop as a result of disease, the distribution of $\dot{V}A/\dot{Q}$ ratios is now heterogeneous. This inequality interferes with exchange of all gases during either uptake or elimination. This interference is manifested as a fall in arterial Po_2 (hypoxemia), an increase in arterial Pco_2 (hypercapnia), and a reduction in both O_2 and CO_2 fluxes ($\dot{V}o_2$ and $\dot{V}co_2$) through the lungs, as shown in Fig. 8. Ventilation/perfusion inequality is seen in all types of lung disease, and is the most common and serious cause of hypoxemia seen clinically.

(4) *Shunting*. While not mentioned until now, if alveoli are *completely unventilated but still perfused with blood, this is called shunting.* It is of particular importance in acute lung diseases where it is common and can be substantial. Such diseases include pneumonia, lung collapse, and pulmonary edema. What is common to these diseases is that the affected alveoli are completely airless because they are either collapsed or filled with cells or fluid. You may choose to think of shunting as the lower limit of $\dot{V}A/\dot{Q}$ (i.e. zero). If such blood, never seeing alveolar gas, passes through the lungs, it will reach the pulmonary veins unchanged and therefore low in O_2 (and high in CO_2). This admixture of venous blood must cause arterial Po_2 to fall. The corresponding increase in Pco_2 is usually quite small however.

2. Summary

The lungs exist for gas exchange — to move O_2 from the air to the blood and CO_2 from the blood to the air. This function is essential to maintenance of cellular respiration, and without it the organism would die. The process of gas exchange takes place by a linked series of transport steps — convective ventilation that exchanges O_2 and CO_2 between the air and the alveolar gas; diffusion that exchanges O_2 and CO_2 between the alveolar gas and the pulmonary capillary blood; and convective blood flow that takes the blood from the pulmonary capillaries to the tissues. These transport steps follow simple physical laws and as a result, the quantitative behavior of the lungs can be predicted quite accurately using a set of simple mass conservation equations. The structure of the lung is uniquely evolved to perform these transport tasks. Diffusion in particular is facilitated by dividing the alveolar volume into a very large number of very small, thin-walled alveoli, but this structure causes a number of potentially serious problems for lung function that have been overcomed by several ingenious mechanisms, including the mucociliary system, surfactant, mechanical interdependence and others. It is surprising that in the face of these problems, gas exchange is nearly perfectly efficient in health. But in lung disease, four separate mechanisms of hypoxemia and hypercapnia may develop, impairing gas exchange and potentially leading to death. Fortunately, several compensatory mechanisms are brought into play to compensate and in most cases allow the lungs to continue to exchange the requisite volume of O_2 and CO_2 despite persisting arterial hypoxemia, and often, also hypercapnia.

References

1. E. P. Hill, G. G. Power and L. D. Longo, Kinetics of O_2 and CO_2 exchange. In: *Bioengineering Aspects of the Lung*, ed. J. B. West (Dekker, New York, 1977), pp. 459–514.
2. M. P. Hlastala, *Physiology of Respiration*, eds. M. P. Hlastala and A. J. Berger (Oxford Press, New York, 2001).
3. A. J. Vander, *Human Physiology: The Mechanisms of Body Function*, eds. E. P. Widmaier, H. Raff and K. T. Strang (McGraw-Hill, Boston, 2006).
4. P. D. Wagner and J. B. West, Ventilation-perfusion relationships. In: *Pulmonary Gas Exchange,* ed. J. B. West (Academic Press, New York, 1980), pp. 219–262.
5. P. D. Wagner and J. B. West, Ventilation, blood flow and gas exchange, In: *Textbook of Respiratory Medicine*, eds. J. F. Murray and J. A. Nadel (W. B. Saunders, Philadelphia, 2005), pp. 51–86.
6. J. B. West, *Respiratory Physiology: The Essentials*, ed. J. B. West (Lippincott Williams & Wilkins, Baltimore, 2000).

CHAPTER 13

ENGINEERING APPROACHES TO UNDERSTANDING THE KIDNEY

Scott C. Thomson

Abstract

In this chapter, kidney physiology is singled out as an exemplar of how engineering methods shape the laws of motion, thermodynamics, and control theory into applicable techniques for the biologist. It is the role of kidneys to stabilize the volume and composition of the body fluids against outside disturbance. This is accomplished by generating a large volume of ultrafiltrate from plasma in the renal glomeruli, then subjecting the ultrafiltrate to extensive modification by the renal tubules to form a final urine that matches the dietary intake. A brief overview is provided of the various kidney functions, followed by a description of how engineering approaches have been central to our understanding of glomerular ultrafiltration, basic tubular reabsorption, the countercurrent multiplier system for enabling a concentrated urine, and internal negative feedback controllers that stabilize kidney function.

1. Introduction

Physiology is a pillar of bioengineering, and engineering contributes greatly to physiology. Engineering approaches are necessary for the full comprehension of biological systems on several scales. At the molecular level, analogs of membranes and sodium pumps are essentially engineering problems. At the whole-organ level, the study of the lung compliance or the aortic pulse also involves engineering principles. Finally, engineering approaches are necessary for predicting how the broad network of physiologic processes that interact to stabilize the body's internal environment will respond to an outside disturbance. In each of these examples, the fundamental elements of physics or control theory have been shaped by engineering into applicable techniques for the biologist. The approach that engineers use for applying principles of physics and mathematics makes it possible for them to think about biological systems in ways that are not possible for the traditional biologist who is mostly trained to identify and classify the components of a complex system rather than to explain how the whole system runs. The purpose of this chapter is to persuade bioengineering students to take an interest in physiology by describing prior successes of the engineering approach to kidney physiology. Kidney physiology is singled out, partly because the author is a kidney physiologist, but also because the kidney is useful for illustrating engineering principles that can be subsequently applied to understand the other organs and the interactions among them.

Science is built up of facts, as a house is with stones. But a collection of facts is no more a science than a heap of stones is a house.

— Henri Poincare

2. Why are there Kidneys?

The lay view is that kidneys mainly exist to remove soluble by-products of metabolism, the prototype being urea. But when someone dies from kidney failure, they do not succumb to the toxicity of urea. Instead, death occurs because of the loss of the means to provide a stable physicochemical environment for the cells in the body. This internal environment is formed from a dilute aqueous solution of salts of electrolytes and minerals. If the volume, ionic composition, osmolarity, or pH of a cell exceeds certain bounds, then it ceases to function and death occurs. It is the main role of the kidney to maintain these parameters within bounds so that the cells can function. Hence, the main role of the kidney is regulatory, not excretory, and the health of a kidney is determined, not by how much it can excrete, but by how efficiently it regulates the volume and composition of the body fluids.

3. What is Physiology?

Physiology is the branch of science that deals with the physical and chemical processes that underpin the activities of life. By attending to *processes,* physiology is distinguished from traditional biology, which mainly attends to *things*. The difference between *processes* and *things* is that *processes happen*, while *things exist*. Just as complex things (such as the genome) are constituted from a few simple things (four nucleic acids), complex processes (such as formation of the urine) are constituted from a few simple processes. In fact, every event that happens in physiology can theoretically be reduced to recurrent combinations of three fundamental processes that reflect the possible fates of fundamental things (i.e. molecules): (1) molecules can move; (2) molecules can be transformed into different molecules; and (3) molecules can bind reversibly to each other.

4. Motion in Physiology

Things move when acted on by forces, and the rate at which something moves is the product of its mobility and the net force acting on it. The forces that apply in physiology include the fundamental gravitational and coulomb forces, and the emergent "forces" related to pressure and diffusion. An object is rendered susceptible to each of these forces by virtue of a specific property. Gravity applies to mass, coulomb force applies to charge, pressure applies to volume, and diffusion applies to "chemical activity," which is usually synonymous with concentration. (It should be noted that diffusion is not a fundamental force. Instead, it is an emergent phenomenon that arises from the random motion of a large number of particles according to the second law of thermodynamics. Diffusion is treated like a force in physiology because it has the mathematical properties of a force.)

In physiology, one is often interested in the passive movement of some entity x from side a to side b across some barrier such as a cell membrane, capillary wall, or layer of epithelial cells, where not enough is known about the structural detail to describe the forces acting inside of the barrier, but it is possible to know the *potential difference* across the barrier. *Potential difference* refers to the amount of energy required to move a unit of x from a to b against the various forces acting on it. For a conservative force field, the average force within the barrier must be equal to the difference in total potential across the barrier divided by its thickness. Often, the thickness of the barrier is not known, but its effect is incorporated, along with the mobility of x within the membrane, to give an overall permeability coefficient of the barrier Px. Then the flux, Jx, of x through the barrier is simply given by the product of Px and the potential difference, $\mu_{(x,b-a)}$;

$$Jx = -Px \cdot \mu_{(x,b-a)}. \tag{1}$$

Relating flux to a potential difference is ubiquitous throughout electrical, chemical, and mechanical engineering, and the foregoing expression reduces to Ohm's law, Fick's first law of diffusion, or Darcy's law, depending on the nature of x.

Describing the flux of x across a barrier is complicated if x has properties that make it susceptible to more than one kind of force. The typical case is that of a small ion being driven across a membrane by both diffusion and coulomb forces. The total transmembrane potential is summed from the electrical potential (voltage • valence of x) and chemical potential (temperature • difference in the log concentration for x). Additional complexity arises when the movement of x is coupled to the movement of some other system element, w. In the prototypical case, x represents a neutral solute and w represents water. The potential for water to cross a barrier is the sum of the pressure and chemical potentials for water. The latter is given by the osmotic pressure provided by the solute and, thus, depends on the concentration of x. Since the body fluids are dilute, we ignore the volume occupied by x and derive the potential for x to cross the barrier from its concentration difference. But a molecule of x interacts with the water molecules that surround it and gets dragged across the membrane along with the water that contains it. If the barrier is more permeable to water than to x, then sieving occurs that causes the concentration of x to be less on the downstream side of the membrane. This simultaneously establishes a concentration gradient that favors the forward diffusion of x and an osmotic gradient that retards the forward flux of water. This theme of solute-solvent coupling appears recurrently in kidney physiology and represents an important application of Onsager's theory of non-equilibrium thermodynamics, which is a tenet of chemical engineering.

5. Transformation

Transformation is the whole class of processes that result in conversion of one molecular species into another. All chemical reactions that involve making or breaking of covalent bonds are included here. These may be catalyzed by enzymes or may occur spontaneously. A specific example is the hydrolysis of ATP to yield ADP and 600 meV of free energy, which then becomes available to perform work. The net flux of x into a compartment is the sum of its entry-departure due to movement and its formation-destruction via the process of transformation, the latter being subject to fundamental rate laws from physics and chemistry that can be described with differential equations.

6. Binding

Binding is the class of processes that consists of two or more molecules combining reversibly to form a single complex. Examples include hormones and neurotransmitters binding to receptors on the cell surface, cAMP binding to the regulatory subunit of protein kinase A, and acid-base buffering. It can reasonably be argued that binding is just a special case of transformation and that there should only be two classes of biological processes. But binding is such a widespread biological phenomenon that it deserves separate recognition. Furthermore, binding and unbinding occurs on a time scale that is much shorter than the time scale we use to describe changes in the local concentration of substances that result from movement or chemical synthesis. In other words, binding processes are in rapid equilibrium, unlike processes of motion and transformation. Therefore, when we use engineering tools to solve problems in physiology, motion and transformations are described with differential equations, whereas binding is often described with algebra.

7. Homeostasis

As already noted, the kidneys are important because they regulate the volume and composition of the body fluids. Relative stability of the body fluids is a primary condition of life and the kidney is responsible for controlling the volume, ionic composition, osmolarity, and pH of the body fluids. The challenge of maintaining this stability is considerable, given that the body operates far from thermodynamic equilibrium, both internally, and with respect to the surrounding environment with which it interfaces. The coordinated physiological processes that maintain steady states in the organism were once deemed so complex and so peculiar to living beings that the term *homeostasis* was coined to designate them. Homeostasis ultimately results from physical and chemical processes. But homeostasis has intrinsic properties that can be described without reference to physics or chemistry. Furthermore, those intrinsic properties are not unique to living things. Instead, they obey the same laws set out in control theory that apply to any cybernetic system. Hence, theories of homeostasis are the purview of engineering.

In classical control theory, stability is achieved through negative feedback. Typically there is a "desired" output of a system, which is set as a point of reference. To ensure that the system follows the reference over time, a closed-loop controller is inserted that adjusts the input based on a comparison between the output and the reference set-point. If an outside disturbance is then imposed on the system to cause a deviation from the set-point, feedback from the controller will mitigate the impact of the disturbance. In physiology, as in engineering, the most interesting features of a feedback system often reside in its dynamics, or the time-dependence of its response. The dynamics of a feedback system are easiest to describe for linear systems, which can be solved in terms of simple functions using Laplace transforms. This standard engineering approach has yielded numerous insights into physiology over the years, but many physiologic systems are decidedly nonlinear and require more modern signal processing techniques, some of which are being developed by engineers doing kidney research.[1,2]

8. Brief Overview of the Kidney Functions

The kidney is constituted from functional units called nephrons, which transit the outer (cortical) and inner (medullary) parts of the kidney. Each human kidney contains about one million nephrons that operate in parallel with some cross-talk. Each nephron consists of a glomerulus and a tubule (Fig. 1). The glomerulus is a tuft of capillaries where the formation of urine begins. The space outside the glomerular capillaries is a port of entry to the tubule. The renal tubule is a hollow structure lined with epithelial cells that transport various solutes in one direction or the other between the tubular lumen and the surrounding interstitium. The tubule is lengthwise heterogeneous and performs specific operations on the tubular fluid in its various segments. The segments include, sequentially, the proximal tubule, descending and ascending limbs of the loop of Henle, distal tubule, and collecting duct. All glomeruli, proximal tubules and distal tubules are confined to the cortex. The loop of Henle begins in the cortex, transits the medulla, and returns to the cortex. The collecting duct begins in the cortex, transits the medulla, and finally drains urine into the renal pelvis from the inner medulla. The cortex is in osmotic

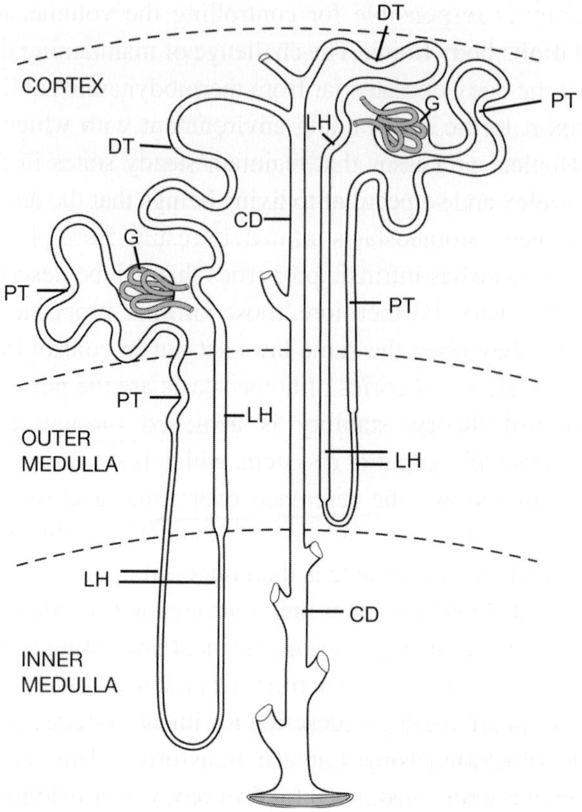

Fig. 1. Anatomy of the nephron. G: Glomerulus, PT: proximal tubule, LH: loop of Henle, DT: distal convoluted tubule and connecting segment, and CD: collecting duct.

equilibrium with the rest of the body. There is an osmotic gradient that begins at the junction of the cortex and medulla and builds toward the inner medulla. This gradient is necessary to form concentrated urine.

The nephrons form urine from the blood plasma in two steps. First, a fraction of the plasma flowing along a glomerular capillary is converted to an ultrafiltrate by sieving across the wall of the capillary. This process, known as *glomerular filtration*, is driven by the blood pressure at a rate further determined by the amount of plasma available to be filtered and by the permeability of the capillary wall. The glomerular capillary wall is a complex structure that forms an effective filtration barrier to cells and to the major blood proteins with molecular diameters exceeding 36Å. Unrestricted passage across the capillary wall is given to smaller solutes such that the concentrations of these in the filtrate are equal to those in the plasma. A pair of human kidneys will typically generate 180 liters per day of glomerular filtrate containing about 27 moles of sodium. This is about 13-fold the total body content of sodium and about 250-fold the typical daily intake of sodium. The second step in the formation of urine occurs as the glomerular filtrate passes along the renal tubule where it is processed to make the final urine.

This processing involves the addition of some substances to the fluid by tubular secretion and the removal of others by tubular reabsorption. In the course of matching the urine to the dietary intake, the tubule usually must reabsorb > 99% of the filtered water and sodium. Approximately 65% of this reabsorption occurs in the proximal tubule, 20% in the loop of Henle, and the remainder in the distal tubule and collecting duct. The proximal tubule is designed for energy-efficient bulk transport, which it can accomplish because it is not required to maintain large potential differences for sodium or osmolarity. Hence, while the proximal tubule reclaims much of the filtered salt and water, it does not much alter the osmolarity and ionic composition of the tubular fluid. The loop of Henle and distal tubule reabsorb salt in excess of water, causing the tubular fluid to become dilute (hypoosmolar) relative to the rest of the body fluids as it enters the collecting duct. When the diet is hypoosmolar, the kidney is called upon to excrete a dilute urine. Under these conditions, the collecting duct maintains low water permeability so that the tubular fluid remains hypoosmolar as it transits the collecting duct and becomes the final dilute urine. When the diet is hyperosmolar, the collecting duct is stimulated to become water-permeable. In turn, water is reabsorbed osmotically from the medullary collecting duct, which maintains a high interstitial osmolarity. This leads to formation of concentrated urine. The body fluids contain about 300 mOsm/kg. By maintaining the medullary osmotic gradient, having a functioning loop of Henle, and modulating the water permeability of the collecting duct, the osmolarity of human urine can be made to vary between 50 and 1200 mOsm/kg.

The collecting duct is also responsible for fine-tuning of the urine solutes and, thus, is required to sustain major gradients between the tubular fluid and surrounding interstitium for the various solutes. To fulfill this requirement, the collecting duct is endowed with a low intrinsic permeability to solutes and a limited overall capacity for transport relative to the proximal tubule.

Free energy is required for tubular secretion or reabsorption and free energy for all transport along the tubule is ultimately traceable to a sodium pump in the abluminal membrane of most tubular cells. This pump uses the energy liberated by ATP hydrolysis to reduce the cell sodium and elevate the cell potassium. Since cells are permeable to potassium, potassium diffuses back out to attain electrochemical equilibrium, causing a negative cell voltage. The negative cell voltage and low cell sodium provide the free energy for a whole array of transport processes, some occurring in the same cell and some occurring distantly. Water reabsorption, wherever it occurs along the nephron, is mainly driven by local osmotic forces, not by hydraulic pressure or pumping. The kidney uses multiple mechanisms to harness energy from the sodium gradient in one portion of the nephron and export that energy for use by other sites along the nephron. One such mechanism employs bicarbonate reabsorption in the early proximal tubule to reduce the overall cost of reabsorbing chloride later in the proximal tubule. Another such mechanism involves a countercurrent multiplier system to generate the medullary osmotic gradient, which becomes the source of energy for raising the osmolarity of the urine above the osmolarity of plasma, as occurs during periods of water deprivation.

Our understanding of these systems is incomplete, but to the degree that we understand them, this understanding owes largely to engineering approaches.

To satisfy conservation of mass, the time-averaged urinary excretion of every substance in the body must perfectly match the difference between the gains from dietary intake plus net formation and the extra-renal losses. Here, we consider how this applies to water, salt, and urea. Salt and urea form the principal osmoles of the urine, while water accounts for most of its volume. The extra-renal losses of salt and water are not part of any feedback loop to regulate the body fluids, so these losses can be subsumed as adjustments in net intake. Urea is formed in the liver from dietary protein and excreted by the kidney to maintain nitrogen balance. If the body fluids are in steady state, the relative proportions of salt, water, and urea that appear in the urine must match their rate of appearance in the body, which is determined solely by the diet. These dietary proportions may vary from one day to the next and they may bear no resemblance to the composition of the body fluids, which are reflected in the glomerular filtrate. Hence, processing by the tubule must be extensive and flexible. Furthermore, the tubule must generate and sustain large concentration gradients for various individual solutes and total osmolarity.

9. Approaching Long-Term Regulation of the Body Fluids

It is an engineering challenge for the kidney to match the excretion to the diet so as to render the body's internal environment relatively insensitive to changes in the diet. To impart a feel for the challenge of fine-tuning the urine, we point out that the tubule must precisely pare down the 180 liters of glomerular filtrate, which contains about 27 moles of salt, into a final urine that matches a diet that may contain 0.5 – 15 liters of water and 0.1 – 1.0 moles of salt. Without invoking any physiologic mechanism for how this occurs, common sense dictates that the long-term average net flux for total body salt and water must be zero, since "long-term" positive or negative flux would lead to an infinite, or infinitely negative, total body content, respectively. Perfect "long-term" balance for salt and water is a conservation-of-mass boundary condition for steady-states that transcends life, but the trajectory taken by the salt and water excretion in passing from one steady-state to another provides important information about the physiology.

It turns out that, following a step change in dietary salt or water, urinary excretion approaches the new intake asymptotically and without overshoot, as if driven purely by the changes in the total body content that gradually accrue after the diet is changed. Given the complex physiologic mechanisms involved, it is truly remarkable that the difference between intake and excretion for both salt and water can be described by simple exponential curves with rate constants for salt and water being about 0.8 $days^{-1}$ and 10 $days^{-1}$, respectively. It is these rate constants that determine the sensitivity of the body's internal environment to outside disturbances. For example, doubling the steady-state salt intake from 100 to 200 mmoles per day will cause the steady state salt content of the body to increase by only 6%. Diluting an isosmolar diet by 50% with 2 liters per day of additional water intake will reduce the steady state osmolarity of the body fluids by < 0.5%. If the kidneys or the feedback systems that regulate them become damaged, then the new steady state is approached more slowly. In other words, the salt and water excretion react less to a given change in total body content. This, in turn, makes the volume and salinity of the body fluids more sensitive to changes in the dietary intake and lessens the flexibility to eat what we want and remain alive. Notably, the exponential decay constants for salt and water balance are valued such that the body more efficiently fends of disturbances in osmolarity than in total body salt. Not surprisingly, challenges to osmolarity are the greater threat to survival.

10. Engineering Approaches to Glomerular Filtration

In the early 1970s, technical innovation made it possible to study the physical basis of glomerular filtration at the single nephron level. A wealth of information about the process was subsequently revealed using engineering approaches combined with the experimental technique of renal micropuncture. This was based on a paradigm for the glomerular capillary as a circular cylinder with uniform hydraulic permeability, *Lp*, and minimal axial resistance to flow[3] (Fig. 2). Water flux across the wall of the cylinder at any point along its length is the product of *Lp* and the effective potential for ultrafiltration, P_{UF}, which is given as the difference between the hydrostatic pressure gradient, ΔP, and the opposing osmotic pressure exerted by the plasma proteins, Π_{GC}. Single nephron glomerular filtration rate, *SNGFR*, is obtained by integrating filtration flux along the entire capillary, where x represents position along the capillary.

$$Jx = -Px \cdot \mu_{(x,b-a)}. \tag{2}$$

Since there is low resistance to flow in the capillary, ΔP is essentially constant along the length, while Π_{GC} rises due to the removal of water. Since Π_{GC} will increase at a

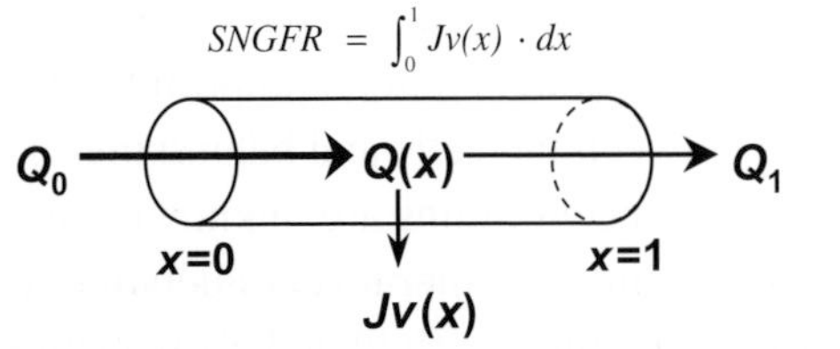

Fig. 2. Schematic representation of filtration by glomerular capillary. *Q*: Plasma flow. *Jv*: Filtration flux, is driven by the ultrafiltration pressure, which declines along the capillary as the plasma proteins become concentrated causing the oncotic pressure to rise. *X*: Position along capillary. SNGFR: Single nephron glomerular filtration rate.

lesser rate when the plasma flow is higher, nephron plasma flow becomes a determinant of *SNGFR*, even though plasma flow does not appear in the equation for flux.

This model for glomerular filtration assumes that the capillary wall is either completely permeable or completely impermeable to any given solute. A completely permeant solute exerts no effective osmotic pressure and a completely impermeant solute will be absent from the urinary space. Thus, the model allows that all osmotic pressure in the system is exerted by non-filterable proteins in the plasma. This approximation is good enough for computing the water flux, but another paradigm is required to study the permselectivity of the capillary wall. The theoretical basis for explaining the molecular size cutoff for glomerular sieving and the passage of minute quantities of proteins across the capillary wall is provided by "pore theory." Pore theory depicts the glomerular capillary wall as a sheet perforated by cylindrical holes (pores) of various sizes. Hydrodynamic theory is applied to the system to predict how the convective and diffusive fluxes of solutes of different sizes depend on the density and geometry of the pores and on the free energies of the solutes and water on both sides of the membrane. The model is applied to data on molecular sieving of tracer molecules obtained from humans or experimental animals to compute the size and distribution of the idealized pores. Pore theory was first envisioned for the glomerulus in the 1960s and applied to human data in the 1970s. It remains a subject of interest to modelers and experimentalists who seek to advance our understanding of the glomerulus, particularly in regard to describing changes that cause the glomerular capillary wall to leak proteins in diseased kidneys.[4,5]

Glomerular filtration can also be described using the finite element method of computational fluid dynamics where pressure and flow field equations are written based on a more realistic structural rendering of the capillary wall.[6]

11. Engineering Approaches to Tubular Transport

Engineering models of renal tubular function have been developed for most of the nephron segments. Work continues in this area according to a basic strategy that was laid out in the past.[7] The simplest models are single equations that represent the overall solute and water fluxes as functions of conditions on both sides of the tubule. More detailed models incorporate a series arrangement of the luminal and abluminal cell membranes arranged in parallel with the intercellular tight junction and lateral intercellular space, which resides between the cells and behind the tight junction. The intracellular and intercellular solute concentrations are unknown variables that are determined by conservation equations. Adding a further level of detail to the model allows transport to affect the concentration of the tubular fluid and changes in the tubular fluid to affect transport. In this case, the conservation of mass is depicted by a system of ordinary differential equations that must be integrated along the length of the tubule. More complex multitubular models have also been developed where transport along the tubule alters the interstitial composition. Computational effort is required to determine the interstitial concentrations compatible with all the tubule fluxes. The purpose of these models has been to illustrate proposed transport mechanisms or to identify specific model parameters, such as the water permeability or the osmotic reflection coefficient of a particular solute, from experimental data. The proximal tubule has been most extensively modeled. But several issues are unresolved, such as how much water passes between the epithelial cells versus across them and why different small electrolytes ions exhibit different osmotic reflection coefficients. Recently, it has been observed that the amount of transport machinery present in the luminal cell membrane is upregulated when shear stress is applied to the membrane. This novel mechanism to explain the flow-dependence of reabsorption in the proximal tubule is the subject of new modeling efforts. An overview of the techniques for modeling tubular transport was published in the mid-1990s, and it remains applicable.

12. Engineering Approaches to the Concentrating Mechanism

The countercurrent system for concentrating the urine has fascinated modelers for over 50 years, with highly detailed mathematical descriptions now available to explain the concentrating mechanism in the outer medulla.[8,9] Yet the inner medullary mechanism remains enigmatic.

What is missing for the inner medulla is adequate energy for the small transverse osmotic gradient that gets multiplied by the countercurrent flow. In the outer medulla, this energy is supplied by abluminal sodium pumps in the ascending loop of Henle. But sodium pumps are absent in the inner medulla, which lacks sufficient oxygen to run them. So the inner medullary loop of Henle passively reabsorbs salt using free energy imported from the cortex where salt reabsorption generates a tubular fluid that is poor in salt and rich in urea. When this urea is recycled to the medulla via the collecting duct, the result is a local decrease in mixing entropy, which provides free energy for passive reabsorption of salt from the loop of Henle. This elegant hypothesis was articulated in mathematical detail during the 1970s,[10] before there were data on the various physical parameters involved. Unfortunately, when these data became available it was learned that the osmotic gradient, as it appears in real life, is steeper than that predicted when the actual membrane permeabilities and diffusion coefficients are entered into the equations. While the search continues for a better explanation, this “solute mixing” model, however imperfect, remains the standard explanation given for the inner medullary osmotic gradient (Fig. 3).

13. Engineering Approaches to Renal Autoregulation

The renal blood flow is subject to many influences from throughout the body. Two of these involve negative feedback control systems intrinsic to the kidney. The first of

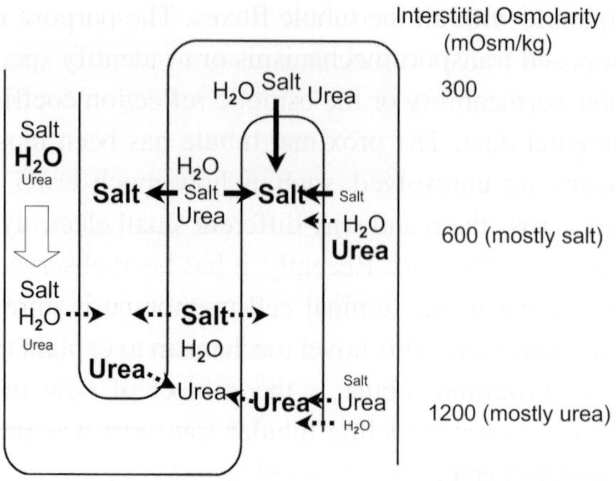

Fig. 3. Qualitative depiction of the countercurrent mechanism under conditions of antidiuresis, where a concentrated urine is being formed. Solid arrows: active transport, dashed arrows: passive transport. The process begins with active reabsorption of salt from the ascending loop of Henle (LOH) in outer medulla. This adds osmoles to the medullary interstitium while generating a tubular fluid that is dilute relative to plasma. The new intersititial osmoles drive osmotic water reabsorption from the descending LOH. Since the tubular fluid leaving the medulla is dilute relative to the fluid that enters from the proximal tubule, the overall osmolarity of the medulla must be higher than the cortex. The descending and ascending LOH form a countercurrent multiplier that augments this difference and causes osmolarity to reach 1200 in the inner medulla. This drives water reabsorption from the collecting duct to concentrate the final urine. In the inner medulla, salt reabsorption is passive and depends on a high ratio of urea : salt osmoles deep in the interstitium The lowering of mixing entropy required to raise the urea : salt ratio is accomplished by active salt reabsorption in other segments. The tubular fluid becomes a concentrated urea solution as water is removed in the collecting duct. Urea is then passively reabsorbed by the most downstream collecting duct to provide osmoles for driving water reabsorption from the descending limb to raise its salt concentration. Salt then diffuses out of the thin ascending limb, which is permeable to salt.

these is a *myogenic* mechanism in the pre-glomerular arterioles, which undergo active constriction in response to increased wall-tension. This has the effect of making vascular resistance a positive function of perfusion pressure, thereby rendering the blood flow relatively insensitive to changes in renal perfusion pressure. The second mechanism is a system of *tubuloglomerular feedback* (*TGF*), which functions within each nephron to stabilize SNGFR and salt delivery to the distal nephron. The TGF system includes a sensor element in the tubule at the downstream end of Henle's loop and a response element in the pre-glomerular arteriole. The tubule senses changes in salt delivery of salt to the early distal nephron and elicits a feedback response from the glomerulus which adjusts SNGFR so as to offset the change in distal salt delivery. Several teleologic arguments have been made to justify the importance of TGF. One of these is that it stabilizes the load to the distal nephron, which facilitates fine-tuning of the salt excretion. The student is referred elsewhere for a simple description of the TGF anatomy and current thoughts as to its mechanism.[11]

The application of engineering principles to TGF has yielded an extensive literature that continues to grow. For instance, a simple negative feedback model has been used by us to quantify the homeostatic efficiency of TGF from micropuncture data.[12] Of greater interest from an engineering standpoint is the array of sophisticated approaches that have been applied to study the dynamical behavior of TGF. Interest in this area began in the 1980s when it was recognized that regular oscillations in the tubular flow and related parameters in normal rats were mediated by TGF and arise due to a switch from a steady state to a limit cycled oscillation (i.e. an oscillation with bounds and certain trajectory) that occurs when the open-loop gain of the TGF response is sufficiently large and the delay in the TGF signal transmission is sufficiently long.[13] Since then, engineering approaches have been exploited to address inter-nephron coupling of the TGF response, coupling of the TGF and myogenic mechanisms, and other issues. Particular interest has focused on the observation that rats with high blood pressure exhibit complexity in their TGF power spectra. Attempts to explain this have invoked bifurcation to deterministic chaos, TGF gain as a stochastic process, and multi-stability with switching between multiple dynamic modes. (Earlier work is cited in this most recent paper on the subject.[14])

14. Summary

Engineering models have been essential to understanding various functions of the kidney including the mechanisms, control, and regulation of glomerular filtration, tubular reabsorption, and the elaboration of a concentrated urine. Imagination, judgment, and mathematical reasoning, are the core attributes of engineering. These will remain essential to kidney physiology, which is charged with making sense of the facts of kidney biology and with explaining the properties and functions that arise from the interacting parts in this complex organ.

Acknowledgments

Funds in support of this project were provided by the US Department of Veterans Affairs Research Service and NIDDK (RO1 DK56248).

References

1. L. Feng, K. Siu, L. C. Moore, D. J. Marsh and K. H. Chon, A robust method for detection of linear and nonlinear interactions: application to renal blood flow dynamics, *Ann. Biomed. Eng.* **34**: 339–353 (2006).

2. H. Zhao, S. Lu, R. Zou, K. Ju and K. H. Chon, Estimation of time-varying coherence function using time-varying transfer function, *Ann. Biomed. Eng.* **33**: 1582–1594 (2005).
3. W. M. Deen, C. R. Robertson and B. M. Brenner, A model of glomerular ultrafiltration in the rat, *Am. J. Physiol.* **223**: 1178–1183 (1972).
4. M. A. Katz, R. C. Schaeffer Jr., M. Gratrix, M. Ducha and J. Carbajal, The glomerular barrier fits a two-pore-and-fiber-matrix model: derivation and physiologic test, *Microvasc. Res.* **57**: 227–243 (1999).
5. D. Venturoli and B. Rippe, Ficoll and dextran vs. globular proteins as probes for testing glomerular permselectivity: effects of molecular size, shape, charge, and deformability (Review), *Am. J. Physiol. Renal Physiol.* **288**: F605–613 (2005).
6. W. M. Deen, M. J. Lazzara and B. D. Myers, Structurual determinants of glomerular permeability, *Am. J. Physiol. Renal Physiol.* **281**: F579–596 (2001).
7. A. M. Weinstein, Mathematical models of tubular transport (Review), *Annu. Rev. Physiol.* **56**: 691–709 (1994).
8. A. T. Layton and H. E. Layton, A region-based mathematical model of the urine concentrating mechanism in the rat outer medulla. I. Formulation and base-case results, *Am. J. Physiol. Renal Physiol.* **289**: F1346–1366 (2005).
9. A. T. Layton and H. E. Layton, A region-based mathematical model of the urine concentrating mechanism in the rat outer medulla. II. Parameter sensitivity and tubular inhomogeneity, *Am. J. Physiol. Renal Physiol.* **289**: F1367–1381 (2005).
10. J. L. Stephenson, Countercurrent transport in the kidney, *Annu. Rev. Biophys. Bioeng.* **7**: 315–339 (1978).
11. S. C. Thomson, Adenosine and purinergic mediators of tubuloglomerular feedback (Review), *Curr. Opin. Nephrol. Hypertens.* **11**: 81–86 (2002).
12. S. C. Thomson and R. C. Blantz, Homeostatic efficiency of tubuloglomerular feedback in hydropenia, euvolemia, and acute volume expansion, *Am. J. Physiol.* **264**(Pt. 2): F930–936 (1993).
13. N. H. Holstein-Rathlou and D. J. Marsh, A dynamic model of the tubuloglomerular feedback mechanism, *Am. J. Physiol.* **258**(Pt. 2): F1448–1459 (1990).
14. A. T. Layton, L. C. Moore and H. E. Layton, Multistability in tubuloglomerular feedback and spectral complexity in spontaneously hypertensive rats, *Am. J. Physiol. Renal Physiol.* **291**: F79–97 (2005).

SECTION V

TISSUE ENGINEERING AND REGENERATIVE MEDICINE

CHAPTER 14

SKELETAL MUSCLE TISSUE BIOENGINEERING

Richard L. Lieber and Samuel R. Ward

Abstract

Skeletal muscle represents a classic biological example of a structure-function relationship. Skeletal muscle anatomy can be determined using a combination of direct tissue dissection and magnetic resonance imaging (MRI). Whole muscle force is determined primarily by the orientation and number of fibers within the muscle, known as skeletal muscle architecture. Within muscle fibers, sarcomere arrangement determines muscle fiber force generated. These sarcomeres are very sensitive to both length and velocity and must be incorporated into any model that attempts to predict muscle force during normal function.

1. Introduction

Skeletal muscle is a highly specialized tissue that is designed to produce force and movement. It is therefore probably not surprising that it has been the subject of extensive experimentation by bioengineers for hundreds of years. Those interested in muscle mechanics, muscle biochemistry, muscle imaging and muscle energetics have implemented a wide array of experimental methods to determine the way in which muscle performs its function. Because muscle functional studies are so closely tied to muscle structural studies, it is probably not an overstatement to say that muscle represents the classic biological example of a structure-function relationship. Therefore, a major goal for bioengineers is to integrate muscle functional data with known structural properties.[1] Conversely, it is important for muscle morphologists to relate the appearance of their "structures" with known functional properties.[2] Finally, by way of introduction, it is important to point out that skeletal muscle as a tissue has provided the world of bioengineering with a number of "lessons" that turn out to apply to all biological systems. For example, the "molecular motors" of muscle were discovered to consist of the filamentous myosin proteins. However, it turns out that molecular motors are found in every cell in the body, but are only polymerized into filaments in the highly specialized striated muscles. Because skeletal muscles can be isolated and retain their function, they are a great model system for performing mechanical, imaging and biological studies.

2. Basic Skeletal Muscle Actions

Skeletal muscles are described based on where they are positioned in the body and on the function they perform when activated. The portion of the body from which the skeletal muscle arises is known as its origin and the location where the muscle ends is known as its insertion. Anatomists thus describe the origin and insertion of every skeletal muscle in the body and this information is used to suggest its action. For example, the biceps brachii is a "two headed" muscle (the Latin meaning of biceps) that extends along the arm (the Latin word for arm is brachium). This muscle crosses the elbow and shoulder joints and thus, when activated, can flex both of them. As another example, the muscles that extend the fingers are named the extensor digitorum communis. This literally translates into "common digital extensors." Similarly, there are two sets of muscles that flex the fingers. These are referred as the flexor digitorum superficialis (FDS) and the flexor digitorum profundus (FDP). As the names imply, the FDS are more superficial or closer to the surface of the body, while the FDP are deeper within the body. Perusal of an anatomy textbook will provide numerous analogous examples.

Many muscles cross only a single joint and perform a single function. Other muscles (for example, those involved in finger manipulation) cross a number of joints and can perform a variety of functions. It is therefore important to integrate skeletal muscle function with the movement of the joint(s) affected.[3] Table 1 provides a brief description of the origin, insertion, and action of many of the most common muscles in the body.

3. The Human Vastus Lateralis Muscle

The most commonly-studied human muscle is the vastus lateralis. Based on direct translation of its name, the vastus lateralis is the "enormous" muscle on the lateral aspect (the "outside") of the leg which one can easily see (Fig. 1A) and feel. The origin of the vastus lateralis is the posterior and lateral aspects of the femur and its insertion is the patellar tendon. In this chapter, we will use the vastus lateralis as the "model muscle" to describe basic muscle structure and function. The reader should be aware that while the vastus lateralis provides some valuable bioengineering lessons, there are a wide range of muscles throughout the human body that can be studied, each one of which provides its own unique lessons.

The origin, insertion, and shape of the vastus lateralis can be appreciated from surface anatomy (Fig. 1A) and understood based on the gross cadaveric human specimen shown in Fig. 1B. (In this context, "gross" simply means large and distinguishes it from

Table 1. Basic muscle groups — the most common limb muscles. Some muscles are listed under multiple categories as they cross multiple joints and thus have variable actions.

Muscle group (common name)	Major component muscles
Hip extensors	Gluteus maximus Adductor magnus Biceps femoris Semitendinosus Semimembranosus
Hip flexors	Psoas major Iliacus Rectus femoris
Knee extensors (quadriceps)	Rectus femoris Vastus lateralis Vastus intermedius Vastus medialis
Knee flexors (hamstrings)	Biceps femoris Semitendinosus Semimembranosus
Ankle flexors (dorsiflexors)	Tibialis anterior Extensor digitorum longus
Ankle extensors (plantarflexors)	Soleus Gastrocnemius
Elbow flexors	Biceps brachii Brachialis
Elbow extensors	Triceps brachii
Wrist flexors	Flexor carpi ulnaris Flexor carpi radialis
Wrist extensors	Extensor carpi ulnaris Extensor carpi radialis

a microscopic picture or micrograph.) The vastus lateralis is an extremely large muscle and, therefore, is very much involved in leg stability. Extending from the vastus lateralis proximally (toward the head) and distally (away from the head) are the glistening white fibers of the tendons. Some muscles have very well defined tendons that extend out from the muscle and can clearly be identified while other muscles have very short tendons that may be embedded within the muscle and barely visible as the muscle inserts onto the bone. Tendons connect muscle fibers to bone.[4] From an engineering

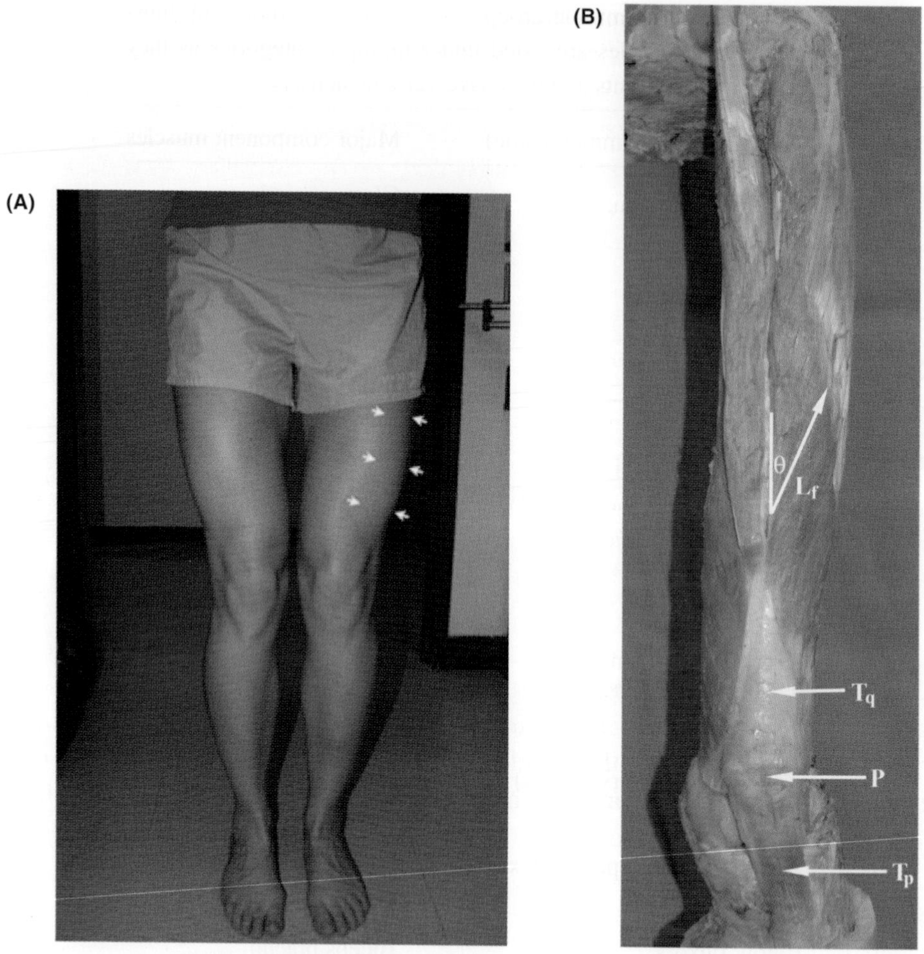

Fig. 1. **(A)** Gross picture of human thigh with surface anatomy of vastus lateralis identified as dotted outlines. **(B)** Gross picture of vastus lateralis muscle with tendons, fibers and pennation angle labeled. T_p = Patellar tendon, P = patella, T_q = quadriceps tendon, θ = pennation angle, L_f = fiber length.

point of view, they provide an intermediate structure between the fairly compliant muscle and the very stiff bone. This prevents tremendous stress concentration, which could result in muscle rupture. At the microscopic level of course, there is a fascinating transition as muscle fibers insert into the external tendon. If one examines the gross photograph a bit more closely (Fig. 1(B)), it can be seen that the muscle tissue has a fibrous appearance. These "fibers" are actually the muscle cells that comprise the muscle tissue itself. Note also that the muscle fibers are oriented at an angle relative to the action of the muscle. Thus, if one drew a straight line between the action of the muscle, which is basically along the femur, and the angle of the fibrous pattern of the vastus lateralis, this so-called "pennation" angle is approximately 30°.[5] Muscle pennation angle is characteristic of a muscle and extremely consistent from person to person. In fact, across a population, the pennation angle of the vastus lateralis is $21 \pm 2°$ in all subjects,[5] regardless of age, gender, or race. This is interesting from a developmental point of view as there is really no information as to how muscles take on this very stereotypic orientation during human growth and development. We will have more to say about pennation angle later in the chapter, but suffice for now to say that pennation of muscle fibers is a strategy used to pack a large number of fibers into a small volume.

3.1 *Magnetic resonance images of the human vastus lateralis muscle*

While surface anatomy (Fig. 1) provides a general view of the vastus, it would be difficult to make measurements or diagnoses from such views. Recent radiological advances permit high-resolution images to be obtained of soft tissues using magnetic resonance imaging (MRI).[6] For decades it has been easy to image bones using X-ray analysis. Now, soft tissues, which are radiolucent (in order words, they barely show up on an X-ray), can be imaged. The MRI method creates an image of tissue based on the behavior of protons within that tissue. Since skeletal muscles contain a fair amount of water they can be easily distinguished from fat and bone using MRI. The power of MRI is that it is possible to obtain three-dimensional views of an entire muscle (and tendon for that matter) and "section" the muscle in any plane. Figure 2 provides a series of MRI views of the vastus lateralis muscle shown in Fig. 1, which is now sectioned into the three "anatomical" planes: the transverse plane (Fig. 2A, perpendicular to the long axis of the femur), the coronal plane (Fig. 2B, parallel to the long axis of the femur and in the same plane as the body), and the sagittal plane (Fig. 2C, parallel to the long axis of the femur, perpendicular to the plane of the body). Spend some time inspecting these images and convince yourself of the size, shape and orientation of the vastus lateralis muscle. Using MRI imaging, the total muscle volume can be reconstructed as shown in Fig. 2D. These reconstructed volumes are currently being used in research and medicine to provide specific information on the changes in muscle after exercise training, disease and injury.

4. Functional Properties of Skeletal Muscle Sarcomeres

We have seen that MRI provides tremendous insight into skeletal muscle structure. Yet interestingly, knowledge of the size and shape of a skeletal muscle provides very little information regarding its function. Why is this? The answer is that it is the orientation and number of skeletal muscle fibers within a muscle, not its size, shape or volume, that determines its function. Thus, there is no way to determine from a simple value of muscle volume (for example, the vastus lateralis has a volume of ~380 cm^3) the actual force generated, excursion, or velocity that could be achieved during contraction. In order to

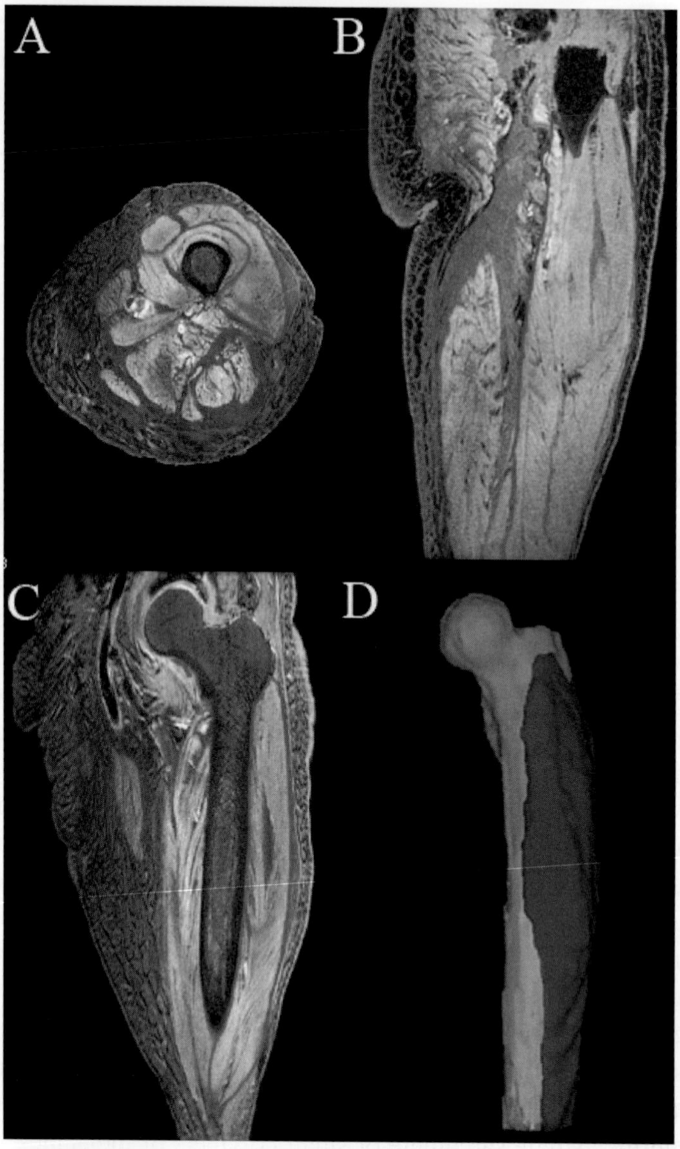

Fig. 2. Magnetic resonance image (MRI) of human vastus lateralis in **(A)** transverse, **(B)** coronal, and **(C)** sagittal planes. **(D)** Three-dimensional reconstruction of the vastus lateralis based on images in **(A)** to **(C)**.

predict muscle function from structural measurements, we must first discuss the basic units of force generation in skeletal muscle, the so-called "sarcomeres." (Note that sarco- and myo- represent the Latin and Greek prefixes respectively for "muscle" so many muscle structures have sarco- and myo- prefixes.) To a first approximation, the functional properties of a muscle fiber are simply an amplified version of the functional properties of a sarcomere. So what are the sarcomere's functional properties?

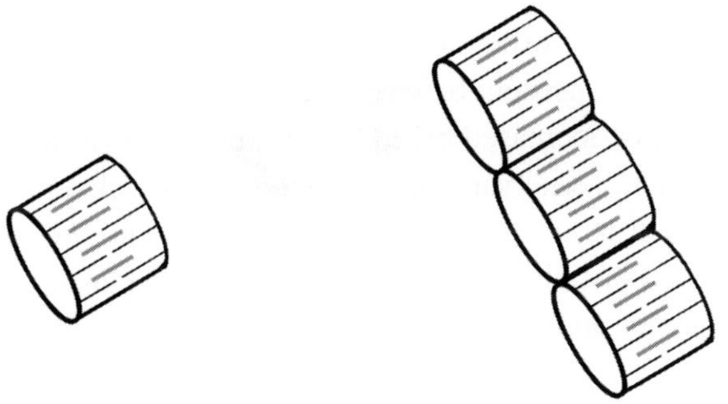

Fig. 3. Graphic illustration of **(A)** sarcomeres in series and **(B)** sarcomeres in parallel. Note that two sarcomeres in series generate two units of excursion and one unit of force while sarcomeres in parallel generate two units of force and one unit of excursion. This series and parallel arrangement of sarcomeres amplify sarcomere properties in whole muscle.

Sarcomeres are protein machines that can generate approximately 50 μN of force over a range of approximately 2 μm. These numbers are approximate and the specific values vary based on the species from which the sarcomere is obtained and slightly based on the muscle within that species. However, if, for purposes of discussion, we simply consider a sarcomere as generating a "unit" of force and a unit of displacement, we can discuss how sarcomere properties are "amplified" into muscle fiber properties. Consider the situation in Fig. 3A, where two individual sarcomeres are placed in series into a "mini" muscle fiber. The properties of this "minifiber" are based on the properties of the individual sarcomeres and since the sarcomeres have been placed in series, one end of the "minifiber" relative to the other end will produce two units of displacement and, because the total cross-sectional area of contractile material is that of one sarcomere, the "minifiber" will produce one unit of force. It can be seen, therefore, that increasing serial sarcomere number is a design strategy used to increase excursion over that which can be obtained by a single sarcomere or small polymers of molecules. Conversely, in Fig. 3B, we show the sarcomeres placed in parallel. Now, in contrast to the serial arrangement, this "minifiber" will produce one unit of excursion, but, since the amount of contractile material in parallel has tripled, this "minifiber" will produce three units of force. Thus, it can be seen that parallel arrangements of sarcomeres is a design strategy to increase contractile force. In both Figs. 3A and 3B, the same amount of contractile material, and thus, the same contractile volume produced quite different functional results as a direct result of the sarcomere arrangement within the fibers. This argument can be extended to any number of sarcomeres in series and in parallel. Human skeletal muscle fibers are 40–70 μm in diameter. Since the diameter of a sarcomere is approximately one micron, this yields a fiber cross-sectional area of 5000–15,000 μm^2 and since each sarcomere produces ~50 μN of force, a muscle fiber will produce 250–750 mN of force. Muscle fiber length, in contrast to muscle fiber diameter, is much more highly variable. Based on the argument presented above, the greater the number of sarcomeres in series the greater the muscle fiber excursion. Even though the number of sarcomeres in series and in parallel is fairly consistent between individuals for a given muscle, sarcomere number in parallel can be increased based on training protocols that create muscle stress (strength training) and serial sarcomere number can be altered with chronic length changes of muscle such as occurs after surgery, tendon rupture, or even after chronic immobilization.[7] The plasticity of skeletal muscle is a fantastic area of study, beyond the scope of this chapter and the reader is referred elsewhere for details.[8–10]

5. Skeletal Muscle Fiber Architecture

The preceding discussion described the anatomical basis for the functional properties of skeletal muscle fibers. At this point, we define a new term, "skeletal muscle architecture." Skeletal muscle architecture is defined as the orientation and number of fibers within a skeletal muscle.[11] As shown above in the example for vastus lateralis, the vastus lateralis is a relatively long muscle spanning most of the length of the thigh. However, the muscle fibers within the vastus lateralis do not run the entire muscle length as shown in Fig. 4. This is true for all skeletal muscles. Muscle fiber length is almost never the same as muscle length but is almost always shorter and, in some cases, much shorter than muscle length. An implication of this point is that, since muscle fiber length determines the excursion of a muscle, muscle length and muscle excursion are not well correlated. Muscle excursion and muscle fiber length are extremely well correlated as has been demonstrated experimentally.[12] Therefore, based on the discussion above (Sec. 4), it should be clear that one strategy that can be used in the body to increase the excursion within a muscle is to make that muscle out of long muscle fibers.

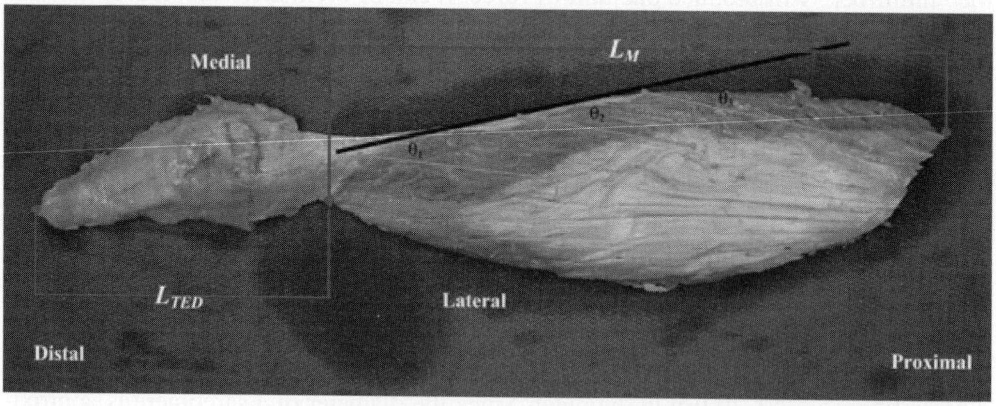

Fig. 4. Photograph of vastus lateralis muscle dissected free from a human cadaver. Labeled are fiber length, muscle length, and pennation angle.

Similarly, if one is to create a very strong muscle, the strategy should be to place a large number of fibers in parallel. One problem with placing a lot of muscles in parallel, is that they physically take up greater width than can be secured to the skeleton. Thus, short fibers are "packed" at an angle (the pennation angle) so that they can be placed onto long, narrow bones. Again, referring to the vastus lateralis muscle, one can see that the vastus lateralis is actually a very strong muscle, not based on its high volume, but based on the large number of muscle fibers that are "packed" in parallel along the muscle length. A survey of the muscles that are placed throughout the human body reveal a huge range of architectural types with some muscles being "designed" for high excursion and others "designed" for high force generation.[5,13–15] One useful index of excursion is the ratio of muscle fiber length to muscle length. The greater this value, the longer the muscle fibers are relative to the whole muscle length. The greatest fiber length : muscle length ratio is observed for the human lumbrical muscles of the hand, muscles designed to control finger movement during fine motor control but which cross many joints. The muscles that have the greatest design for stability are the interossei muscles of the hand that are also involved in hand control. It is interesting that most highly developed human features, namely fine motor manipulation, is accomplished in part by highly specialized skeletal muscles.

The preceding discussion makes it clear that muscle fiber length dominates muscle function and muscle volume, as would be obtained from an image or by direct dissection, is not very useful. What number can we derive experimentally that will predict muscle function? The parameter most often calculated is the so-called muscle "physiological cross-sectional area" (PCSA). The equation for PCSA is given by equation (1) as:

$$\text{PCSA}\,(\text{mm}^2) = \frac{M\,(\text{g}) \bullet \cos\theta}{\rho\,(\text{g}/\text{m}^3) \bullet L_f\,(\text{mm})} \tag{1}$$

where M is muscle mass, θ is pennation angle, ρ is muscle density and L_f is fiber length. The easiest way to understand the basis for this equation is to assume that the muscle is a cylinder. The equation for the volume of a cylinder is $\pi r^2 h$ where πr^2 is the area of the cylinder and h is cylinder length. Based on the geometry of the muscle, we can consider muscle fiber length similar to the value h, and can thus define the cross-sectional area of the muscle as volume divided by fiber length as shown below in equation (2):

$$\text{PCSA}\,(\text{mm}^3) = \frac{\text{Volume}\,(\text{mm}^2)}{L_f\,(\text{mm})}. \tag{2}$$

Of course, muscles are not cylinders and this PCSA value represents a "virtual" area, not an actual area that can be defined anatomically. However, experimental measurements have demonstrated that the area calculated as volume divided by fiber length is an excellent approximation of the total area of muscle fibers acting in parallel, and this term has been referred as the "physiological" cross-sectional area or PCSA.[16] The term physiological cross-sectional area distinguishes it from anatomical cross-sectional area, which could be measured directly from some type of image or directly from the tissue itself. Therefore, if one is to predict the force a particular muscle would generate, one divides volume by fiber length. In human studies, especially, non-invasive studies, volume can be obtained directly from high resolution MRI. In experimental studies, volume is most easily determined by weighing the sample and, dividing by muscle density.[17] Muscle force produced is PCSA multiplied by the specific muscle force of 250 kPa for mammalian muscle.[16]

In contrast to PCSA, there is not yet a useful non-invasive method for determining fiber length. This is because, even though the ultrasound and MRI imaging methods are good, it is still not possible to determine muscle fiber length unequivocally because sarcomere length cannot be defined. Since muscle excursion is proportional to fiber length, the only way to obtain this value, using current methods, is to perform direct microdissection on fixed tissues. This direct measurement method is subject to bias based on the position of the limb that occurred during fixation. For example, as we were discussing the vastus lateralis, if vastus lateralis fiber length were measured with the knee highly flexed, fiber length would be apparently very long, whereas if it were measured with the knee fully extended, fiber length would be relatively short. How does one compensate for such variability? The easiest method is to normalize all experimentally measured fiber lengths to a single sarcomere length. Since the muscle fiber can be considered as a number of sarcomere in series, if fiber length increases, sarcomere length increases proportionately. If fiber length decreases, sarcomere length decreases proportionately. Therefore, one must perform microdissection, measure sarcomere length and then normalize all fiber lengths to a standard sarcomere length using equation (3):

$$L_f\ (\text{mm}) = \frac{L_f'\ (\mu\text{m}) \bullet L_s\ (\mu\text{m})}{L_s'\ (\mu\text{m})} \qquad (3)$$

where L_f is normalized fiber length, L_f' is experimentally measured (raw) fiber length, L_s is some standard sarcomere length (usually the "optimal length" for that species) and L_s' is the experimentally measured sarcomere length at the experimentally measured fiber length.[18]

6. Skeletal Muscle Fiber Properties

6.1 *The skeletal muscle sarcomere*

Having summarized the anatomical factors that govern the maximum muscle force generated and the extent of muscle excursion, it is now of interest to delve deeper into skeletal muscle function by defining skeletal muscle fibers properties at the cellular

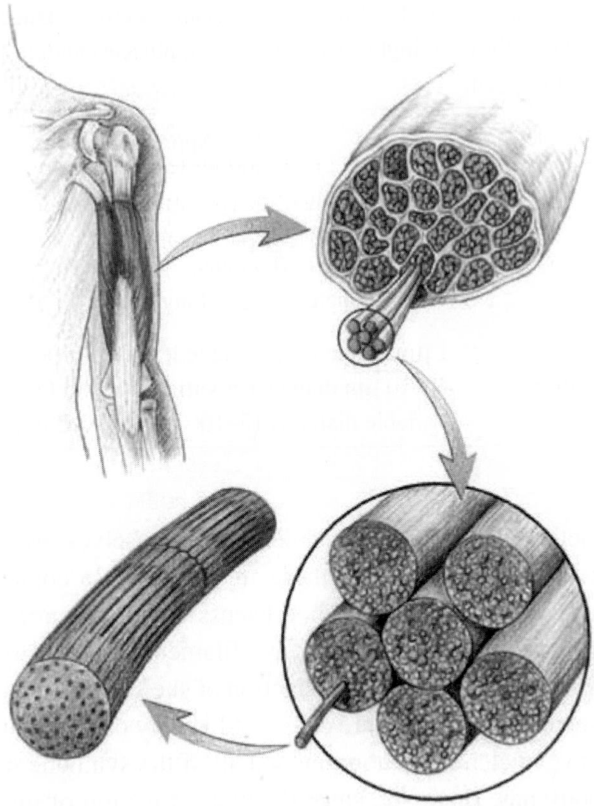

Fig. 5. Hierarchy of skeletal muscle structure. Whole muscles are composed of muscle fibers which are composed of myofibrils which are composed of sarcomeres which are composed of myofilaments which are composed of contractile proteins. Figure used by permission from Lieber.[28]

level. Muscle cells can be considered polymers of polymers of polymers, etc. It is known that the skeletal muscle contractile filaments (Fig. 5) are themselves organized by a molecular polymerization process. The molecular motor, myosin, is polymerized in a specific fashion to form a myosin filament, which, in vertebrate skeletal muscle, is always 1.65 μm long. The actin molecule is also polymerized in an ATP-dependent fashion into a long filament, that varies in length depending on the particular vertebrates species. Actin filament lengths can vary from 2.0 μm (in frogs) to 2.7 μm (in humans). It is the interaction between actin and myosin filaments that forms the structural basis for force generation in skeletal muscle. Interestingly, while it can be shown that the myosin molecule can indeed "power" muscle contraction, the myosin molecule itself only generates a few pN of force and only moves a few nm of distance. Thus, how is it possible that human skeletal muscles can generate tremendous forces, even though the molecules themselves generate such small forces? The answer lies in the fact that the molecules are arranged in series and in parallel such that the individual molecular events add tremendously. The actin and myosin filaments, which as we stated are composed of

Table 2. Size of the major muscle structural components. Sizes listed as "variable" are presented as such because they are highly variable among muscles and, in these cases, a range for human muscles is provided.

Structure	Approximate size
Myosin motor molecule	5 nm × 15 nm
Myosin filament	10 nm × 1.65 μm
Actin filament	5 nm × 2.7 μm (humans)
Sarcomere	1 μm diameter × 3 μm long
Myofibril	1 μm diameter × variable length (5–100 mm)
Myofiber (=muscle fiber)	40–70 μm diameter × variable length (5–100 mm)
Muscle	Variable diameter (5–100 mm^2) × variable length (5–100 mm)

polymers of the actin and myosin molecules, are themselves arranged into parallel arrays of filaments that interdigitate and these are known as sarcomeres (Fig. 5). There are numerous other proteins that we will not discuss that mechanically stabilize the filaments within the sarcomere and that interface filaments with the rest of the cell.[19] It must be remembered that while the major function of skeletal muscle is to produce force and movement, skeletal muscle cells are cells just like any other and, in addition to contractile proteins, have nuclei, mitochondria, and all of the synthetic machinery that any other cell in the body has. However, since the major function of muscle is to produce force and movement, the tissue itself is packed with contractile proteins. This is one of the reasons that when trying to build up a muscle, either through athletics or through rehabilitation, a relatively high protein diet is often recommended.

Sarcomeres are arranged within the cell in long threads known as myofibrils. The diameter of a sarcomere is approximately 1 μm, the spacing between myofilaments is approximately 8 nm and therefore a sarcomere is composed of approximately 2000 myosin filaments. The myofibril itself forms a continuous thread along the length of the fiber. While myofibrillar diameter is approximately 1 μm, many myofibrils are packed in parallel to form a muscle fiber whose size in mammals varies from 40–70 μm. Thus, the structural hierarchy of muscle proceeds from the motor molecule itself to the muscle tissue with the order being: myosin molecule, myosin filament, sarcomere, myofibril, fiber, and muscle. Table 2 lists the sizes of these various structures. An understanding of this order is critical for understanding the structural basis of movement and for performing skeletal muscle modeling.

6.2 *The sarcomere length-tension relationship*

What are the mechanical properties of the muscle fiber? In a series of elegant experiments in the 1960s and 1970s, it was demonstrated that the muscle fiber is essentially an amplified sarcomere.[20–22] There are subtle exceptions to this rule, but to a first approximation, if one understands sarcomere properties, one can understand muscle cell properties. When a muscle cell is held a fixed length and activated, this is referred to as an isometric ("same length") contraction. During an isometric contraction, skeletal muscle generates force and the amount of force generated depends on the amount of interaction between actin and myosin filaments. Recall above that we stated that a sarcomere represents an array of actin filaments interdigitated into an array of myosin filaments. Because the physical connection that occurs during contraction is myosin physically connecting to actin, the more actin sites that are available to myosin, the more force is generated by the sarcomere. Since sarcomere length is very much related to fiber length, muscle force generation is a strong function of muscle fiber length. This relationship has been experimentally determined and is relatively simple. When muscle fibers are highly stretched to the point where there is no interaction between actin and myosin filaments, no force is generated (position #3 on Fig. 6). As the muscle fiber is shortened from this long length, there is a linear increase in the number of interactions possible between the actin and myosin filaments and therefore, force increases linearly. At the point where there are the maximum number of interactions between actin and myosin (i.e. "optimal length") maximum force is generated (position #2 on Fig. 6). As shortening continues, actin filaments from one side of the sarcomere interfere with actin-myosin interaction on the opposite side of the sarcomere and force begins to decrease (position #1 on Fig. 6). Finally, as shortening continues, even though the muscle may be activated, there is no way that force can be generated. Interestingly, this can

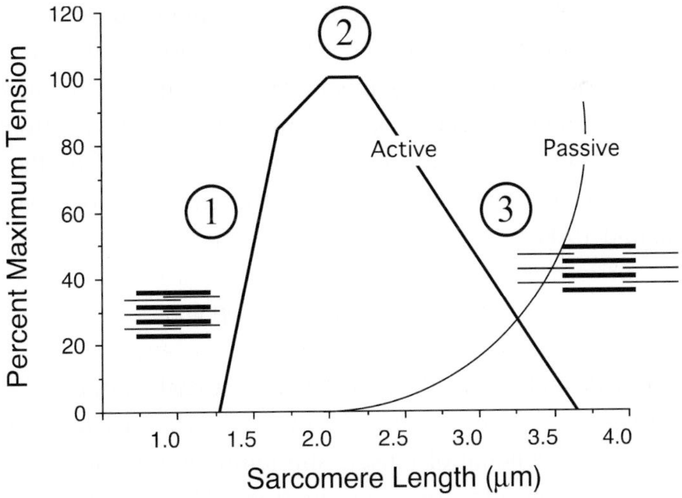

Fig. 6. Relationship between sarcomere length and force generated by a skeletal muscle. Note that, at very long lengths, muscle force is low due to unfavorable interaction between myofilaments. At optimal length (2.7 μm in this example) the maximum number of interactions between actin and myosin filaments occurs.

be accomplished even in humans in the physiological range. An understanding of the range over which specific muscles operate can be used clinically to diagnose muscle diseases and injuries.

6.3 *Estimating muscle force using the sarcomere length-tension relationship*

While Fig. 6 illustrates the relation between the sarcomere length and tension, for a whole muscle, the important parameter to be modeled is muscle force as a function of muscle length. Since the amount of stress generated by mammalian skeletal muscle is relatively constant (250 kPa), muscle force generated equals muscle physiological cross-sectional area multiplied by 250 kPa. If one is to predict whole muscle excursion, one must have an estimate of fiber length and thus the number of sarcomeres in series.[23] From these data, muscle length-tension curves will have the same shape as the sarcomere length curve, but will simply be scaled appropriately.[24]

6.4 *The sarcomere force-velocity relationship*

Having described the relationship between muscle force and muscle length during an isometric contraction, what happens to force when a muscle begins to move? Even though muscle force and muscle length are highly correlated as described above, muscles are even more sensitive to velocity changes than they are to length changes. Thus, if one is to hold muscle at a constant length, activate it and allow it to shorten, force drops dramatically as shortening occurs.[25–27] This is because, as filaments begin to slide pass one another, the probability of attachment between the actin and myosin filaments decreases, the number of cross-bridge connections decreases and therefore muscle force decreases. This is shown graphically in Fig. 7 and is described by the classic force-velocity equation (Eq. (4)) elucidated primarily by one of the original bioengineers, A.V. Hill, in the mid-1900s:

$$(P + a) \bullet v = b \bullet (P_o + P) \tag{4}$$

where a and b are constants derived experimentally (usually about 0.25), P is muscle force, v is muscle velocity and P_o is maximum isometric muscle tension. This relationship reveals that any amount of muscle shortening will be accompanied by a relatively rapid loss in muscle force production. This obviously has powerful implications for any kind of modeling involving movement. Based on a muscle's sensitivity to length and velocity, it is not enough to know how large a muscle is or its dimensions in order to predict its function. One must be able to estimate muscle velocity to predict muscle force generated and this is typically very difficult. A second

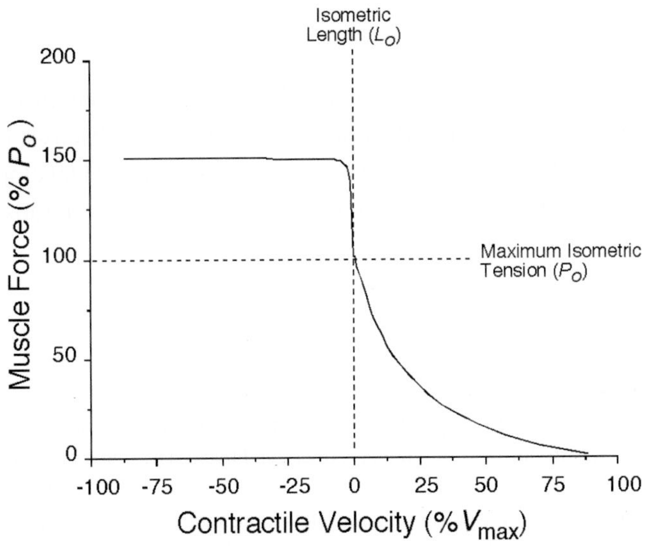

Fig. 7. Relationship between muscle velocity and force generated by a skeletal muscle. As muscles shorten (positive velocity), muscle force decreases rapidly. When a muscle is forced to length (negative velocity), force increases precipitously.

point that is obvious based on inspection of Fig. 7 is that velocity can be positive or negative. By convention, positive velocity refers to shortening, but muscles can also be forced to lengthen. This type of action is common in everyday movement when muscles are forced to lengthen, for example during walking. In this case, forced muscle lengthening results in a high increase in force borne by the muscle. As a rule of thumb, muscles can bear loads that are approximately 75% greater than the forces they themselves can generate isometrically and these actions represent very potent strengthening stimuli. There are some schools of thought in exercise science that strengthening of a muscle must involve forced lengthening of the muscle while it is activated to create such high stresses.

7. Conclusions

Hopefully, this brief presentation has convinced the student that virtually every aspect of skeletal muscle function (skeletal muscle anatomy, function, length and velocity measurement) can be studied using the modern tools of bioengineering that include imaging, mechanics, molecular biology, cell biology and bioinformatics. It should be appreciated that this chapter represents only a brief presentation of the very basic properties of muscle and that bioengineering laboratories throughout the world are constantly testing and expanding the limits of our understanding of skeletal muscle structure and function.

Acknowledgments

This laboratory has been supported generously by the National Institutes of Health, National Institutes of Arthritis and Musculoskeletal and Skin Diseases (NIAMS) grants AR40050, AR40539, and National Institutes of Child Health and Development (NICHD) grants HD048501 and HD44822 as well as the Department of Veterans Affairs.

References

1. A. F. Huxley, Muscular contraction, *J. Physiol.* **243**: 1–43 (1974).
2. B. R. Eisenberg, Quantitative ultrastructure of mammalian skeletal muscle. In: *Skeletal Muscle*, eds. L. D. Peachey, R. H. Adrian and S. R. Geiger (American Physiological Society, Baltimore, MD, 1983), pp. 73–112.
3. F. E. Zajac and M. E. Gordon, Determining muscle's force and action in multi-articular movement. In: *Exercise and Sport Sciences Review*, ed. J. Holloszy (Wiilaims & Wilkins, Baltimore, MD, 1989), pp. 187–230.
4. J. G. Tidball, Myotendinous junction: morphological changes and mechanical failure associated with muscle cell atrophy, *Exp. Mol. Pathol.* **40**: 1–12 (1984).
5. T. L. Wickiewicz, R. R. Roy, P. L. Powell and V. R. Edgerton, Muscle architecture of the human lower limb, *Clin. Orthon. Relat. Res.* **179**: 275–283 (1983).
6. R. A. Meyer and B. M. Prior, Functional magnetic resonance imaging of muscle, *Exerc. Sport Sci. Rev.* **28**: 89–92 (2000).
7. S. W. Herring, A. F. Grimm and B. R. Grimm, Regulation of sarcomere number in skeletal muscle: a comparison of hypotheses, *Muscle Nerve* **7**: 161–173 (1984).
8. S. Salmons and J. Henriksson, The adaptive response of skeletal muscle to increased use, *Muscle Nerve* **4**: 94–105 (1981).
9. D. Pette, *The Dynamic State of Muscle Fibers* (Walter de Gruyter & Company, Berlin, 1990).
10. D. Pette, *Plasticity of Muscle* (Walter de Gruyter & Company, New York, 1980).
11. R. L. Lieber and J. Fridén, Functional and clinical significance of skeletal muscle architecture, *Muscle Nerve* **23**: 1647–1666 (2000).
12. S. C. Bodine, R. R. Roy, D. A. Meadows, R. F. Zernicke, R. D. Sacks, M. Fournier and V. R. Edgerton, Architectural, histochemical, and contractile characteristics of a unique biarticular muscle: the cat semitendinosus, *J. Neurophysiol.* **48**: 192–201 (1982).
13. P. W. Brand, R. B. Beach and D. E. Thompson, Relative tension and potential excursion of muscles in the forearm and hand, *J. Hand Surg. [Am.]* **3A**: 209–219 (1981).
14. R. L. Lieber, M. D. Jacobson, B. M. Fazeli, R. A. Abrams and M. J. Botte, Architecture of selected muscles of the am and forearm: anatomy aid implications for tendon transfer, *J. Hand Surg. [Am.]* **17A**: 787–798 (1992).
15. R. L. Lieber, B. M. Fazeli and M. J. Botte, Architecture of selected wrist flexor and extensor muscles, *J. Hand Surg. [Am.]* **15A**: 244–250 (1990).
16. P. L. Powell, R. R. Roy, P. Kanim, M. Bello and V. R. Edgerton, Predictability of skeletal muscle tension from architectural determinations in guinea pig hindlimbs, *J. Appl Physiol.* **57**: 1715–1721 (1984).
17. S. R. Ward and R. L. Lieber, Density and hydration of fresh and fixed skeletal muscle, *J. Biomech,* **38**: 2317–2320 (2005).
18. A. Felder, S. R. Ward and R. L. Lieber, Sarcomere length measurement permits high resolution normalization of muscle fiber length in architectural studies, *J. Exp. Biol.* **208**: 3275–3279 (2005).

19. J. Squire, *The Structural Basis of Muscular Contraction* (Plenum Press, New York, 1981).
20. A. F. Huxley and R. M. Simmons, Proposed mechanism of force generation in vertebrate striated muscle, *Nature* **233**: 533–538 (1971).
21. K. Edman, The relation between sarcomere length and active tension in isolated semitendinosus fibers of the frog, *J. Physiol.* **183**: 407–417 (1966).
22. A. M. Gordon, A. F. Huxley and F. J. Julian, The variation in isometric tension with sarcomere length in vertebrate muscle fibers, *J. Physiol.* **184**: 170–192 (1966).
23. S. L. Delp, J. P. Loan, M. G. Hoy, F. E. Zajac, E. L. Topp and J. M. Rosen, An interactive graphics-based model of the lower extremity to study orthopaedic surgical procedures, *IEEE Trans. Biomed. Eng.* **37**: 757–767 (1990).
24. F. E. Zajac, Muscle and tendon: properties, models, scaling, and application to biomechanics and motor control, *Crit. Rev. Biomed. Eng.* **17**: 359–411 (1989).
25. A. V. Hill, *First and Last Experiments in Muscle Mechanics* (Cambridge University Press, New York, NY, 1970).
26. A. V. Hill, The mechanics of active muscle, *Proc. R. Soc. Lond. Ser. B. Biol. Sci.* **141**: 104–117 (1953).
27. B. Katz, The relation between force and speed in muscular contrtaction, *J. Physiol.* **96**: 45–64 (1939).
28. R. Lieber, *Skeletal Muscle Structure, Function, and Plasticity*, 2nd ed. (Lippincott Williams & Wilkins, Philadelphia, PA, 2002).

CHAPTER 15

MULTI-SCALE BIOMECHANICS OF ARTICULAR CARTILAGE

Won C. Bae and Robert L. Sah

Abstract

Articular cartilage is a connective tissue that covers the ends of bones in the body, bearing and transmitting load while allowing low-friction and low-wear joint articulation. The biomechanical functions of articular cartilage have been examined at multiple length scales, ranging from intact joints to cellular and molecular components. The objective of this chapter is to provide an introduction to (1) the composition, structure, and function of articular cartilage, (2) biomechanical tests of articular cartilage, and consideration of the uses, configurations, and length scales of such tests, and (3) mathematical analysis of tissue deformation and strain. The knowledge gained from such multi-scale biomechanical studies facilitates the understanding of cartilage function in growth, aging, health and disease.

1. Composition, Structure, and Functions of Articular Cartilage

Articular cartilage is a connective tissue that covers the ends of certain bone surfaces. Cartilage is present in articulating joints such as the knee and the hip. It provides a load-bearing surface for low-friction and wear-resistant joint motion. Adult cartilage is composed of cells, known as chondrocytes, sparsely distributed (~1%–10% by volume) within a fluid-filled extracellular matrix that is composed mainly of collagens and proteoglycans. The collagen network of normal articular cartilage is strong in tension,[1] whereas the proteoglycans, due to a high density of negatively-charged chemical moieties, swell in physiologic solutions and provide resistance to compression.[2,3] The balance between the swelling pressure of proteoglycans and restraining function of the collagen network is important for the maintenance of normal cartilage function and compromised in arthritic disease.[4] (Fig. 1)

Adult articular cartilage, appearing relatively uniform at a first glance, is a complex material, being both inhomogeneous and anisotropic. It is inhomogeneous in that its properties vary with position, with location in joints and with depth relative to the articular

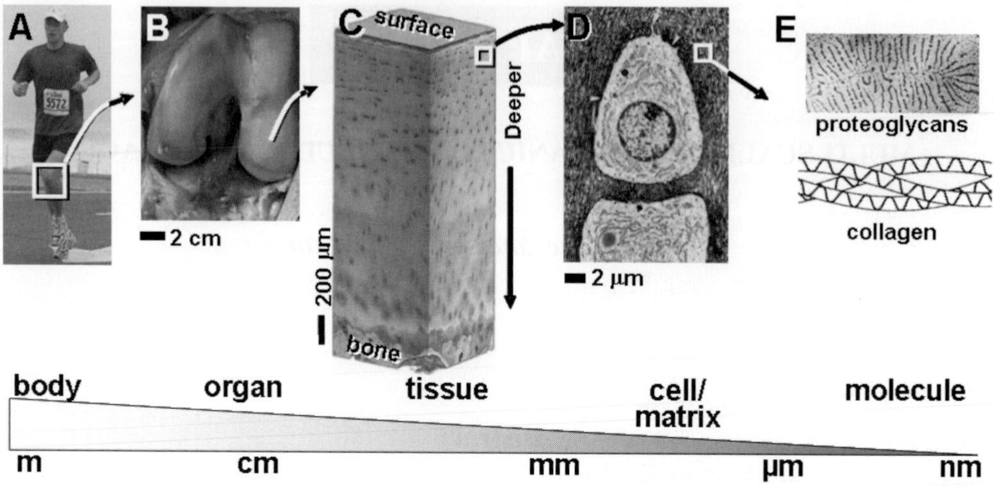

Fig. 1. Articular cartilage at different length scales. **(A)** Human knee joints are covered with **(B** and **C)** articular cartilage containing **(D)** chondrocytes sparsely embedded in extracellular matrix, largely consisting of **(E)** proteoglycans and collagen and hydrated in synovial fluid. (Photo is provided by T. J. Klein. Micrographs and collagen schematic were adapted from Refs. 66 to 68.)

surface. For example, the proteoglycan content of cartilage is inhomogeneous, increasing markedly with depth from the surface to a peak in the middle zone.[2] In addition, collagen fibrils are generally fine and densely packed near the surface, and thicker in the deep layers,[5] while the collagen content decreases slightly with depth from the surface into the deep zone.[6,7] Articular cartilage is also anisotropic, i.e. its properties vary with tissue orientation. Collagen fibrils are long and thin, and thereby impart anisotropy; in the superficial zone, collagen is oriented parallel to the articular surface. Cartilage is inhomogeneous with respect to collagen orientation, since orientation varies with depth and becomes perpendicular to the surface in the deep zone.[8,9] The orientation of collagen fibrils has been speculated to be reflected by the split-line direction, the direction in which the tissue preferentially splits open when pierced with an awl.[6]

The inhomogeneity and anisotropy of articular cartilage is evident not only in its structure and composition, but also in its biomechanical properties. Cartilage exhibits large depth-associated variations, both in compression and tension. As described below, the compressive modulus,[10] a measure of intrinsic compressive stiffness, is low in the superficial layer and increases with depth from the articular surface.[11] This variation in compressive modulus can be attributed in large part to the variation in proteoglycan content.[2,3,12] In contrast, tensile stiffness decreases with depth[6] and as the axis of extension deviates from the split-line direction.[6] This variation in tensile modulus, and strength, has been ascribed to the primary direction of the collagen network.

Understanding cartilage biomechanical properties and behavior is important in understanding health and diseases of diarthrodial joints. With advancing age in adults, the articular cartilage of humans often undergoes degeneration and progresses to osteoarthritis, a degenerative joint disease that affects over 60 million Americans[13] with a large economic impact of over US$50 billion annually.[14] A hallmark of aging and cartilage degeneration is biomechanical softening and weakening.[6,15–17] Additionally, the characteristics of tissue strain have implications for cartilage biology, including regulation of cell metabolism and causation of cell injury. Relatively low strain magnitudes of ~5%, applied dynamically in compression[18,19] or shear,[20] stimulate chondrocyte biosynthesis. However, higher strain magnitudes under certain conditions can be injurious to cartilage and cells.[21–25]

2. Confined Compression and Indentation Biomechanical Tests of Articular Cartilage at Different Length Scales

There are a variety of ways to study the biomechanics of cartilage, depending on the objectives of the observer. One such objective is to gain a better understanding of tissue biomechanical properties and function of cartilage, which allows for description of cartilage as a material in a geometry-independent fashion and subject to the complex loading conditions of the body or test configuration. Another objective is to obtain a non-destructive biomechanical index of tissue function, such as for diagnosing disease or determining the efficacy of a potential therapy. In this case, a method to assess biomechanical response of cartilage non-destructively and rapidly would be desired, even without determination of cartilage material properties.

Traditionally, biomechanical properties of cartilage have been elucidated using several types of mechanical tests (Fig. 2). The confined compression test[10,11,26] utilizes a fluid-impermeable chamber that surrounds the side and the bottom of the sample, and a porous upper platen that presses against one surface of the sample (Fig. 2A). This configuration has the advantage of providing a deformation that is primarily one-dimensional and thereby simplifies the biomechanical analysis.[10,27] The indentation test[28–32] utilizes an indenter tip of a certain size and shape (typically a plane-ended cylinder or a sphere)

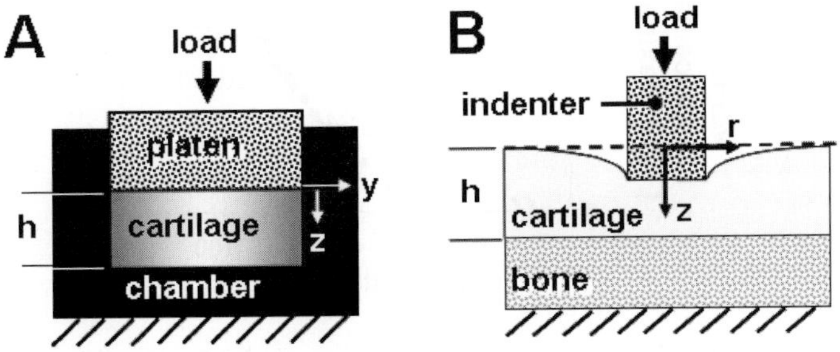

Fig. 2. Methods to biomechanically assess articular cartilage include: **(A)** confined compression and **(B)** indentation tests.

that presses against the cartilage surface, causing it to dimple (Fig. 2B). Each test can be performed at length scales ranging from full-thickness regions of cartilage over the joint surface, down to cellular and extracellular regions of cartilage tissue. Cartilage biomechanical behavior varies markedly with the length scale of analysis, as described below.

2.1 *Macro-scale biomechanical testing*

The classical biomechanical experiments on cartilage have relied on assessing the relationship between load and displacement, applied or measured at the surfaces of a cartilage sample. From load-displacement data, physical properties which are dependent on sample and test geometries can be determined. Also, from these experiments and deduced stress-strain data, material properties which are independent of sample and test configurations can be computed, based on models that often employ simplifying assumptions such as homogeneity (i.e. mechanical properties do not vary with position).

Determination of a load-displacement relationship requires controlling one quantity and measuring the other, over a duration of time. In the common situation where load is controlled and displacement is measured, known as the creep test, a constant step-load is applied (Fig. 3A) and resultant displacement (Fig. 3B) is measured. In this

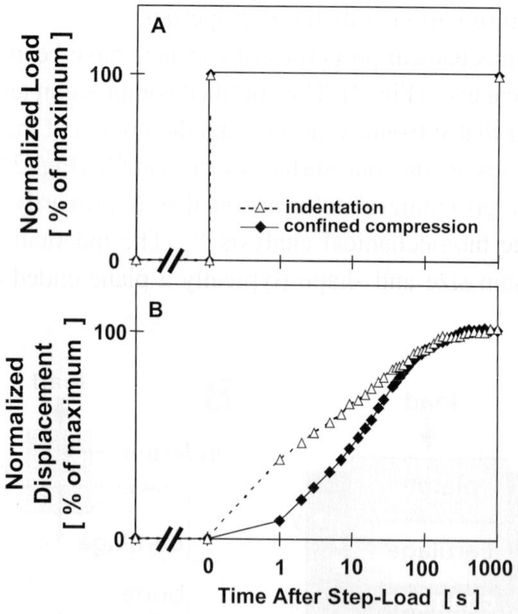

Fig. 3. Time-dependent deformation of cartilage, measured using traditional testing methods of (diamond) confined compression and (triangle) indentation tests. Normalized **(A)** applied load and **(B)** the resultant displacement of the platen or the indenter. Adapted from Ref. 33.

protocol, an increase in displacement with time (i.e. consolidation of cartilage) to an equilibrium is observed. Due to the hydrated and porous nature of the cartilage, fluid is slowly forced out until the solid portion of the cartilage bears all of the load. The converse protocol, known as the stress-relaxation test, applies and holds a displacement and the resultant load is measured. Here, a decrease in load with time to equilibrium also occurs with fluid exudation or redistribution. For either type of dynamic protocol, the equilibrium and transient data can be analyzed[10,27,32] to derive material properties assumed to be homogeneous. Such properties include the aggregate modulus (H_A, a measure of compressive modulus at equilibrium), and hydraulic permeability (k_p, a measure of resistance to fluid exudation). In the theoretical model of this situation, a characteristic time constant is given by $\tau = h^2/(H_A \bullet k_p)$ where h is the characteristic length. This transient nature of cartilage biomechanical behavior is important to note since both confined compression and indentation tests can require a long duration (~hr) to reach equilibrium, as shown in Fig. 3B.

Experimentally, macro-scale testing of cartilage requires planning for the range of loads and displacements that will be applied and measured. Given that cartilage is a thin (~mm) tissue, a mechanical tester with high sensitivity and fine motion control is desired. The expected range of load and displacement should be determined prior to any testing, using constitutive relationships such as:

$$\sigma = H_A \cdot \varepsilon \tag{1}$$

where σ is the stress (load divided by the relevant cross-sectional area), H_A is aggregate modulus (~1 MPa), and ε is strain (displacement divided by sample thickness). For example, if 100 kPa compressive stress is to be applied during a creep test, one can expect a strain of approximately –0.1 (or compression of ~10% of sample thickness). If the sample is 1 mm thick, the testing system should be able to resolve displacements <<100 μm. For a stress relaxation test, the maximum load at the onset of the applied displacement can be several folds higher than the equilibrium load. Such calculations should be used as a guideline for planning experiments and specifying an appropriate testing apparatus.

2.1.1 *Confined compression test — macroscopic*

Cartilage samples from many different species of animals can be prepared for a confined compression test. Often, a sheet of cartilage is cut with a microtome to remove it from underlying bone and then punched to yield a cartilage disk of known diameter. The disk is measured for thickness at multiple locations and averaged, and then placed into a confining chamber (Fig. 2A). Alternatively, cartilage can be tested *in situ* (i.e. remaining on the joint),[33] using a confining ring that cuts into the cartilage and bone, providing a confinement similar to that (Fig. 2A) for excised tissue. The confined sample in either case is topped with a porous platen (Fig. 2A), which allows for fluid flow and exudation out of the platen. The entire setup is often submerged or irrigated with physiologic saline to prevent dehydration of the sample during testing, since such would likely alter the test results.

Then a prescribed load or displacement is applied axially onto the platen to initiate the test. In one study of rabbit cartilage samples *in situ*, each sample (1.8 mm in diameter by 0.3 mm thick) was topped with a porous frit and compressed with a 0.13 N step-load (~50 kPa area-averaged stress, Fig. 3A), using a benchtop mechanical tester. The resultant displacement (e.g. Fig. 3B) increased with time (i.e. becoming compacted), until an equilibrium was reached, after ~1000 sec. The displacement did not change instantaneously at the onset of the load, since (1) cartilage is confined by the chamber, (2) fluid in cartilage cannot exude instantaneously, and (3) the fluid is incompressible under the testing condition. Overall (averaged) tissue strain is typically calculated as the displacement divided by the initial sample (cartilage) thickness.

2.1.2 *Indentation test — macroscopic*

The indentation test (Fig. 2B) is another widely used testing configuration for cartilage biomechanics, due in part to its non-destructive and rapid nature and its potential usefulness as a diagnostic tool. For example, while confined compression testing requires cartilage samples to be cut to certain shapes and sizes, indentation can be performed directly on an intact surface of cartilage and has been used to assess cartilage arthroscopically. Indentation testing at the macro-scale requires instrumentation similar to that for confined compression testing in order to control and measure load and displacement at the surface of the sample (Sec. 2.1.1). The indentation behavior of a sample is affected by size and shape of the indenter tip,[29] cartilage thickness[29] and boundary conditions.[34] Also, unlike confined compression which results in displacement at the sample surface that is generally uniform, displacement of the tissue during indentation depends on position (e.g. r and z, Fig. 2B) relative to the indenter. Therefore, indentation test results are often described in terms of indenter displacement (or the displacement of cartilage under the indenter) and measured or applied load.

During indentation, the indenter tip is aligned orthogonal to the cartilage surface, and then a loading protocol is initiated. The loading protocol may be identical to those used for confined compression (i.e. long-duration creep or stress-relaxation), or may be rapid, whereby a short duration application of load or displacement is used. Short-duration protocols, more clinically advantageous than long protocols, are feasible for indentation tests since the tissue is not confined and the cartilage surface deflects "instantaneously" with applied load and yields a short-term load-displacement relation. In an indentation study (Fig. 3),[33] a 0.56 mm diameter plane-ended cylindrical tip was used to compress rabbit cartilage with a ~70 kPa stress. The result (Fig. 3B) was an initial jump in displacement, followed by a gradual increase in displacement to an equilibrium.

Both the initial and time-dependent responses during indentation tests have been modeled assuming homogeneity and a number of other assumptions,[29,32] which allows for determination of material properties from force-displacement-time (or stress-strain-time) experimental data. If a rapid indentation test is performed, the result may be shown as indenter force (initial force for an applied displacement),[35] modulus (initial force divided by indenter area), or the apparent stiffness (force divided by displacement).[30]

2.2 *Micro-scale biomechanical testing*

The biomechanical properties that are discerned are dependent on the length scale at which cartilage is examined (Fig. 1). Cartilage can be analyzed at length scales ranging from intact joints to full-thickness regions of cartilage over the joint surface, layers of cartilage tissue from different depths, cellular and extracellular regions of cartilage tissue, molecular constituents within these regions, and beyond. One set of biomechanical properties can be determined if articular cartilage is assumed to be homogeneous.[10] Another set of very different behaviors can be observed if cartilage is analyzed at the finer length scale of tissue layers at varying depths from the articular surface.[11]

In order to discern deformation through the depth of cartilage, microscopy (Fig. 4A) has been widely used along with two-dimensional biomechanical testing configurations

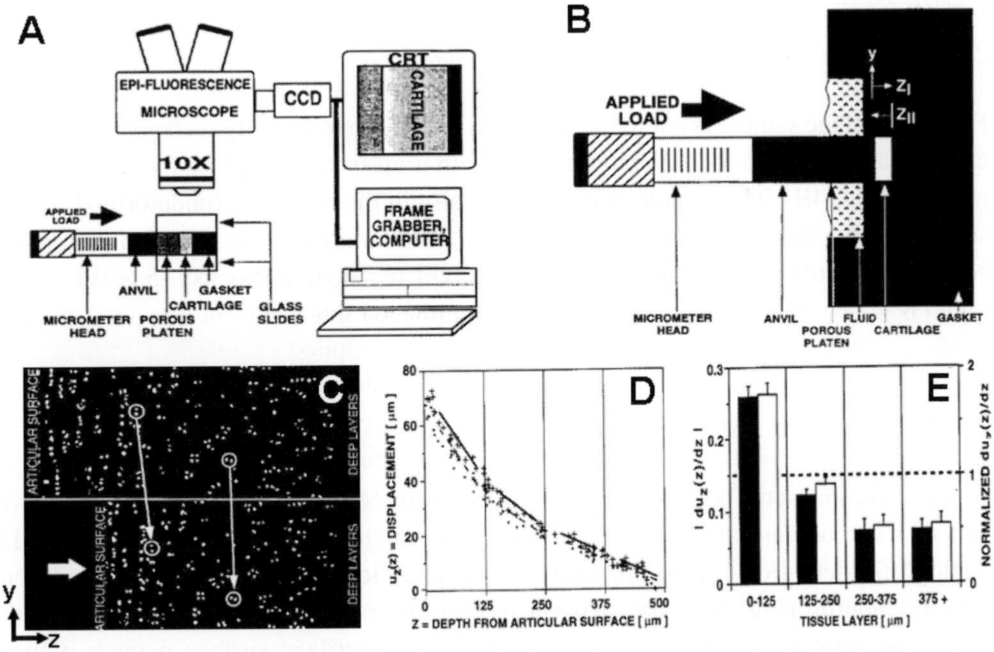

Fig. 4. **(A)** Apparatus to perform cell-level confined compression test. **(B)** Confining chamber is viewed from the microscope to obtain **(C)** images of cartilage deformation, analyzed to yield axial **(D)** displacement and **(E)** strain as a function of tissue-depth. Adapted from Ref. 36.

similar to their three-dimensional counterparts. The basic concept is to impose deformation by loading tissue surfaces using test configurations similar to those used in macro-scale tests (Figs. 4A and 4B) and record multiple images (Fig. 4C) for analysis of deformation (Figs. 4D and 4E). For example, confined compression testing has been performed while microscopically imaging cartilage simultaneously.[11,12,36] Typically, fluorescence or reflected light imaging is used. With fluorescence microscopy, nucleus-binding dyes such as Hoechst 33258 and propidium iodide are applied in order to highlight the indwelling cells which serve as intrinsic fiducial markers (e.g. Fig. 4C).[31,36] Alternatively, cartilage surfaces can be impregnated with dark materials (e.g. black enamel[37]) which provide sufficient contrast when illuminated with a light source and serve as markers.

Images of undeformed and deformed cartilage can be analyzed in a number of ways to assess intra-tissue deformation. Computer-assisted visual identification of applied markers or fluorescent cell nuclei in the samples[11,36,37] provide direct but time-consuming means of tracking tissue deformation as the 2-D displacement of the markers and nuclei. Such cell nuclei are typically identified as objects of ~5 μm diameter and location that can be resolved to ~0.2 μm (by calculating the centroid). Semi- or fully-automated methods, such as video image correlation[38] are available, and have been used for strain analysis of various non-biological,[39] and biological materials such as bone,[40] retina,[41] and cartilage.[31]

2.2.1 *Confined compression test — microscopic*

Here, a cartilage sample is cut to the desired size and shape (usually rectangular or half-cylindrical) to fit into a confining chamber with a glass-wall to allow for microscopic imaging (Fig. 4B). The sample is imaged initially in the reference (undeformed) state, subjected to compression, and then re-imaged after an appropriate duration (e.g. after equilibration following stress-relaxation).[36] Small rectangular blocks, $\sim 5 \times 0.7 \times 0.5$ mm^3 ($W \times L \times H$), of bovine cartilage has been imaged this way.

2.2.2 *Indentation test — microscopic*

Using a similar setup as that used for micro-scale confined compression (Figs. 4A and 4B), osteochondral (cartilage-on-bone) samples have been imaged during a 2-D analog of indentation testing. To study 3-D indentation by a plane-ended cylinder, a 2-D indentation (Fig. 5, insert) using a rectangular prismatic tip can be performed. (Here, an axial plane during 3-D indentation of a cartilage surface using a plane-ended cylinder reveals rectangular cut-planes for both the indenter and the sample.) With this analog, the main features of the 3-D indentation can be studied with microscopic viewing.

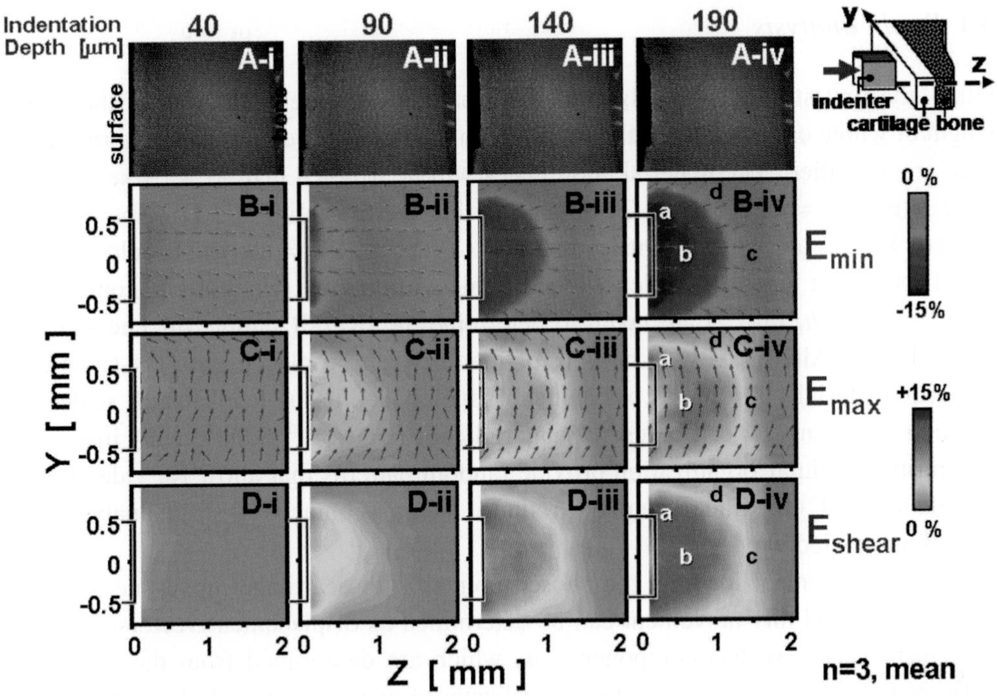

Fig. 5. Two-dimensional microscopic indentation test (insert) was used to obtain **(A)** micrographs of deforming cartilage at depths of (i) 40, (ii) 90, (iii) 140, and (iv) 190 mm, which was analyzed using video image correlation and strain calculation to yield principal strain maps. **(B)** The minimum and **(C)** the maximum principal strain components, and **(D)** the maximum shear strain components were plotted and (arrows) principal directions were overlaid. Adapted from Ref. 31.

In one study,[31] human osteochondral blocks, ~10 × 10 × 4 mm^3 ($W \times L \times H$), were indented with rectangular prismatic tip (contact area of 1 mm wide by 4 mm tall) dynamically, to a depth of ~200 μm (at ~130 μm/s). Digital images (Fig. 5A) were taken at indentation depths of 40, 90, 140, and 190 μm (~2.2%, 4.9%, 7.7%, and 10.4% of the 1.83 ± 0.15 mm cartilage thickness).

3. Tissue Deformation and Strain Within Articular Cartilage

Cartilage deformation can be quantified in a number of ways, including with measures of strain. Strain is a normalized measure of change in length. For example, if a one-dimensional bar shortens by 1% of the initial length, its strain is ~0.01, depending on the type of strain being calculated. For cartilage deformation, infinitesimal or Lagrangian descriptions of strain have been used widely for small and large deformations, respectively. (Readers are referred to Fung[42] for a thorough introduction to definition of strain and analysis methods.)

3.1 *Strain analysis*

Often, the first step in the analysis of deformation is the determination of the initial length of whole or segments of the body. To do this, material coordinates at a reference state are identified. The material coordinates will be denoted here as X_R, where $R = 1$, 2, or 3 (e.g. $X_R = (X_1, X_2, X_3)$, Cartesian coordinates of a point in 3-D). Note that the numeric indices are arbitrary; instead of $X_R = (X_1, X_2, X_3)$, $X_R = (X, Y, Z)$ could be used, in 3-D cases. At large, tissue-level length scales, often only two material coordinates (i.e. $z = 0$ and h, Fig. 2) are taken to determine the overall sample thickness or the gauge length. Alternatively, a number of markers can be placed onto the sample to serve as material coordinates. At a much smaller, cell/tissue-level length scale, the location of cell nuclei can be useful material coordinates.[11,37] Additionally, images of the intrinsic cellular organization provide random patterns that allow for video image correlation (VIC) method, recently introduced for non-contact 2-D strain analysis,[43] in a semi-automated way.

After mechanical loading, the chosen points on the cartilage move due to tissue deformation, and this movement can be determined as displacement vectors. The displacement vector, **u**, has components, u_i, which are determined from the difference between the current and the reference coordinates of material points. If the current coordinates are denoted as x_i, then

$$u_i = x_i - X_i. \quad (i = 1,\ 2,\ \text{and}\ 3) \tag{2}$$

For the cases where several material coordinates are chosen, a displacement field (displacement as a function of spatial position) is obtained. It is possible at this stage to introduce various interpolation schemes to re-map the chosen coordinates to desired spatial positions (e.g. to obtain even spacing of coordinates).

The displacement gradient tensor, **G** (with components G_{iR}), can also be determined and subsequently used to determine strain. A gradient is a measure of "slope" of the quantity of interest at a particular position. For displacement, the gradient is determined as:

$$G_{iR} = \frac{\partial u_i}{\partial X_R}. \quad (i,\ R = 1,\ 2,\ \text{and}\ 3) \tag{3}$$

For example, for a 2-D case (i, R = 1 and 2), there are four components of displacement gradients, i.e.

$$\frac{\partial u_1}{\partial X_1}, \frac{\partial u_1}{\partial X_2}, \frac{\partial u_2}{\partial X_1}, \text{and}\ \frac{\partial u_2}{\partial X_2}.$$

Strains of various types can be determined directly from displacement gradients. For small displacements (relative to initial length), infinitesimal strains are typically calculated as follows:

$$\varepsilon_{ij} = \frac{1}{2}\left[\frac{\partial u_j}{\partial X_i} + \frac{\partial u_i}{\partial X_j}\right]. \tag{4}$$

Explicitly for a 2-D case, the three unique components (due to symmetry) of the infinitesimal strain are:

$$\varepsilon_{11} = \partial u_1/\partial X_1 \tag{4a}$$

$$\varepsilon_{22} = \partial u_2/\partial X_2 \tag{4b}$$

$$\varepsilon_{12} = \varepsilon_{21} = 0.5\,(\partial u_1/\partial X_2 + \partial u_2/\partial X_1). \tag{4c}$$

For large displacements (relative to initial length), the Lagrangian description of strain is often used:

$$E_{ij} = \frac{1}{2}\left[\frac{\partial u_j}{\partial X_i} + \frac{\partial u_i}{\partial X_j} + \frac{\partial u_\alpha}{\partial X_i}\frac{\partial u_\alpha}{\partial X_j}\right], \tag{5}$$

where repeated indices indicate summation. Explicitly for a 2-D case, the components of the Lagrangian strain are:

$$E_{11} = \frac{1}{2}\left[2\frac{\partial u_1}{\partial X_1} + \left(\frac{\partial u_1}{\partial X_1}\right)^2 + \left(\frac{\partial u_2}{\partial X_1}\right)^2\right] \tag{5a}$$

$$E_{22} = \frac{1}{2}\left[2\frac{\partial u_2}{\partial X_2} + \left(\frac{\partial u_1}{\partial X_2}\right)^2 + \left(\frac{\partial u_2}{\partial X_2}\right)^2\right] \tag{5b}$$

$$E_{12} = E_{21} = \frac{1}{2}\left[\frac{\partial u_1}{\partial X_2} + \frac{\partial u_2}{\partial X_1} + \left(\frac{\partial u_1}{\partial X_1}\frac{\partial u_1}{\partial X_2}\right) + \left(\frac{\partial u_2}{\partial X_1}\frac{\partial u_2}{\partial X_2}\right)\right]. \tag{5c}$$

E_{11}, E_{22}, and E_{12} represent normal (in 1- and 2-directions) and shear strain components, respectively, in the current frame of reference. It is sometimes useful to determine principal strain components, which are independent of the frame of reference. Principal strains are obtained when the frame of reference is rotated such that the shear strain vanishes; for example, pure shear can be decomposed into a combination of stretch in one direction and collapse in the orthogonal direction. Principal strains are useful for a number of engineering applications, e.g. determination of material failure and for facilitating comparison of strain data from different sets of experiments. The minimum principal, the maximum principal, and the maximum shear strain components are determined[44] respectively as:

$$E_{\min} = \frac{1}{2}(E_{11} + E_{22}) - \frac{1}{2}\sqrt{(E_{11} - E_{22})^2 + 4E_{12}^2} \tag{6A}$$

$$E_{\max} = \frac{1}{2}(E_{11} + E_{22}) + \frac{1}{2}\sqrt{(E_{11} - E_{22})^2 + 4E_{12}^2} \tag{6B}$$

$$E_{\text{shear}} = \frac{1}{2}\sqrt{(E_{11} - E_{22})^2 + 4E_{12}^2}. \tag{6C}$$

The maximum and minimum principal directions are determined[44] as follows:

$$\tan(2\theta_{\max}) = \frac{2E_{12}}{E_{11} - E_{22}} \quad \text{and} \tag{7a}$$

$$\theta_{\min} = \theta_{\max} + \frac{\pi}{2}. \tag{7b}$$

3.2 *Intra-tissue strain during confined compression testing*

Images are processed to accentuate cell nuclei (Fig. 4C), and nuclei from different depths of the tissue are selected for tracking (arrows, Fig. 4C). The displacement profile is measured (Fig. 4D) and axial strain is calculated in sequential 125–250 μm thick cartilage layers (Fig. 4E), using linear regression fitting of the displacement values in each layer. The slope of the regression fits are taken as the infinitesimal strain.

The cell nuclei in images are typically easily identified (Fig. 4C). Measurement of the change in nucleus position between the reference (i.e. uncompressed) and compressed images allow for quantitation of displacement, $u(y, z)$, which is primarily parallel to the loading axis (z). The displacement in the z-direction, $u_z(z)$, varies nonlinearly with depth in adult cartilage. Note that a homogeneous material at equilibrium would be predicted to exhibit a linear $u_z(z)$ profile.

In the study on adult bovine cartilage (Fig. 4),[36] the strain magnitude decreased with depth from the articular surface. The strains were: layer 1, 0.365 ± 0.017; layer 2,

0.150 ± 0.007; layer 3, 0.067 ± 0.007; and layer 4, 0.060 ± 0.009. Those strains were statistically different from each other, with the exception of strains in layers 3 and 4. Qualitatively, the strain in layer 2 was similar to the applied compression of 15% (i.e. dashed line, Fig. 4E), whereas that for layer 1 was greater than, and for layers 3 and 4 was less than, the applied compression. The results indicated that the compressive modulus of articular cartilage varies markedly with depth, and that this feature of cartilage may affect joint function during compressive loading.

3.3 *Intra-tissue strain during indentation testing*

Images are analyzed to determine a displacement field using direct tracking or an automated method such as video image correlation as described in Sec. 2.2. Video image correlation compares digital images of the undeformed and deformed samples in small patches that contain unique patterns of pixel intensities (provided by stained nuclei). Assuming that there exists a one-to-one correspondence between the pixels in the reference and the deformed subsets, and that the subset region deforms linearly, the displacement of the reference subset can be computed.[38] Multiple discrete subsets are analyzed in this way to determine the displacement of the center point of each subset, and collectively, a displacement map, $u_i(z, y)$, relative to the reference image. While convenient, any automated method such as this requires careful validation against a direct tracking method.

Using displacement gradients [equation (3)], Lagrangian principal strains of E_{min}, E_{max}, and E_{shear} are calculated [equations (5) and (6)] along with the principal direction of θ_{min} and θ_{max} [equation (7)]. In the study of human articular cartilage,[31] strain profiles during 2-D indentation were determined as described above. Strain profiles varied greatly as a function of position relative to the indenter (Fig. 5). E_{min} was compressive, generally high in magnitude in the areas of cartilage under and near the indenter (areas **a** and **b**, Fig. 5B-iv), and the greatest in area **a** directly under the indenter edges (–0.18 ± 0.02 at [$Y = 0.5$ mm, $Z = 0.1$ mm], Fig. 5B-iv). For the same samples, E_{min} values diminished with cartilage depth, being in the range of –0.05 to 0 in area **c** near the bone (–0.03 ± 0.01 at [0 mm, 1.5 mm], Fig. 5B-iv), and in area **d**, far away from the centerline (–0.05 ± 0.01 at [0.75 mm, 0.75 mm], Fig. 5B-iv). In addition, E_{max} was tensile and high in magnitude (0.07–0.10) in areas **a** and **b** (Fig. 5C-iv) while being relatively low (~0.03) in areas **c** and **d** (Fig. 5C-iv), with a spatial profile similar to that of E_{min}. However, the range of strain magnitudes was not as large as that seen in the E_{min} profile. The spatial variations in E_{shear} were also similar to that of E_{min}, with the maximum values found in area **a** (0.14 ± 0.01 at [0.5 mm, 0.1 mm], Fig. 5D-iv) and diminishing with distance away from the indenter and centerline.

The principal directions (arrows in Figs. 5B and 5C) indicated that E_{min}, compressive strain, was mainly in the axial direction, while E_{max}, tensile strain, was mainly in the lateral direction. The profiles of intra-tissue strain along with the in-plane principal directions suggested that as the indenter tips compressed the cartilage surface, the bulk tissue compressed vertically and expanded laterally while also undergoing a shear deformation.

These results have several implications for clinical indentation tests. The relatively high strain amplitudes near the surface and directly under or adjacent to the indenter indicate that the mechanical properties of these tissue regions will dominate the indentation response. Also, the region-specific strain values help to identify potential effects of indentation on cartilage biology.

4. Summary and Conclusions

Biomechanical testing methods for articular cartilage (Sec. 2) and the theoretical formulation (Sec. 3) to determine strain at different length scales were presented. For both confined compression and indentation tests, the tissue-level testing required the application and measurement of surface force and displacement, while the cell-level testing required microscopic imaging of cell nuclei and sophisticated methods to determine a displacement map. These quantities were then used to determine tissue strain.

Mechanical testing at tissue- and cell-level length scales provides useful measures for basic and clinically-oriented research of articular cartilage. For basic understanding of cartilage mechanics, biomechanical "material properties" of cartilage can be deduced from stress and strain (and time), along with theoretical models.[10,29,32,45,46] These deduced properties, or even simple measures such as stiffness, may be related to other non-biomechanical measures such as gross- and histo-morphology,[47,48] biochemical content,[49,50] and cell metabolism,[51,52] to elucidate relationships between function, structure, and composition of articular cartilage and how those modulate cellular responses. For clinical usefulness, non-destructive and rapid indentation test provides measures of cartilage stiffness that have been correlated with other measures of cartilage health and disease, such as gross degeneration,[16,17] histological grading,[53–55] and biochemical content.[54,55]

There are numerous methods for mechanical testing of cartilage and its constituents, and at other length scales than those presented here. At greater length scales, deformation and stresses in cartilage in whole joints have been determined using MRI[56,57] and mechanical methods.[58–60] At smaller length scales, deformation of individual chondrocytes have been assessed using bright-field,[61] confocal,[62] and atomic force microscopy.[63] At the molecular length scale, collagen molecule strain was measured using an X-ray diffraction technique,[64] while electro-mechanical interactions between proteoglycan molecules have been studied using atomic force microscopy.[65] Combined, these studies of articular cartilage at a wide variety of length scales provide an increasingly complete picture of articular cartilage, which will help to advance our understanding of cartilage, function in health and disease.

Acknowledgments

This work was supported in part by grants from the National Institutes of Health, National Science Foundation, Musculoskeletal Transplant Foundation, and National Football League Charities, and also a grant from the Howard Hughes Medical Institute Professors Program to University of California-San Diego (in support of RLS) and a Faculty Fellow Award from the University of California (to WCB). The authors thank Gregory Williams for proofreading the chapter.

References

1. V. C. Mow, W. Zhu and A. Ratcliffe, Structure and function of articular cartilage and meniscus. In: *Basic Orthopaedic Biomechanics*, eds. V. C. Mow and W. C. Hayes (Raven Press, New York, 1991), pp. 143–198.
2. A. Maroudas, Physico-chemical properties of articular cartilage. In: *Adult Articular Cartilage*, eds. M. A. R. Freeman (Pitman Medical, Tunbridge Wells, England, 1979), pp. 215–290.
3. M. D. Buschmann and A. J. Grodzinsky, A molecular model of proteoglycan-associated electrostatic forces in cartilage mechanics, *J. Biomech. Eng.* **117**: 179–192 (1995).
4. A. Maroudas, Balance between swelling pressure and collagen tension in normal and degenerate cartilage, *Nature* **260**: 808–809 (1976).
5. H. Muir, P. Bullough and A. Maroudas, The distribution of collagen in human articular cartilage with some of its physiological implications, *J. Bone Joint Surg. Br.* **52**: 554–563 (1970).
6. G. E. Kempson, H. Muir, C. Pollard and M. Tuke, The tensile properties of the cartilage of human femoral condyles related to the content of collagen and glycosaminoglycans, *Biochim. Biophys. Acta* **297**: 456–472 (1973).
7. M. F. Venn, Variation of chemical composition with age in human femoral head cartilage, *Ann. Rheum. Dis.* **37**: 168–174 (1978).
8. R. M. Aspden and D. W. L. Hukins, Collagen organization in articular cartilage, determined by x-ray diffraction, and its relationship to tissue function, *Proc. R. Soc. Lond. B* **212**: 299–304 (1981).
9. N. D. Broom, The collagen framework of articular cartilage: its profound influence on normal and abnormal load-bearing function. In: *Collagen: Chemistry, Biology, and Biotechnology*, ed. M. E. Nimni (CRC Press, Inc., Boca Raton, FL, 1988), pp. 243–265.
10. V. C. Mow, S. C. Kuei, W. M. Lai and C. G. Armstrong, Biphasic creep and stress relaxation of articular cartilage in compression: theory and experiment, *J. Biomech. Eng.* **102**: 73–84 (1980).
11. R. M. Schinagl, D. Gurskis, A. C. Chen and R. L. Sah, Depth-dependent confined compression modulus of full-thickness bovine articular cartilage, *J. Orthop. Res.* **15**: 499–506 (1997).
12. S. S. Chen, Y. H. Falcovitz, R. Schneiderman, A. Maroudas and R. L. Sah, Depth-dependent compressive properties of normal aged human femoral head articular cartilage, *Osteoarthr. Cartil.* **9**: 561–569 (2001).
13. T. E. Andreoli, J. C. Bennett, C. C. J. Carpenter, F. Plum and L. H. J. Smith, *Cecil Essentials of Medicine* (W. B. Saunders Company, Philadelphia, PA, 1993), p. 921.
14. A. Praemer, S. Furner and D. P. Rice, *Musculoskeletal Conditions in the United States* (American Academy of Orthopaedic Surgeons, Park Ridge, IL, 1992), p. 199.
15. L. Sokoloff, Elasticity of aging cartilage, *Proc. Fedn. Am. Socs. Exp. Biol.* **25**: 1089–1095 (1966).
16. S. Roberts, B. Weightman, J. Urban and D. Chappell, Mechanical and biochemical properties of human articular cartilage in osteoarthritic femoral heads and in autopsy specimens, *J. Bone Joint Surg. Br.* **68–B**: 278–288 (1986).

17. G. E. Kempson, C. J. Spivey, S. A. Swanson and M. A. Freeman, Patterns of cartilage stiffness on normal and degenerate human femoral heads, *J. Biomech.* **4**: 597–609 (1971).
18. R. L. Sah, Y. J. Kim, J. H. Doong, A. J. Grodzinsky, A. H. K. Plaas and J. D. Sandy, Biosynthetic response of cartilage explants to dynamic compression, *J. Orthop. Res.* **7**: 619–636 (1989).
19. Y. J. Kim, R. L. Sah, A. J. Grodzinsky, A. H. Plaas and J. D. Sandy, Mechanical regulation of cartilage biosynthetic behavior: physical stimuli, *Arch. Biochem. Biophys.* **311**: 1–12 (1994).
20. M. Jin, E. H. Frank, T. M. Quinn, E. B. Hunziker and A. J. Grodzinsky, Tissue shear deformation stimulates proteoglycan and protein biosynthesis in bovine cartilage explants, *Arch. Biochem. Biophys.* **395**: 41–48 (2001).
21. M. Thibault, A. R. Poole and M. D. Buschmann, Cyclic compression of cartilage/bone explants *in vitro* leads to physical weakening, mechanical breakdown of collagen and release of matrix fragments, *J. Orthop. Res.* **20**: 1265–1273 (2002).
22. B. Kurz, M. Jin, P. Patwari, D. M. Cheng, M. W. Lark and A. J. Grodzinsky, Biosynthetic response and mechanical properties of articular cartilage after injurious compression, *J. Orthop. Res.* **19**: 1140–1146 (2001).
23. D. D. D'Lima, S. Hashimoto, P. C. Chen, M. K. Lotz and C. W. Colwell, Jr., Cartilage injury induces chondrocyte apoptosis, *J. Bone Joint Surg. Am.* **83–A**(Suppl. 2): 19–21 (2001).
24. P. Patwari, V. Gaschen, I. E. James, E. Berger, S. M. Blake, M. W. Lark, *et al.*, Ultrastructural quantification of cell death after injurious compression of bovine calf articular cartilage, *Osteoarthr. Cartil.* **12**: 245–252 (2004).
25. T. M. Quinn, R. G. Allen, B. J. Schalet, P. Perumbuli and E. B. Hunziker, Matrix and cell injury due to sub-impact loading of adult bovine articular cartilage explants: effects of strain rate and peak stress, *J. Orthop. Res.* **19**: 242–249 (2001).
26. A. K. Williamson, A. C. Chen and R. L. Sah, Compressive properties and function-composition relationships of developing bovine articular cartilage, *J. Orthop. Res.* **19**: 1113–1121 (2001).
27. E. H. Frank and A. J. Grodzinsky, Cartilage electromechanics-II. A continuum model of cartilage electrokinetics and correlation with experiments, *J. Biomech.* **20**: 629–639 (1987).
28. G. E. Kempson, M. A. R. Freeman and S. A. V. Swanson, The determination of a creep modulus for articular cartilage by indentation tests of the human femoral head, *J. Biomech.* **4**: 239–250 (1971).
29. W. C. Hayes, L. M. Keer, K. G. Herrmann and L. F. Mockros, A mathematical analysis for indentation tests of articular cartilage, *J. Biomech.* **5**: 541–551 (1972).
30. W. C. Bae, A. W. Law, D. Amiel and R. L. Sah, Sensitivity of indentation testing to step-off edges and interface integrity in cartilage repair, *Ann. Biomed. Eng.* **32**: 360–369 (2004).
31. W. C. Bae, C. W. Lewis, M. E. Levenston and R. L. Sah, Indentation testing of human articular cartilage: effects of probe tip geometry and indentation depth on intra-tissue strain, *J. Biomech.* **39**: 1039–1047 (2006).
32. A. F. Mak, W. M. Lai and V. C. Mow, Biphasic indentation of articular cartilage-I. Theoretical analysis, *J. Biomech.* **20**: 703–714 (1987).
33. J. Dounchis, F. L. Harwood, A. C. Chen, W. C. Bae, R. L. Sah, R. D. Coutts, *et al.*, Cartilage repair with autogenic perichondrium cell and polylactic acid grafts, *Clin. Orthop. Rel. Res.* **377**: 248–264 (2000).
34. J.-K. Suh and R. L. Spilker, Indentation analysis of biphasic articular cartilage: nonlinear phenomena under finite deformation, *J. Biomech. Eng.* **116**: 1–9 (1994).
35. T. Lyyra, I. Kiviranta, U. Vaatainen, H. J. Helminen and J. S. Jurvelin, *In vivo* characterization of indentation stiffness of articular cartilage in the normal human knee, *J. Biomed. Mater. Res.* **48**: 482–487 (1999).
36. R. M. Schinagl, M. K. Ting, J. H. Price and R. L. Sah, Video microscopy to quantitate the inhomogeneous equilibrium strain within articular cartilage during confined compression, *Ann. Biomed. Eng.* **24**: 500–512 (1996).

37. D. A. Narmoneva, J. Y. Wang and L. A. Setton, Nonuniform swelling-induced residual strains in articular cartilage, *J. Biomech.* **32**: 401–408 (1999).
38. M. A. Sutton, S. R. McNeill, J. D. Helm and Y. J. Chao, Advances in two-dimensional and three-dimensional computer vision. In: *Photomechanics*, ed. P. K. Rastogi (Springer, New York, 2000), pp. 323–372.
39. T. C. Chu, W. F. Ranson, M. A. Sutton and W. H. Peters, Applications of digital image correlation techniques to experimental mechanics, *Exp. Mech.* **25**: 232–244 (1985).
40. B. K. Bay, Texture correlation: a method for the measurement of detailed strain distributions within trabecular bone, *J. Orthop. Res.* **13**: 258–267 (1995).
41. W. Wu, W. H. Peters, 3rd and M. E. Hammer, Basic mechanical properties of retina in simple elongation, *J. Biomech. Eng.* **109**: 65–67 (1987).
42. Y. C. Fung, *A First Course in Continuum Mechanics* (Prentice-Hall, Englewood Cliffs, 1977), p. 340.
43. W. H. Peters and W. F. Ranson, Digital imaging techniques in experimental stress analysis, *Opt. Eng.* **21**: 427–432 (1982).
44. A. C. Eringen, *Mechanics of Continua* (Wiley, New York, 1967).
45. V. C. Mow, M. C. Gibbs, W. M. Lai and W. B. Zhu and K. A. Athanasiou, Biphasic indentation of articular cartilage-II. A numerical algorithm and an experimental study, *J. Biomech.* **22**: 853–861 (1989).
46. C. G. Armstrong, W. M. Lai, V. C. Mow, An analysis of the unconfined compression of articular cartilage, *J. Biomech. Eng.* **106**: 165–173 (1984).
47. D. H. Collins, *The Pathology of Articular and Spinal Disease* (Arnold, London, 1949).
48. K. D. Jadin, B. L. Wong, W. C. Bae, K. W. Li, A. K. Williamson, B. L. Schumacher, *et al.* Depth-varying density and organization of chondrocyte in immature and mature bovine articular cartilage assessed by 3-D imaging and analysis, *J. Histochem. Cytochem.* **53**: 1109–1119 (2005).
49. R. W. Farndale, C. A. Sayers and A. J. Barrett, A direct spectrophotometric microassay for sulfated glycosaminoglycans in cartilage cultures, *Connect. Tissue. Res.* **9**: 247–248 (1982).
50. J. F. Woessner, The determination of hydroxyproline in tissue and protein samples containing small proportions of this imino acid, *Arch. Biochem. Biophys.* **93**: 440–447 (1961).
51. R. L. Sah, Y. J. Kim and A. J. Grodzinsky, The effect of mechanical compression on cartilage metabolism. In: *Methods in Cartilage Research*, eds. A. Maroudas and K. E. Kuettner (Academic Press, New York, 1990), pp. 116–119.
52. Y. Kim, *Physical Regulation of Cartilage Metabolism: Effects of Compression on Specific Matrix Molecules and Their Spatial Distribution* (Massachusetts Institute of Technology, 1993).
53. H. J. Mankin and L. Lipiello, Biochemical and metabolic abnormalities in articular cartilage from osteoarthritic human hips, *J. Bone Joint Surg. Am.* **52-A**: 424–434 (1970).
54. T. Franz, E. M. Hasler, R. Hagg, C. Weiler, R. P. Jakob and P. Mainil-Varlet, *In situ* compressive stiffness, biochemical composition, and structural integrity of articular cartilage of the human knee joint, *Osteoarthr. Cartil.* **9**: 582–592 (2001).
55. W. C. Bae, M. M. Temple, D. Amiel, R. D. Coutts, G. G. Niederauer and R. L. Sah, Indentation testing of human cartilage: sensitivity to articular surface degeneration, *Arthritis Rheum.* **48**: 3382–3394 (2003).
56. F. Eckstein, B. Lemberger, C. Gratzke, M. Hudelmaier, C. Glaser, K. H. Englmeier, *et al.*, *In vivo* cartilage deformation after different types of activity and its dependence on physical training status, *Ann. Rheum. Dis.* **64**: 291–295 (2005).
57. F. Eckstein, B. Lemberger, T. Stammberger, K. H. Englmeier and M. Reiser, Patellar cartilage deformation *in vivo* after static versus dynamic loading, *J. Biomech.* **33**: 819–825 (2000).
58. R. C. Haut, Contact pressures in the patellofemoral joint during impact loading on the human flexed knee, *J. Orthop. Res.* **7**: 272–280 (1989).
59. W. A. Hodge, R. S. Fijan, K. L. Carlson, R. G. Burgess, W. H. Harris and R. W. Mann, Contact pressures in the human hip joint measured *in vivo*, *Proc. Natl. Acad. Sci. USA* **83**: 2879–2883 (1986).

60. T. D. Brown and D. T. Shaw, *In vitro* contact stress distributions in the natural human hip, *J. Biomech.* **16**: 373–384 (1983).
61. W. R. Jones, H. P. Ting-Beall, G. M. Lee, S. S. Kelley, R. M. Hochmuth and F. Guilak, Alterations in the Young's modulus and volumetric properties of chondrocytes isolated from normal and osteoarthritic human cartilage, *J. Biomech.* **32**: 119–127 (1999).
62. F. Guilak, A. Ratcliffe and V. C. Mow, Chondrocyte deformation and local tissue strain in articular cartilage: a confocal microscopy study, *J. Orthop. Res.* **13**: 410–421 (1995).
63. K. D. Costa and F. C. P. Yin, Analysis of indentation: implications for measuring mechanical properties with atomic force microscopy, *J. Biomech. Eng.* **121**: 462–471 (1999).
64. N. Sasaki and S. Odajima, Stress-strain curve and Young's modulus of a collagen molecule as determined by the x-ray diffraction technique, *J. Biomech.* **29**: 655–658 (1996).
65. L. Ng, A. J. Grodzinsky, P. Patwari, J. Sandy, A. Plaas and C. Ortiz, Individual cartilage aggrecan macromolecules and their constituent glycosaminoglycans visualized via atomic force microscopy, *J. Struct. Biol.* **143**: 242–257 (2003).
66. E. B. Hunziker, Articular cartilage structure in humans and experimental animals. In: *Articular Cartilage and Osteoarthritis*, eds. K. E. Kuettner, R. Schleyerbach, J. G. Peyron and V. C. Hascall (Raven Press, New York, 1992), pp. 183–199.
67. J. A. Buckwalter and L. C. Rosenberg, Structural changes during development in bovine fetal epiphyseal cartilage, *Coll. Relat. Res.* **3**: 489–504 (1983).
68. M. Doherty, ed., *Color Atlas and Text of Osteoarthritis* (Times Mirror International Publishers Limited, London, 1994), p. 208.

CHAPTER 16

DESIGN AND DEVELOPMENT OF AN *IN VIVO* FORCE-SENSING KNEE PROSTHESIS

Darryl D. D'Lima and Peter C. Y. Chen

Abstract

Understanding the origin and distribution of tibiofemoral forces is essential in the design and development of total knee replacement prostheses because these forces determine the fit, wear, and long-term outcome of the prosthesis. A total knee replacement tibial tray component with four embedded force transducers and a telemetry system was developed to directly measure tibiofemoral compressive forces *in vivo*. The design underwent many modifications since 1993 and has been tested extensively in the laboratory to determine performance, accuracy, and safety. Trial surgical implantation was performed to demonstrate feasibility and to test the utility of the prosthesis as a dynamic ligament balancing device, which is an essential measure for proper alignment and success of the operation. Dr. Clifford Colwell Jr. performed the first permanent surgical implantation on February 27, 2004 and *in vivo* tibial forces were recorded for the first time in history. Knee forces were monitored during recovery and rehabilitation in the early period after surgery, as well as for activities of daily living and exercise over the first 12 months after surgery. These data were supplemented with video motion analysis, electromyography, and ground reaction force measurement. The data will provide the necessary information for the orthopedic scientific community to improve knee prostheses designs.

1. Introduction

Total knee arthroplasty or replacement is the treatment of choice for end-stage arthritis of the knee. Approximately 500,000 knee replacements were performed in the United States in 2006 and are projected to reach 3.5 million in 2030. As indicated in Lee's chapter on entrepreneurship, this operation is considered to be one of the 24 key medical and biological innovations voted by the fellows of AIMBE in 2005 to the AIMBE Hall of Fame. The surgery involves removal of diseased bone and cartilage lining the knee joint.

The removed articular surfaces are replaced by artificial components (prosthetic implants), typically made of metal and polyethylene. The popular materials in use today are alloys of cobalt-chrome or titanium articulating against ultra high molecular weight polyethylene. The lower end of the femur is replaced with a convex metal implant, while the upper end of the tibia is replaced with a concave polyethylene implant (Figs. 1a and 1b). The articular surface of the patella is also often replaced with a convex polyethylene implant. The surgery provides dramatic relief from arthritic pain; however, normal

(a)

(b)

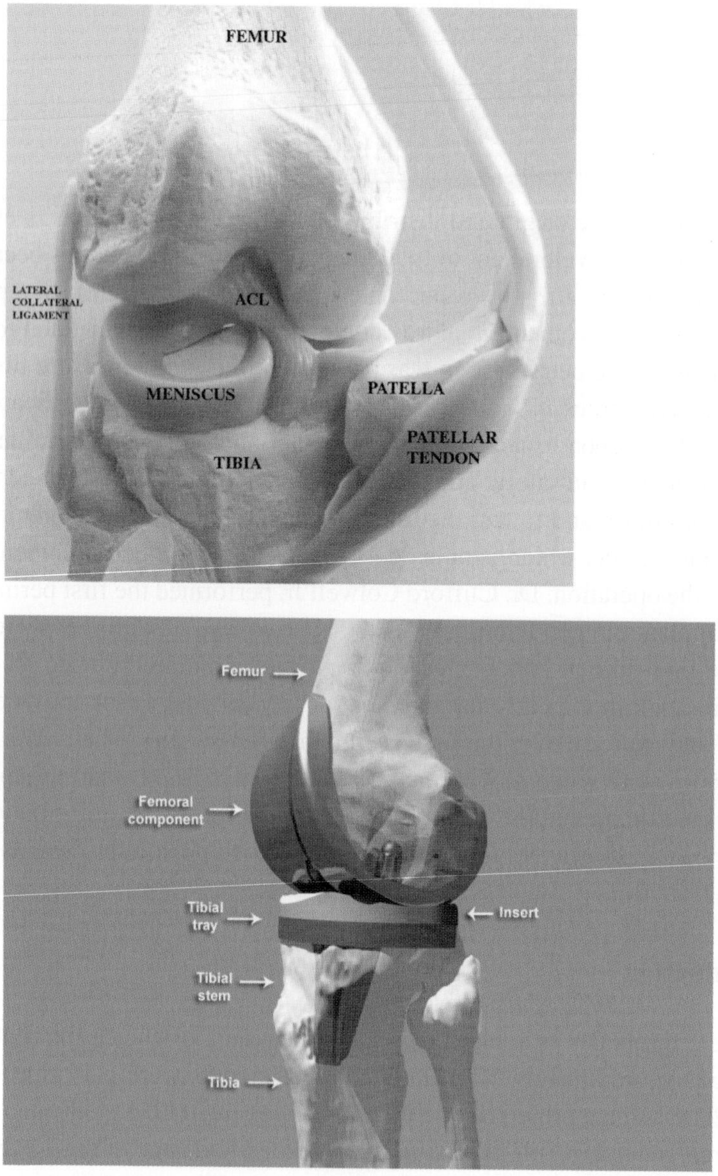

Fig. 1. (**a**) Model of a normal knee. (**b**) After installation of a typical knee implant.

knee function may not always be completely restored. In addition, the lifetime of the prosthesis is limited (typically 15 to 20 years). Important factors that determine the durability of the arthroplasty include wear and fatigue of the polyethylene component and loosening of the metallic components from their fixation in the bone.

Knowing the distribution of tibiofemoral forces is essential in the development, understanding, and optimal results of total knee arthroplasty. These forces determine wear and permanent plastic deformation in the polyethylene component, stress distribution in the implant and in the implant-bone interface, and stress transfer to the underlying bone. Tibial prostheses have been instrumented with force transducers to measure tibial forces *in vitro*;[1] however, direct measurement of tibial forces *in vivo* has not been reported. Telemetry has been shown to be a safe and accurate means of obtaining force data from implanted transducers and has been used to measure *in vivo* forces in the hip and the femur.[2–12] Hip forces measured directly through telemetry have typically been lower than those predicted mathematically. The knee is more difficult to model than the hip because of its complex geometry, six degrees of freedom, and the fact that soft tissues play a major role in stabilizing and directing knee function. Knee function is further complicated by the high degree of mechanical redundancy due to the numerous muscles involved in knee motion, including some that act on the hip or the ankle joint. Theoretical estimates of tibiofemoral forces are variable and not reliable depending on the mathematical models used and on the type of activity analyzed.[13–20]

This chapter describes the ongoing design, development, testing, and implantation of an electronic knee prosthesis that was started in 1993. A knee arthroplasty tibial tray with force transducers and a telemetry system was developed to measure directly tibiofemoral compressive forces *in vivo*. The tibial tray design with embedded transducers was reported in 1996 for *in vitro* experimental use.[1] A working prototype of the telemetry and remote powering was initially presented in 1999.[21] Subsequent years were spent refining manufacturing techniques, improving durability, and safety testing. Intraoperative trial implantation was also performed to demonstrate feasibility and to test the utility of the prosthesis as a dynamic ligament-balancing device. Ligament balancing is an essential process in the alignment of the knee during surgery and is currently determined by the surgical skill of the surgeon. On February 27, 2004 the electronic knee replacement prosthesis was implanted in the first patient. Forces were monitored during recovery and rehabilitation in the early postoperative period as well as for activities of daily living and exercise over the first 12 months after surgery. These measurements were supplemented by video motion analysis, electromyography, which is used to assess muscle contraction, and ground reaction force measurement.

2. Design of an Implantable Electronic Tibial Tray

Transducers: A titanium alloy revision tibial prosthesis design was instrumented with force transducers, a microtransmitter, and an antenna. The tibial tray consists of upper

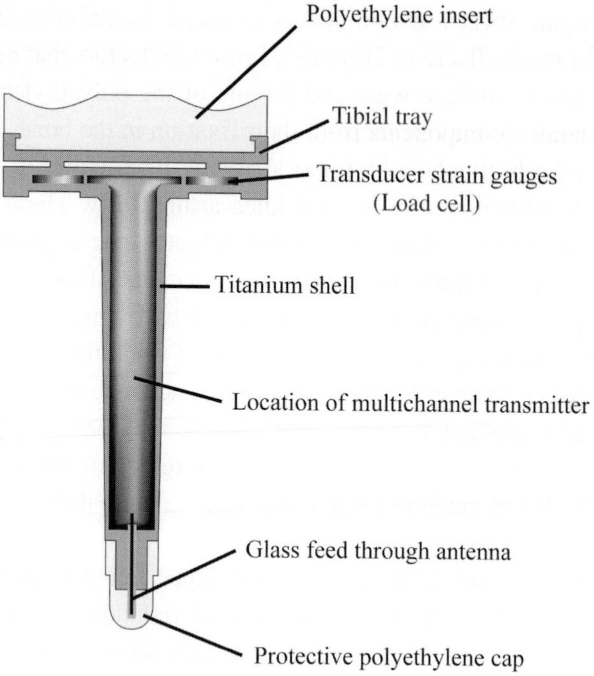

Fig. 2. Diagram of implant with instrumentation.[43]

and lower halves separated by support posts, below which lie the load cells (Fig. 2). The load cells are located at the four quadrants of the tibial tray as described by Kaufman *et al.*[1] The lower tibial tray is thinner at the junction with the support posts, thus forming a flexural membrane. Loads are transmitted from the upper tibial tray through the support posts directly to the flexural membranes. The force in each of the membranes is measured with 8 strain gauges. This design is based on the principle that electrical resistance changes when a conductor material is stressed. Each membrane instrumented with strain gauges constitutes a load cell. The load cells are calibrated individually and in combination, using a loading device traceable to the National Institute of Standards and Technology. By measuring the force on each load cell and the total axial load, the location of the center of pressure can be determined. In addition, the distribution of forces between the medial and the lateral compartments can be calculated. If the tibiofemoral contact points are known, moments along the anteroposterior and the mediolateral axes can also be determined. Figure 2 displays the location of the multichannel transmitter and the hermetic feedthrough antenna.

Telemetry: The microtransmitter receives the analog signal through leads from the load cells. The microtransmitter was developed by Microstrain (Burlington, VT) entirely from off-the-shelf surface-mount integrated circuits. The microprocessor filters and converts the analog signal to a digital signal and transmits the data using pulse code modulated radiofrequency (RF).

Transmitting antenna: Since RF signals do not pass through a sealed titanium shell, a single-pin hermetic feedthrough, medical grade, transmitting antenna was welded at the distal tip of the stem and connected to the microtransmitter.

Receiver: The receiving antenna captures the transmitted RF signal and the receiver generates RS-232 signals, which are directly read using custom software on a PC.

Power: Powering the implantable system was challenging. Because batteries contain toxic chemicals, have a limited lifespan, and occupy valuable space, we selected remote powering using magnetic near-field coupling. Approximately 40 milliwatts of power could be generated continuously in the internal coil, which was adequate to power the telemetry system. Details of design features of the microtransmitter, the antenna, the receiver, and the remote powering system can be found in D'Lima *et al.*[21]

3. *In Vitro* Validation and Cadaver Tests

After assembly and sealing, the electronic prosthesis was recalibrated and was checked for thermal drift. The integrity of the RF signal was then tested with the prosthesis cemented in five cadaver knees. The knees were mounted on a dynamic quadriceps-driven knee extension simulator (Oxford knee rig). To test the accuracy of the transducers, compressive loads of varying magnitudes were applied at different locations on a multiaxial testing machine (Force 5™, AMTI, Watertown, MA). To test the effect of the presence of an insert on the tray, loads were applied with and without an 8-mm polyethylene insert that matched the size of the tibial tray. The externally applied loads were compared with those measured by the tibial transducers.

In vitro cadaver tests demonstrated that the implant force transmitted reliably through bone, cement, and soft tissue. After a stable signal was obtained (based on the error detection of the checksum byte), the receiving antenna could be moved between three and five meters before loss of signal. The mean absolute error in measuring applied loads was 1.2%. The mean absolute error between the applied load location and the measured load location was < 1.5 mm.

4. Intraoperative Feasibility Testing

Cadaveric implantation did not reveal any significant issues with surgical technique and instrumentation. The feasibility of live surgical implantation was tested in the operating room. We chose the first intraoperative trial implantation in a revision case, which would not necessitate the removal of additional bone. Other goals of intraoperative implantation were to identify potential for sensitivity of the telemetry system to electronic and wireless equipment in close proximity in the operating room. The adjustment of external coil placement to maximize the efficiency of the power induction and the orientation of the receiving antenna to optimize signal integrity was also assessed in living tissue.

A 65-year-old male patient was recruited for this procedure after appropriate Institutional Review Board approval. This patient was scheduled for revision total knee arthroplasty following an infected primary arthroplasty that necessitated removal of the original tibial and femoral components. The tibial prosthesis was implanted in the tibial canal after preparation with manufacturer-provided instrumentation. A standard femoral component was press fit onto the original femoral bone cuts. Trial reduction with a 10-mm insert gave adequate stability in flexion and extension (bending and straightening of the knee). The RF signal integrity was tested over various knee positions during static and dynamic maneuvers. Tibial forces were recorded during passive knee flexion and forced varus (bowleg) and valgus (knock-knee). Excellent transmission of RF signal was obtained during all maneuvers. The intraoperative measurement of tibial forces *in vivo* provided further proof of concept of using telemetry, which served to validate the ability of the implant to transmit data through living tissue and bone cement. The intraoperative testing also determined that the implant was sensitive to the variations in forces seen in passive knee flexion and was able to pick up changes in knee forces introduced by manually applying varus and valgus moments to the knee.

5. Dynamic Ligament Balancing

In the hip joint, the opposing articular surfaces approximate a ball and socket joint and are highly conformed. In the knee the conformity between opposing articular surfaces is low and the ligaments and joint capsule provide the primary stabilizing support. The success of knee arthroplasty therefore depends on preserving the stability of these soft tissues. If the soft tissues are left too tight, function may be restricted, whereas lax soft tissues can give rise to instability. Further, an imbalance in the tightness of the soft tissues, for example, tightness in the medial soft tissues and laxity in the lateral soft tissues, can generate abnormal forces in the joint leading to early failure of the components. Several complications after total knee arthroplasty such as malalignment, instability, excessive wear, and loosening have been attributed to poor soft-tissue balance after surgery.[22–25] However, soft-tissue balance during surgery is an art, largely dependent on the surgeon's experience. Wide variations in knee kinematics have been reported after total knee arthroplasty among patients implanted with the same prosthesis design.[26,27] One major cause of this variation could be variation in soft-tissue balance.

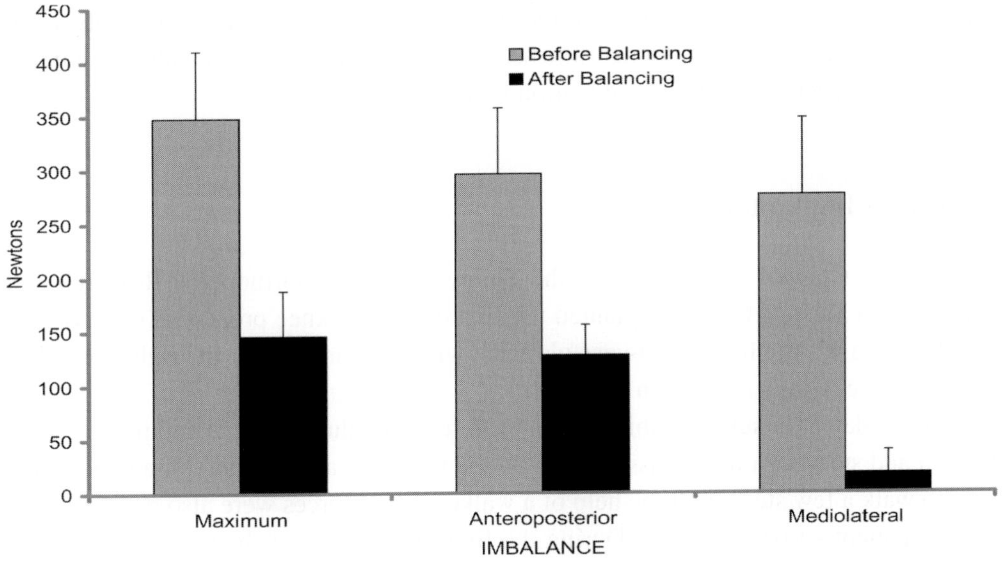

Fig. 3. Tibial tray forces were recorded before and after ligament balancing with the help of standard surgical instrumentation.[44]

The electronic tibial prosthesis was used to measure dynamic soft-tissue balance. The objective of these experiments was to monitor soft-tissue balance during routine total knee arthroplasty. A secondary objective was to determine the sensitivity of the device to imbalance. Fresh-frozen cadaver knees were used in this study. The articular surfaces were removed as in a typical knee arthroplasty. The instrumented tibial component was inserted along with a trial femoral component and a tibial insert. Forces on the tibial tray were recorded three times while the knee was passively ranged between 0° and 90° flexion. Next, soft-tissue balance and fine-tuning of the cut surfaces of the bone were performed as is typically done during real-life surgery. The soft-tissue balancing was performed subjectively and trial components were tested until the surgeon was satisfied. Then the instrumented tibial prosthesis was inserted one more time and the forces were recorded.

The instrumented tibial component was also used as an intraoperative trial prosthesis in patients during total knee arthroplasty. A similar approach was followed as described above for soft-tissue balancing. Tibial tray forces were recorded before and after balancing with the help of standard surgical instrumentation (Fig. 3)

The results of testing confirmed that the electronic knee prosthesis was very sensitive to soft-tissue imbalance. In addition, changes in the bone cuts that were made to improve ligament balance could then be assessed quantitatively. Precisely measuring forces during surgery may help quantify knee tightness, which can then be correlated with postoperative outcome.

Computer-aided surgical navigation systems have recently gained popularity. These systems have been reported to increase the accuracy of cutting the bone and aligning the prosthetic implants.[28–31] However, these systems cannot directly address soft-tissue balance and knee tightness. An instrumented tibial prosthesis can be a valuable adjunct to further enhance the value of navigation tools.

6. *In Vivo* Implantation

Dr. Clifford Colwell Jr., director of the Shiley Center for Orthopedic Research and Education at Scripps Clinic, implanted the first electronic knee prosthesis on February 27, 2004 in a 67-kg (148-lb) 80-year-old male with osteoarthritis of the right knee. The postoperative X-rays are shown in Fig. 4.

On the day after surgery, tibial forces were recorded during active and passive knee flexion and standing up with assistance. On the third day after surgery, the patient was able to walk a few steps with the help of a walker. Tibial forces were also recorded during the patient's first few steps. During the first three weeks of the patient's recovery,

(a) (b)

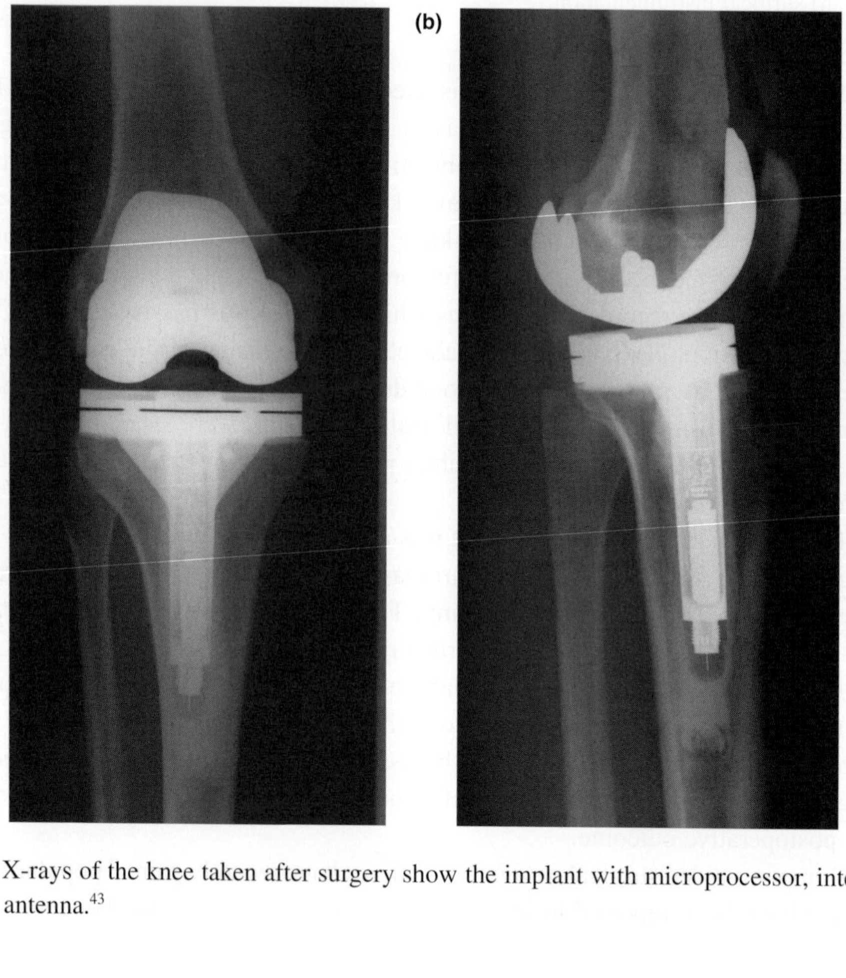

Fig. 4. X-rays of the knee taken after surgery show the implant with microprocessor, internal coil, and antenna.[43]

tibial forces were monitored during walking, stair climbing, rising from a chair, and standing on both legs with and without support.

Early postoperative rehabilitation: The patient was able to stand and bear weight on the operated limb with the help of a walker on postoperative day 1 and peak tibial forces of 1.17 ± 0.09 times body weight were recorded. Normalizing by body weight allows for a more accurate comparison of forces between patients with different weights. When the attending surgeon lifted the patient's leg, the prosthesis measured 0.34 ± 0.04 times body weight. When the patient actively lifted his leg with the knee fully extended, forces averaged 0.84 ± 0.06 times body weight.

Level walking: Tibial forces ranged from 1.26 ± 0.09 times body weight on postoperative day 3 to 2.17 ± 0.20 times body weight at the six-week measurement (Fig. 5). The mean location of the center of pressure recorded during heel strike, midstance, and toe off is shown in Fig. 6.

Stair climbing: Tibial forces peaked at 2.5 times body weight between 35° and 50° of knee flexion (Fig. 7). These forces were substantially higher than the vertical ground reaction force recorded under the same foot, which was 0.95 times body weight at the above flexion angles.

Chair rise: Peak tibial forces recorded reached 1.5 times body weight at 101° knee flexion. The vertical ground reaction force under the right foot was less than 0.5 times body weight for almost the entire chair-rise activity (Fig. 8), while the average ground

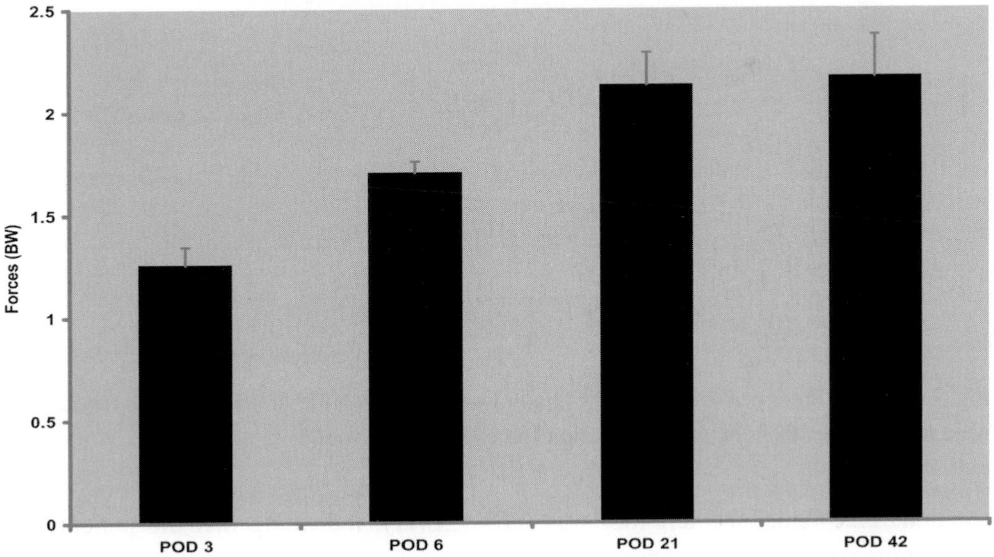

Fig. 5. Peak tibial forces during walking increased rapidly during the first three weeks after surgery (POD = postoperative day).[43]

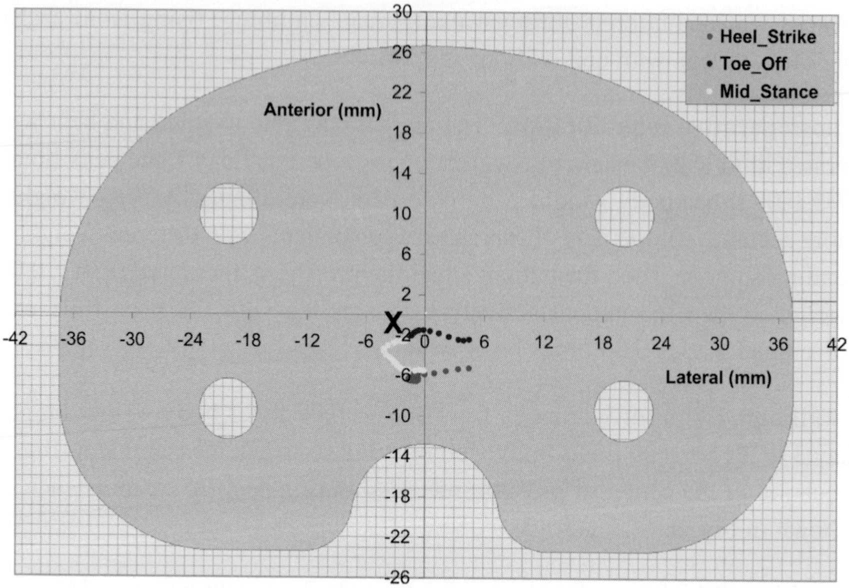

Fig. 6. The center of pressure during the different phases of walking stayed close to the midline.[43]

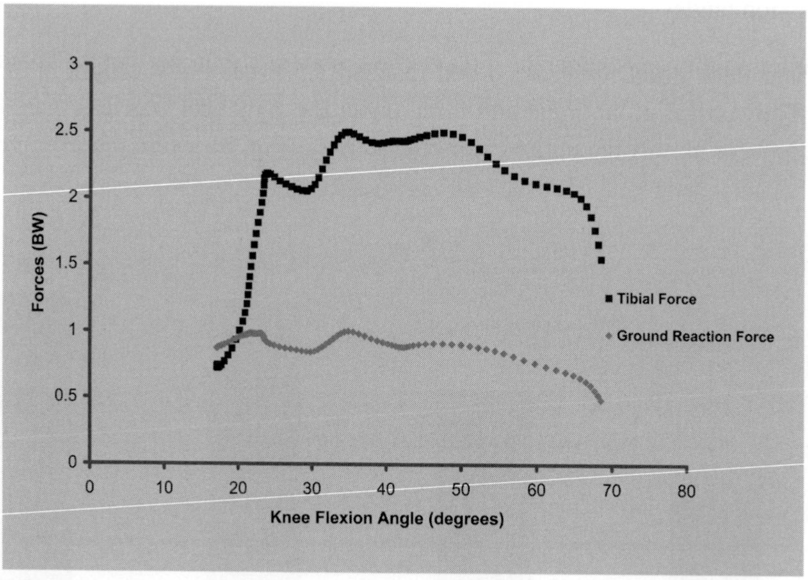

Fig. 7. Tibial forces peaked at 2.5 times body weight between 35° and 50° of knee flexion and were much higher than the ground reaction forces below the foot.[43]

reaction force under the left foot was 0.69 (±0.11), indicating that the patient still favored his right (operated) knee.

This was the first *in vivo* direct force measurement at the tibial tray after total knee arthroplasty. Stair climbing and rising from a chair are activities that involve higher

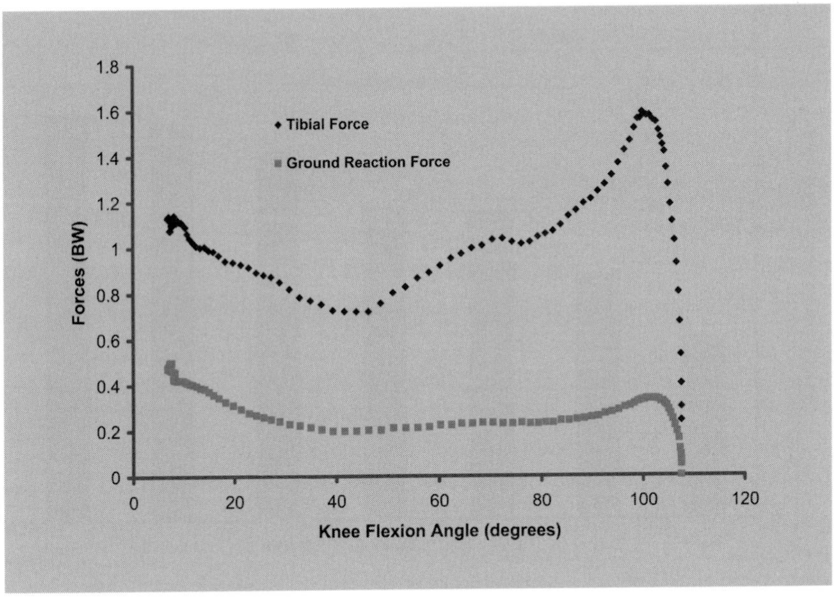

Fig. 8. Peak tibial forces while rising from a chair were again higher than ground reaction forces but not as high as forces measured during stair climbing.[43]

knee flexion angles and generate higher knee moments than walking. Stair climbing generated the highest peak tibial forces among all the activities studied (2.5 times body weight). However, forces were still lower than reports of tibial forces predicted during stair climbing.[32] While the knee flexed to a greater degree during the chair-rise activity (107°) than the stair-climbing activity (69°), tibial forces were substantially lower. Because the chair-rise activity involved double stance, the instrumented knee was loaded at lower levels.

7. Tibial Forces During Activities of Daily Living and Rehabilitation Over the First 12-Month Postoperative Period

We continued to monitor the tibial forces at regular intervals after surgery. This section reports on the tibial forces recorded during the first year after implantation. Activities of daily living were measured: level walking, rising from a chair, and stair ascent and descent under laboratory and natural conditions. During laboratory data collection, ground reaction forces were also monitored and synchronized with the tibial force data. At 11-months postoperative, tibial forces were measured at the Shiley Pavilion, Scripps Center for Integrative Medicine during treadmill walking (Star Trak 7600 Pro, Irvine, CA) at speeds ranging from 1.5–3 miles/hour, exercising on a stair-climbing machine (StairMaster® 4600, StairMaster Health and Fitness Products Inc, Tulsa, OK) and on a stationary bicycle (Spin Cycle, Reebok, Canton, MA) at various levels of difficulty.

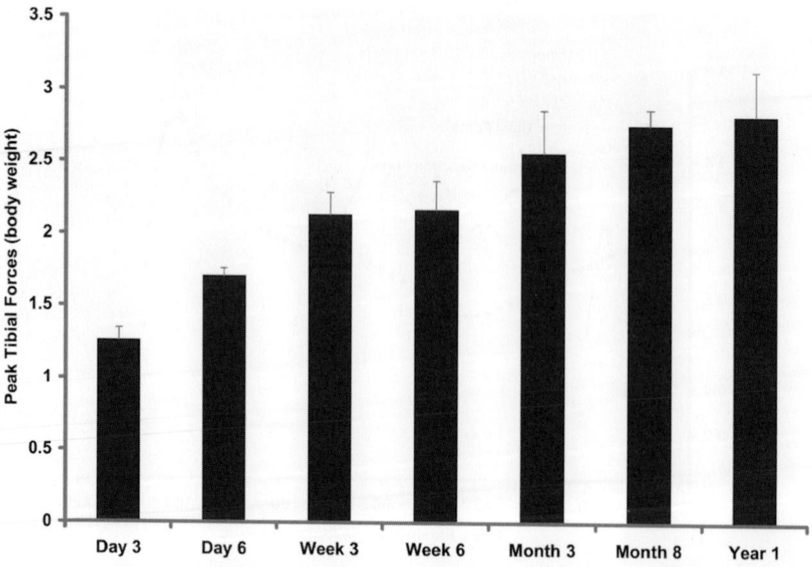

Fig. 9. After the first three weeks, tibial forces during walking increased gradually but steadily over the rest of the first year.[45]

Tibial forces were averaged over a sequence of cycles (typically six cycles) for each activity at each time point.

Walking: Peak tibial forces increased substantially over the first three weeks during rehabilitation and then more gradually over the rest of the year (Fig. 9). At the one-year follow-up, peak forces averaged 2.8 times body weight. Tibial forces were also recorded during treadmill walking (Fig. 10). The overall peak tibial forces remained lower for treadmill walking than for over-ground gait (between 1.75 and 2.03 times body weight).

Chair rise: By the six-week follow-up, the patient was able to rise from a chair without arm support. Tibial forces during chair rise increased from six-weeks postoperative (1.56 times body weight) to the one-year follow-up (2.64 times body weight). Force plates were placed under each foot during the chair-rise activity to monitor the distribution of ground reaction forces between instrumented and contralateral limbs. At the 12-month follow-up, the force plates registered even distribution between the two limbs (48% to 52% of total ground reaction force). Sitting down (stand-to-sit activity) generated more tibial forces than rising from a chair (sit-to-stand activity).

Stair climbing: Peak forces generated during stair ascent increased from a mean of 2.6 times body weight at six weeks to a mean of 2.9 times body weight at one year. Descending stairs generated significantly ($p = 0.004$) larger tibial forces than ascending stairs.

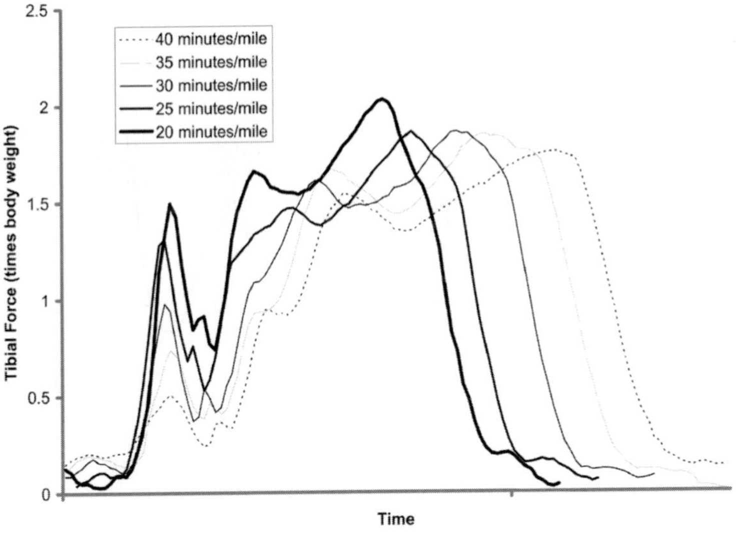

Fig. 10. Forces recorded during treadmill walking (up to 3 miles/hour) remained lower than forces recorded for over-ground gait.[45]

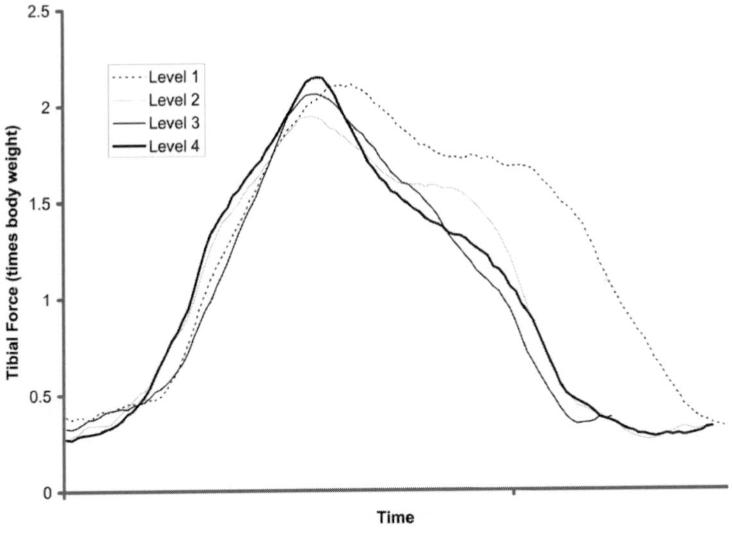

Fig. 11. Forces were recorded while exercising on a StairMaster machine at different levels of difficulty.[45]

When the patient exercised on a StairMaster, peak tibial forces remained close to two times body weight even when the level of difficulty was increased from level 1 to level 4 (Fig. 11). Stationary cycling did not generate high tibial forces (Fig. 12). Tibial forces ranged from a mean of 0.80 times body weight at 50 rpm at level 2, to a mean of 0.92 times body weight at 90 rpm at level 5.

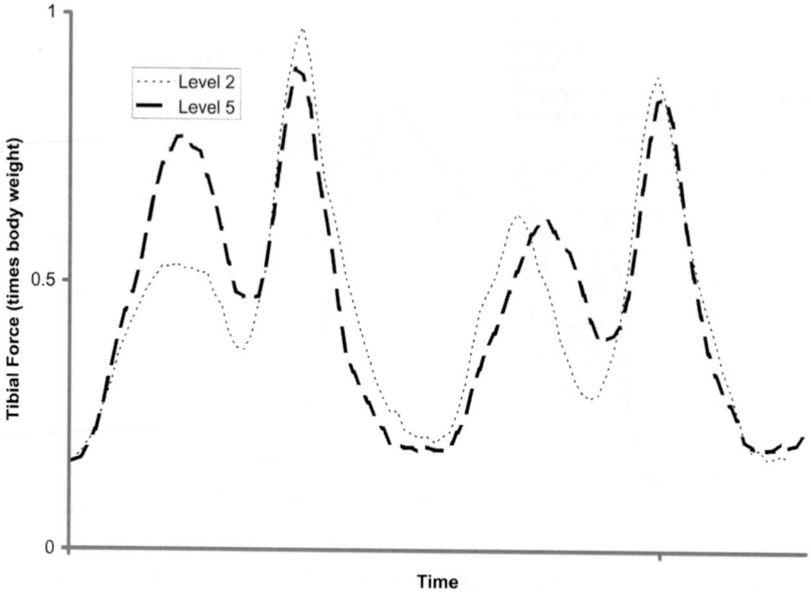

Fig. 12. Forces remained low during stationary cycling at various levels of pedaling cadence and difficulty.[45]

8. Discussion

Walking is the most common dynamic activity of daily living in typical joint arthroplasty patients and has been reported to average one to two million cycles/year.[33–35] By the third postoperative day the peak forces in the knee were greater than the patient's body weight while he walked with the help of a walker. Peak tibial forces increased rapidly over the first three weeks and then more gradually over the next 11 months. These peak tibial forces were within the predicted range (two to four times body weight, Morrison[36]), slightly lower (3 to 3.3 times body weight, Taylor *et al.*[32]) or substantially lower (7.1 times body weight, Seireg and Arvikar[37]) than reported mathematical models.

Walking on a treadmill generated even lower forces at the tibia (two times body weight or less) than walking on the laboratory floor. The ground reaction force was not tested for this activity and the treadmill walking surface may have absorbed more of the ground reaction force than the typical gait lab floor. The above data support the consensus that walking is a relatively safe activity for knee arthroplasty patients.

Knee arthroplasty patients sometimes have difficulty rising from a chair, another common activity of daily living. In addition, high knee moments are typically reported during the chair-rise activity.[38,39] Tibial forces increased substantially (to 2.6 times body weight) at the one-year follow-up. Further, the stand-to-sit activity generated even more tibial forces than the sit-to-stand activity. The increase in tibial force over time can be attributed in part to the distribution of ground reaction force between the patient's lower limbs. At the six-week follow-up, ground reaction forces recorded under the instrumented limb were as low as 20% of the total forces. At the one-year follow-up the ground reaction forces were almost evenly distributed between the lower limbs.

Stair climbing is another common activity of daily living, with the ratio of stair climbing cycles to walking cycles averaging around 1:25 in hip arthroplasty patients.[35] Tibial forces during stair climbing increased to a mean of three times body weight at one year. Stair descent generated even larger tibial forces than stair ascent, which is consistent with reports of increased ground reaction forces[40] and knee moments[41] during stair descent compared with stair ascent. In the present study, using a StairMaster, which exercises the muscles involved in stair climbing, generated lower tibial forces than those during regular stair climbing.

Stationary bicycling is often recommended for patients with knee disorders because it is a low impact activity. Knee joint moments have been shown to vary with pedaling rate.[42] In our study the tibial forces generated were relatively low, remaining below one time body weight through a range of pedaling rates (50 to 90 rpm) and levels of difficulty. These results support the consensus that bicycling is a relatively safe exercise for total knee arthroplasty patients.

9. Significance and Clinical Relevance

Tibial forces are directly related to the transmission of stresses in the implant. These include contact stresses generated at the bearing surface and subsurface, stresses at the implant–cement–bone interface, and stresses transmitted to underlying bone. Stresses at the bearing surface are a major factor in generating wear and fatigue, which determine the life of the implant. Stresses at the implant–cement interface have been correlated with aseptic loosening, implant migration, and the generation of third-body wear particles. Stresses transmitted to underlying bone affect remodeling, stress shielding, and osteoporosis. By directly measuring these forces, the stresses generated during common activities of daily living can be calculated.

Tibial forces generated *in vivo* were directly measured intraoperatively, during activities of daily living and during rehabilitation, which is invaluable in determining the safety of activities and exercises. This information can also be used to develop more effective rehabilitation. We have also shown that the device can be used effectively as a dynamic soft-tissue balancing tool during knee arthroplasty. This information can be a valuable adjunct in developing the next generation of surgical navigation systems for the knee, which can combine careful ligament balance with accurate prosthesis alignment.

These data are being made available to the orthopedic scientific community to validate existing models of the knee that estimate these forces or to develop more accurate computational models. Once validated, these models can provide valuable information that may lead to design changes that can improve the function and longevity of total knee prostheses.

Given the current trends in the increase in older population groups and the increase in the rates of total knee arthroplasty, a significant positive impact on clinical outcomes, longevity, and function is anticipated.

References

1. K. R. Kaufman, N. Kovacevic, S. E. Irby and C. W. Colwell, Jr., Instrumented implant for measuring tibiofemoral forces, *J. Biomech.* **29**: 667–671 (1996).
2. F. Graichen, G. Bergmann and A. Rohlmann, Hip endoprosthesis for *in vivo* measurement of joint force and temperature, *J. Biomech.* **32**: 1113–1117 (1999).
3. E. J. Bassey, J. J. Littlewood and S. J. Taylor, Relations between compressive axial forces in an instrumented massive femoral implant, ground reaction forces, and integrated electromyographs from vastus lateralis during various "osteogenic" exercises, *J. Biomech.* **30**: 213–223 (1997).
4. R. A. Brand, D. R. Pedersen, D. T. Davy, G. M. Kotzar, K. G. Heiple and V. M. Goldberg, Comparison of hip force calculations and measurements in the same patient, *J. Arthroplasty* **9**: 45–51 (1994).
5. G. Bergmann, F. Graichen and A. Rohlmann, Hip joint loading during walking and running, measured in two patients, *J. Biomech.* **26**: 969–990 (1993).
6. G. M. Kotzar, D. T. Davy, V. M. Goldberg, K. G. Heiple, J. Berilla, K. G. J. Heiple, R. H. Brown and A. H. Burstein, Telemeterized *in vivo* hip joint force data: a report on two patients after total hip surgery, *J. Orthop. Res.* **9**: 621–633 (1991).
7. D. T. Davy, G. M. Kotzar, R. H. Brown, K. G. Heiple, V. M. Goldberg, K. G. Heiple, Jr., J. Berilla and A. H. Burstein, Telemetric force measurements across the hip after total arthroplasty, *J. Bone Joint Surg. Am.* **70**A: 45–50 (1988).
8. W. A. Hodge, K. L. Carlson, R. S. Fijan, R. G. Burgess, P. O. Riley, W. H. Harris and R. W. Mann, Contact pressures from an instrumented hip endoprosthesis, *J. Bone Joint Surg. Am.* **71A**: 1378–1386 (1989).
9. T. W. Lu, J. J. O'Connor, S. J. Taylor and P. S. Walker, Validation of a lower limb model with *in vivo* femoral forces telemetered from two subjects, *J. Biomech.* **31**: 63–69 (1998).
10. S. J. Taylor, P. S. Walker, J. S. Perry, S. R. Cannon and R. Woledge, The forces in the distal femur and the knee during walking and other activities measured by telemetry, *J. Arthroplasty* **13**: 428–437 (1998).
11. S. J. Taylor, J. S. Perry, J. M. Meswania, N. Donaldson, P. S. Walker and S. R. Cannon, Telemetry of forces from proximal femoral replacements and relevance to fixation, *J. Biomech.* **30**: 225–234 (1997).
12. T. W. Lu, S. J. Taylor, J. J. O'Connor and P. S. Walker, Influence of muscle activity on the forces in the femur: an *in vivo* study, *J. Biomech.* **30**: 1101–1106 (1997).
13. G. E. Lutz, R. A. Palmitier, K. N. An and E. Y. Chao, Comparison of tibiofemoral joint forces during open-kinetic-chain and closed-kinetic-chain exercises, *J. Bone Joint Surg. Am.* **75**A: 732–739 (1993).
14. R. Nisell, M. O. Ericson, G. Nemeth and J. Ekholm, Tibiofemoral joint forces during isokinetic knee extension, *Am. J. Sports Med.* **17**: 49–54 (1989).
15. J. J. Collins, The redundant nature of locomotor optimization laws, *J. Biomech.* **28**: 251–267 (1995).
16. K. E. Wilk, R. F. Escamilla, G. S. Fleisig, S. W. Barrentine, J. R. Andrews and M. L. Boyd, A comparison of tibiofemoral joint forces and electromyographic activity during open and closed kinetic chain exercises, *Am. J. Sports Med.* **24**: 518–527 (1996).

17. G. Li, K. Kawamura, P. Barrance, E. Y. Chao and K. Kaufman, Prediction of muscle recruitment and its effect on joint reaction forces during knee exercises, *Ann. Biomed. Eng.* **26**: 725–733 (1998).
18. G. Li, K. R. Kaufman, E. Y. Chao and H. E. Rubash, Prediction of antagonistic muscle forces using inverse dynamic optimization during flexion/extension of the knee, *J. Biomech. Eng.* **121**: 316–322 (1999).
19. A. Seireg and R. J. Arvikar, A mathematical model for evaluation of forces in lower extremities of the musculo-skeletal system, *J. Biomech.* **6**: 313–326 (1973).
20. M. I. Ellis, B. B. Seedhom and V. Wright, Forces in the knee joint whilst rising from a seated position, *J. Biomed. Eng.* **6**: 113–120 (1984).
21. D. D. D'Lima, C. P. Townsend, C. W. Arms, B. A. Morris and C. W. Colwell, Jr., An implantable telemetry device to measure intra-articular tibial forces, *J. Biomech.* **38**: 299–304 (2005).
22. P. A. Lotke and M. L. Ecker, Influence of positioning of prosthesis in total knee replacement, *J. Bone Joint Surg. Am.* **59**A: 77–79 (1977).
23. E. L. Feng, S. D. Stulberg and R. L. Wixson, Progressive subluxation and polyethylene wear in total knee replacements with flat articular surfaces, *Clin. Orthop. Relat. Res.* **299**: 60–71 (1994).
24. R. E. Windsor, G. R. Scuderi, M. C. Moran and J. N. Insall, Mechanisms of failure of the femoral and tibial components in total knee arthroplasty, *Clin. Orthop. Relat. Res.* **248**: 15–19 (1989).
25. P. F. Sharkey, W. J. Hozack, R. H. Rothman, S. Shastri and S. M. Jacoby, Insall Award paper. Why are total knee arthroplasties failing today?, *Clin. Orthop. Relat. Res.* **404**: 7–13 (2002).
26. S. A. Banks, G. D. Markovich and W. A. Hodge, *In vivo* kinematics of cruciate-retaining and substituting knee arthroplasties, *J. Arthroplasty* **12**: 297–304 (1997).
27. D. A. Dennis, R. D. Komistek, M. R. Mahfouz, B. D. Haas and J. B. Stiehl, Multicenter determination of *in vivo* kinematics after total knee arthroplasty, *Clin. Orthop. Relat. Res.* **416**: 37–57 (2003).
28. M. Sparmann, B. Wolke, H. Czupalla, D. Banzer and A. Zink, Positioning of total knee arthroplasty with and without navigation support. A prospective, randomised study, *J. Bone Joint Surg. Br.* **85**: 830–835 (2003).
29. B. Stockl, M. Nogler, R. Rosiek, M. Fischer, M. Krismer and O. Kessler, Navigation improves accuracy of rotational alignment in total knee arthroplasty, *Clin. Orthop. Relat. Res.* **426**: 180–186 (2004).
30. R. Hart, M. Janecek, A. Chaker and P. Bucek, Total knee arthroplasty implanted with and without kinematic navigation, *Int. Orthop.* **27**: 366–369 (2003).
31. H. Bathis, L. Perlick, M. Tingart, C. Luring, D. Zurakowski and J. Grifka, Alignment in total knee arthroplasty. A comparison of computer-assisted surgery with the conventional technique, *J. Bone Joint Surg. Br.* **86**: 682–687 (2004).
32. W. R. Taylor, M. O. Heller, G. Bergmann and G. N. Duda, Tibio-femoral loading during human gait and stair climbing, *J. Orthop. Res.* **22**: 625–632 (2004).
33. M. Silva, E. F. Shepherd, W. O. Jackson, F. J. Dorey and T. P. Schmalzried, Average patient walking activity approaches 2 million cycles per year: pedometers under-record walking activity, *J. Arthroplasty* **17**: 693–697 (2002).
34. T. P. Schmalzried, E. S. Szuszczewicz, M. R. Northfield, K. H. Akizuki, R. E. Frankel, G. Belcher and H. C. Amstutz, Quantitative assessment of walking activity after total hip or knee replacement, *J. Bone Joint Surg. Am.* **80**: 54–59 (1998).
35. M. Morlock, E. Schneider, A. Bluhm, M. Vollmer, G. Bergmann, V. Muller and M. Honl, Duration and frequency of every day activities in total hip patients, *J. Biomech.* **34**: 873–881 (2001).
36. J. B. Morrison, The mechanics of the knee joint in relation to normal walking, *J. Biomech.* **3**: 51–61 (1970).
37. A. Seireg and R. J. Arvikar, The prediction of muscular load sharing and joint forces in the lower extremities during walking, *J. Biomech.* **8**: 89–102 (1975).
38. M. W. Rodosky, T. P. Andriacchi and G. B. Andersson, The influence of chair height on lower limb mechanics during rising, *J. Orthop. Res.* **7**: 266–271 (1989).

39. T. Kotake, N. Dohi, T. Kajiwara, N. Sumi, Y. Koyama and T. Miura, An analysis of sit-to-stand movements, *Arch. Phys. Med. Rehabil.* **74**: 1095–1099 (1993).
40. B. J. McFadyen and D. A. Winter, An integrated biomechanical analysis of normal stair ascent and descent, *J. Biomech.* **21**: 733–744 (1988).
41. T. P. Andriacchi, G. B. Andersson, R. W. Fermier, D. Stern and J. O. Galante, A study of lower-limb mechanics during stair-climbing, *J. Bone Joint Surg. Am.* **62A**: 749–757 (1980).
42. R. Redfield and M. L. Hull, On the relation between joint moments and pedalling rates at constant power in bicycling, *J. Biomech.* **19**: 317–329 (1986).
43. D. D. D'Lima, S. Patil, N. Steklov, J. E. Slamin and C. W. Colwell Jr., Tibial forces measured *in vivo* after total knee arthroplasty, *J. Arthroplasty* **21**: 255–262 (2006).
44. D. D. D'Lima, S. Patil, N. Steklov and C. W. Colwell Jr., An ABJS Best Paper Award: Dynamic intra-operative ligament balancing for total knee arthroplasty, *Clin. Orthop. Relat. Res* [Epub ahead of print] (2007). PubMed ID: 17693867.
45. D. D. D'Lima, S. Patil, N. Steklov, J. E. Slamin and C. W. Colwell Jr., The Chitranjan Ranawat Award: *In vivo* knee forces after total knee arthroplasty, *Clin. Orthop. Relat. Res.* **440**: 45–49 (2005).

CHAPTER 17

THE IMPLANTABLE GLUCOSE SENSOR IN DIABETES: A BIOENGINEERING CASE STUDY

David A. Gough

Abstract

Diabetes is a devastating disease that is rapidly becoming more prevalent worldwide. All approaches to therapy in diabetes are based on management of blood glucose, and there is a need for a broadly acceptable sensor to continuously monitor glucose. This chapter describes the rationale for new glucose sensors and how they would be used, and reviews our recent work on development of a long-term implantable sensor. Diabetes is an area where there are many opportunities to employ the powerful tools of bioengineering.

1. Introduction

Since my previous review on this subject that appeared in *Introduction to Bioengineering* in 2000,[1] there have been many changes in the field. Most importantly, diabetes has become much more prevalent, especially type 2 diabetes, and the incidence of new cases worldwide has reached epidemic proportions. This has led to widespread concerns that the consequences of the disease, if unabated, may begin to over-tax health care resources in the near future. At the same time, there has also been remarkable progress toward the development of new technologies for addressing the disease. Nevertheless, there remain numerous opportunities for bioengineers, some of which are reviewed here. This article focuses on the rationale for glucose monitoring and advances made at the University of California, San Diego and is an abridged version of a more extensive recent review.[2]

2. The Case for New Glucose Sensors

Glucose assay is arguably the most common of all medical measurements. Billions of glucose determinations are performed each year by laypeople with diabetes based on "fingersticking" and by health care professionals based on blood samples. However,

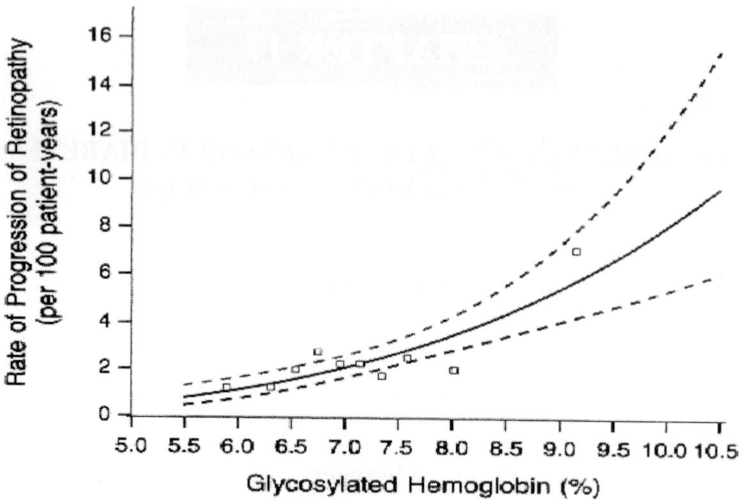

Fig. 1. The results of the DCCT.[4] Results show that improved glucose control, measured by a reduction in the fraction of glycosylated hemoglobin, leads to reduced long-term complications of diabetes. Copyright 1993, Massachusetts Medical Society.

new types of sensors capable of continuous glucose monitoring are nearing clinical introduction. Continuous or near-continuous glucose sensors may make possible new and fundamentally different approaches to the therapy of the disease.

The objective of all forms of therapy for diabetes is the maintenance of blood glucose control near normal levels.[3] The Diabetes Control and Complications Trial (or DCCT) and counterpart studies such as the United Kingdom Prevention of Diabetes Study (UKPDS) have clearly demonstrated that lower mean blood glucose levels resulting from aggressive treatment can lead to a reduced incidence and progression of retinopathy, nephropathy, and other complications of the disease.[4,5] These prospective studies showed definitively that there exists a cause-and-effect relationship between poor blood glucose control and the complications of diabetes (Fig. 1). As convenient means for frequent glucose assay were not available at the time, glucose control was assessed in these trials by glycosylated hemoglobin levels (Hb_{A1c}), which indicate blood glucose concentrations averaged over the previous three-month period. Although Hb_{A1c} levels are useful for assessment of longitudinal blood glucose control, the values indicate only *averaged* blood glucose, rather than blood glucose *dynamics* (i.e., how blood glucose changes with time), and cannot be used for immediate adjustment of therapy. Normalization of blood glucose dynamics may be of equal or greater importance than normalization of average blood glucose. The results of the DCCT point to the need for practical new approaches to achieve control.

The primary need for a new type of glucose sensor is to facilitate improved treatment of type 1 diabetes. In this form of the disease, the insulin producing ability of the pancreas has been partially or fully destroyed due to a misdirected autoimmune process, making insulin replacement essential. The sensor would help avoid the long-term complications associated with hyperglycemia (i.e., above-normal blood glucose) by providing information to specify more timely and appropriate insulin administration. It is now becoming widely appreciated that a new sensor could also be beneficial for people with the more common type 2 diabetes, where a progressive resistance of peripheral tissues to insulin develops, leading to glucose imbalances that can eventually produce long-term clinical consequences similar to type 1 diabetes. Type 2 diabetes is related to obesity, lifestyle and inherited traits. In recent years, the incidence of type 2 diabetes has increased at extraordinary rates in many populations, especially among young people, to the point of becoming a worldwide epidemic.[6] It is estimated that within ten years, the prevalence of diabetes may approach 210 million cases worldwide.[7]

In addition, an automatic or continuous sensor may also have an important role in preventing hypoglycemia (i.e., below-normal blood glucose). Hypoglycemia is caused primarily by a mismatch between the insulin dosage used and the amount of insulin actually needed to return the blood glucose level to normal. Many people with diabetes can reduce the mean blood glucose by adjustment of diet, insulin, and exercise, but when aggressively attempted, this has led to a documented increase in the incidence of hypoglycemia.[8] Below-normal glucose values can rapidly lead to cognitive lapses, loss of consciousness, and life-threatening metabolic crises. A continuous glucose sensor that does not depend on user initiative could be part of an automatic alarm system to warn of hypoglycemia and provide more confidence to the user to lower mean blood glucose, in addition to preventing hypoglycemia by providing improved insulin dosages. Hypoglycemia prevention may be the most important application of a continuous glucose sensor.

Alternatives to sensor-based therapies for diabetes are more distant. Several biological approaches to diabetes treatment have been proposed, including pancreatic transplantation, islet transplantation, genetic therapies, stem cell-based therapies, beta cell preservation, and others. These alternatives require substantial basic research and discovery, and are not likely to be available until far into the future, if eventually feasible.

Although new glucose sensors have the advantage of being closer to clinical introduction, there are certain other advantages as well. Real-time monitoring may lead to an automatic hypoglycemia warning device and entirely new means to implement several present therapies. In addition, the sensor is key to the implementation of the mechanical artificial beta cell. This device would have an automatic glucose sensor, a refillable insulin pump, and a controller containing an algorithm to direct automatic pumping of insulin based on information provided by the sensor. Development of an acceptable glucose sensor has thus far been the most difficult obstacle to implementation of the mechanical artificial beta cell.

3. The Ideal Glucose Sensor

For the glucose sensor to be a widely accepted innovation, the user must have full confidence in its accuracy and reliability, yet remain uninvolved in its operation and maintenance. Sensor systems under development have yet to reach this ideal, but some promising aspirants are described below. Short of the ideal, several intermediate sensing technologies with limited capabilities may find some degree of clinical application and, if used effectively, may lead to substantial improvements in blood glucose control. Nevertheless, the most reliable, unobtrusive sensors will lead to the broadest adoption by users.

4. Present Glucose Sensing Methods

The present method for glucose monitoring is based on samples collected by finger-sticking. In this method, sample collection involves the use of a lancet to puncture the skin of the fingertip or forearm to produce a small volume of blood and tissue fluid, followed by collection of the fluid on a reagent-containing strip and analysis by a handheld meter. The standard of care requires glucose determination by this method six times per day, but it is widely conceded that most people with diabetes sample only about once a day on average. When sampling is not sufficiently frequent, undetected blood glucose excursions can occur between samples. It has been shown that blood glucose measurements must be obtained every ten minutes to detect all blood glucose excursions in the most severe diabetic subjects,[9] although less frequent sampling may be sufficient for many people with diabetes. The fact that the sampling frequency required to detect all glycemic excursions is not realistically feasible at present indicates that the dynamic control of blood glucose is not actually practiced in diabetes management.

Several hundred physical principles for monitoring glucose have been proposed since the 1960s. Many are capable of glucose measurement in simple solutions, but have encountered limitations when used with blood, employed as implants, or tested in clinically relevant applications. Nevertheless, a few sensors have progressed toward clinical applications.

5. The UCSD Glucose Sensor Principle

Our sensor approach is based on the immobilized enzymes glucose oxidase and catalase. The enzymes catalyze the following reaction:

$$glucose + 1/2O_2 \rightarrow gluconic\ \ acid$$

The enzymes are immobilized within a gel membrane in contact with the electrochemical oxygen sensor. Excess oxygen not consumed by the enzymatic process is detected by an oxygen sensor and, after comparison with a similar background oxygen sensor without enzymes, produces a differential signal current that is related to glucose concentration.

Enzyme electrode sensors must contact the sample fluid to be assayed and therefore require either sensor implantation or sample extraction (as in the case of fingerstick devices). By employing the enzyme, sensors can have a significant advantage over non-enzymatic sensors of being *specific* for glucose rather than just selective. However, the benefits of enzyme specificity may not be fully realized unless the sensor is properly designed. To achieve the best performance, enzyme electrode sensors must include design features to address enzyme inactivation, biological oxygen variability, mass transfer dependance, and other effects. These issues are engineering challenges that have been addressed in my previous review.[1]

From the perspective of biocompatibility, sensors can be implanted either in direct contact with blood or with tissues. Biocompatibility in contact with blood depends on the surface properties of the sensor as well as flow characteristics at the implant site. Implantation in an arterial site, where the pressure and fluid shear rates are high, poses the threat of blood clotting and is rarely justified. Central venous implantation is considerably safer, and there are examples of other successful implants in this site (e.g. pacemaker leads, catheters).

Implantation of the sensor in a tissue site is safer, but involves other challenges. The sensing objective is to infer blood glucose concentration from the tissue sensor signal, and factors that affect glucose mass transfer from nearby capillaries to the implanted sensor must be taken into account. These factors include: the pattern and extent of perfusion of the local microvasculature, regional perfusion of the implant site, the heterogeneous distribution of substrates within tissues, and the availability of oxygen. There are also substantial differences in performance between short- and long-term implant applications. In the short term, a dominant wound healing response prevails, whereas in the long term, encapsulation may occur. Definitive studies are needed to establish the real-time accuracy of implanted sensors and determine when recalibration is necessary.

This approach has several unique features.[10] Electrochemical interference and electrode poisoning from endogenous biochemicals are prevented by a pore-free silicone rubber membrane between the electrode and the enzyme layer. This material is permeable to oxygen but completely impermeable to polar molecules that cause electrochemical interference. Appropriate design of the sensor results in sufficient supply of oxygen to the enzyme region to avoid a stoichiometric oxygen deficit.[11] The differential oxygen measurement system can also readily account for variations in oxygen concentration and local perfusion, which may be particularly important for accurate function of the implant in tissues. Excesses of immobilized glucose oxidase can be incorporated to extend the effective enzyme lifetime of this sensor, a feature not feasible with peroxide and conductive polymer-based sensors. Co-immobilization of catalase can further prolong the lifetime of glucose oxidase by preventing peroxide-mediated enzyme inactivation, the main cause of reduced enzyme lifetime.[12] This sensor design also avoids current passage through the body and hydrogen peroxide release into the tissues. These features are not found in most other glucose sensor designs.

5.1 *The long-term central venous sensor*

Based on this principle, we developed a long-term sensor as a central venous implant[10] (Fig. 2). The sensor functioned with implanted telemetry[13] in dogs for >100 days and did not require recalibration during this period (Fig. 3). The central venous site permitted direct exposure of the sensor to blood, which allowed simple verification of the sensor function without mass transfer complications.

These results have lead to several unanticipated conclusions. Although native glucose oxidase is intrinsically unstable, if the sensor is designed appropriately, the apparent catalytic lifetime of the immobilized enzyme can be substantially extended.[14] Also, the potentiostatic oxygen sensor has been shown to be remarkably stable[15] and the oxygen deficit, once thought to be an insurmountable barrier, can be easily overcome by design of the enzyme-containing membrane.[11] Moreover, the central venous implant

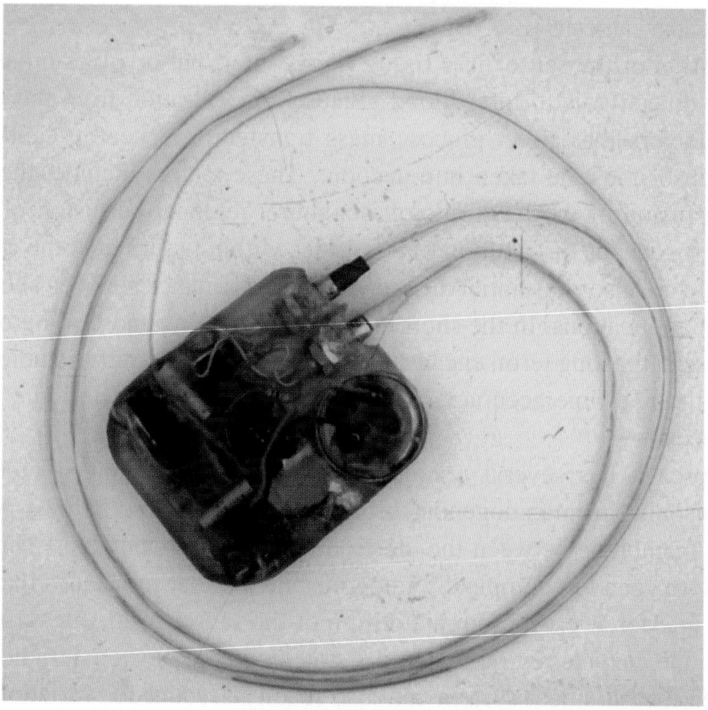

Fig. 2. Animal prototype long-term central venous glucose sensor with implanted telemetry.[13] Glucose and oxygen sensors are at the end of the catheters. The telemetry antenna emerges from the top, left. The telemetry body is 2 × 2.5 inches, and is encapsulated in epoxy. This implantable telemetry unit, designed and built by UCSD students, provides a means for reporting all the same measurements from the sensor (i.e., signal currents, electrode voltages, battery status, temperature, etc.) that could be obtained with non-implanted instrumentation. The implant can be reprogrammed to transmit at different rates by passing a magnet in certain sequences. The device is a powerful experimental tool in its own right.

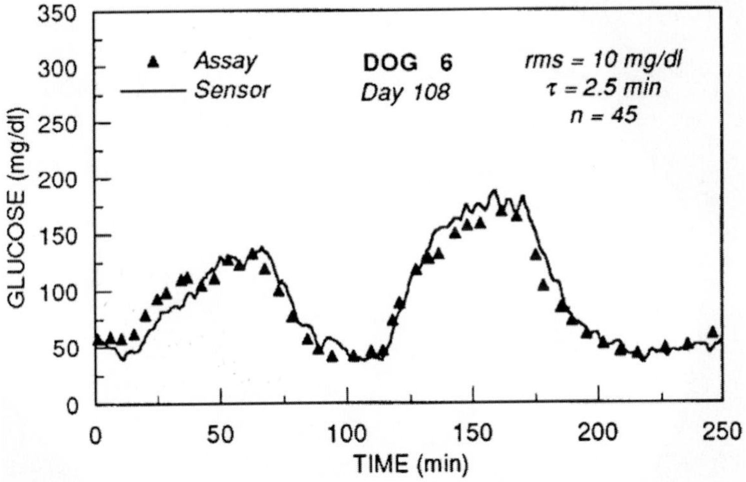

Fig. 3. Response of an implanted intravenous sensor to glucose challenges on day 108 after implantation in a dog.[10] The solid line is the sensor signal and triangles are venous blood glucose assays. Blood glucose excursions with initial rates of 0.28 mM/min were produced by infusions of sterile glucose solutions through an intravenous catheter in a foreleg vein. (Note: 90 mg/dl glucose = 5.0 mM.) Copyright 1990, American Diabetes Association.

site, which is uniquely characterized by slow, steady flow of blood, allows for sufficient long-term biocompatibility with blood.

This arrangement provided an ideal testbed to document the long-term stability and function of the sensor in animals.[10] In human clinical trials, a version of this system has been reported[16] to function continuously for >500 days in humans with <25% change in sensitivity to glucose over that period. This achievement represents a world record for long-term, stable, implanted glucose sensor operation, although certain hurdles still exist related to commercialization of the sensor.

5.2 *The long-term tissue glucose sensor*

Notwithstanding the success of the central venous sensor, there is a potential for blood clotting events. Although this potential is small, it may become significant over many years in individual users. This suggests reservations that may limit clinical acceptance and provides motivation for development of a potentially safer long-term sensor implant in tissues. The successful central venous sensor cannot be simply adopted for use in a safer tissue site, but certain design features of that sensor which promote long-term function, such as immobilized enzyme design, the stable potentiostatic oxygen sensor, and membrane design to eliminate the oxygen deficit, can be incorporated.

The tissue glucose sensor must be designed further to function in the unique environment of tissues. An array of sensors (Fig. 4) from which signals can be averaged is

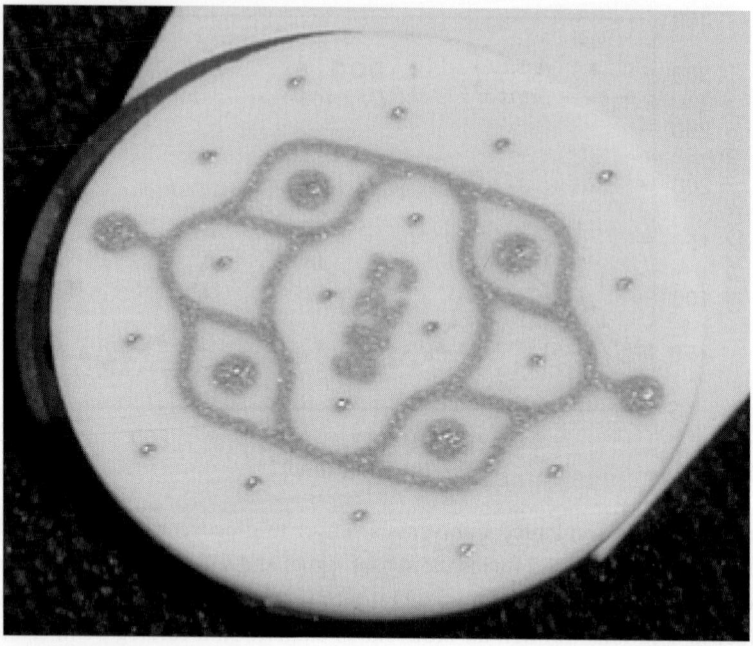

Fig. 4. Close-up view of tissue glucose and oxygen sensor array.[17] Sensor array with small (125-μm diameter) independent platinum working electrodes, large (875-μm diameter) common platinum counter electrodes, and a curved common Ag/AgCl reference electrode. The enzyme-containing membrane is not shown. Copyright 2003, American Physiological Society.

needed to address the heterogeneity of tissues, and there must be signal processing methods to account for variations in microvascular blood flow in tissues.

A systematic approach is required to validate sensor function, based on quantitative experimentation, mass transfer analysis, and accounting for properties of tissues that modulate glucose signals. Several new tools and methods have been developed. A tissue window chamber has been developed that allows direct optical visualization of implanted sensors in rodents, with surrounding tissue and microvasculature, while recording sensor signals[17] (Fig. 5). This facilitates determination of the effects of microvascular architecture and perfusion on the sensor signal. A method has been devised for sensor characterization in the absence of mass transfer boundary layers[18] that can be carried out before implantation and after explantation to infer stability of the implanted sensor. This allows quantitative assessment of mass transfer resistance within the tissue and the effects of long-term tissue changes. A sensor array having multiple glucose and oxygen sensors has also been developed that shows the range of variation of sensor responses within a given tissue.[19] This provides a basis for averaging sensor signals for quantitative correlation to blood glucose concentration.

There are opportunities for research on the tissue sensor array. There is a need to understand the effects of physiologic phenomena such as local perfusion, tissue

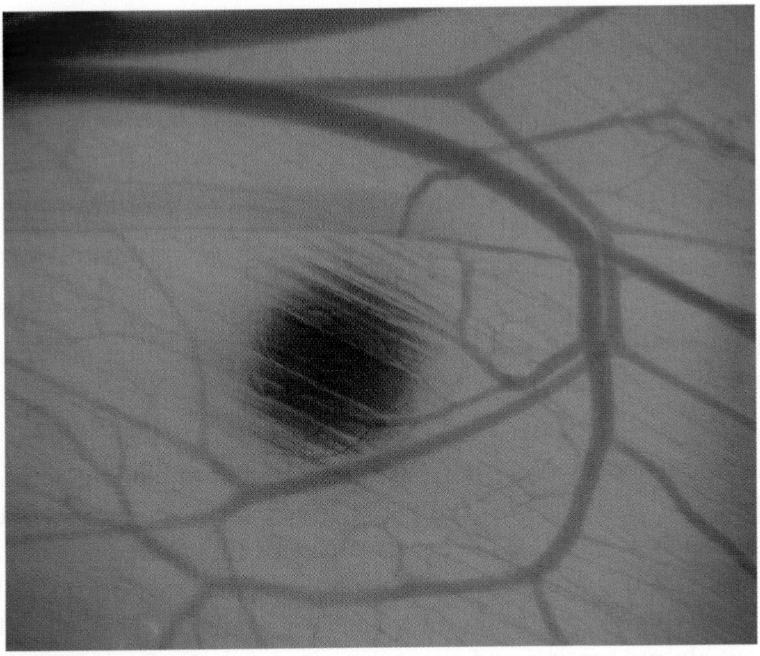

Fig. 5. An implanted glucose sensor and nearby microvasculature.[2] Optical image taken in a hamster window chamber. Sensor diameter is 125 μm.

variability, temperature and movement that modulate sensor responses to glucose and affect measurement accuracy. A detailed understanding of these effects and their dynamics is needed for a full description of the glucose sensing mechanism. It must be shown that the sensor array produces a reliable determination of glucose during exercise, sleeping and other daily life conditions.

A complete explanation for the response of every sensor is being sought, whether it is producing "good" or "bad" results, as more can often be learned for sensor improvement from sensors that produce equivocal results than from those that produce highly predictable signals.[20] As sensors should be useful for hypoglycemia detection, sensor function must be validated in the hypoglycemic state.

6. Blood Glucose Prediction

The ability to monitor blood glucose in real-time has major advantages over present methods based on sample collection that provide only sparse, historical information. There exists, however an additional possibility of using sensor information to *predict* future blood glucose values. It been demonstrated that blood glucose dynamics are not random and that blood glucose values can be predicted using autoregressive moving average (ARMA) methods, at least for the near future, from frequently sampled

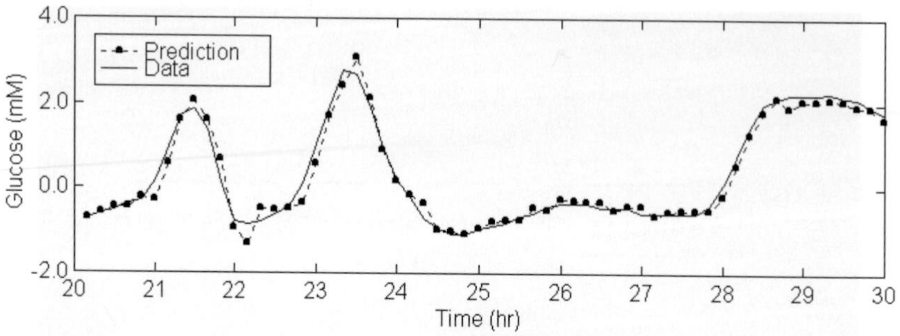

Fig. 6. Blood glucose prediction based on recently sampled values.[21] Ten-minute prediction in a non-diabetic, average rms error = 0.2 mM. Copyright 1999, American Diabetes Association.

previous values[21] (Fig. 6). Prediction based only on recent blood glucose history is particularly advantageous because there is no need to involve models of glucose and insulin distribution, with their inherent requirements for detailed accounting of glucose loads and vascular insulin availability. This capability may be especially beneficial to children. Glucose prediction can potentially amplify the considerable benefits of continuous glucose sensing, and may represent an even further substantial advance in blood glucose management.

7. Closing the Loop

Glucose control is an example of a classical control system. To fully implement this system, there is a need to establish a programmable controller based on continuous glucose sensing, having control laws or algorithms to counter hyper- and hypoglycemic excursions, identify performance targets for optimal insulin administration, and employ insulin pumps (Fig. 7). The objective is to restore optimal blood glucose control while avoiding over-insulinization by adjusting the program, a goal that may not be possible to achieve with alternative cell- or tissue-based insulin replacement strategies.

Programmable external pumps that deliver insulin to the subcutaneous tissue are now widely used and implanted insulin pumps may soon become similarly available. At present, these devices operate mainly in a pre-programmed or *open-loop* mode, with occasional adjustment of the delivery rate based on fingerstick glucose information. However, experimental studies in humans have been reported utilizing *closed-loop* systems based on implanted central venous sensors and intra-abdominal insulin pumps in which automatic control strategies were employed over periods of several hundred days.[22] There is a need to expand development of such systems for broad acceptance. Reviews of pump development can be found elsewhere.[23]

These results demonstrate that an implantable artificial beta cell is potentially feasible, but more effort is required to incorporate a generally acceptable glucose sensor, validate the system extensively, and demonstrate its robust response.

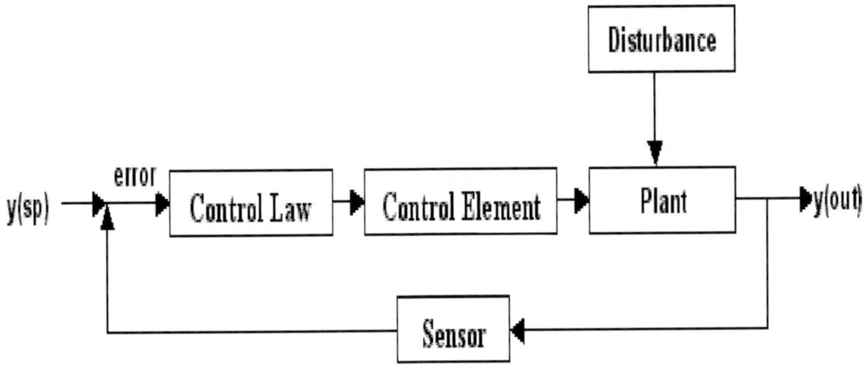

Fig. 7. A simple control system for blood glucose. *y(out)* is the blood glucose concentration, *y(sp)* is the desired blood glucose, the natural sensor is in the pancreatic beta cell, the plant is the body over which glucose is distributed, and the disturbance is absorption of glucose from the gut via digestion. The control element can be an insulin pump. The control law is an algorithm that directs the pump in response to the difference between measured and target blood glucose.

8. Conclusions and Opportunities for Student Involvement

The need for new glucose sensors in diabetes is now greater than ever. Although development of an acceptable, continuous and automatic glucose sensor has proven to be a substantial challenge, progress over the past several decades has defined sensor performance requirements and has focused development efforts on a limited group of promising candidates. The advent of new glucose sensing technologies could facilitate fundamentally new approaches to the therapy of diabetes. The developments at UCSD provide excellent opportunities for bioengineering students to get involved in research.

Acknowledgments

The author holds equity interest in GlySens, Inc., a company dedicated to the development of a new glucose sensor. This arrangement that has been approved by the University of California, San Diego in accordance with its conflict of interest policies.

References

1. D. A. Gough, The implantable glucose sensor: an example of bioengineering design. In: *Introduction to Bioengineering*, ed. Y. C. Fung (World Scientific Publishing, Co., Singapore, 2000), pp. 57–74.
2. F. Rahaghi and D. A. Gough, Glucose sensors. In: *Encyclopedia of Medical Devices and Instrumentation*, Vol. 3, ed. J. G. Webster (John Wiley & Sons, Hoboken, NJ, 2005), pp. 393–406.
3. G. F. Cahill Jr., L. D. Etzwiler and N. Freinkel, "Control" and diabetes (Editorial), *N. Engl. J. Med.* **294**(18): 1004–1005 (1976).

4. The Diabetes Control and Complications Trial Research Group, The effect of intensive treatment of diabetes on the development and progression of long-term complications in insulin-dependent diabetes mellitus, *N. Engl. J. Med.* **329**: 977–986 (1993).
5. I. M. Stratton, *et al.*, Association of glycaemia with macrovascular and microvascular complications of type 2 diabetes (UKPDS 35): prospective observational study, *Br. Med. J.* **321**: 405–412 (2000).
6. J. S. Skyler and C. Oddo, Diabetes trends in the USA, *Diabetes Metab. Res. Rev.* **18**(Suppl. 3): S21–26 (2002).
7. P. Zimmet, K. G. Alberti and J. Shaw, Global and societal implications of the diabetes epidemic, *Nature* **414**: 782–787 (2001).
8. J. F. Rovet and R. M. Ehrlich, The effect of hypoglycemic seizures on cognitive function in children with diabetes: a 7-year prospective study, *J. Pediatr.* **134:** 503–506 (1999).
9. D. A. Gough, K. Kreutz-Delgado and T. M. Bremer, Frequency characterization of blood glucose dynamics, *Ann. Biomed. Eng.* **31**: 91–97 (2003).
10. J. C. Armour, *et al.*, Application of chronic intravascular blood glucose sensor in dogs, *Diabetes* **39**: 1519–1526 (1990).
11. D. A. Gough, J. Y. Lucisano and P. H. Tse, Two-dimensional enzyme electrode sensor for glucose, *Anal. Chem.* **57**: 2351–2357 (1985).
12. P. H. S. Tse, J. K. Leypoldt and D. A. Gough, Determination of the intrinsic kinetic constants of immobilized glucose oxidase and catalase, *Biotechol. Bioeng.* **29**: 696–704 (1987).
13. B. D. McKean and D. A. Gough, A telemetry-instrumentation system for chronically implanted glucose and oxygen sensors, *IEEE Trans. Biomed. Eng.* **35**: 526–532 (1988).
14. D. A. Gough and T. Bremer, Immobilized glucose oxidase in implantable glucose sensor technology, *Diabetes Technol. Ther.* **2**: 377–380 (2000).
15. J. Y. Lucisano, J. C. Armour and D. A. Gough, *In vitro* stability of an oxygen sensor, *Anal. Chem.* **59**: 736–739 (1987).
16. Medtronic/MiniMed presentation at the Diabetes Technology and Therapeutics Conference, San Francisco, CA (2003).
17. M. T. Makale, *et al.*, Tissue window chamber system for validation of implanted oxygen sensors, *Am. J. Physiol. Heart Circ. Physiol.* **284**: H2288–2294 (2003).
18. M. T. Makale, M. C. Jablecki and D. A. Gough, Mass transfer and gas-phase calibration of implanted oxygen sensors, *Anal. Chem.* **76**: 1773–1777 (2004).
19. M. T. Makale, P. C. Chen and D. A. Gough, Variants of the tissue/sensor array chamber, *Am. J. Physiol. Heart Circ. Physiol.* (Cover figure) **289**: H57–65 (2005).
20. D. A. Gough and J. C. Armour, Development of the implantable glucose sensor. What are the prospects and why is it taking so long?, *Diabetes* **44**: 1005–1009 (1995).
21. T. Bremer and D. A. Gough, Is blood glucose predictable from previous values? A solicitation for data, *Diabetes* **48**: 445–451 (1999).
22. G. M. Steil, A. E. Panteleon and K. Rebrin, Closed-loop insulin delivery-the path to physiological glucose control, *Adv. Drug Deliv. Rev.* **56**: 125–144 (2004).
23. C. D. Saudek, *et al.*, Implantable pumps. In: *International Textbook of Diabetes Mellitus*, 3rd ed. (John Wiley & Sons, New Jersey, 2004).

CHAPTER 18

STEM CELLS IN REGENERATIVE MEDICINE

Shu Chien and Lawrence S. B. Goldstein

Abstract

Stem cells can either self-renew for long periods without differentiation or can become differentiated under specific conditions into specialized cells. They have great potential to treat disease someday by regenerating the dysfunctional cells or by providing novel ways to develop either drugs or other therapies. Embryonic stem cells from blastocysts are pluripotent in that they can differentiate into all types of cells in any organ/tissue. Adult stem cells, which are present in small proportions in organs/tissues after birth, are thought to be multipotent in that they differentiate only into the types of cells that exist in the organ/tissue in which they reside. There is some evidence that adult stem cells might become pluripotent and trans-differentiate into cells of other organs/tissues, but this evidence need replication. Currently, human embryonic stem cells hold great promise for medical advances in treating diseases, but there have been some objections to using cells from a blastocyst due to moral and religious considerations. The critical issues are whether undifferentiated blastocysts should be thought of as being a person and how do we balance potential life and existing adult life. New methods are being developed to produce pluripotent stem-cell lines with the aim of circumventing such objections based on religious, ethical and/or political grounds. There is much to be done in research on stem cells in order to realize their maximum benefits in clinical applications. Bioengineers can play a significant role in fostering the advance of stem cell research and applications.

1. Introduction

Stem cell research is an important subject for basic science and clinical applications, but it also involves societal and ethical considerations. It has become a household term, and bioengineers can contribute to its progress by applying engineering principles and techniques to this rapidly developing frontier area. The aim of this chapter is to provide a fundamental introduction to stem cells and how they may be used to benefit humankind.

Many reference materials on this subject can be found on the Internet, e.g. The National Institutes of Health (http://stemcells.nih.gov/info/basics), the US National Academies (http://dels.nas.edu/bls/stemcells), and the American Association for the Advancement of Sciences (http://www.aaas.org/spp/sfrl/projects/stem/index.shtml), as well as many books (a few examples are given in Refs. 1 to 3).

Animal life starts with a fertilized egg, which can divide (grow) into multiple cells. At specific stages of growth, most cells begin to differentiate to become specialized in structure and function, e.g. muscle cells, nerve cells, bone cells, adipose cells, etc. Stem cells are cells that remain unspecialized and can renew themselves through cell division for long periods of time without differentiation. Under specific physiological or experimental conditions, however, stem cells can be induced to differentiate into specialized cell types. Thus, the two distinct characteristics of stem cells are (1) they can self-renew by division, perhaps without limit, and (2) they can undergo differentiation to replenish the various types of specialized cells under specific conditions.

When a stem cell divides, each new cell has the potential to either remain a stem cell or become another type of cell with a more specialized structure and function. The remarkable ability of stem cells to develop into virtually every cell type in the body make them extremely valuable in serving as potential repair system for many different degenerative diseases and injuries.

2. Classification and Sources of Stem Cells

There are two primary types of stem cells from animals and humans that are used in scientific studies: embryonic stem cells and adult (tissue) stem cells; they have different characteristics and functions as outlined below.

2.1 *Embryonic stem cells*

Cell divisions of a fertilized egg lead to the formation of a blastocyst, which is a small sphere of cells surrounding a central cavity that contains the "inner cell mass" (Fig. 1). Embryonic stem cells are harvested from the inner cell mass of the blastocyst when it is about five to six days old and consists of approximately 200 cells.

Embryonic stem cells were first obtained from early mouse embryos in the 1980s. As a result of studying the biology of mouse stem cells, it became possible in 1998 to isolate stem cells from human blastocysts and grow human embryonic stem cells in the laboratory. The blastocysts that are generally used to obtain human embryonic stem cells are those developed from eggs fertilized *in vitro* for infertility purposes (in an *in vitro* fertilization clinic) but no longer needed for that purpose; they are then donated

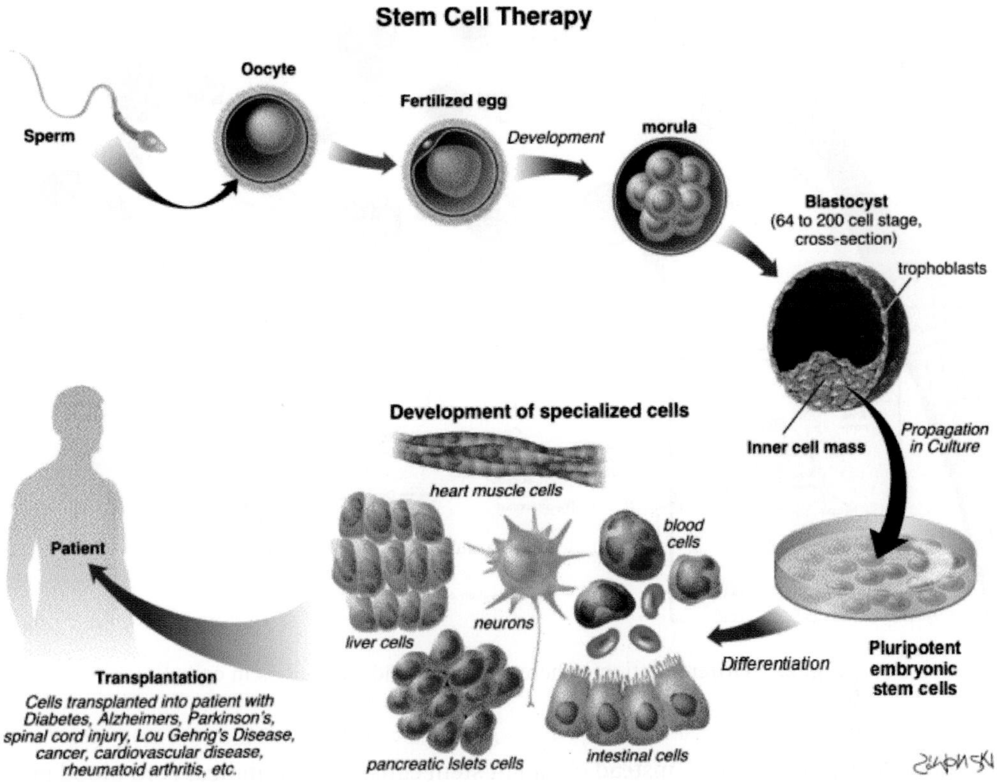

Fig. 1. Schematic drawing showing the development of embryonic stem cells from a fertilized egg and a blastocyst with an inner cell mass. Also shown is the differentiation of the pluripotent embryonic stem cell into specialized cells for potential stem cell therapy. Courtesy of Fred H. Gage and Jamie Simon at the Salk Institute.

for research with the informed consent of the donor. They are not derived from eggs fertilized in a woman's body.

Embryonic stem cells are pluripotent, i.e. they can differentiate under appropriate conditions into all types of cells to generate the cell types in any organ or tissue in the body.

2.2 *Adult stem cells*

Adults have stem cells in many different tissues in the body, which are necessary for our survival. Examples are skin stem cells, which renew and repair our skin, and bone marrow stem cells, which generate the different cell types in our blood. The adult stem cells are unspecialized cells found in a tissue or organ that appear to be able to yield all the specialized types of cells of that tissue or organ. Since fetuses, babies, and children also have such stem cells, some scientists have suggested using the term "somatic stem

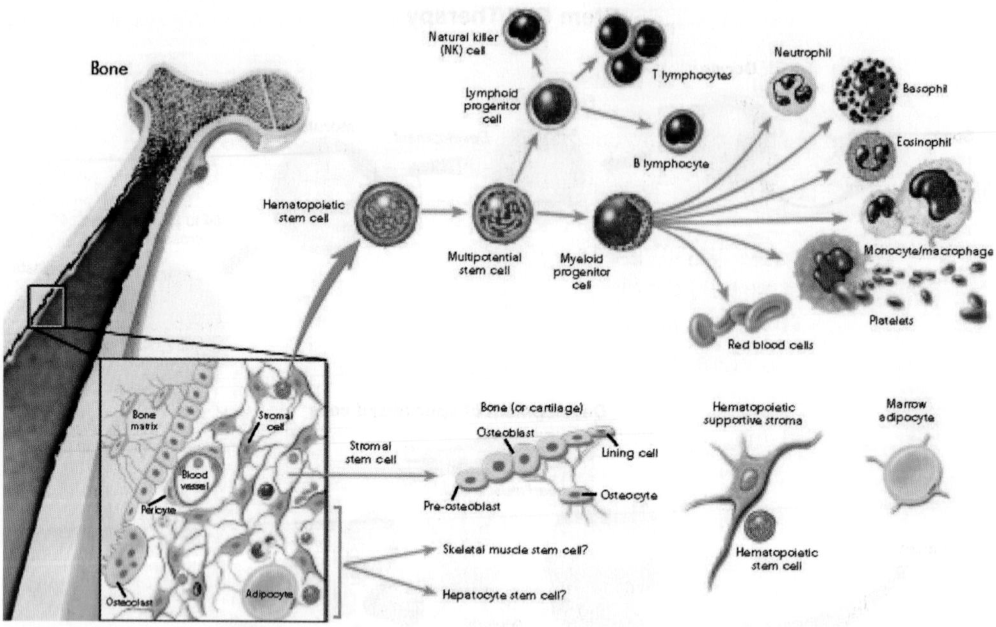

Fig. 2. Differentiation of hematopoietic and stromal stem cells.[17]

cells" or "tissue stem cells," instead of "adult stem cells." The umbilical cord blood provides another source of such "adult" stem cells.

Adult stem cells are undifferentiated cells, which exist in a small proportion among mostly differentiated cells in a tissue or organ. Adult stem cells may remain quiescent (non-dividing) for long periods until they are activated by signals due to turnover, disease, or tissue injury to repair the tissue in which they reside.

The adult tissues that have been found to contain stem cells include brain, bone marrow, skeletal muscle, skin, liver, and many others. Examples of differentiation pathways of adult stem cells are given below and in Fig. 2.

- Hematopoietic stem cells give rise to all types of blood cells, i.e. red blood cells, different white blood cells (including lymphocytes, granulocytes, and monocytes-macrophages), and platelets.
- Bone marrow stromal cells (mesenchymal stem cells) give rise to bone cells, cartilage cells, fat cells, and other types of connective tissue cells such as those in tendons.
- Neural stem cells in the brain give rise to nerve cells (or neurons) and the non-neuronal cells (astrocytes and oligodendrocytes).
- Epithelial stem cells in the lining of deep crypts of the digestive tract give rise to absorptive cells, intestinal endocrine cells, and several other cell types.
- Skin stem cells are found in the basal layer of the epidermis and at the base of hair follicles. The epidermal stem cells give rise to keratinocytes, which migrate to the

skin surface to form a protective layer. The follicular stem cells give rise to the hair follicle and the epidermis.

A single adult stem cell can generate a line of genetically identical cells. Such a clone of cells then gives rise to all the appropriate differentiated cell types of the tissue or organ in which they reside. Thus, adult stem cells are generally multipotent rather than pluripotent, which is a characteristic of embryonic stem cells. However, some recent studies suggest that, under certain conditions, adult stem cells may be able to give rise to cell types of completely different tissues, a phenomenon known as trans-differentiation or plasticity. Reports of adult stem cell plasticity include the differentiation of:

- Hematopoietic stem cells in the bone marrow into major types of brain cells (neurons, oligodendrocytes, and astrocytes); skeletal muscle cells; cardiac muscle cells; and liver cells.
- Bone marrow stromal cells into cardiac muscle cells and skeletal muscle cells.
- Brain stem cells into blood cells and skeletal muscle cells.

These studies are, at present, controversial, some have not been reproducible, and some have been subject to technical artifacts. Thus, more work is needed to explore this important issue. The possibility that stem cells from one tissue may be controlled to differentiate to cell types of a completely different tissue has generated a great deal of research interests and activities. If the mechanisms of trans-differentiation can be identified and controlled, existing stem cells from a healthy tissue might be induced to repopulate and repair a diseased tissue, thus making it possible to use adult stem cells for cell-based therapies.

There are a number of key questions about adult stem cells that remain to be answered. These include:

- What is their origin? Why do they remain in an undifferentiated state among all the differentiated cells around them?
- Do adult stem cells normally exhibit plasticity? What are the signals that regulate their proliferation (renewal) versus differentiation? Can adult stem cells be manipulated to trans-differentiate into cells of any organ or tissue?
- How can the proliferation of adult stem cells be enhanced to generate sufficient tissues for transplants?
- Are these factors that stimulate stem cells to target to sites of injury or damage?

2.3 *Comparison between embryonic and adult stem cells*

As pointed out above, adult and embryonic stem cells differ in the number and types of cells into which they can differentiate. Embryonic stem cells are pluripotent and can become all cell types. Adult stem cells are generally multipotent and are limited to differentiating into the cell types of their tissue of origin, though there is some evidence of plasticity of adult stem cells. Human embryonic and adult stem cells each have advantages and disadvantages regarding their potential use for cell-based regenerative therapies.

Large numbers of embryonic stem cells can be relatively easily grown in culture, while adult stem cells are rare in mature tissues and methods for expanding their numbers in cell culture have generally not yet been worked out. It is important to be able to obtain large numbers of cells that are needed for stem cell replacement therapies.

A potential advantage of using adult stem cells is that the patient's own cells could be expanded in culture and then re-introduced into the same individual. The use of the patient's own adult stem cells would avoid rejection of the cells by the immune system. This represents a significant advantage since immune rejection is a difficult problem that can only be partially circumvented with immunosuppressive drugs. However, whether the recipient would reject donor embryonic stem cells has not been determined in human experiments.

3. Cultivation of Stem Cells

Human embryonic stem cells are isolated by transferring the inner cell mass into a plastic laboratory culture dish that contains a culture medium. The cells divide and spread over the surface of the dish. The inner surface of the culture dish is typically coated with mouse embryonic fibroblasts that have been treated so they will not divide. This coating layer of cells is called a feeder layer. The reason for having the mouse cells in the bottom of the culture dish is to give the inner cell mass cells a sticky surface to which they can attach and to provide nutrients into the culture medium. Recently, scientists have begun to devise ways of growing embryonic stem cells without the mouse feeder cells by using human feeder cells or without using any feeder cells. This is a significant advancement because it can avoid the risk that viruses or other macromolecules in the mouse cells may be transmitted to the human cells.

Over the course of several days, the cells of the inner cell mass proliferate and begin to crowd the culture dish. These cells are then removed gently and plated (subcultured) into several fresh culture dishes. The process of subculturing is repeated many times and for many months, with each cycle referred to as a passage. After six months or more, the original cells of the inner cell mass would yield millions of embryonic stem cells. Embryonic stem cells that have proliferated in cell culture for six or more months without differentiating are pluripotent and appear genetically normal; they are referred to as an embryonic stem cell line. Once such cell lines are established, or even before that stage, batches of them can be frozen and shipped to other laboratories for further culture and experimentation.

4. Identification of Embryonic Stem Cells

At various points during the process of generating embryonic stem cell lines, the process of characterization needs to be carried out to test whether the cells exhibit the fundamental properties that make them embryonic stem cells.

Scientists have not yet agreed on a standard battery of tests that measure the human embryonic stem cells' fundamental properties. Also, it is acknowledged that many of the tests used may not be good indicators of the cells' most important biological properties and functions. Nevertheless, laboratories that grow human embryonic stem cell lines use the following tests:

- Growing and subculturing the stem cells for many months. This ensures that the cells are capable of long-term self-renewal. Scientists inspect the cultures through a microscope to see that the cells look healthy and remain undifferentiated.
- Using techniques to determine the presence of specific markers, e.g. Oct-4, which is a transcription factor found only in undifferentiated cells and is important in cell differentiation and embryonic development.
- Examining the chromosomes under a microscope to assess whether the chromosomes are damaged or if the number of chromosomes has changed.
- Determining whether the cells can be subcultured after freezing, thawing, and replating.
- Testing whether the human embryonic stem cells are pluripotent by (1) allowing the cells to differentiate spontaneously in cell culture; (2) manipulating the cells so they will differentiate to form specific cell types; or (3) injecting the cells into an immuno-suppressed mouse to test for the formation of a benign tumor called a teratoma, which typically contains a mixture of many differentiated or partly differentiated cell types.

5. Research on Stem Cells

Research on stem cells is needed to elucidate the essential properties that make them different from specialized cell types and to learn how to use them in clinical applications such as developing cell-based therapies, screening new drugs and toxins, and understanding birth defects. Some of the basic research activities involve: (1) determining how embryonic stem cells can remain proliferative for years in the laboratory without differentiating and (2) identifying the factors in living organisms that normally regulate proliferation and self-renewal of stem cells and the signals that cause them to become specialized cells.

A starting population of stem cells can proliferate for many months in the laboratory to yield millions of unspecialized cells, i.e. long-term self-renewal. There is a critical need to elucidate the specific factors, signals, and conditions that cause a stem cell population in a mature organism to proliferate and remain unspecialized until the cells are needed for repair of a specific tissue. Answers to these questions will help us to understand how cell proliferation is regulated during normal embryonic development or during the abnormal cell division that leads to cancer.

The signals that trigger stem cells to differentiate into specialized cells are only partially understood. The signals generated inside the cell are controlled by its genes. The external signals include chemicals secreted by other cells, physical contact with neighboring cells, molecules in the microenvironment, and external mechanical forces such as shear stress. It is not known whether the internal and external signals required for cell differentiation are similar for all kinds of stem cells, and whether specific sets of signals promote differentiation into specific cell types. Answers to these questions may lead to finding new ways of controlling stem cell differentiation in the laboratory, thereby growing cells or tissues that can be used for specific purposes including cell-based therapies. Some examples of potential treatments include replacing the dopamine-producing cells in the brains of Parkinson's patients or developing insulin-producing cells for type I diabetes (Fig. 3), as well as repairing damaged heart muscle following a heart attack with cardiac muscle cells, among many others.

As pointed out above, the embryonic stem cells in culture can remain undifferentiated for a long time. If these cells are allowed to form embryoid bodies, they begin to differentiate spontaneously to form specific cell types such as muscle cells, nerve cells, etc. Although spontaneous differentiation is a good indication that a culture of embryonic stem cells is healthy, it is not an efficient way to produce cultures of specific cell types, and controlled differentiation is needed.

Controlled differentiation of embryonic stem cells to generate cultures of specific types of cells can be achieved by a number of manipulations, e.g. changing the chemical composition of the culture medium, altering the surface of the culture dish, or modifying the cells by inserting specific genes. Some basic protocols or "recipes" have been established for the directed differentiation of embryonic stem cells into some specific cell types.

Research efforts on developing ways to reliably direct the differentiation of embryonic stem cells into specific cell types will be extremely valuable for the treatment of certain diseases.

6. Uses of Stem Cells to Treat Disease

Each type of cell makes a particular product(s) or provides a particular function, e.g. beta cells in the pancreas produce insulin to help modulate blood sugar level, heart cells contract to pump blood, and brain cells send electrical signals to produce thought or movement. Diseases such as diabetes, heart failure, or Alzheimer's disease are largely caused by a breakdown in functions or death of these specific cells. Stem cells are

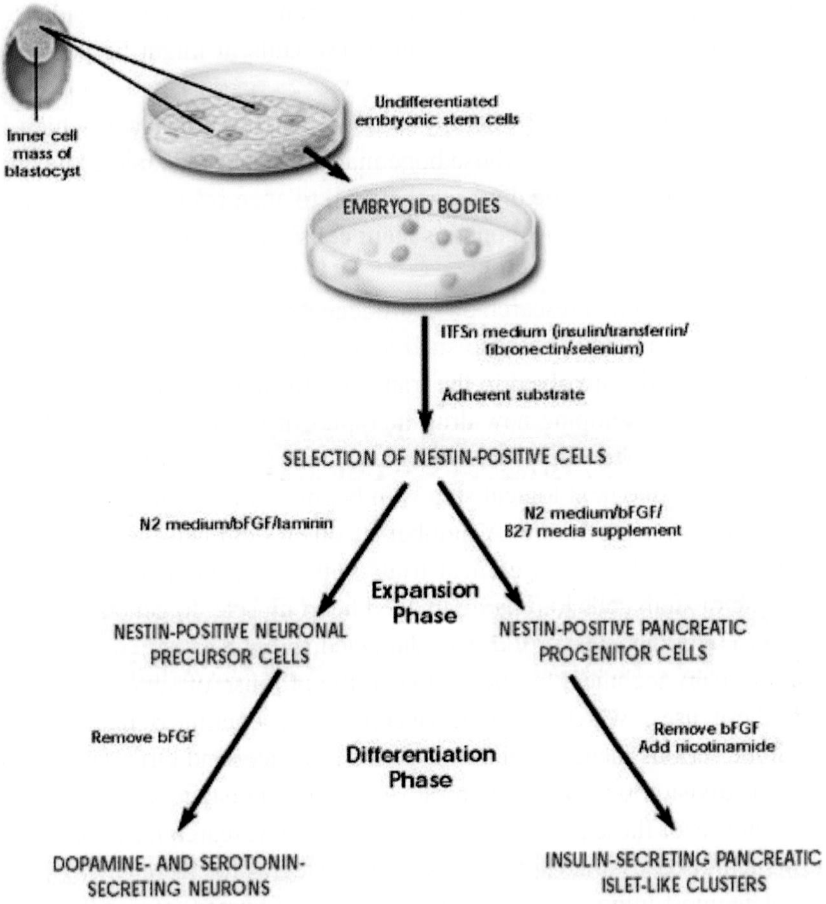

Fig. 3. Potential stem cell treatment in which embryonic stem cells can be selected, expanded, and induced to differentiate into dopamine- and serotonin-secreting neurons or insulin-secreting pancreatic islet-like clusters.[18]

important because they may be used to replace the damaged cells or tissues, as well as helping us better understand diseases or develop drug therapies.

Two main avenues are being explored for using stem cells to treat disease.

6.1 *Use of stem cells in "regenerative medicine" (cell-based therapy)*

Discrete populations of adult stem cells can generate replacements for cells that are lost through normal wear and tear, injury, or disease. A wide range of diseases (diabetes, heart disease, Parkinson's disease, motor neuron disease, etc.) may be amenable to stem cell therapy if stem cells can be made to go to the appropriate place in the body and become the appropriate cell type. For example, if stem cells could be made to target to an injured spinal cord and differentiate into nerve cells, it might be possible to cure paralysis.

The most common stem cell therapy is bone marrow transplant, which has been used for decades to treat patients whose bone marrow has been damaged by chemotherapy, or who have certain genetic diseases that lead to anemia or immune deficiency. As pointed out above, bone marrow contains stem cells that can give rise to all cell types in the blood.

Much basic and applied research on neurodegenerative and other human diseases is performed in animal models, because such work can rarely be done on people. There are, however, difficulties involved in the transition from animal research to diseases in humans, whether in developing new drug therapies or testing new ideas. In some diseases, such as amyotrophic lateral sclerosis (Lou Gehrig's disease), advancements have been made such that the next logical step is to begin working with human cells, which have a lot of hope for the future. A number of other stem cell therapies have shown promise in animal studies or in clinical trials with small numbers of humans. These include the use of stem cells to repair damaged heart muscle, to generate cells that produce insulin to treat diabetes, or to repair damaged spinal nerves.

There are many technical hurdles between the promise of stem cells and the realization of these uses, which need be overcome by continued intensive stem cell research. Some serious medical conditions, such as cancer and birth defects, are due to abnormal cell division and differentiation. A better understanding of the genetic and molecular controls of these processes through stem cell research may yield information about how such diseases arise and suggest new strategies for therapy.

Perhaps the most important potential application of human stem cells is the generation of cells and tissues that could be used for cell-based therapies. Today, donated organs and tissues are often used to replace ailing or destroyed tissue, but the need for transplantable tissues and organs far outweighs the available supply. Stem cells, directed to differentiate into specific cell types, offer the possibility of a renewable source of replacement cells and tissues to treat diseases including Parkinson's disease, spinal cord injury, stroke, burns, heart disease, diabetes, osteoarthritis, and rheumatoid arthritis.

For example, it may become possible to generate healthy heart muscle cells in the laboratory and then transplant those cells into patients with chronic heart disease. Other recent studies in cell culture systems indicate that it may be possible to direct the differentiation of embryonic stem cells or adult bone marrow cells into heart muscle cells.

In people who suffer from type I diabetes, the cells of the pancreas that normally produce insulin are destroyed by the patient's own immune system. New studies indicate that it may be possible to direct the differentiation of human embryonic stem cells in cell culture to form insulin-producing cells that eventually could be used in transplantation therapy for diabetics.

To fully realize the promise of novel cell-based therapies for these debilitating diseases, it is essential to be able to easily and reproducibly manipulate stem cells so that they develop the necessary characteristics for successful differentiation, transplantation and engraftment. The following is a list of steps in successful cell-based treatments that are needed to achieve precise control and to bring such treatments to the clinic. Thus, stem cells must be reproducibly made to:

- Proliferate extensively and generate sufficient quantities of tissue.
- Differentiate into the desired cell type(s).
- Survive in the recipient after transplant.
- Integrate into the surrounding tissue after transplant.
- Function appropriately for the duration of the recipient's life.
- Avoid harming the recipient in any way.

Also, to avoid the problem of immune rejection, it is necessary to develop strategies to generate tissues that will not be rejected.

6.2 *Use of stem cells to develop drug therapies*

It is possible to make stem cells that are genetically identical to the cell type that is defective in a patient with a disease such as Lou Gehrig's disease. By studying these cells, researchers can gain insight into what is the molecular derangement in the disease at the molecular level. These cells can also be used to test drugs that might block the progression of the disease.

Human stem cells could be used to test new drugs. For example, new medications could be tested for safety on cells differentiated from human pluripotent cell lines. The availability of pluripotent stem cells would allow drug testing in a wider range of cell types. Because the effective screening of drugs requires the conditions to be identical when comparing different drugs, the state of differentiation of stem cells into the specific cell type must be precisely controlled, and this requires knowledge of the signals controlling differentiation, as mentioned above.

Therefore, the promise of stem cells in regenerative medicine and in drug development for novel therapies is very exciting, but significant technical hurdles remain that can be overcome only through years of intensive research.

7. Public Policy and Ethics Related to Stem Cell Research

Stem cell research involves a close interplay between science and public policy. Some key issues in public policy related to stem cells involve morals, religions, and ethics. The scientific data indicate that we need research with both adult stem cells and embryonic stem cells to improve our understanding of diseases and to deliver new treatments that are urgently needed to save patients. The moral and religious problem is that some of the most valuable kinds of cells come from early human embryos at the blastocyst stage, when each embryo is a tiny ball of approximately 200 cells, with no organs such as brain and heart, nor blood. The blastocyst can be frozen for five to ten years and then thawed out successfully. Some people argue that such an early-stage embryo (the blastocyst) is the same as a person or a baby, and hence are opposed to human embryonic stem cell research. Most people, however, do not think that such a ball of cells is the same as a person. If one does not equate these frozen embryos to an actual person, then we should be able to use these cells in research, because we have a duty to understand and treat important human diseases and save human lives. Scientists are citizens and members of the community, as everybody else. Many of them become frustrated when a patient passed away due to disease conditions that could someday be corrected with appropriate treatment based on stem cell research if we had moved faster in this direction. However, we must be sensitive to all points of view on this delicate issue.

When scientific issues such as stem cell research are debated in the public realm, scientists should participate by providing the scientific information and viewpoints in order for the public to have an educated discussion. Since the funding for most scientific work is provided by the public, scientists also have a responsibility to provide the public with the best information possible. People may not agree at the end of a discussion on such a complex problem, but they would gain some knowledge about the scientific issues. Therefore, such discussions should be viewed as community service, and they may help to de-mystify science.

Ethics is an important issue in stem cell research. The critical questions cannot be settled by science alone and the answers depend on the way we view human life and the society. If such debates should result in moratoria or bans on research on human embryonic stem cells, the medical advances that would come from such work will be held in abeyance, and patients with a narrow window of opportunity for treatment may be lost, i.e. human lives could be lost.

8. Government Policies on Human Embryonic Stem Cell Research

8.1 *Federal policy on stem cell research and NIH's role*

The NIH website (http://stemcells.nih.gov/policy/NIHFedPolicy.asp) states the following regarding Federal policy and NIH's role:

"On August 9, 2001, President George W. Bush announced that federal funds may be awarded for research using human embryonic stem cells if the following criteria are met:

- *The derivation process (which begins with the destruction of the embryo) was initiated prior to 9:00 P.M. EDT on August 9, 2001.*

- *The stem cells must have been derived from an embryo that was created for reproductive purposes and was no longer needed.*
- *Informed consent must have been obtained for the donation of the embryo and that donation must not have involved financial inducements.*

The NIH, as the Federal government's leading biomedical research organization, is implementing the President's policy. The NIH funds research scientists to conduct research on existing human embryonic stem cells and to explore the enormous promise of these unique cells, including their potential to produce breakthrough therapies and cures.

Investigators from ten laboratories in the United States, Australia, India, Israel, and Sweden have derived stem cells from 71 individual, genetically diverse blastocysts. These derivations meet the President's criteria for use in federally funded human embryonic stem cell research. The NIH has consulted with each of the investigators who have derived these cells. These scientists are working with the NIH and the research community to establish a research infrastructure to ensure the successful handling and the use of these cells in the laboratory.

To review the list of the international research organizations that have derived cell lines that are eligible for funding under the President's policy, see http://stemcells.nih.gov/research/registry/eligibilitycriteria.asp.

To review a list of those eligible cell lines that are currently available for shipping to researchers, see http://stemcells.nih.gov/research/registry."

Thus, research on human embryonic stem cells in laboratories in the United States that receive federal research funding is currently limited to the 71 approved cell lines. In practice, however, many fewer of these lines are in sufficiently good condition for research purpose.

8.2 *Proposition 71 in the State of California*

To enhance the output from human embryonic stem cell research beyond the Federally approved cell lines to generate new ways to treat human diseases through regenerative medicine and development of drug therapy, the State of California placed on its 2004 State election ballot Proposition 71 on "Stem Cell Research. Funding. Bonds. Initiative Constitutional Amendment and Statute" (http://www.lao.ca.gov/ballot/2004/71_11_2004.htm). The measure authorizes the state to sell US$3 billion in general obligation bonds to provide funding for stem cell research and research facilities in California, to establish a new state medical research institute to use the bond funds for awarding grants and loans for stem cell research and research facilities, and to manage stem cell research activities funded by this measure within California. Priority for research grant funding would be given to stem cell research that met the institute's criteria and was unlikely to receive federal funding. In some cases, funding could also be provided for other types of research that were determined to cure or provide new types of treatment of diseases and injuries. The institute would not be allowed to fund research on human reproductive cloning. Consistent with current statute, this measure would make conducting stem cell research a state constitutional right.

In November 2004, the Proposition was passed by a large majority of the voting citizens of the State of California. Thereafter, an Independent Citizens Oversight Committee (ICOC) was formed to establish medical and scientific accountability standards for the conduct of stem cell research in accordance with the California constitution, thus assuring that the research is conducted safely, in accordance with the highest ethical standards, and in compliance with the state and national policies that protect patient safety, patient rights, and patient privacy. In April 2006, the State formed a California Institute of Regenerative Medicine (CIRM) (http://www.cirm.ca.gov/) to formally launch its scientific strategic planning process and to guide the scientific programs over the coming years. CIRM hosts a series of Scientific Conferences for the ICOC and the public to discuss the scientific strategies and approaches to funding stem cell research and the development of related therapies, as well as conducting public hearings at which members of the public may speak.

To honor the public mandate to pursue expeditiously the human embryonic stem cell research prior to the favorable decision on a litigation that has delayed the issuance of the bonds, private citizens of California have raised money for CIRM training grants, and Governor Arnold Schwarzenegger has directed the California Department of Finance to provide loans to CIRM to begin the funding of research grants. Grants for the training of doctoral students, postdoctoral fellows and physician scientists and for the conduct of research on human embryonic stem cells have been awarded to many California institutions after rigorous review. Since the legality and constitutionality of Proposition 71 were upheld by the California Superior and Appeals Courts and the decline by the Supreme Court of California to hear any further appeals on May 16, 2007, CIRM has been able to begin selling bonds to fund large numbers of stem cell grants in 2007 and beyond. Requests for Proposals have been issued for grants including New Faculty Awards, Major Facilities, Disease Team Planning Grants, and Generation of New Cell Lines. Such timely funding support will enable California to pursue actively this most promising area of medical science, i.e. stem cell research, and hopefully this will set a trend for the whole nation.

9. Alternative Approaches to Human Embryonic Stem Cells

As mentioned above, despite the great promise of research on embryonic stem cells for understanding and treating disease, there are people who oppose such research based on religious and ethical grounds. There have been efforts to develop new methods for producing pluripotent stem-cell lines with the aim of circumventing such objections based on religious, ethical and/or political grounds.[4]

9.1 *Removing one cell from the eight-cell stage before blastocyst formation*

Chung *et al.*[5] in Lanza's group adapted a method commonly used during *in vitro* fertilization (IVF) or assisted-reproduction clinics for preimplantation genetic diagnosis (PGD) of genetic defects. This involves the removal of one cell from the eight-cell stage of development of the fertilized egg before the formation of the blastocyst (Fig. 4) and using it to produce embryonic stem cell lines, thus not compromising the embryo from which the blastomere was obtained. The single blastomere cells are co-cultured with established embryonic stem cell lines, and then separated from them to form fully competent, pluripotent embryonic stem cell lines originating from that one blastomere.

The embryonic stem cells thus produced would have the same genes as the embryo, essentially a mix from the two parents, just as the cells undergoing *in vitro* fertilization treatment. One of the goals for stem cell research is to produce pluripotent cells that represent the full genetic diversity of humans, or that are genetically identical to a particular donor (e.g. a patient with a genetic disorder), in order to study the cellular and genetic bases of disease development.[6] Stem cell lines generated by this approach might be manipulated in culture by replacing the defective gene with healthy copies, and thereby one could validate the role of particular drug targets or the efficacy of certain therapies. In the long term, healthy cells derived from repaired stem cells might aid the regeneration of tissues from the donor patient.

The cells are "pluripotent," i.e. they can grow into the three major tissue types.[7] Since the method does not involve destroying embryos, it is hoped that this will solve

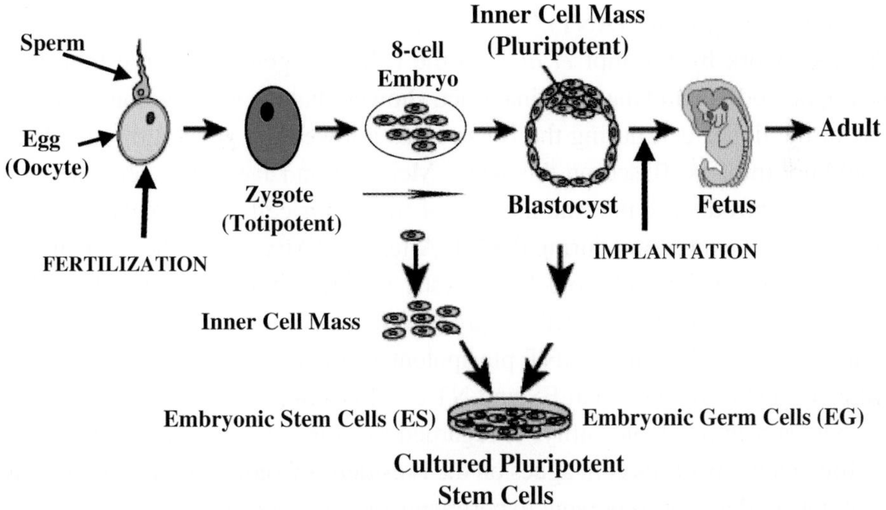

Fig. 4. Cell removal as an alternative to destroying embryos. One cell was from the eight-cell stage of development of the fertilized egg before the formation of the blastocyst and used to produce embryonic stem cell lines. Adapted from www.nih.gov/news/stemcell/fig3.gif.

the political impasse that resulted in the federal banning of research on human ESCs. While the IVF embryos that have been biopsied for PGD have grown into normal babies, some embryos do not survive. Until the safety issues have been examined more closely, it is felt that the procedure should only be used on a cell already taken from an embryo for PGD.[8]

9.2 *Nuclear transfer*

Nuclear transfer (NT) involves removing the nucleus from a donor body cell (e.g. a skin cell) and injecting it into an egg cell that has had its own chromosomes removed. This egg cell is then subjected to procedures to facilitate the formation of an embryoid (embryo-like) blastocyst. During this process, the body-cell nucleus undergoes "reprogramming" to change the expression of its genes from those of the skin to more pluripotent genes. Embryoid blastocysts have an inner cell mass like normal blastocysts, and these cells can become pluripotent stem cells. Such NT stem cells, as ES cells, can self-renew or differentiate to become most types of mature body cell.

Animal tests, however, show that reprogramming of embryoid blastocysts may not complete all the developmental stages to birth; this failure can injure or even kill the mother bearing it in later stages of pregnancy. Furthermore, because human NT stem cells come from embryoid blastocysts, their derivation still raised objections on political, ethical and religious grounds. A possible solution to the controversy is the invention of a process that would produce an entity that cannot implant in the uterus, termed alternative nuclear transfer (ANT).[9]

Previous work by Strumpf *et al.*[10] on the role of a gene cdx2 in establishing the mouse trophectoderm and the intestinal tract, showed that the suppression of cdx2 in the nucleus of the donor cell during the NT process allowed the generation of NT entities that could not implant. Based on this work, Meissner and Jaenisch[11] generated ANT in the mouse by introducing into the nucleus of the donor cell a gene encoding an RNA that inhibits cdx2 expression during the NT. After the derivation of the ANT pluripotent stem-cell line from the embryoid blastocysts, the inhibitor gene was clipped out to enable the resulting ANT stem cells to produce mature intestinal epithelia when given the right cues (Fig. 5). These ANT pluripotent stem-cell lines can form many other mature cells, just as the classical ES and NT cell lines do.

Such a "non-implantable entity" is regarded by Hurlbut as non-viable for becoming human life, but many of his colleagues on the President's Council on Bioethics disagree.[9]

Attempts to delay or to prevent experiments on human embryonic stem cells derive from the belief that the preimplantation embryos, though they may have little or no potential to form a functioning organism, are human, and thus have the same rights as born humans. Although the alternative approaches to generate human embryonic stem cells might be regarded by some as a diversion of good science by politics, there is the

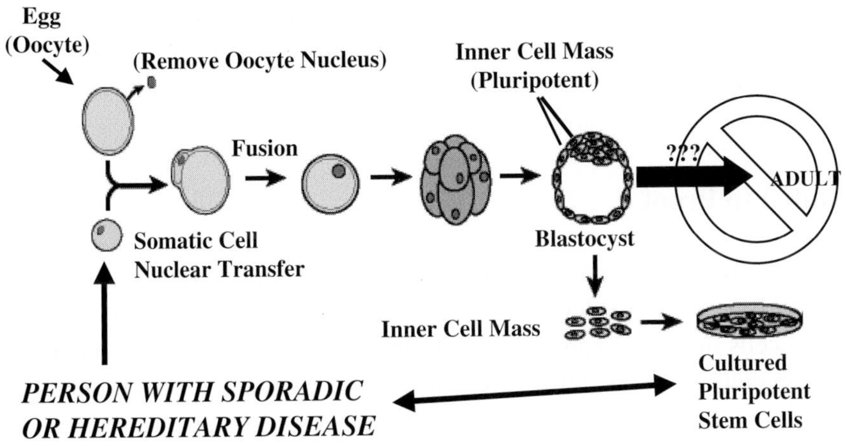

Fig. 5. Nuclear transfer (NT) technology. Adapted from www.nih.gov/news/stemcell/fig4B.gif.

general belief that all of these attempts to advance and translate medical science should be pursued in parallel.[4]

10. Role of Bioengineering in Stem Cell Research

Investigations on stem cells involve interdisciplinary research that needs to combine many biological fields with natural sciences, engineering, and clinical medicine. All cells, including stem cells, respond to physico-chemical changes in their micro-environment. While most of the studies have been focused on the roles of the chemical factors such as growth factors and matrix proteins, physical factors have also been shown to play a significant role. Thus, mechanical forces such as shear and stretch, as well as geometrical factors and mechanical properties of the extracellular matrix can modulate the differentiation of stem cells.[12–15]

Bioengineering can make important contributions to this frontier area by providing technological innovation, e.g. high-throughput screening of the optimum physico-chemical environment for desired differentiation.[16] With the high-throughput approach, the amount of experimental data will be vast, and they will need to be analyzed with the aid of bioinformatics and systems biology. In Sec. VII of this book, there are five chapters on these subjects, and the approaches there can be very valuable to the identification and elucidation of the genetic, molecular and cellular interactions in bringing about the specific pathways of self-renewal and differentiation, the two major characteristics of stem cells.

Some of the other bioengineering approaches that are valuable for stem cell research are medical imaging (including molecular imaging), cell separation and characterization, transport processes for targeted delivery, biomaterials, large scale culturing and testing, computational modeling, development of novel instrumentation, etc.

Through the application of these methods and principles, bioengineering can make valuable contributions and form an essential component of interdisciplinary stem cell research.

Note Added to Proof

The field of stem cell research progresses very rapidly. On the day of proof-reading of this Chapter (November 20, 2007), two teams of scientists reported that they had turned human skin cells into what appear to be pluripotent stem cells without having to make or destroy an embryo. If this procedure proves to be practical, it might quell the ethical debate troubling the field. These papers are Takahashi *et al.*[20] and Yu *et al.*[21]

References

1. Comm. Biol. Biomed. Appl. Stem Cell Res., NRC, *Stem Cells and the Future of Regenerative Medicine*, Board of Life Sciences (National Research Council, Washington, DC, 2001).
2. A. Bongso and E. H. Lee, *Stem Cells: From Bench to Bedside* (World Scientific Publishing Co., New Jersey, 2005).
3. A. Kiessling and S. C. Anderson, *Human Embryonic Stem Cells* (Jones & Bartlett, Boston, 2003).
4. I. L. Weissman, Medicine: politic stem cells, *Nature* **439**: 145–147 (2006).
5. Y. Chung, I. Klimanskaya, S. Becker, J. Marh, S. J. Lu, J. Johnson, L. Meisner and R. Lanza, Embryonic and extraembryonic stem cell lines derived from single mouse blastomeres, *Nature* **439**: 216–219 (2006).
6. I. L. Weissman, Stem cells — scientific, medical, and political issues, *N. Engl. J. Med.* **346**: 1576–1579 (2002).
7. R. Lanza, *Nature*, DOI: 10.1038/nature05142.
8. R. Hooper, Embryonic stem cells without embryo death, *New Sci.* **2566** (23 August 2006): 10.
9. W. Hurlbut, *Monitoring Stem Cell Research* (President's Council on Bioethics, Washington DC, 2004).
10. D. Strumpf, C. A. Mao, Y. Yamanaka, A. Ralston, K. Chawengsaksophak, F. Beck and J. Rossant, Cdx2 is required for correct cell fate specification and differentiation of trophectoderm in the mouse blastocyst, *Development* **132**: 2093–2102 (2005).
11. A. Meissner and R. Jaenisch, Generation of nuclear transfer-derived pluripotent ES cells from cloned Cdx2-deficient blastocysts, *Nature* **439**: 212–215 (2006).
12. A. Lemmon, N. H. Sniadecki, S. A. Ruiz, J. L. Tan, L. H. Romer and C. S. Chen, Shear force at the cell-matrix interface: enhanced analysis for microfabricated post array detectors, *Mech. Chem. Biosyst.* **2**: 1–16 (2005).
13. A. Margariti, L. Zeng and Q. Xu, Stem cells, vascular smooth muscle cells and atherosclerosis, *Histol. Histopathol.* **21**: 979–985 (2006).
14. D. Shumann, R. Kujat, M. Nerlich and P. Angele, Mechanobiological conditioning of stem cells for cartilage tissue engineering, *Biomed. Mater. Eng.* **16**(4 Suppl.): S37–52 (2006).
15. A. J. Engler, S. Shen, H. L. Sweeney and D. E. Discher, Matrix elasticity directs stem cell lineage specification, *Cell* **122**: 677–689 (2006).

16. S. Flaim, S. Chien and S. Bhatia, An extracellular matrix microarray for probing cellular differentiation, *Nat. Methods* **2**: 119–125 (2005).
17. Stem Cell Basics: What are adult stem cells? In: *Stem Cell Information* (National Institutes of Health, US Department of Health and Human Services, Bethesda, MD, 2006). Available at http://stemcells.nih.gov/info/basics/basics4.
18. Appendix B: Mouse embryonic stem cell cultures. In: *Stem Cell Information* (National Institutes of Health, US Department of Health and Human Services, Bethesda, MD, 2006). Available at http://stemcells.nih.gov/info/scireport/appendixb.
19. Stem Cell Basics: What are the potential uses of human stem cells and the obstacles that must be overcome before these potential uses will be realized? In: *Stem Cell Information* (National Institutes of Health, US Department of Health and Human Services, Bethesda, MD, 2006). Available at http://stemcells.nih.gov/info/basics/basics6.
20. K. Takahashi, K. Tanabe, M. Ohnuki, M. Narita, T. Ichisaka, K. Tomoda and S. Yamanaka, Induction of pluripotent stem cells from adult human fibroblasts by defined factors, *Cell* **131**: 1–12 (2007).
21. J. Yu, M. A. Vodyani, K. Smuga-Otto, J. Antosiewicz-Bourget, J. L. Frane, S. Tian, J. Nie, G. A. Jonsdottir, V. Ruotti, R. Stewart, I. I. Slukvin and J. A. Thomson, Induced pluripotent stem cell lines derived from human somatic cells. *Science* DOI: 10.1126/science.1151526 (published online November 20, 2007).

SECTION VI

NANOSCIENCE AND NANOTECHNOLOGY

CHAPTER 19

ENGINEERING COMPOUNDS TARGETED TO VASCULAR ZIP CODES

Erkki Ruoslahti

Abstract

The blood vessels in different tissues carry specific molecular markers. Various disease processes, such as cancer, inflammation or atherosclerosis, express their own molecular markers on the vasculature. These tissue- and disease-specific molecules create a "zip code" system of vascular addresses. The vascular addresses for tissues and disease processes reside in the inner lining of the blood vessels, the endothelium, and are thus readily accessible from the blood stream. The screening of phage-displayed peptide libraries *in vivo* has been a particularly effective way of identifying vascular zip codes. Targeting a drug to these addresses can enhance the efficacy of the drug while reducing its side effects. The greatest potential of the vascular targeting technology may be in constructing smart nanodevices.

1. Introduction

The targeting concept discussed in this chapter is based on the "magic bullet" idea, which has been around for 100 years: a specific binding molecule, such as an antibody, is used to ferry a therapeutic molecule to a specific target in the body. For example, antibodies that recognize tumor cells have been used to deliver radioactive substances into tumors. The radioactive substance is chemically linked to the antibody; because the antibody binds to tumor cells, but not to other cells, the radioactivity becomes concentrated in the tumor. This approach has worked well in lymphomas and leukemias where tumor cells circulate in the blood, but the results with solid tumors have generally been disappointing, mostly because tumor cells in solid tumor are not readily accessible to circulating drugs.

A new twist to the magic bullet concept is to use specific addresses in the blood vessels for the delivery. Blood vessels, particularly the endothelial cells that form the inner lining of the vessel wall, are readily available to a therapeutic agent that circulates in the blood, circumventing the problem with access to the tissue outside the vessels. Fortunately, the vasculature provides many opportunities for targeted delivery. Tumors put their signature on blood vessels as do other pathological processes, and even individual normal tissues each have their own marker profile.[1] These marker molecules can serve as targets to which to direct diagnostic and therapeutic agents, including nanodevices.

Targeted delivery could potentially make it possible to design "smart" nanoparticles that seek out the site of disease. First-generation targeted nanoparticles have already been designed that bind to tumor vessels and passively release a drug or provide a signal that can be detected by a diagnostic instrument. Some of these nanodevices are in or close to entering clinical trials.[2–5] In the future, we are likely to see nanodevices that perform more functions than the current ones and do so with much greater sophistication. This chapter provides a review of some of the recent developments in vascular zip codes and discussions of their use in the engineering of targetable diagnostic and therapeutic agents.

2. Vascular Zip Codes

2.1 *Tissue-specific zip codes in blood vessels*

Most, perhaps all, normal tissues put on their vasculature a tissue-specific molecular signature defined by specific molecular markers. Antibodies prepared against endothelial cells isolated from a given tissue have been used to identify some of these vascular markers.[6,7] Moreover, analyzing mRNAs expressed by endothelial cells from two tissues (lung and kidney) has revealed extensive differences in their gene expression patterns.[8] Also, a genomic promoter sequence that allows gene expression in endothelial cells from one tissue, but not from another, has been described.[9] However, most of the vascular heterogeneity known at this time has been discovered using phage display.

Phage display is one of the methods in which large libraries of compounds are screened to identify compounds with useful properties. In phage display, a large collection (up to billions) of peptides or protein fragments is expressed at the surface of a phage, which is a virus that infects bacteria. Each phage particle expresses one peptide, and the sequence of the peptide is encoded in the phage's genome as a piece of inserted DNA and can be retrieved by sequencing the DNA insert. Phage library screening has been adapted for *in vivo* use to discover specific vascular markers (Fig. 1). The phage library is injected into the tail vein of a mouse and the phage is rescued from the tissue of interest, such as a tumor, yielding a phage pool enriched in phage that selectively binds to the target tissue. By repeating the procedure a number of times, the enrichment can be brought to a level that allows the isolation of individual clones of homing phage, and from the DNA of these clones, identification of the homing peptides. All normal tissues that have been analyzed by *in vivo* phage display so

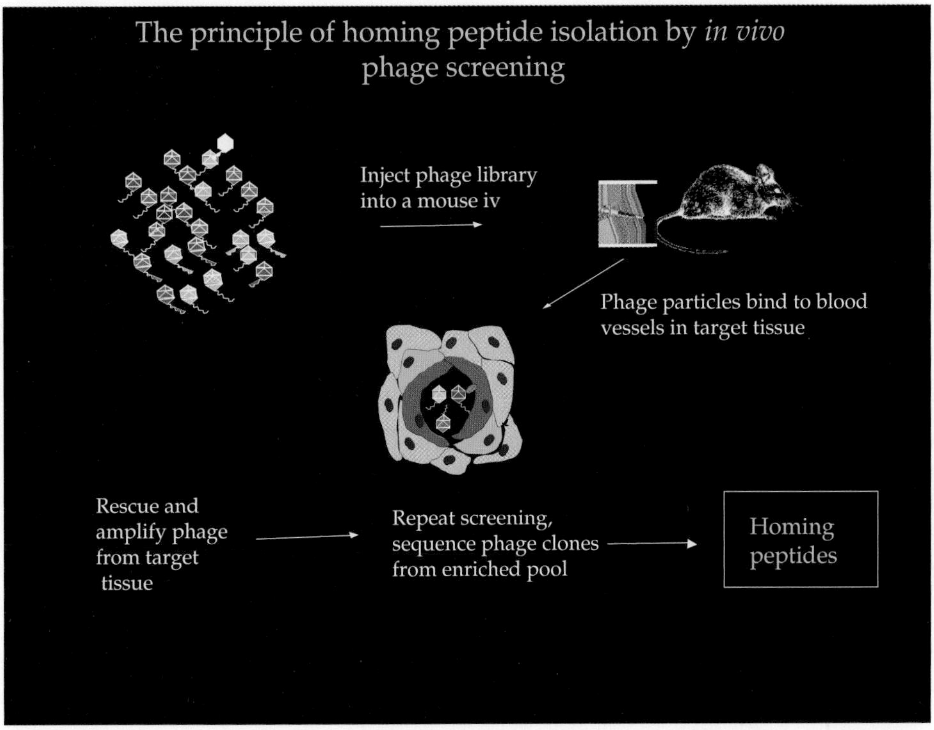

Fig. 1. Schematic representation of the protocol for *in vivo* phage screening and homing peptide characterization. A library containing as many as several billion peptides and expressed on the surface of bacteriophage is intravenously injected into a mouse. A few minutes later, the target tissue is removed, and a phage pool enriched in phage carrying homing peptides for the target tissue is rescued from it by infecting bacteria. A highly enriched pool is obtained by repeating the process several times. The part of the phage's DNA that encodes the peptide insert is then sequenced from a sample of phage clones in the final enriched pool. These DNA sequences can then be translated into the sequences of the peptides displayed by the phage clones. Those peptide sequences that are repeated several times (i.e. have become enriched) are the candidate homing peptides.

far have turned out to express tissue-specific endothelial markers. These tissues include both major organs, such as the brain, lungs, heart, and kidneys, and small ones such as the prostate.[1]

2.2 *Vascular zip codes in pathological conditions*

Angiogenesis. Inflammation causes changes in adjacent vessels. The endothelial cells of these vessels turn on the expression of adhesion molecules that capture circulating leukocytes to the endothelial surface. The attached leukocytes subsequently penetrate through the vessel wall to fight the agent that caused the inflammation. The blood vessels

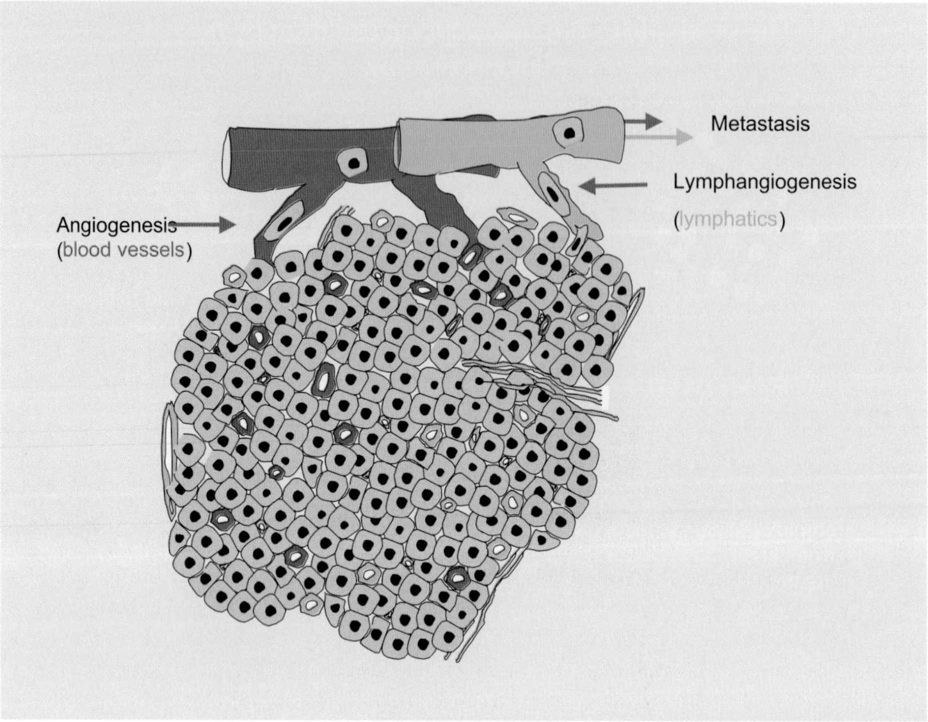

Fig. 2. Blood and lymphatic circulation in tumors. Tumors induce the formation of new blood and lymphatic vessels. Both serve as conduits of metastatic spreading of cancer.

that nurture tumors are also different from normal vessels.[1] Many of the changes in tumor vessels relate to the growth of new vessels into the tumor and are shared by vessels in inflammatory tissues and in tissues undergoing regeneration. The sprouting of new blood vessels induced by growing tissues is called angiogenesis, or when it affects lymphatic vessels, lymphangiogenesis (Fig. 2). The activated endothelial cells in these new vessels express molecules that are not expressed, or are expressed at much lower levels, in normal, established vessels.

The markers of angiogenesis include increased expression of the $\alpha v\beta 3$, $\alpha v\beta 5$ and $\alpha 5\beta 1$ integrins.[10,11] These integrins, particularly $\alpha v\beta 3$ and $\alpha v\beta 5$, are only expressed at low levels in normal tissues, but are up-regulated in growing tissues, including blood vessels undergoing angiogenesis. Blocking these integrins with peptides or other compounds has an anti-angiogenic effect. As discussed below, such anti-angiogenic compounds can also serve as guidance molecules for therapeutics targeted to tumor blood vessels.

Other proteins selectively expressed in angiogenic vessels include aminopeptidase N; its involvement in angiogenesis was established by *in vivo* phage display.[12] The form of aminopeptidase N expressed in tumor vasculature may be distinct from those expressed by normal cells.[13] A similar situation exists with an oncofetal splicing

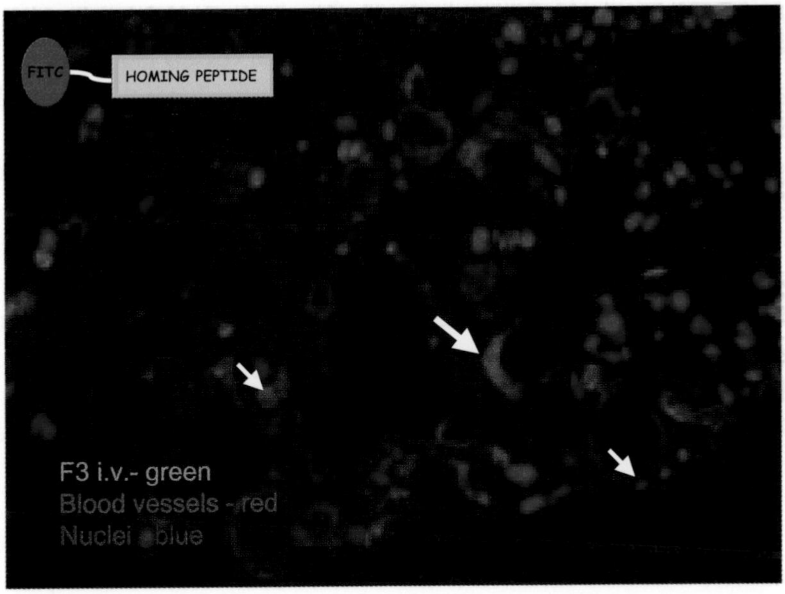

Fig. 3. An example of homing to blood vessels. A fluorescein conjugate of a tumor-homing peptide designated as F3 accumulates in the nuclei of endothelial cells and tumor cells in tumor produced by injecting human tumor cells in an immunosuppressed mouse (a xenograft tumor) (Reproduced from Porkka *et al.*[15])

variant of fibronectin that is selectively expressed in angiogenic blood vessels, whereas fibronectin, in general, is ubiquitous.[14] The variant fibronectin illustrates the fact that the extracellular matrix of tumor vessels can also contain tumor-specific vascular markers.

In some cases an angiogenesis marker is created by expression of a molecule at an aberrant location. Three examples of this are known: nucleolin, which is primarily an intracellular protein, is expressed at the cell surface by angiogenic endothelial cells (and tumor cells), but not by resting endothelial cells of normal vessels (Fig. 3).[15,16]

A similar aberrant cell surface expression has been shown for annexin 1.[17] Moreover, the polarity of phospholipids in cell membrane is perturbed in angiogenic endothelial cells. Anionic phospholipids, such as phosphatidyl serine, are generally confined to the inner leaflet of the lipid bilayer, but appear in the outer leaflet in angiogenesis[18] (curiously enough, this same change is found in apoptotic cells). Comparison of mRNA levels in endothelial cells from tumor vessels to vessels in the adjacent normal tissue has revealed a number of novel angiogenesis markers.[19] Strikingly, various collagen mRNAs turn out to be strongly over-expressed in angiogenic vessels. Perhaps the most interesting angiogenesis marker that has emerged from these studies is one of the receptors for the anthrax toxin. Some groups are attempting to change the toxin so that it would only recognize the angiogenesis-related receptor, with the hope that the toxin could then be used to destroy tumor vessels.

Anti-angiogenic treatments. Controlling angiogenesis is emerging as an important therapeutic strategy; promoting the growth of new blood vessels facilitates tissue regeneration, while destroying the vasculature in a tumor can starve the tumor. The FDA recently approved the first anti-angiogenic drug, an antibody against vascular endothelial growth factor, for cancer therapy. However, a phenomenon that limits the effectiveness of this drug was soon noted, i.e. it preferentially destroys blood vessels that are poorly constructed and do not contribute much to the blood supply of the tumor. The few vessels with a normal or near-normal structure tend to be resistant to anti-angiogenic therapies, and it is these vessels that account for most of the blood supply into the tumor tissue. Hence the phenomenon has been referred to as "normalization" of tumor vessels as a result of anti-angiogenic therapy. To fully realize the potential of anti-angiogenic cancer therapy, it is obviously important to devise ways of extending the effect of the therapy to the "normal" vessels in tumors.

Vascular changes associated with tumor type and stage. In addition to angiogenesis-related markers, which are common to all tumors, tumor vessels also express changes that are specific for a given tumor type. These latter markers are expressed in one tumor type or in a limited range of tumors. It seems that the tumor tissue sends the endothelium the signals necessary for the expression of specific molecules. Moreover, tumor stage-specific vascular markers also exist; homing peptides can distinguish the vasculature of premalignant lesions from normal vessels and from the vessels in a subsequently developing malignant tumor.[20,21] While the existence of tumor type- and tumor stage-specific vascular markers can be inferred from the binding specificity of homing peptides, the molecular nature of those markers remains to be elucidated.

Tumor-specific lymphatic vessel markers. Like tumor blood vessels, lymphatic vessels in tumors are also distinct. The main function of lymphatic vessels is to collect tissue fluid that has leaked from blood capillaries into tissue space, and return it to the blood circulation. Poor lymphatic function (or excessive leakage of fluid from blood vessels) causes tissue swelling, edema. Lymphatic vessels within tumors are generally poorly functional, and this is thought to be the reason why the interstitial pressure in tumors is higher than that in normal tissues. The high tissue pressure in turn makes tumors poorly permeable to drugs; high molecular weight compounds, such as antibodies, are particularly difficult to deliver into tumors.[22] The lymphatics also have an important role in lymphocyte trafficking and immunity. Finally, lymphatic vessels are an important conduit of metastasis, the spreading of tumor cells to distant sites in the body. The first site of metastasis is typically the lymph node to which the lymphatics from the tumor area drain. Hence the "sentinel" lymph node is the place where the surgeon looks for signs of metastasis in a cancer operation.

Like tumor blood vessels, tumor lymphatics grow with the tumor. The process is known as lymphangiogenesis. The newly formed and growing lymphatic vessels are dependent on specific growth factors of the vascular endothelial growth factor family and can be destroyed by inhibiting the activity of these growth factors. The result is suppression of metastasis in various experimental tumor models. Thus, destroying tumor lymphatics may have a place in multifaceted cancer therapy.[23]

Endothelial cells of tumor lymphatics express markers that are not present in normal lymphatics or in the blood vessels of the same tumor.[24,25] The lymphatic markers also provide a second "zip code" system for tumor therapies that are based on physical targeting of drugs to tumors. Targeting drugs to both blood vessels and lymphatics may allow a two-pronged attack on tumors. The areas in tumors that are rich in lymphatics tend to have relatively few blood vessels and *vice versa*, making the two-pronged approach potentially quite attractive. It is likely that lymphatic vessels also express tissue-specific markers and markers that are selectively expressed in diseases other than cancer, but such markers remain to be discovered.

In vivo phage display has revealed an unexpected degree of diversity in the vasculature of tumors, both in tumor blood vessels and lymphatics. This method and similar analyses have not been extensively applied to the study of the vasculature in diseases other than cancer, but it is likely that vascular markers for lesions in other diseases will be identified in the near future. Initial studies on atherosclerotic plaques have already shown promise in this regard.[26,27] Diseases that are vascular in nature, or have a vascular component (e.g. Alzheimer's disease), are very likely to express specific markers that could be identified with these methods in animal models and that could then form the basis of targeted therapies.

2.3 *Vascular zip codes in diagnosis and therapy*

There are two principal ways vascular zip codes can be used to improve existing diagnostic and therapeutic procedures, and to develop new ones. First, any molecule that is specifically expressed in a biological process is likely to play a functional role in that process. Thus, angiogenesis markers are likely to be functionally important in angiogenesis. As discussed below, recent results have shown that this is so. Second, a homing peptide or other binding molecule that binds to a vascular zip code molecule can concentrate a payload at a targeted site, such as a tumor. The payload can be a diagnostic probe, a drug, some other therapeutic or diagnostic agent, or a nanodevice. The selective delivery to the targeted site can be expected to increase the efficacy of the therapy at the target site, while decreasing side effects elsewhere.

The functional involvement of angiogenesis markers in the growth of new blood vessels is illustrated by integrins and aminopeptidase N. The integrins that are selectively up-regulated in angiogenic vessels, $\alpha v\beta 3$, $\alpha v\beta 5$ and $\alpha 5\beta 1$, all bind to their extracellular matrix ligands through the RGD (arginine-glycine-aspartic acid) tripeptide shared by many extracellular matrix proteins.[28] Small peptides containing the RGD sequence, compounds that mimic RGD, or antibodies that block the RGD-binding site can be used to block the function of these integrins. Such compounds have a number of

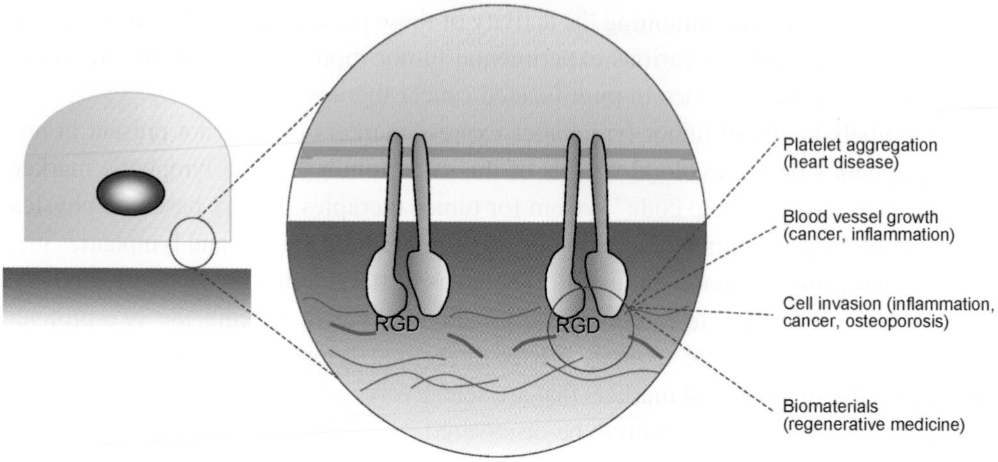

Fig. 4. The RGD cell attachment site. An RGD (arginine-glycine-aspartic acid) governs the attachment of many types of cells to an extracellular matrix. Short synthetic peptides containing the RGD sequence and RGD-mimicking compounds are gaining broad utility in medicine.

proven and prospective medical uses, including an anti-angiogenic activity that indicates a role for integrins in angiogenesis[1] (Fig. 4). Anti-angiogenic activities of aminopeptidase N antibodies and chemical inhibitors of the enzyme[12] similarly suggest a functional role for aminopeptidase N in angiogenesis.

2.4 *Delivery of diagnostic and therapeutic agents to vascular targets*

Here, we are particularly interested in the second mode of using vascular zip codes: the delivery of materials to specific vascular sites, such as tumors. Homing peptides and antibodies that bind to specific vascular markers can carry a payload to the sites where that marker is expressed. When the payload is a drug, the delivery will concentrate the drug at the target site, and increased efficacy and lesser side effects should result. Peptides that specifically bind to the $\alpha v\beta 3$ and $\alpha v\beta 5$ integrins (RGD motif peptide) or to aminopeptidase N (NGR motif peptide) have been used to target doxorubicin, tumor necrosis factor α, and pro-apoptotic/anti-bacterial peptides to tumors.[1] In each case, inhibition of tumor growth was obtained.

Experiments with a pro-apoptotic/anti-bacterial peptide particularly vividly illustrate the power of vascular targeting: an amphipathic peptide that disrupts bacterial and mitochondrial membranes and induces apoptosis if internalized by mammalian cells has been conjugated to homing peptides and delivered by intravenous injection to tumors, arthritic synovium or normal prostate as a conjugate with homing peptides. The effects were selective and dependent on which homing peptide was used; conjugates with tumor-homing peptide produce inhibition of tumor growth, conjugates with peptides that home to sites of angiogenesis suppress inflammation in arthritic synovium, and conjugates with a peptide that homes to the vasculature of normal mouse prostate cause destruction of prostate tissue and delay the development of prostate cancer in mice transgenic for a prostate-expressed oncogene.[1] Thus, a non-specifically toxic peptide can be engineered into a specific drug by coupling it with an appropriate homing peptide partner.

Internalization and subcellular targeting. The targeted apoptosis-inducing peptide conjugates described above also illustrate another point: the ability to be internalized by the target cell is an important characteristic in a homing peptide. Internalization made the homing peptide-pro-apoptotic peptide conjugates effective, and internalization greatly increases the accumulation of a peptide and its payload at the target. An example of a particularly effective internalizing peptide is shown in Fig. 5.

This peptide concentrates its payload (fluorescein) so effectively from an intravenous injection that a mammary fat pad tumor can be visualized on a light table without any other equipment.[29] Moreover, this and some other homing peptides take their payload all the way to the nucleus of the target cells. In an important distinction from well-known cell-penetrating peptides such as the Tat peptide, which internalize into all

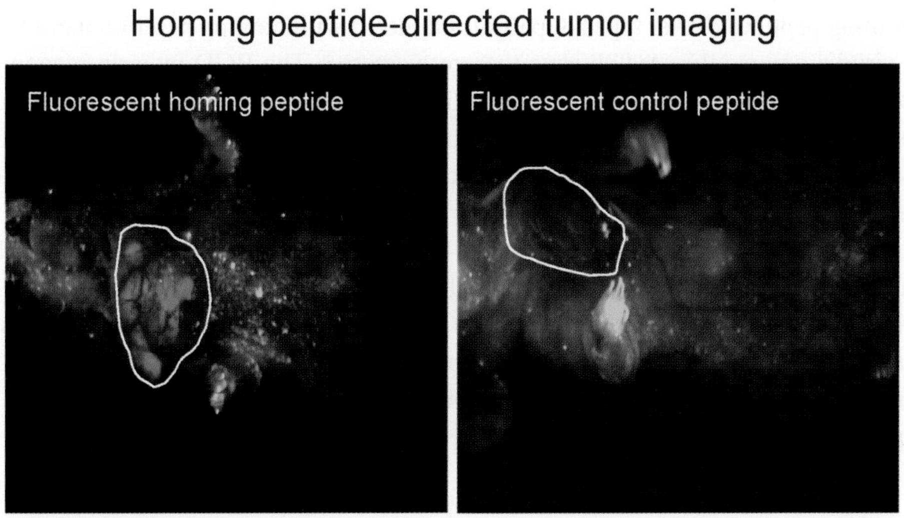

Fig. 5. Homing peptide-directed tumor imaging: a homing peptide that recognizes tumor lymphatics and tumor cells accumulates in tumor tissue. Mice with MDA-MB-435 human breast cancer xenografts were injected with tumor-homing LyP-1 peptide (left) or a control peptide (right), both labeled with fluorescein. The mice were examined under blue light 20 hours later for fluorescence. Yellow autofluorescence outlines the mice; the tumor (circled in white) in the LyP-1-injected mouse is brightly fluorescent (green), whereas no specific fluorescence is seen in the tumor of the mouse injected with the control peptide. (Modified from Laakkonen *et al.*[24])

cells, the homing peptides are cell type-specific. They only enter specific target cells, for example, in tumor vasculature. Some vascular homing peptides bind to the tumor cells in addition to the endothelial cells in the tumor. This further increases the utility of the peptides, as the payload is also specifically delivered into tumor cells.

2.5 *Nanoparticle targeting*

Nanotechnology is predicted to bring about the next revolution in medicine, and vascular zip codes are likely to contribute to this revolution. The promise of nanotechnology is that nanodevices have the potential of being more versatile and effective than even the best drugs. This is because unlike drugs, nanodevices can incorporate multiple functions in one entity. For example, a "smart" nanoparticle could zero in on a site of disease, send a signal from that site for diagnostic purposes, release a drug that treats the disease, and measure the effect. Vascular zip codes provide a system that can supply nanoparticles with the ability to zero in on a target (and could also contribute some sensor functions).

The bacteriophage is a virus that infects bacteria (but not eukaryotic cells). As the phage is a nanoparticle (diameter about 40 nm), the very fact that it can be used to identify vascular zip codes shows that nanoparticles can be targeted to a vascular zip code by homing peptides. Other types of particles targeted with homing peptides include animal viruses commonly used as gene therapy vectors.[30] The RGD integrin recognition sequence has been particularly popular in this regard, but work on other homing peptides is underway. Chemical mimetics of the RGD sequence have been used to target tumor vasculature with lipid-based nanoparticles designed for imaging purposes, and as gene therapy vectors.[2–4]

Homing peptides can also direct the homing of inorganic nanoparticles. Thus, exquisitely specific homing of intravenously injected quantum dots into target tissues has been demonstrated. For example, it has been possible to label tumor blood vessels in one color and the lymphatics in another by using quantum dots that emit light at different wave lengths in conjunction with homing peptide selective for blood vessels or lymphatics of the same tumor.[31]

3. Future Directions

The homing peptide studies described above indicate that the vasculature of each tissue expresses unique cell surface marker molecules. These marker molecules represent a diverse array of proteins ranging from proteases to integrins, and to growth factor receptors and proteoglycans. Some (perhaps all) of these molecules are functionally important (for example, markers of angiogenic vasculature are needed for angiogenesis to take place). New vascular functions and new drug targets are likely to emerge from identification of the target molecules for homing peptides that have not yet been matched with a receptor.

The vascular zip code molecules that have been identified in tumors include angiogenesis-associated molecules and molecules that are expressed in the vasculature of a certain tumor type. Tumor type-specific molecular markers would be potentially advantageous in targeted delivery of therapies because sites of beneficial angiogenesis would not be affected. Markers that are specific for premalignant lesions could perhaps be used to detect and eradicate developing cancers. Tumor lymphatics differ from their normal counterpart vessels, providing another readily accessible target for the delivery of therapies.

Finally, disease-associated changes in the vasculature may make it possible to develop modularly designed nanodevices that can incorporate more functions than currently available drugs or diagnostic probes (Fig. 6).

Targeting capabilities provided by homing peptides could be used in a number of ways. For example, one can envision diagnostic nanodevices that recognize a vascular target through homing peptide binding, become altered as a result of the encounter with

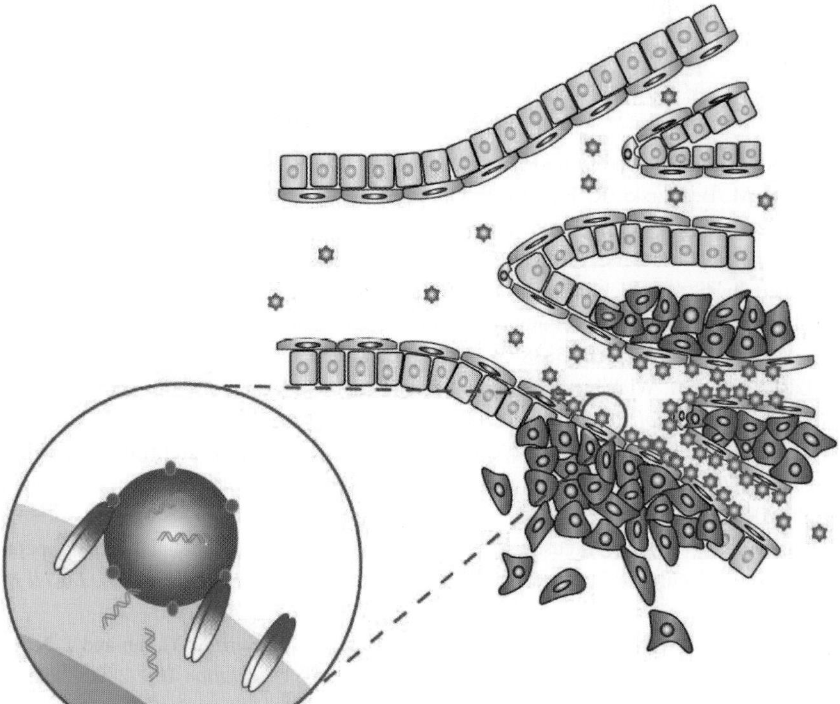

Fig. 6. Schematic illustration of nanoparticle design for the recognition of tumor vasculature and additional functions. Circulating nanoparticles bind a specific marker on the endothelial cells of tumor vessels, but do not bind to the vessels in normal tissue. The bound particles discharge a cargo into the tumor. (Reproduced with permission from Ruoslahti.[1])

the target, and report back on the encounter. Externally accessible tissues could be patrolled by nanodevices administered through other routes. For example, orally administered devices could survey the gut, while particles introduced into the urinary bladder, or delivered there by filtration from the blood, could do the same for bladder cancer, and so on. Nanodevices capable of recognizing premalignant lesions could extend the diagnostic capabilities to include the all-important early detection of cancer. Adding responsiveness to an external signal, such as a magnetic field, would allow remote control of nanodevices that have identified their target in the body, allowing the physician to trigger the release of a therapeutic at an optimal time and dose. One of the challenges of bioengineers is to design these new devices; major advances in medical practice are likely to ensue.

Acknowledgments

The author's work was supported by the NIH grants CA 082713, CA 104898, CA115410 and Cancer Center Support Grant CA 30199, and the DAMD 17-02-1-0315 Innovator Award from the Department of Defense.

References

1. E. Ruoslahti, Specialization of tumour vasculature, *Nat. Rev. Cancer* **2**: 83–90 (2002).
2. J. D. Hood, M. Bednarski, R. Frausto, S. Guccione, R. A. Reisfeld, R. Xiang and D. A. Cheresh, Tumor regression by targeted gene delivery to the neovasculature, *Science* **296**: 2404–2407 (2002).
3. P. M. Winter, S. D. Caruthers, A. Kassner, T. D. Harris, L. K. Chinen, J. S. Allen, E. K. Lacy, H. Zhang, J. D. Robertson, S. A. Wickline and G. M. Lanza, Molecular imaging of angiogenesis in nascent Vx-2 rabbit tumors using a novel alpha(nu)beta3-targeted nanoparticle and 1.5 tesla magnetic resonance imaging, *Cancer Res.* **63**: 5838–5843 (2003).
4. G. M. Lanza and S. A. Wickline, Targeted ultrasonic contrast agents for molecular imaging and therapy, *Curr. Probl. Cardiol.* **28**: 625–653 (2003).
5. K. A. Kelly, J. R. Allport, A. Tsourkas, V. R. Shinde-Patil, L. Josephson and R. Weissleder, Detection of vascular adhesion molecule-1 expression using a novel multimodal nanoparticle, *Circul. Res.* **96**: 327–336 (2005).
6. R. C. Johnson, D. Zhu, H. G. Augustin-Voss and B. U. Pauli, Lung endothelial dipeptidyl peptidase IV is an adhesion molecule for lung metastatic rat breast and prostate carcinoma cells, *J. Cell Biol.* **121**: 1423–1432 (1993).
7. B. S. Ding, Y. J. Zhou, X. Y. Chen, J. Zhang, P. X. Zhang, Z. Y. Sun, X. Y. Tan and J. N. Liu, Lung endothelium targeting for pulmonary embolism thrombolysis, *Circulation* **108**: 2892–2898 (2003).
8. J.-T. Pai and E. Ruoslahti, Identification of genes up-regulated in endothelia, *Gene* **347**: 21–33 (2005).
9. W. C. Aird, J. M. Edelberg, H. Weiler-Guettler, W. W. Simmons, T. W. Smith and R. D. Rosenberg, Vascular bed-specific expression of an endothelial cell gene is programmed by the tissue microenvironment, *J. Cell Biol.* **138**: 1117–1124 (1997).
10. B. P. Eliceiri and D. A. Cheresh, The role of alphav integrins during angiogenesis: insights into potential mechanisms of action and clinical development, *J. Clin. Invest.* **103**: 1227–1230 (1999).

11. S. Kim, K. Bell, S. A. Mousa and J. A. Varner, Regulation of angiogenesis *in vivo* by ligation of integrin $\alpha5\beta1$ with the central cell-binding domain of fibronectin, *Am J. Pathol.* **156**: 1345–1462 (2000).
12. R. Pasqualini, E. Koivunen, R. Kain, J. Lahdenranta, M. Sakamoto, A. Stryhn, R. A. Ashmun, L. H. Shapiro, W. Arap and E. Ruoslahti, Aminopeptidase N is a receptor for tumor-homing peptides and a target for inhibiting angiogenesis, *Cancer Res.* **60**: 722–727 (2000).
13. F. Curnis, G. Arrigoni, A. Sacchi, L. Fischetti, W. Arap, R. Pasqualini and A. Corti, Differential binding of drugs containing the NGR motif to CD13 isoforms in tumor vessels, epithelia, and myeloid cells, *Cancer Res.* **62**: 867–874 (2002).
14. C. Halin, S. Rondini, F. Nilsson, A. Berndt, H. Kosmehl, L. Zardi and D. Neri, Enhancement of the antitumor activity of interleukin-12 by targeted delivery to neovasculature, *Nature Biotechnol.* **20**: 264–269 (2002).
15. K. Porkka, P. Laakkonen, J. A. Hoffman, M. Bernasconi and E. Ruoslahti, A fragment of the HMGN2 protein homes to the nuclei of tumor cells and tumor endothelial cells *in vivo*, *Proc. Natl. Acad. Sci. USA* **99**: 7444–7449 (2002).
16. S. Christian, J. Pilch, K. Porkka, P. Laakkonen and E. Ruoslahti, Nucleoilin expressed at the cell surface is a marker of endothelial cells in tumor blood vessels, *J. Cell Biol.* **163**: 871–878 (2003).
17. P. Oh, Y. Li, J. Yu, E. Durr, K. M. Krasinska, L. A. Carver, J. E. Testa and J. E. Schnitzer, Subtractive proteomic mapping of the endothelial surface in lung and solid tumours for tissue-specific therapy, *Nature* **429**: 629–635 (2004).
18. X. Huang, M. Bennett and P. E. Thorpe, A monoclonal antibody that binds anionic phospholipids on tumor blood vessels enhances the antitumor effect of docetaxel on human breast tumors in mice, *Cancer Res.* **65**: 4408–4416 (2005).
19. A. Nanda and B. St. Croix, Tumor endothelial markers: new targets for cancer therapy, *Curr. Opin. Oncol.* **16**: 44–49 (2004).
20. J. A. Hoffman, E. Giraudo, M. Singh, M. Inoue, K. Porkka, D. Hanahan and E. Ruoslahti, Progressive vascular changes in a transgenic mouse model of squamous cell carcinoma, *Cancer Cell* **4**: 383–391 (2003).
21. J. A. Joyce, P. Laakkonen, M. Bernasconi, G. Bergers, E. Ruoslahti and D. Hanahan, Stage-specific vascular markers revealed by phage display in a mouse model of pancreatic islet tumorigenesis, *Cancer Cell* **4**: 393–403 (2003).
22. R. K. Jain, Normalization of tumor vasculature: an emerging concept in antiangiogenic therapy, *Science* **307**: 58–62 (2005).
23. K. Alitalo, S. Mohla and E. Ruoslahti, Lymphangiogenesis and cancer: meeting report, *Cancer Res.* **64**: 9225–9229 (2004).
24. P. Laakkonen, K. Porkka, J. A. Hoffman and E. Ruoslahti, A tumor-homing peptide with a lymphatic vessel-related targeting specificity, *Nat. Med.* **8**: 743–751 (2002).
25. L. Zhang, E. Giraudo, J. A. Hoffman, D. Hanahan and E. Ruoslahti, Lymphatic zip codes in premalignant lesions and tumors, *Cancer Res.* **66**: 5696–5706. (2006).
26. P. Houston, J. Goodman, A. Lewis, C. J. Campbell and M. Braddock, Homing markers for atherosclerosis: applications for drug delivery, gene delivery and vascular imaging, *FEBS Lett.* **492**: 73–77 (2001).
27. C. Liu, G. Bhattacharjee, W. Boisvert, R. Dilley and T. Edgington, *In vivo* interrogation of the molecular display of atherosclerotic lesion surfaces, *Am. J. Pathol.* **163**: 1859–1871 (2003).
28. E. Ruoslahti, The RGD story: a personal account. A landmark essay. *Matrix Biol.* **22**: 459–465 (2003).
29. P. Laakkonen, M. E. Akerman, H. Biliran, M. Yang, F. Ferrer, T. Karpanen, R. M. Hoffman and E. Ruoslahti, Antitumor activity of a homing peptide that targets tumor lymphatics and tumor cells, *Proc. Natl. Acad. Sci. USA* **101**: 9381–9386 (2004).
30. T. J. Wickham, Targeting adenovirus, *Gene Ther.* **7**: 110–114 (2000).
31. M. E. Akerman, W. C. Chan, P. Laakkonen, S. N. Bhatia and E. Ruoslahti, Nanocrystal targeting *in vivo*, *Proc. Natl. Acad. Sci. USA* **99**: 12617–12621 (2002).

CHAPTER 20

THE STRUCTURE OF THE CENTRAL NERVOUS SYSTEM AND NANOENGINEERING APPROACHES FOR STUDYING AND REPAIRING IT

Gabriel A. Silva

Abstract

Nanotechnologies involve materials and devices with an engineered functional organization at the nanometer scale. Applications of nanotechnology to cell biology and physiology provide targeted interactions at a fundamental molecular level. In neuroscience, this entails specific interactions with neurons and glial cells, which complements the cellular and molecular organization and structure of the nervous system. Examples of current work include technologies designed to better interact with neural cells, advanced molecular imaging technologies, applications of materials and hybrid molecules for neural regeneration and neuroprotection, and targeted delivery of drugs and small molecules across the blood-brain barrier.

1. Introduction

Nanotechnology and nanoengineering stand to produce significant scientific and technological advances in diverse fields including medicine and physiology. In a broad sense, they can be defined as the science and engineering involved in the design, synthesis, characterization, and application of materials and devices whose smallest functional organization in at least one dimension is on the nanometer scale, ranging from a few to several hundred nanometers. A nanometer is one billionth of a meter, or three orders of magnitude smaller then a micron, roughly the size scale of a molecule itself (e.g. a DNA molecule is about 2.5 nm long, while a sodium atom is about 0.2 nm). To give an appreciation of just how significant an order of magnitude is, let alone three orders when going from micron to nanometer scales, consider that no one would ever walk from New York to San Diego; but, with a single order of magnitude change in speed, the equivalent of changing speed from walking to driving, you would get to San Diego in a few days. Flying, which would be two orders of magnitude faster than driving, would get you there in a few hours, and three orders faster than walking would take you minutes. (Walking a straight line between the two cities would take about 42 days at an average speed of three miles per hour.)

The potential impact of nanotechnology stems directly from the spatial and temporal scales being considered: materials and devices engineered at the nanometer scale imply controlled manipulation of individual constituent molecules and atoms in how they are arranged to form the bulk macroscopic substrate. This, in turn, means that nanoengineered substrates can be designed to exhibit very specific and controlled bulk chemical and physical properties as a result of the control over their molecular synthesis and assembly.

For applications to medicine and physiology, including neuroscience, these materials and devices can be designed to interact with cells and tissues at a molecular (i.e. sub-cellular) level with a high degree of functional specificity, thus allowing a degree of integration between technology and biological systems not previously attainable. It should be appreciated that nanotechnology is not in itself a single emerging scientific discipline, but a meeting of traditional sciences such as chemistry, physics, materials science and biology in order to bring together the required collective expertise needed to develop these novel technologies. Bionanotechnology applications to the central nervous system (CNS) are designed to interact with cells and tissues at a subcellular molecular level, the functional building block level associated with the constituent protein elements that make up the functional cellular unit, such as cell surface receptors, transmembrane proteins, ion channels, and the cell's cytoskeleton. The complexity associated with the cellular heterogeneity, structure, and functional organization of the CNS present some unique challenges for designing and using bionanotechnologies, but the potential offered by the unique properties associated with nanoengineered materials and devices complement other neurobiological approaches and provide a significant opportunity to advance our basic understanding of cellular neurobiology and neurophysiology, and provide novel clinical treatments for neurological disorders.

2. The Micro- and Nanoscale Structure of the Central Nervous System (CNS)

2.1 *The organization of the CNS*

As with all body systems, the fundamental functional scale of the central nervous system is at the micro- and nanoscales. This section begins with a review of the structure and organization of the CNS, working down in scale to the protein (i.e. nanoscale) level. For general reviews and further reading on neurophysiology the reader is referred to any of the several excellent texts which cover the subject to varying degrees and in different ways.[1]

The CNS essentially refers to three gross anatomical structures: the brain, spinal cord, and neural retina, which is an extension of the brain itself. The entire CNS is separated from the rest of the body by the blood brain barrier (BBB), and in the case of the retina the blood retinal barrier. This makes the CNS immunologically privileged, in the sense that specific immune cells and factors (e.g. antibodies) circulating in the periphery do not have access to CNS structures. This also makes the local extracellular CNS environment, the cerebral spinal fluid (CSF), quite chemically unique. The CSF surrounds the brain and spinal cord and provides a degree of cushioning from the bony structures which protect them (i.e. the skull and vertebrae). The BBB, which controls the rate at which factors cross in and out of the CNS, is made up by the endothelial cells of the small blood vessels that feed it. The BBB provides both a mechanical mechanism of restricted access mediated by tight junctions between adjacent cells, and physiological mechanisms, mediated by specialized active transport processes. In general, lipid soluble factors are able to cross the BBB much more readily than less lipid soluble factors, a fact that has been an important consideration for drug delivery into the CNS and is an important consideration for nanotechnological approaches that strive to do the same. Between the neural tissues and the bone that protect them are a series of membranous covers collectively referred to as the meninges. The thick dura matter is closest to the bone, the arachnoid is next, and the very thin pia matter is immediately adjacent to the neural tissue. The subarachnoid space, between the arachnoid and the pia matter is where the CSF is.

During development the brain forms its main subdivisions, which consist of the cerebrum, diencephalon, cerebellum, and brainstem. The brain also has four cavities in which CSF is produced called the ventricles. The cerebrum and diencephalon together make up the forebrain, which include other key brain structures such as the basal ganglia, an area important in various aspects of behavior and movement, the thalamus, which is where much of the neural information passing on to the cortex is integrated, the hypothalamus, which is a small but critical region coordinating much of brain's functions, and the evolutionary primitive and complex limbic system, which lies deep in the brain and is involved in various aspects of learning, experience, behavior, and emotions. The cerebrum itself is where most of the complex neural processing occurs. The corpus collosum refers to the massive group of fibers that connect the two hemispheres of the brain. The cerebellum, which lies at the base of the brain towards the back, is involved in the coordination of posture, balance, and movement. The brainstem, which consists of the midbrain, the pons, and medulla oblongata, acts as the relay between the spinal cord and brain, but also has critical roles in basic life sustaining functions such as breathing. The spinal cord is made up of many complex ascending and descending pathways to the brain (afferent and efferent pathways, respectively) with specific offshoots along its length (the dorsal and ventral roots) that innervate all the structures outside the CNS that are controlled by the CNS. Most sensory and control information is shuttled through the spinal cord, with the exception of structures innervated by the 12 cranial nerves which connect to the brain directly. One of these nerves, the optic nerve, is made up by bundles of nerve fibers (properly called axons) that form the output from the neural retina on their way to the brain. The retina consists of several distinct anatomical and physiological layers of cells, and is where all visual sensory information begins as a result of a process called phototransduction in photoreceptor neurons where incoming light is transduced into a neuro-chemical signal. The axons formed by the last layer of neurons in the retina, the retinal ganglion cells, form the optic nerve. The retina is also responsible for a significant degree of pre-processing of neural information before it goes on to the brain. Neural structures outside of the CNS make up, by definition, the peripheral nervous system (PNS).

2.2 *The cellular and molecular structure of the CNS*

Each of the CNS structures described above consists of distinct anatomical and/or functional sub-structures that serve specific functions. All of these sub-structures are composed of a very complex interconnection of cells, which are the functional units responsible for all the information processing that occurs in the CNS (Fig. 1).

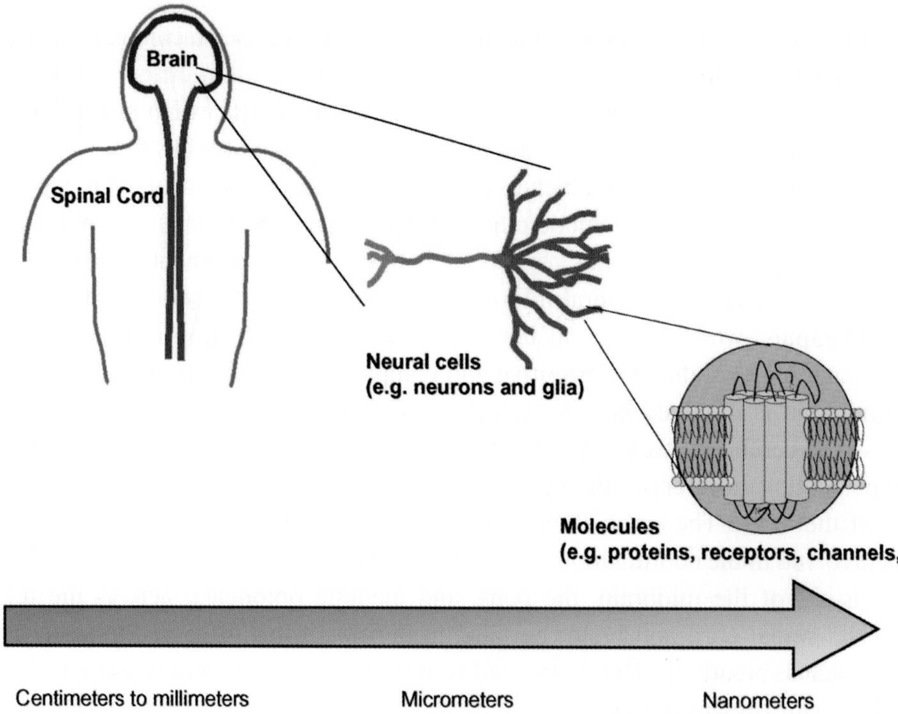

Fig. 1. Overview of the structure of the central nervous system (CNS), from organs (i.e. the brain and spinal cord) to cells to molecules. The size scale progressively decreases from centimeter and millimeter sized anatomical and functional structures to micrometer sized cells, to nanometer sized molecules.

There are many different types and sub-types of cells in the CNS, each a highly specialized terminal phenotype cell designed to carry out a specific function. In general, there are two main classes of cells in the nervous system: neurons and glial cells. Neurons are the fundamental functional (i.e. information processing) unit of the nervous system. Neurons come in a tremendous variety of morphologies which reflect the vast number of specialized roles they are designed for. In general, however, an idealized neuron consists of dendrites, a cell body, an axon, and a synaptic terminal. The dendrites collect sub-threshold inputs from other neurons (i.e. presynaptic neurons) called excitatory postsynaptic potentials (EPSP's) or inhibitory postsynaptic potentials (IPSP's). If the temporal and/or spatial summation of these inputs reach a specific threshold level at the neuron's axon hillock, a specialized zone where the axon meets the cell body, an electrical event called an action potential is generated that travels down the length of the axon and terminates at the synaptic terminal where neurotransmitter molecules are released by the neuron and signal other downstream neurons by contributing to those neuron's EPSP's and IPSP's. Some neurons have up to 400,000 dendrites, and the complexity of the summating postsynaptic inputs both in space (e.g. as a function of where on the dendritic tree they are) and in time is staggering and represents the principal mechanism of computational processing in the CNS. The cell body, which contains the neuron's nucleus, carries out many of the typical life-supporting processes of all cells, although in the neuron this is also where chemical-electrical processes result in an action potential being generated (or not). The axon is an elongated cellular process containing many intermediate filaments and microtubules that transport vesicles and other organelles up and down its length. It is also through which the action potential travels to signal a downstream neuron. Some neurons have very short axons, such as the inter-neurons in the brain and spinal cord or the photoreceptors in the retina, while some axons can be almost a meter long, such as the spinal cord motor neurons that innervate the muscles of the leg.

The action potential is a self-propagating, self-renewing chemical-electric event that begins at the axon hillock and travels the length of the axon uninterrupted. The molecular basis of the action potential is the movement of ions down strong electrochemical and diffusion gradients between the inside and outside of the neuron separated by the cell membrane of the axon. Na^+, which is actively pumped out of the neuron and is at much higher concentrations extracellularly, enters the cell through Na^+ specific ion channels while K^+, whose situation is reversed, flows out. Other ions, such as Cl^- are also involved. The ion channels are voltage sensitive, so that the activation of channels induces a voltage change to open adjacent channels, thereby creating a self-propagating event. The amplitude of the action potential is constant and its duration is very short, so that in engineering terms it can be regarded as a traveling or propagating delta function. It is important to appreciate that changes in only a very thin shell of ions on either side of the axon's cell membrane are required to produce an action potential, so that the restoration of ionic compositions by active pumping of ions by the transmembrane Na^+-K^+-ATPase pumps are only required over long time scales, but not over the millisecond time scale of an individual action potential. In fact, several thousand action potentials can occur following the chemical inactivation of the Na^+-K^+-ATPase pumps before any significant effect on the amplitude of the action potential can be measured. At the end of it all is the synaptic terminal of the presynaptic neuron which synapses with a dendrite of a postsynaptic neuron. We will pay particular attention to the molecular details of the synapse below as an example of a naturally occurring form of a nanoengineered structure. The action potential and synapse constitute the fundamental mechanisms by which information is transmitted through the nervous system. It should be noted that the term "nerve", which everyone is familiar with, properly speaking refers to a bundle of axons running through the PNS. A similar bundle of axons running through the CNS is called a tract. Similarly, a collection of neuron cell bodies in the PNS is referred to as a ganglion (e.g. the dorsal root ganglia of the spinal cord which sit just outside the vertebral bodies); while in the CNS they are called a nucleus (e.g. the lateral geniculate nucleus in the brain which receives inputs from the retina). Note that this use of the term "nucleus" is not to be confused with the nucleus of a cell. Some axons are "insulated" by the processes of a type of specialized non-neuronal cell wrapping around them that give the axon a white appearance due to a high lipid content in these processes. This is referred to as myelination and results in increased speeds of action potential transmission by a process called saltatory conduction, since the action potential "jumps" from one spot on the axon to the next spot not covered by myelin, roughly equally spaced down the length of the axon (the nodes of Ranvier). The cells that do this insulating in the CNS are called oligodendrocytes, while their counterparts in the PNS are called Schwann cells. Both are types of glial cells, the second major cell type in the nervous system, described below. In contrast, unmyelinated axons and cell bodies appear grayish in color. This is the physical basis for the gray matter and white matter of the CNS. In the brain the gray matter is found on the surface, giving the brain a gray color, while the heavily myelinated tracts are hidden beneath. In the spinal cord it is the reverse, the white matter is in the periphery of the cord, with the butterfly shaped gray matter in the center.

The second major cell type in the nervous system are glial cells, which include several distinct types of cells. As introduced above, oligodendrocytes in the CNS and Schwann cells in the PNS are primarily responsible for the myelination of axons. A breakdown in this process compromises the entire nervous system and is the pathological basis of multiple sclerosis. Another type of glial cell is the microglia, the CNS's phagocytic immune cells that are in part responsible for removing waste products. The third major classes of glial cells are the astrocytes in the brain and spinal cord and astrocytes and Muller cells in the neural retina. From a functional standpoint these cells are remarkably interesting. They were classically regarded as housekeeping cells that existed to support neurons by secreting growth factors and other trophic factors and maintaining a homeostatic extracellular environment.

Relatively recently, however, work by several groups have shown that astrocytes in particular are not just housekeeping cells, but also directly influence and participate in the regulation and transmission of information in the CNS by interacting with neurons, to the point that the concept of a synapse involving just pre- and postsynaptic neurons is being redefined to include an astrocytic process.[2,3] These cells form highly complex signaling networks with each other and with neurons, although the molecular details are very different from those of the action potential. Astrocyte-astrocyte communication is mediated by intracellular calcium waves that spread throughout an astrocyte and results in the release of various extracellular signaling factors, including adenosine triphosphate among others, to signal both other nearby and relatively far away astrocytes. Astrocytes are able to signal neurons and neurons are able to signal astrocytes by a variety of different mechanisms that include direct cell-cell gap junctional coupling and over longer distances by chemical diffusion. Because a single astrocyte can have multiple processes associated with many neurons, neuronal-glial signaling adds a significant layer of computational complexity to information processing and flow in the CNS since these cells are able to "short circuit" neuronal connections.

2.3 *Membrane proteins and receptors*

All cells, in a very real sense, are exquisitely engineered nanoscale machines designed to carry out many complex tasks in a coordinated way that respond, signal, and adapt to the environment in which they find themselves. The fundamental functional units that make up the cell are proteins and other molecules, the principal targets which nanoengineered materials and devices are designed to interact with. This can be achieved in one of two ways: one is to develop bionanotechnologies that target ubiquitous components of cellular and biochemical signaling systems. Depending on the specific cell with which the device interacts with, targeting a ubiquitous signaling pathway would affect one or more specific downstream events. A good example would be targeting protein phosphorylation sites, which is a ubiquitous mechanism for modifying and altering the function of proteins. Targeting phosphorylation sites on β-tubulin III would produce specific affects in neurons, since it is a neuron-specific microtubule, while doing the same thing on glial fibrillary acidic protein (GFAP) phosphorylation sites in astrocytes, a macroglial-specific intermediate filament would produce very different results. The second way would be to develop bionanotechnologies designed to target a cell-specific signaling process in order to affect a known functional end point. This approach requires significant expertise in the physiological or biological system under consideration, and as such will most likely require interdisciplinary cross-training between fields and/or highly interdisciplinary collaborations. It should be appreciated that the former approach of developing nanotechnologies that target general or ubiquitous cellular processes would

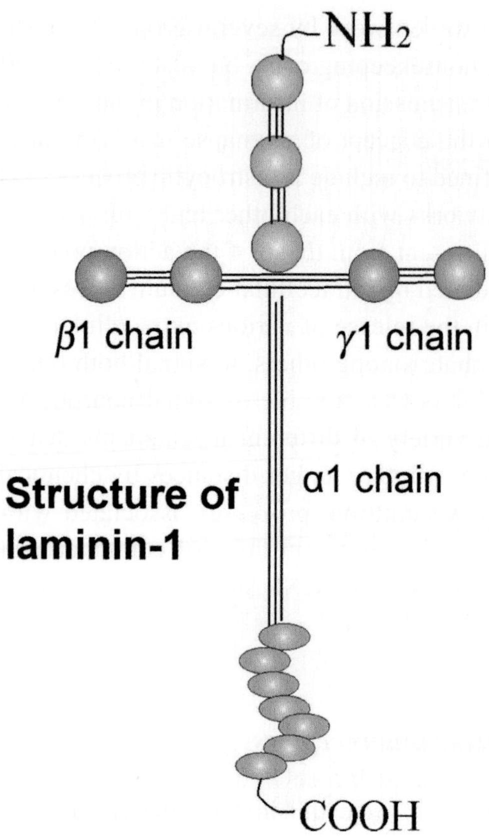

Fig. 2. Schematic of the structure of laminin-1 showing its α1, β1, and γ1 component chains.

ultimately still require interdisciplinary collaborations to produce meaningful clinical or biological applications.

A particularly significant potential target for nanotechnologies designed to interact with neurons, for example, is the laminin family of proteins.[4,5] The laminins have been used in micro- and nanoengineered systems with neurons both *in vitro* and *in vivo*, particularly within the context of neural regeneration. Laminins are large multi-domain trimeric proteins made up of α, β, and γ chains (Fig. 2), of which there exist various isotypes due to proteolytic cleavage and alternative splicing post-translational modifications.[6,7]

The expression of different laminin isoforms varies both spatially and temporally between tissues and within a specific organ (e.g. the brain) during development and adulthood. In this way, the laminins play key roles in signaling and coordinating cell specific events from the outside world. The failure of laminin expression or their incorrect expression, either spatially or temporally, results in numerous pathologies of varying severity. For example, mutations in the laminin-5 genes LAMα3, LAMβ3, and LAMα2 produce skin blisters, while mutations in the LAMα2 gene results in muscular dystrophy.[8] A loss-of-function deletion of LAMα1 produces embryonic death in mice null mutants.[9]

To date, five different α chains have been identified, labeled α1 to α5, three different β chains, β1 to β3, and three different γ chains, γ1 to γ3. This allows 45 possible different combinations of trimeric isoforms, although in reality there are restrictions on the number of actual combinations that occur naturally, with specific chains interacting only with other specific chains. Twelve laminins have been identified, laminin 1 to 12, with laminin 1 being the most extensively studied. Ultrastructurally, laminin 1 has a cross-shaped structure with the long arm ending in a G domain formed by the C terminus of the α1 chain. The N terminus of the β1 and γ1 domains bend and form the short arms of the cross, while the N terminus of the α1 chain extends to form the top of the cross. Beyond the point where the three chains meet they form an α helix that ends at the G domain. The α1 chain is 400 kDa while the β1 and γ1 chains are 200 kDa each. In general, this is the structure of most of the laminins.[6,10] Neurons in particular respond very strongly to the laminins and in particular to laminin 1, displaying strong cell adhesion and neurite outgrowth properties that are cell-type and peptide dependent.[11] A large variety of specific peptide sequences have been identified that affect neurons in different ways. Several groups, including our own, have recognized the advantages of engineering materials that not only provide mechanical and structural support to cellular systems, but also express functional chemistry that promotes favorable cellular responses by incorporating extracellular matrix (ECM) signals.[11,12] Nanoengineered systems provide an ideal opportunity to incorporate these molecular signals into novel materials that are designed to specifically interact with target cells.

2.4 *Example of the nanoengineered CNS: self-assembly of the presynaptic particle web*

A particularly striking example of biological "nanoengineering" in the CNS was described by Philips and colleagues[13] with regards to the self-assembly of the presynaptic particle web, a dense arrangement of proteins and cytoskeletal structures in presynaptic neurons. The neuronal synapse, both presynaptically and postsynaptically, is an amazing nanoscale machine of exquisite complexity and specialization, the full details of which are beyond the scope of this chapter (the reader is referred to Refs. 14 and 15 on the topic). The synapse is made up of pre- and postsynaptic membranes in direct opposition to each other separated by a very narrow gap of about 25 nm. Across this gap extend projections from both the pre- and postsynaptic cells that hold the synapse tightly together, so much so that it is difficult to physically separate pre- and postsynaptic terminals. On the postsynaptic side is the postsynaptic density (PSD), an ultrastructurally dense collection of intracellular structures, proteins, and vesicles that can be easily seen with electron microscopy. The PSD, a highly studied structure, has classically been regarded as the principal adherence component of the synapse, responsible for maintaining its structural integrity, a sort of synaptic "glue." On the presynaptic side there exists a presynaptic cytoskeletal "web" made up of electron dense particles about 50 nm in diameter arranged in a structured network. This structure is much less studied then the PSD. It is partly thought to guide presynaptic vesicles containing neurotransmitters to the synaptic terminal for release into the synaptic cleft, but is also the presynaptic terminal's contribution to adherent stabilization. Phillips *et al.* were able to experimentally determine the conditions under which synaptic junctional complexes can be isolated and purified, and demonstrated that the presynaptic web particles directly connect to the PSD through nanosize filamentous tethers, presumably participating in cell-cell adhesion at the synapse. Self-assembly is essentially responsible for how the synapse is put together (for an excellent review see Ref. 16). What is particularly striking is that these authors were able to show that, given the right physical and chemical conditions, the elements of the presynaptic web can undergo soluble disassembly and re-assembly just by changing the pH of its environment from physiological pH to a pH of 8 and then back to 6. Indeed, using mass spectrometry they were able to show that at least some of the proteins that make up the web were known proteins that participate in vesicle recycling and membrane fusion.

3. Synthetic and Engineering Methods in Nanotechnology

Different methods for the synthesis of nanoengineered materials and devices can accommodate precursors from solid, liquid, or gas phases, and encompass a tremendously varied set of experimental techniques. In general, however, most synthetic methods can be classified into two main approaches: "top down" and "bottom up" approaches, and combinations thereof.

"Top down" techniques begin with a macroscopic material or group of materials and incorporate smaller scale details into them. The best known example of a "top down" approach are the photolithography techniques used by the semiconductor industry to create integrated circuits by etching patterns in silicon wafers. This process generally starts by covering a piece of silicon with some kind of photoresist, a photochemical sensitive polymer that hardens upon exposure to laser light of specific wavelengths. Patterns are then "drawn" into the photoresist coated silicon with a laser, so that the untreated photoresist can be washed away and the exposed silicon treated so it can act as an electrode. Aluminum wires are then placed down in the etched patterns after removing the hardened photoresist to yield transistors. A similar approach has been used to develop micro-scale connected wells in agar using poly(dimethylsiloxane) molds in order to study neuron-astrocyte communication. Cell cultures are set up where neurons are in one well and astrocytes in an adjacent well connected by a channel that allows the diffusion of soluble factors.[17] This is an example of a lithographic technique applied to cell biology. Other types of nanolithographic techniques such as dip-pen nanolithography[18,19] and electrostatic atomic force microscope nanolithography,[20] where individual molecules are deposited or moved, respectively, into desired configurations, are able to produce true nanoscale features in various materials.

"Bottom up" approaches, on the other hand, begin by designing and synthesizing custom-made molecules that have the ability to self-assemble or self-organize into higher order mesoscale and macroscale structures.[21,22] The challenge is to synthesize molecules that spontaneously self-assemble upon the controlled change of a specific chemical or physical trigger, such as a change in pH, the concentration of a specific solute, or the application of an electric field. The physical mechanisms that produce self-assembly, i.e. the driving forces that push these molecules to self-assemble into organized structures, are due to thermodynamics and competing molecular interactions, including hydrophobic/hydrophilic forces, hydrogen bonding, and van Der Waals interactions, that aim to minimize energy states for different molecular configurations. The trick is to design systems that self-assemble into macroscopic higher order structures that display desirable chemical and/or physical properties not displayed by the constituent molecules themselves.

A related synthetic approach relevant to biological applications are templating or scaffolding techniques which use pre-existing structures to guide the nucleation and growth of a nanostructured material. Good examples of this are biomineralization-type applications such as the growth of artificial bone biomimetics.[23,24] This is achieved by inducing the formation of organoapetite on (surgically implantable) titanium structures such as foils, meshes, or porous cylinders by the pre-adsorption onto the bare metal of poly-L-lysine or poly-L-glutamic acid, which when in contact with the metal form a polyionic bilayer. It is thought that the capture of tiny embryonic crystals and subsequent nucleation of these seed crystals are what lead to the actual growth of the apetite on the metal. Other techniques are attempting to mimic the ultrastructure of bone without the requirement of deposition on pre-existing metal structures, by inducing the self-assembly and formation of molecular nanofibers interspaced with mineralized hydroxyapetite crystals, structures which are similar in spirit to the collagen ultrastructure of bone.[25,26] As these technologies develop, it is expected that these advanced materials will not only provide desirable mechanical properties, but also incorporate functional cell signaling properties such as neurotrophic support or anti-inflammatory effects.

4. Applications of Nanoengineering and Nanotechnology to the CNS

4.1 *Applications to basic neuroscience*

Molecular deposition and lithographic patterning of neuronal specific molecules with nanometer resolutions[27,28] are an extension of micro-patterning approaches (Fig. 3).[29,30] By depositing proteins and other molecules that promote and support neuronal adhesion

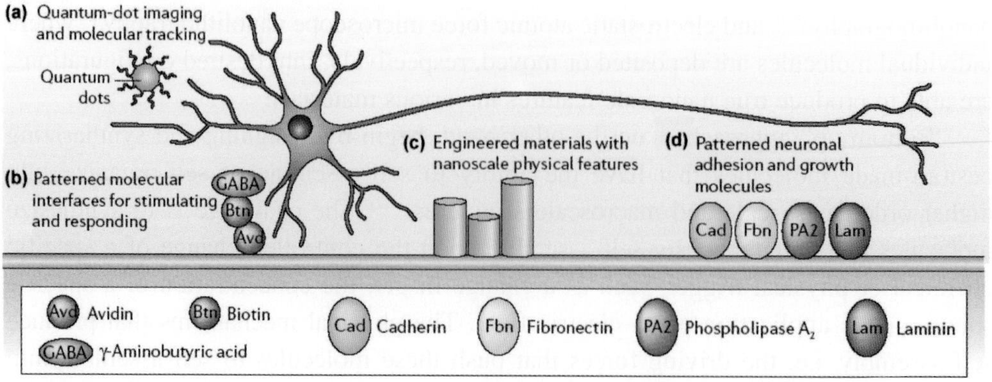

Fig. 3. Applications of nanotechnology to basic neuroscience include materials and devices designed to interact with neurons and glia at molecular scales in order to influence and respond to cellular events. In all cases, these engineered technologies allow controlled interactions at cellular and sub-cellular scales. Examples of these include: chemically functionalized fluorescent quantum dot nanocrystals used to visualize ligand–target interactions **(a)**; functionalized surfaces with neurotransmitter ligands in order to induce controlled signaling **(b)**; engineered materials with nanoscale physical features that produce ultrastructural morphological changes **(c)**; and surfaces and materials functionalized with different neural specific effector molecules to induce controlled cellular adhesion and growth **(d)**. Reproduced with permission from *Nature Reviews Neuroscience*.[72]

and growth on surfaces, geometric selective patterning and growth of neurons (for example, controlled neurite extension) can be achieved. This allows studying cellular communication and signaling and provides a test system for investigating the effects of drugs and other molecules.

The ability to control this process at nanometer, as opposed to micron, resolutions provides the ability to investigate how neurons respond to anisotropic physical and chemical cues. Micron-scale patterning can provide a functional boundary for controlling and influencing cellular behavior, but ultimately the neuron sees a stimulus (or stimuli in the case of multiple signals) that averages its effects over the whole cell. Nanotechnology approaches present sub-cellular stimuli that can vary from one part of the neuron to another. For example, photolithography and layer-by-layer self-assembly have been used to pattern phopholipase A2, which promotes neuronal adhesion, on a background of poly(diallyldimethylammonium chloride) (PDDA).[31] This approach provides the ability for nanoscale patterning at resolutions that can yield complex functional architectures that are tailored to the needs of a particular experiment. Layer-by-layer self-assembly has also been used on silicon rubber to pattern alternating laminin and poly-D-lysine or fibronectin / poly-D-lysine ultra-thin layers, which are 3.5 – 4.4 nm thick that support the growth of cerebellar neurons.[27] These studies suggest that bioactive ultra-thin layers could coat electrodes designed for long-term implants in order to promote cell adhesion and limit immune responses. In a different approach, electrodes coated with nano-porous silicon increased neurite outgrowth from PC 12 cells compared to uncoated electrodes, and displayed a decreased glial response, thereby limiting the insulating effects of the glial scar.[28] Coating electrodes with ultra-thin bioactive layers might have other advantages, such as limiting increases in the thickness of the electrodes and thereby minimizing local trauma due to their insertion and resultant cellular responses. Other work has shown the effects of nanoscale physical features on neuronal behavior. Substania nigra neurons cultured on SiO_2 surfaces with different nanoscale topographies demonstrated differential cell adhesion properties.[32] Neurons cultured on surfaces with physical features (i.e. surface roughness) between 20–70 nm adhered and grew better than neurons cultured on surfaces with features less than 10 nm or greater than 70 nm. These neurons also displayed normal morphologies and normal production of tyrosine hydroxylase, a marker of metabolic activity (Fig. 3).

Another emerging area of neuroscience nanotechnology are materials and devices designed to interact, record, and/or stimulate neurons at molecular scales.[33,34] Recent work has demonstrated the feasibility of functionalizing mica or glass tethered with the inhibitory neurotransmitter γ-aminobutryic acid (GABA) and its analog muscimol (5-aminomethyl-3-hydroxyisoxazole) through biotin-avidin binding interactions.[33,34] The functional integrity of the bound version of the neurotransmitter, which *in vivo* acts not as a bound ligand but as a diffusible signal across the synaptic cleft, was demonstrated electrophysiologically *in vitro* by eliciting an agonist response to cloned GABA(A) and GABA(C) receptors in Xenopus oocytes.[33] Such sophisticated systems, although still in the early conceptual and testing phases, may provide powerful molecule-based platforms for testing drugs and for neural prosthetic devices. Currently, all neural prostheses (including neural retinal prosthesis) rely on micron-scale features and cannot interact with the nervous system in a controlled way at molecular scales, which is a significant drawback. Other work is focusing on achieving nanoscale measurements of cellular responses. Atomic force microscopy (AFM) has been used to measure local nanometer morphological responses to micro-electrode array stimulation of neuroblastoma cells.[35] AFM is a technique that, among other capabilities, allows measuring height changes in the topography of a surface (for example, a living cell or a synthetic material) with nanometer resolution.[36,37] It is essentially a nanoscale cantilever that can measure surface topologies at the atomic level. This technique can measure cross-sectional changes in cell height (between 100–300 nm), which are produced by biphasic pulses at a frequency of 1 Hz, thereby providing information on ultra-fine morphological changes to electrical stimuli in neurons which cannot be achieved by other technologies (Fig. 3).

An area of nanotechnology that holds significant promise for probing the details of molecular and cellular processes in neural cells are functionalized semi-conductor quantum dot nanocrystals (Figs. 3 and 4).[38,39] Quantum dots are nanometer-sized particles composed of a heavy metal core of materials such as cadmium selenium or cadmium telluride, with an intermediate unreactive zinc sulfide shell and an outer

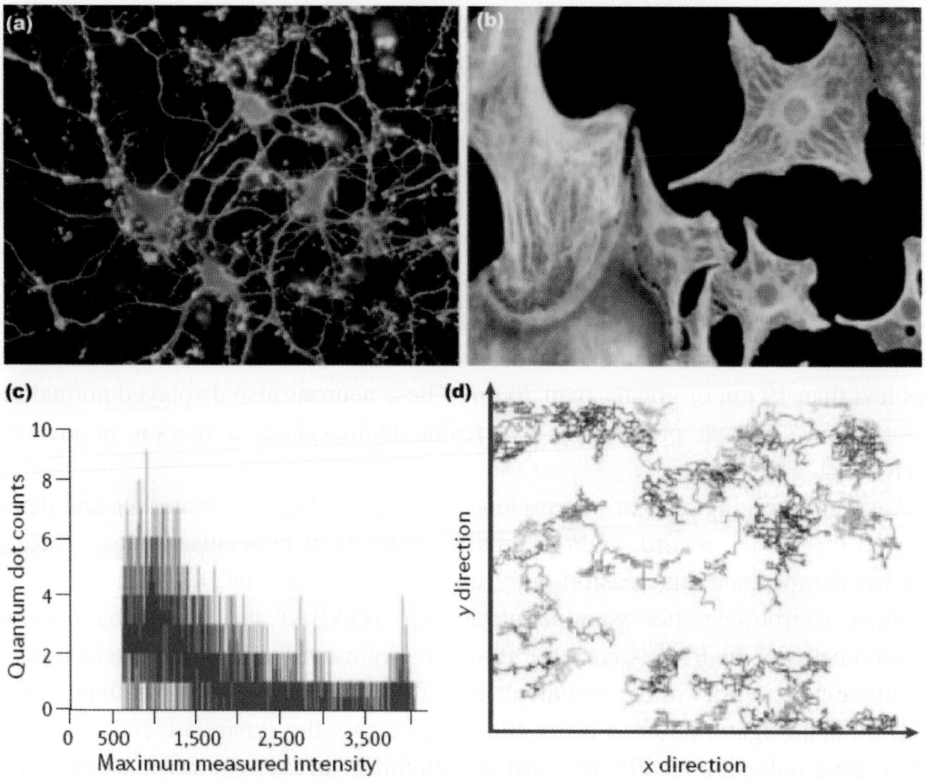

Fig. 4. The quantum dot toolbox. Fluorescent quantum dots are nanoscale particles which can be chemically functionalized with a large variety of biological molecules attached on their outer surface (e.g. antibodies, peptides, trophic factors, etc.). This allows specific molecular interactions in both live and fixed target cells, which can be visualized at high resolutions taking advantage of quantum dots' unique optical properties, such as their prolonged photo-stability (i.e. minimal photobleaching), large excitation absorption spectra and extremely narrow emission spectra. Specific examples of applications of quantum dots to neuroscience can be found in Refs. 39, 48 and 49. **(a)** Primary rat cortical neurons labeled with anti-β-tubulin III antibody conjugated quantum dots. β-tubulin is a neuronal specific intermediate filament protein. **(b)** Primary rat astrocytes labeled with anti-GFAP antibody conjugated quantum dots. Glial fibrillary acidic protein (GFAP) is a glial specific intermediate filament protein. **(c)** One of the main advantages of quantum dot nanotechnology is that qualitative observations as well as quantitative data can be obtained, which provides detailed molecular and biophysical information about the biological system being investigated. For example, by using computational and morphometric tools, individual quantum dots can be counted across a sample image to yield information on the distribution and number of ligand–target interactions. **(d)** Another example is the ability to perform single particle tracking (SPT) of ligand–target pairs, such as tracking the motion of a receptor in a cell membrane. Illustration of the trajectory of a field of 55 quantum dots undergoing Brownian diffusion, with each color corresponding to a different quantum dot. The small size, photochemistry and bioactivity of functionalized quantum dots therefore provide an extensive new toolbox for investigating molecular and cellular processes in neurons and glia. Reproduced with permission from *Nature Reviews Neuroscience*.[72]

coating composed of selective bioactive molecules tailored to a particular application. The physical nature of quantum dots gives them unique and highly stable fluorescent optical properties that can be changed by altering their chemistry and physical size.

Quantum dots can be tagged with fluorescent proteins of interest using different chemical approaches similar to fluorophore immunocytochemistry. However, quantum dots have significant advantages: they experience minimal photobleaching and can have much higher signal-to-noise ratios, resulting in dramatically improved signal detection. This is because quantum dots have broad absorption spectra but very narrow emission spectra.[40,41] In addition, they can be used for single-particle tracking of target molecules in live cells, such as tracking ligand-receptor dynamics in the cell membrane.[42,43] Quantum dot labeling of both fixed and live cells is well established now, and has been used in a wide variety of cell types — mostly *in vitro* or *in situ*,[44,45] with some examples *in vivo*.[46,47] Despite the growing literature on the applications of quantum dots to various cell types, their applications to neurons[38,39] and glia[48] have been slower to develop, and care must be taken to validate labeling methods specific for neural cells, as methods for labeling other cell types do not necessarily work for neural cells.[48] Recent work has illustrated the potential of this technology in neuroscience. The real-time dynamics of glycine receptors in spinal neurons have been tracked and analyzed using single-particle tracking over periods of seconds to minutes.[39] These investigators characterized the dynamics of receptor diffusion, which differed as a function of the spatial localization of the receptors relative to the synapse, depending on whether they were in synaptic, perisynaptic or extrasynaptic regions. In another example, immobilized quantum dots that were conjugated with β-nerve growth factor (β-NGF) were shown to interact with TrkA receptors in PC12 cells and regulate their differentiation in a controlled way.[49] This may provide new tools for studying neuronal signaling processes. Although most applications of quantum dots in neuroscience have taken place *ex vivo*, *in vivo* microangiography of mouse brains has been achieved using serum that has been labeled with quantum dots.[38] New functionalization and labeling methods of quantum dots have been developed and tested by labeling AMPA neurotransmitter receptors,[50] and by measuring the cytotoxicity of hippocampal neurons.[51] Although quantum-dot nanotechnology, when used correctly, has little cytotoxic effects on cells *in vitro* (so that experimental results are not affected), their *in vivo* applications have different challenges due to the possibility of local and systemic toxicity. To address these issues, the safety of quantum dot applications to both neural cells and other cell types continues to be an active area of research.[52,53]

4.2 *Applications to clinical neuroscience*

Applications of nanotechnology which aim at limiting and reversing neurological disorders by promoting neural regeneration and achieving neuroprotection, are active areas

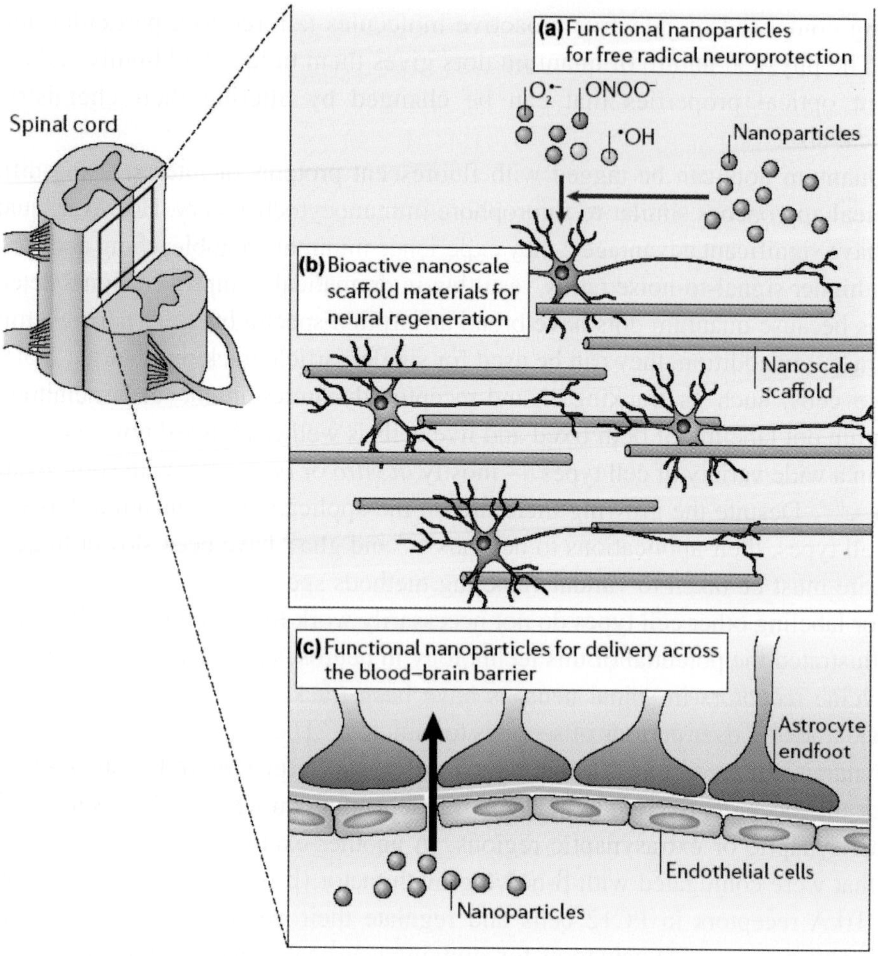

Fig. 5. Applications of nanotechnology to clinical neuroscience intended to limit and/or reverse neuropathological disease processes at a molecular level or facilitate and support other approaches intended to do so. These include: nanoparticles that promote neuroprotection by limiting the effects of free radicals produced following trauma (e.g. such as those produced by central nervous system secondary injury mechanisms) **(a)**; the development and use of nanoengineered scaffold materials that mimic extracellular matrix and provide a physical and/or bioactive environment for neural regeneration **(b)**; and nanoparticles designed to allow the transport of drugs and small molecules across the blood brain barrier **(c)**. Reproduced with permission from *Nature Reviews Neuroscience*.[72]

of research (Fig. 5). The development of nanoengineered scaffolds that support and promote neurite and axonal growth are evolving from tissue engineering approaches based on the manipulation of bulk materials. Examples of micron-scale tissue engineering include poly-(L-lactic) acid (PLLA) and other synthetic hydrogels that have engineered micro-scale features, and scaffolds derived from naturally occurring materials such as collagen.[54,55]

One example of a nanoengineered system extended from this type of work are PLLA scaffolds with an ultrastructure consisting of cast PLLA fibers, which have a diameter of 50–350 nm and a porosity of about 85%.[56] The scaffolds were constructed using liquid–liquid phase separation by dissolving PLLA in tetrahydrofuran (THF) rather than casting them on glass. When cultured in the scaffolds, neonatal mouse cerebellar progenitor cells were able to extend neurites and differentiate into mature neurons. A fundamentally different approach for the development of a nanomaterial that promotes and supports neural regeneration is the self-assembly of nanofiber networks composed of peptide amphiphile molecules (Fig. 6).[5]

In this study, peptide amphiphile molecules, which consisted of a hydrophobic carbon tail and a hydrophilic peptide head group, self-assembled into a dense network of nanofibers when exposed to physiological ionic conditions, which trapped the surrounding water molecules and macroscopically form a weak self-supporting gel. The hydrophilic peptide head groups, which formed the outside of the fibers, consisted of the bioactive laminin-derived peptide IKVAV, which promotes neurite sprouting and growth.[11,57] Encapsulation of neural progenitor cells from embryonic mouse cortex in the nanofiber networks resulted in fast and robust neuronal differentiation (30% and 50% of neural progenitor cells differentiated into neurons at 1 and 7 days *in vitro*, respectively), with minimal astrocytic differentiation (1% and 5% of neural progenitor cells differentiated into astrocytes at 1 and 7 days *in vitro*, respectively). This approach may therefore promote neuronal differentiation at an injury site while potentially limiting the effects of reactive gliosis and glial scarring, which are ubiquitous neuropathological disease processes (Fig. 5).

Applications of nanotechnologies for neuroprotection have focused on limiting the damaging effects of free radicals generated after injury, which is a key neuropathological process that contributes to central nervous system ischemia, trauma and degenerative disorders.[58,59] Fullerenols, which are derivatives of hydroxyl functionalized fullerenes (molecules composed of regular arrangements of carbon atoms[60,61]), have been shown to have antioxidant properties and act as free radical scavengers, which can lead to a reduction in the extent of excitotoxicity and apoptosis induced by glutamate, NMDA, AMPA and kainite.[62,63] Fullerenol-mediated neuroprotection has been demonstrated *in vitro* by limiting excitotoxicity and apoptosis of cultured cortical neurons, and *in vivo* by the delayed onset of motor degeneration in a mouse model of familial amyotrophic lateral sclerosis. The neuroprotective effect of fullerenols might be partly mediated by inhibition of glutamate receptors, as it had no effect on GABA(A) or taurine receptors. It also lowered glutamate-induced elevations in intracellular calcium, which is an important mechanism of neuronal excitotoxicity.[62,63]

Another clinically relevant area of intense research is nanotechnology approaches for the delivery of drugs and small molecules across the blood-brain barrier (BBB) using functionalized nanoparticles.[64,65] The goal is to design nanoparticles that can be administered systemically and can deliver drugs and other molecules across the BBB,

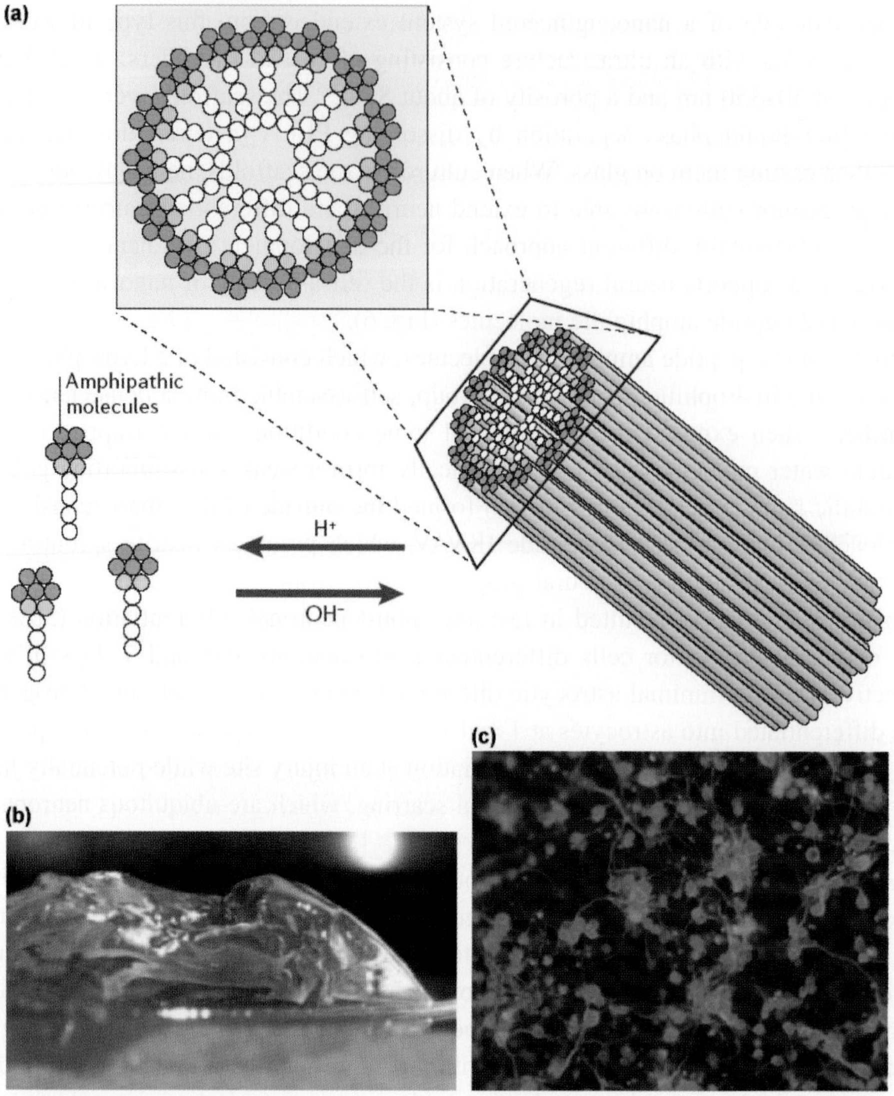

Fig. 6. Example of a nanoengineered material for neural regeneration. Engineered nanomaterials provide highly specific ways to induce controlled cellular interactions that can promote desired neurobiological effects. **(a)** In this example, peptide amphiphile molecules can be coaxed to self-assemble into elongated micelles that produce a dense nanofiber matrix.[25,26] Under physiological conditions, the self-assembly process traps the surrounding aqueous environment and macroscopically produces a self-supporting gel in which neural progenitor cells and stem cells can be encapsulated. **(b)** An example of a peptide, amphiphile nanogel on a 12 mm glass coverslip. **(c)** The surface of the nanofibers consists of laminin-derived, neuronal specific pentapeptides, which are encountered by the encapsulated cells at high concentrations, resulting in robust differentiation into neurons while suppressing astrocyte differentiation.[5] Neurons (in green) are stained for β-tubulin III, with astrocytes stained for glial fibrillary acidic protein (GFAP; although none are present). All nuclei were stained with a non-specific nuclear Hoescht stain. Reproduced with permission from *Nature Reviews Neuroscience*.[72]

which is a major clinical objective for treating a wide range of neurological disorders. To achieve this, a variety of materials and synthetic approaches are being investigated. Oligonucleotides have been delivered in gels of cross-linked poly(ethylene glycol) and polyethylenimine.[66] Charge differences in the electrostatic forces between the gel and the oligonucleotides provide a reversible delivery mechanism that can be used for shuttling molecules across the BBB and then releasing them from the delivery system. Neuropeptides (such as enkephalins), the NMDA receptor antagonist MRZ 2/576, and the chemotherapeutic drug doxorubicin have been absorbed onto the surface of poly(butylcyanoacrylate) nanoparticles coated with polysorbate 80.[67,68] The polysorbate on the surface of the nanoparticles adsorb apolipoproteins B and E from the blood, and are taken up by brain capillary endothelial cells via receptor-mediated endocytosis.[69] Nanoparticles that target tumors in the CNS may be a particularly important application of this technology due to the high morbidity and mortality of often aggressive neoplasms in the physically confined spaces of the cranium and spinal canal.

5. Challenges Associated with Nanotechnology Applications to the CNS

In general, nanotechnology applications aimed at the CNS are lagging behind other areas of medicine and biology, such as orthopedic applications, DNA/genomic sensors, and novel drug and gene delivery approaches (see for example Refs. 70 and 71). This may be due in part to the inherent complexity of the CNS and in part due to its difficult and restricted anatomical access. The CNS represents an extremely heterogeneous cellular and molecular environment, with many different types and sub-types of cells. This degree of cellular heterogeneity underlies the nervous system's anatomical and functional "wiring" that is the basis of the system's extremely complex information processing. Nanotechnologies designed to interact with CNS cells and processes *in vivo* must take this complexity into consideration, even if they are designed to target a specific cell type or physiological process. Failure to do so may result in unforeseen "side effects" or interactions. Our ability to chemically synthesize and engineer nanotechnologies with highly specific chemical and/or physical properties tailored to a particular molecular or cellular target is far from perfect, and these are very active areas of research that have seen tremendous improvements and advancements in recent years and will continue to improve in the years to come. But the reality is that at present, both top down (e.g. lithographic) and bottom up (e.g. self-assembling) chemistry and nanoengineering approaches and methods are very basic compared to physiological nanoengineering. One would have to anticipate that as the materials science and synthesis methods improve, the greater synthetic sophistication and ability to control desired mesoscale and macroscale properties of materials and devices will translate into more improved and more specific interactions with the CNS.

A second major consideration besides physiological complexity is that any nanotechnology that aims to study and understand an underlying process *in vivo* must take into account the highly anatomically restrictive nature of the CNS. CNS structures are highly protected from mechanical and physical injury, and are immunologically privileged behind the BBB, creating a unique molecular and cellular environment unique to the CNS. This is due to the essential and critical roles the CNS plays in regulating, maintaining, and coordinating sensory and motor functions and the multitude of essential life preserving functions. Any nanotechnology designed to interact with CNS cells in the intact organism must be first delivered to their target location. This in itself may present technical challenges as difficult as any others. The physical and functional isolation of the CNS means that careful consideration needs to be paid to the delivery vehicle or mechanism by which nanoengineered materials or devices are meant to be delivered to their target sites. Care must be taken to disrupt the BBB as little as possible during delivery. In essence, these are the same issues associated with drug delivery into the CNS, although one can think of some unique obstacles specific to nanotechnologies. With drugs the challenge is to get them safely and effectively across the BBB at sufficient concentrations, but once in it is assumed that the drug will carry out its effects by diffusing through the cerebral spinal fluid or vasculature and binding with its target receptors. With nanotechnologies these issues are just as relevant and challenging, but with the additional consideration that the proper functioning of the nanomaterial or device may depend on other chemical or physical processes in addition to diffusion, such as electrochemical reactions or cellular endocytosis. The delivery mechanism must be biologically compatible while at the same time ensuring the functional integrity of the delivered technology. The simplest and most straightforward delivery approach is sterotaxic injection directly to the target site, but this is of course highly invasive. More sophisticated approaches may provide a mechanism for targeted delivery across the BBB from the periphery. A truly sophisticated nanotechnology would be able to cross the BBB in one configuration, sense and respond to its environment, and subsequently carry out its primary functional effects within the CNS in a second configuration.

Taking these issues into consideration, three things must occur simultaneously for CNS bionanotechnology to be developed and have meaningful scientific and clinical impacts:

(1) There must be continued advancements in fundamental chemistry and materials science. As these advancements occur, they will lead to ever more sophisticated synthetic and characterization approaches for nanoengineered materials and devices that are able to control multiple chemical and/or physical parameters that target multiple molecular, cellular, and physiological processes in the biological system.
(2) There must be continued advancements in our understanding of the molecular biology, biochemistry, neurophysiology, and neuropathology of the CNS. This is necessary to provide ever more specific functional targets for novel nanotechnologies.

The more we know about and understand how the healthy CNS works and how it fails in disease, the better we can design and apply bionanotechnologies that are highly specific to one or more CNS processes.

(3) The design and integration of specific nanoengineered technologies for the CNS must take advantage of the first two points, and will require significant interdisciplinary collaborations between physical and life scientists, as well as clinicians. As these areas develop in an integrated and parallel fashion, bionanotechnology applications to the CNS will emerge.

6. Summary and Conclusions

Applications of nanotechnology to neuroscience are already having a significant impact, which will continue to grow in the foreseeable future. Short-term progress has benefited *in vitro* and *ex vivo* studies of neural cells, often supporting or augmenting standard technologies. These advances contribute to both our basic understanding of cellular neurobiology and neurophysiology, and to our understanding and interpretation of neuropathology. Although nanotechnologies designed to interact with the nervous system *in vivo* are slow and challenging to develop, they will have significant, direct clinical implications. Nanotechnologies aimed at supporting cellular or pharmacological therapies or to facilitate direct physiological effects *in vivo* will make significant contributions to clinical care and prevention. The reason for the tremendous potential that nanotechnology applications can have in biology and medicine in general and neuroscience in particular stems from the ability of these technologies to specifically interact with cells at a molecular scale.

Acknowledgments

This work was supported in part by grants from the Stein Institute for Research on Aging, the Culpepper Biomedical Pilot Initiative, the Whitaker Foundation, and the National Institute for Neurological Diseases and Stroke at the National Institutes of Health (RO1 NS054736-01). This work was adapted from Refs. 72 to 74.

References

1. E. R. Kandel, J. H. Schwartz and T. M. Jessell, *Principles of Neural Science* (McGraw-Hill Health Professions Division, New York, 2000).
2. P. Bezzi and A. Volterra, A neuron-glia signalling network in the active brain, *Curr. Opin. Neurobiol.* **11**: 387–394 (2001).
3. A. Araque, G. Carmignoto and P. G. Haydon, Dynamic signaling between astrocytes and neurons, *Annu. Rev. Physiol.* **63**: 795–813 (2001).

4. J. M. Frade, J. R. Martinez-Morales and A. Rodriguez-Tebar, Laminin-1 selectively stimulates neuron generation from cultured retinal neuroepithelial cells, *Exp. Cell Res.* **222**: 140–149 (1996).
5. G. A. Silva, C. Czeisler, K. L. Niece, E. Beniash, D. A. Harrington, J. A. Kessler and S. I. Stupp, Selective differentiation of neural progenitor cells by high-epitope density nanofibers, *Science* **303**: 1352–1355 (2004).
6. P. Tunggal, N. Smyth, M. Paulsson and M. C. Ott, Laminins: structure and genetic regulation, *Microsc. Res. Tech.* **51**: 214–227 (2000).
7. E. Hohenester and J. Engel, Domain structure and organisation in extracellular matrix proteins, *Matrix Biol.* **21**: 115–128 (2002).
8. U. M. Wewer and E. Engvall, Merosin/laminin-2 and muscular dystrophy, *Neuromuscul. Disord.* **6**: 409–418 (1996).
9. N. Smyth, H. S. Vatansever, P. Murray, M. Meyer, C. Frie, M. Paulsson and D. Edgar, Absence of basement membranes after targeting the LAMC1 gene results in embryonic lethality due to failure of endoderm differentiation, *J. Cell Biol.* **144**: 151–160 (1999).
10. Y. S. Cheng, M. F. Champliaud, R. E. Burgeson, M. P. Marinkovich and P. D. Yurchenco, Self-assembly of laminin isoforms, *J. Biol. Chem.* **272**: 31525–31532 (1997).
11. S. K. Powell, J. Rao, E. Roque, M. Nomizu, Y. Kuratomi, Y. Yamada and H. K. Kleinman, Neural cell response to multiple novel sites on laminin-1, *J. Neurosci. Res.* **61**: 302–312 (2000).
12. E. Freire, F. C. Gomes, R. Linden, V. M. Neto and T. Coelho-Sampaio, Structure of laminin substrate modulates cellular signaling for neuritogenesis, *J. Cell Sci.* **115**: 4867–4876 (2002).
13. G. R. Phillips, J. K. Huang, Y. Wang, H. Tanaka, L. Shapiro, W. Zhang, W. S. Shan, K. Arndt, M. Frank, R. E. Gordon *et al.*, The presynaptic particle web: ultrastructure, composition, dissolution, and reconstitution, *Neuron* **32**: 63–77 (2001).
14. N. Kasthuri and J. W. Lichtman, Structural dynamics of synapses in living animals, *Curr. Opin. Neurobiol.* **14**: 105–111 (2004).
15. V. N. Murthy and P. De Camilli, Cell biology of the presynaptic terminal, *Annu. Rev. Neurosci.* **26**: 701–728 (2003).
16. Z. Li and M. Sheng, Some assembly required: the development of neuronal synapses, *Nat. Rev. Mol. Cell Biol.* **4**: 833–841 (2003).
17. H. Takano, J. Y. Sul, M. L. Mazzanti, R. T. Doyle, P. G. Haydon and M. D. Porter, Micropatterned substrates: approach to probing intercellular communication pathways, *Anal. Chem.* **74**: 4640–4646 (2002).
18. K. B. Lee, S. J. Park, C. A. Mirkin, J. C. Smith and M. Mrksich, Protein nanoarrays generated by dip-pen nanolithography, *Science* **295**: 1702–1705 (2002).
19. B. P. Liu, A. Fournier, T. GrandPre and S. M. Strittmatter, Myelin-associated glycoprotein as a functional ligand for the Nogo-66 receptor, *Science* **297**: 1190–1193 (2002).
20. S. F. Lyuksyutov, R. A. Vaia, P. B. Paramonov, S. Juhl, L. Waterhouse, R. M. Ralich, G. Sigalov and E. Sancaktar, Electrostatic nanolithography in polymers using atomic force microscopy, *Nat. Mater.* **2**: 468–472 (2003).
21. S. I. Stupp, V. V. LeBonheur, K. Walker, L. S. Li, K. E. Huggins, M. Keser and A. Amstutz, Supramolecular materials: self-organized nanostructures, *Science* **276**: 384–389 (1997).
22. S. I. Stupp and P. V. Braun, Molecular manipulation of microstructures: biomaterials, ceramics, and semiconductors, *Science* **277**: 1242–1248 (1997).
23. S. I. Stupp and G. W. Ciegler, Organoapatites: materials for artificial bone. I. Synthesis and microstructure, *J. Biomed. Mater. Res.* **26**: 169–183 (1992).
24. S. I. Stupp, G. C. Mejicano and J. A. Hanson, Organoapatites: materials for artificial bone. II. Hardening reactions and properties, *J. Biomed. Mater. Res.* **27**: 289–299 (1993).
25. J. D. Hartgerink, E. Beniash and S. I. Stupp, Self-assembly and mineralization of peptide-amphiphile nanofibers, *Science* **294**: 1684–1688 (2001).

26. J. D. Hartgerink, E. Beniash and S. I. Stupp, Peptide-amphiphile nanofibers: a versatile scaffold for the preparation of self-assembling materials, *Proc. Natl. Acad. Sci. USA* **99**: 5133–5138 (2002).
27. H. Ai, H. Meng, I. Ichinose, S. A. Jones, D. K. Mills, Y. M. Lvov and X. Qiao, Biocompatibility of layer-by-layer self-assembled nanofilm on silicone rubber for neurons, *J. Neurosci. Methods* **128**: 1–8 (2003).
28. K. A. Moxon, N. M. Kalkhoran, M. Markert, M. A. Sambito, J. L. McKenzie and J. T. Webster, Nanostructured surface modification of ceramic-based microelectrodes to enhance biocompatibility for a direct brain-machine interface, *IEEE Trans. Biomed. Eng.* **51**: 881–889 (2004).
29. T. H. Park and M. L. Shuler, Integration of cell culture and microfabrication technology, *Biotechnol. Prog.* **19**: 243–253 (2003).
30. B. C. Wheeler, J. M. Corey, G. J. Brewer and D. W. Branch, Microcontact printing for precise control of nerve cell growth in culture, *J. Biomech. Eng.* **121**: 73–78 (1999).
31. J. S. Mohammed, M. A. DeCoster and M. J. McShane, Micropatterning of nanoengineered surfaces to study neuronal cell attachment *in vitro*, *Biomacromolecules* **5**: 1745–1755 (2004).
32. Y. W. Fan, F. Z. Cui, S. P. Hou, Q. Y. Xu, L. N. Chen and I. S. Lee, Culture of neural cells on silicon wafers with nano-scale surface topograph, *J. Neurosci. Methods* **120**: 17–23 (2002).
33. T. Q. Vu, S. Chowdhury, N. J. Muni, H. Qian, R. F. Standaert and D. R. Pepperberg, Activation of membrane receptors by a neurotransmitter conjugate designed for surface attachment, *Biomaterials* **26**: 1895–1903 (2005).
34. U. Saifuddin, T. Q. Vu, M. Rezac, H. Qian, D. R. Pepperberg and T. A. Desai, Assembly and characterization of biofunctional neurotransmitter-immobilized surfaces for interaction with postsynaptic membrane receptors, *J. Biomed. Mater. Res. A* **66**: 184–191 (2003).
35. M. B. Shenai, K. G. Putchakayala, J. A. Hessler, B. G. Orr, M. M. Banaszak Holl and J. R. Baker, Jr., A novel MEA/AFM platform for measurement of real-time, nanometric morphological alterations of electrically stimulated neuroblastoma cells, *IEEE Trans. Nanobioscience* **3**: 111–117 (2004).
36. H. G. Hansma, K. Kasuya and E. Oroudjev, Atomic force microscopy imaging and pulling of nucleic acids, *Curr. Opin. Struct. Biol.* **14**: 380–385 (2004).
37. N. C. Santos and M. A. Castanho, An overview of the biophysical applications of atomic force microscopy, *Biophys. Chem.* **107**: 133–149 (2004).
38. M. J. Levene, D. A. Dombeck, K. A. Kasischke, R. P. Molloy and W. W. Webb, *In vivo* multiphoton microscopy of deep brain tissue, *J. Neurophysiol.* **91**: 1908–1912 (2004).
39. M. Dahan, S. Levi, C. Luccardini, P. Rostaing, B. Riveau and A. Triller, Diffusion dynamics of glycine receptors revealed by single-quantum dot tracking, *Science* **302**: 442–445 (2003).
40. J. L. West and N. J. Halas, Engineered nanomaterials for biophotonics applications: improving sensing, imaging, and therapeutics, *Annu. Rev. Biomed. Eng.* **5**: 285–292 (2003).
41. W. C. Chan and S. Nie, Quantum dot bioconjugates for ultrasensitive nonisotopic detection, *Science* **281**: 2016–2018 (1998).
42. M. J. Saxton and K. Jacobson, Single-particle tracking: applications to membrane dynamics, *Annu. Rev. Biophys. Biomol. Struct.* **26**: 373–399 (1997).
43. S. Bonneau, M. Dahan and L. D. Cohen, Single quantum dot tracking based on perceptual grouping using minimal paths in a spatiotemporal volume, *IEEE Trans. Image Process.* **14**: 1384–1395 (2005).
44. F. Tokumasu and J. Dvorak, Development and application of quantum dots for immunocytochemistry of human erythrocytes, *J. Microsc.* **211**: 256–261 (2003).
45. J. K. Jaiswal, H. Mattoussi, J. M. Mauro and S. M. Simon, Long-term multiple color imaging of live cells using quantum dot bioconjugates, *Nat. Biotechnol.* **21**: 47–51 (2003).
46. M. E. Akerman, W. C. Chan, P. Laakkonen, S. N. Bhatia and E. Ruoslahti, Nanocrystal targeting *in vivo*, *Proc. Natl. Acad. Sci. USA* **99**: 12617–12621 (2002).
47. X. Michalet, F. F. Pinaud, L. A. Bentolila, J. M. Tsay, S. Doose, J. J. Li, G. Sundaresan, A. M. Wu, S. S. Gambhir and S. Weiss, Quantum dots for live cells, *in vivo* imaging, and diagnostics, *Science* **307**: 538–544 (2005).

48. S. Pathak, E. Cao, M. C. Davidson, S. Jin and G. A. Silva, Quantum dot applications to neuroscience: new tools for probing neurons and glia, *J. Neurosci.* **26**: 1893–1895 (2006).
49. T. Q. Vu, R. Maddipati, T. A. Blute, B. J. Nehilla, L. Nusblat, and T. A. Desai, Peptide-conjugated quantum dots activate neuronal receptors and initiate downstream signaling of neurite growth, *Nano. Lett.* **5**: 603–607 (2005).
50. M. Howarth, K. Takao, Y. Hayashi and A. Y. Ting, Targeting quantum dots to surface proteins in living cells with biotin ligase, *Proc. Natl. Acad. Sci. USA* **102**: 7583–7588 (2005).
51. H. Fan, E. W. Leve, C. Scullin, J. Gabaldon, D. Tallant, S. Bunge, T. Boyle, M. C. Wilson and C. J. Brinker, Surfactant-assisted synthesis of water-soluble and biocompatible semiconductor quantum dot micelles, *Nano. Lett.* **5**: 645–648 (2005).
52. L. Braydich-Stolle, S. Hussain, J. J. Schlager and M. C. Hofmann, *In vitro* cytotoxicity of nanoparticles in mammalian germline stem cells, *Toxicol. Sci.* **88**: 412–419 (2005).
53. E. B. Voura, J. K. Jaiswal, H. Mattoussi and S. M. Simon, Tracking metastatic tumor cell extravasation with quantum dot nanocrystals and fluorescence emission-scanning microscopy, *Nat. Med.* **10**: 993–998 (2004).
54. F. Stang, H. Fansa, G. Wolf and G. Keilhoff, Collagen nerve conduits — assessment of biocompatibility and axonal regeneration, *Biomed. Mater. Eng.* **15**: 3–12 (2005).
55. T. T. Yu and M. S. Shoichet, Guided cell adhesion and outgrowth in peptide-modified channels for neural tissue engineering, *Biomaterials* **26**: 1507–1514 (2005).
56. F. Yang, R. Murugan, S. Ramakrishna, X. Wang, Y. X. Ma and S. Wang, Fabrication of nano-structured porous PLLA scaffold intended for nerve tissue engineering, *Biomaterials* **25**: 1891–1900 (2004).
57. M. Nomizu, B. S. Weeks, C. A. Weston, W. H. Kim, H. K. Kleinman and Y. Yamada, Structure-activity study of a laminin alpha 1 chain active peptide segment Ile-Lys-Val-Ala-Val (IKVAV), *FEBS Lett.* **365**: 227–231 (1995).
58. E. H. Lo, T. Dalkara and M. A. Moskowitz, Mechanisms, challenges and opportunities in stroke, *Nat. Rev. Neurosci.* **4**: 399–415 (2003).
59. R. J. Gagliardi, Neuroprotection, excitotoxicity and NMDA antagonists, *Arq. Neuropsiquiatr.* **58**: 583–588 (2000).
60. D. Gust, T. A. Moore and A. L. Moore, Photochemistry of supramolecular systems containing C60, *J. Photochem. Photobiol. B* **58**: 63–71 (2000).
61. J. L. Segura and N. Martin, New concepts in tetrathiafulvalene chemistry, *Angew. Chem. Int. Ed. Engl.* **40**: 1372–1409 (2001).
62. L. L. Dugan, E. G. Lovett, K. L. Quick, J. Lotharius, T. T. Lin and K. L. O'Malley, Fullerene-based antioxidants and neurodegenerative disorders, *Parkinsonism Relat. Disord.* **7**: 243–246 (2001).
63. H. Jin, W. Q. Chen, X. W. Tang, L. Y. Chiang, C. Y. Yang, J. V. Schloss and J. Y. Wu, Polyhydroxylated C(60), fullerenols, as glutamate receptor antagonists and neuroprotective agents, *J. Neurosci. Res.* **62**: 600–607 (2000).
64. P. R. Lockman, R. J. Mumper, M. A. Khan and D. D. Allen, Nanoparticle technology for drug delivery across the blood-brain barrier, *Drug Dev. Ind. Pharm.* **28**: 1–13 (2002).
65. U. Schroeder, P. Sommerfeld, S. Ulrich and B. A. Sabel, Nanoparticle technology for delivery of drugs across the blood-brain barrier, *J. Pharm. Sci.* **87**: 1305–1307 (1998).
66. S. V. Vinogradov, E. V. Batrakova and A. V. Kabanov, Nanogels for oligonucleotide delivery to the brain, *Bioconjug. Chem.* **15**: 50–60 (2004).
67. R. N. Alyaudtin, A. Reichel, R. Lobenberg, P. Ramge, J. Kreuter and D. J. Begley, Interaction of poly(butylcyanoacrylate) nanoparticles with the blood-brain barrier *in vivo* and *in vitro*, *J. Drug Target* **9**: 209–221 (2001).
68. J. Kreuter, Nanoparticulate systems for brain delivery of drugs, *Adv. Drug. Deliv. Rev.* **47**: 65–81 (2001).

69. J. Kreuter, D. Shamenkov, V. Petrov, P. Ramge, K. Cychutek, C. Koch-Brandt and R. Alyautdin, Apolipoprotein-mediated transport of nanoparticle-bound drugs across the blood-brain barrier, *J. Drug Target* **10**: 317–325 (2002).
70. T. G. Drummond, M. G. Hill and J. K. Barton, Electrochemical DNA sensors, *Nat. Biotechnol.* **21**: 1192–1199 (2003).
71. S. Zhang, Fabrication of novel biomaterials through molecular self-assembly, *Nat. Biotechnol.* **21**: 1171–1178 (2003).
72. G. A. Silva, Neuroscience nanotechnology: progress, opportunities and challenges, *Nat. Rev. Neurosci.* **7**: 65–74 (2006).
73. G. A. Silva, Small neuroscience: the nanostructure of the central nervous system and emerging nanotechnology applications, *Curr. Nanosci.* **3**: 225–236 (2005).
74. G. A. Silva, Introduction to nanotechnology and its applications to medicine, *Surg. Neurol.* **61**: 216–220 (2004).

CHAPTER 21

CELLULAR BIOPHOTONICS: LASER SCISSORS (ABLATION)

Michael W. Berns

Abstract

This chapter examines the use of light (photons) at the tissue and cellular levels and is focused towards achieving an understanding and application of light to bioengineering and medical problems. Initial discussion focuses on the mechanisms of photon interaction at the tissue and cellular levels. These are of a thermal, mechanical, and chemical nature. Examples are given for the eye because it is the organ that is the most studied and manipulated by light. The use of light for the diagnosis and treatment of cancer particularly related to the use of light-sensitive photochemical agents is presented (photodynamic therapy). At the cellular and sub-cellular level the use of a laser microbeam is discussed. Particular interest is focused on the use of the laser microbeam to manipulate the organelles of the dividing cell. The advent of the GFP gene-fusion proteins and their use in facilitating the visualization and targeting of sub-cellular structures is recognized. The combined use of cell tracking and robotics in the devlopment of an Internet-based laser microscope is also discussed.

1. Introduction

Even though this chapter will focus on the cellular application of photons in an "ablative mode" it should be recognized that this is only one small part of a much more inclusive field of biophotonics that involves non-destructive laser trapping,[1–3] and a myriad of cellular and tissue-level imaging modalities.[4]

The *absorption* of the photons by the tissue/cell can lead to a variety of controlled (and uncontrolled) responses. These may be *photothermal*, *photochemical*, *photoablative*, *photodisruptive*, *re-emissive (fluorescence)*, *and multi-photonic* (Figs. 1A and 1B).

In addition, alterations in cell/tissue structure can be achieved by *nonlinear* highly dynamic physical events caused by the high electric fields and plasmas generated in a tightly focused pulsed laser beam.[5] These events are generally not wavelength

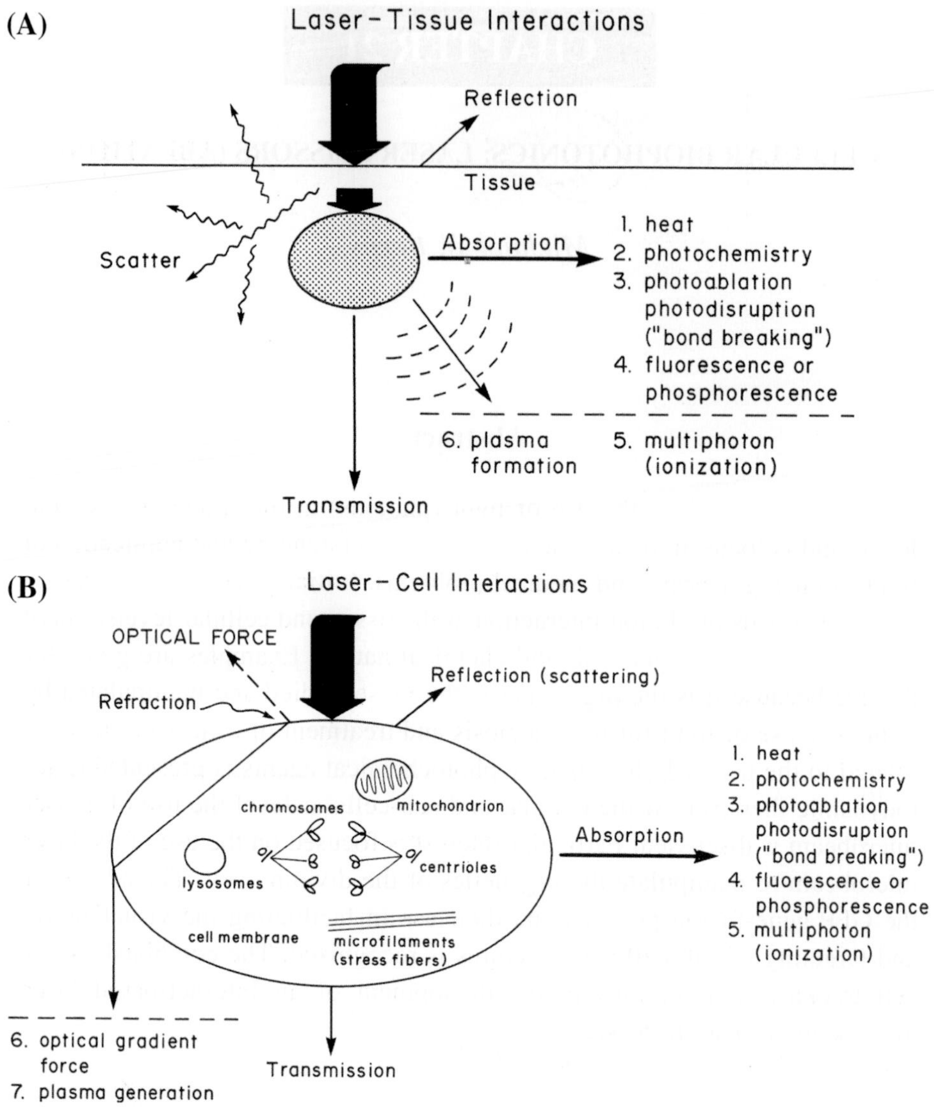

Fig. 1. **(A)** Mechanisms of photon interactions with tissue. **(B)** Mechanisms of photon interactions with cells. Individual organelles are diagrammed. Laser microbeams can be focused on to any one of these organelles in order to study the organelle structure and function.

dependent and result from the high intensity of photons in a small focal volume. Such disruptive events have been demonstrated in both sub-cellular irradiation[6–8] and in the disruption of tissue for human medical use (see Niemz[5] for a good review). In general, it is possible to relate the irradiance (W/cm^2) in the cell/tissue with the duration of light exposure (femtoseconds to seconds), and the amount of energy delivered (joules/cm^2). This has been done both for tissue[8,9] (Fig. 2A) and cells[8,10] (Fig. 2B).

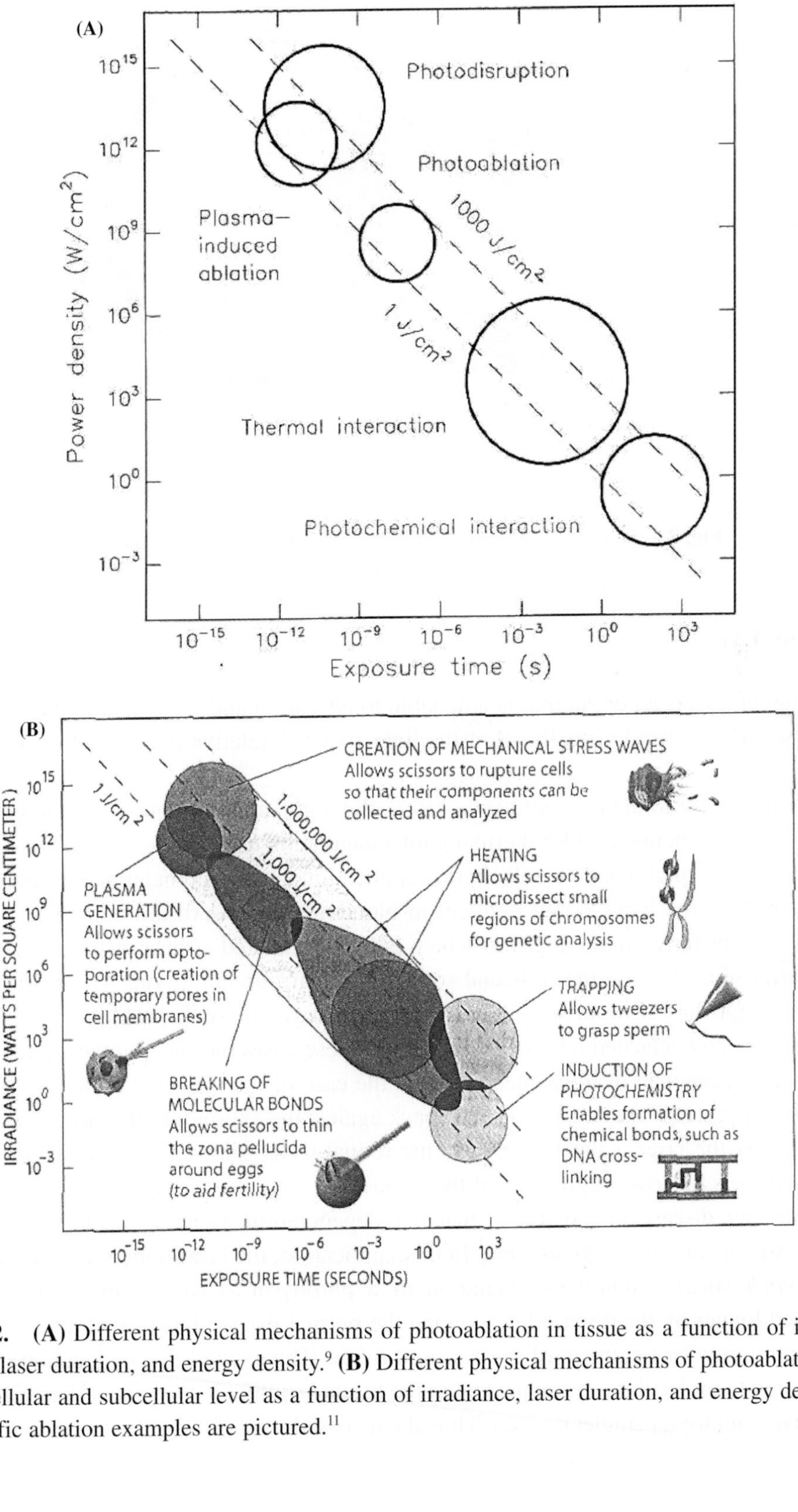

Fig. 2. **(A)** Different physical mechanisms of photoablation in tissue as a function of irradiance, laser duration, and energy density.[9] **(B)** Different physical mechanisms of photoablation at the cellular and subcellular level as a function of irradiance, laser duration, and energy density. Specific ablation examples are pictured.[11]

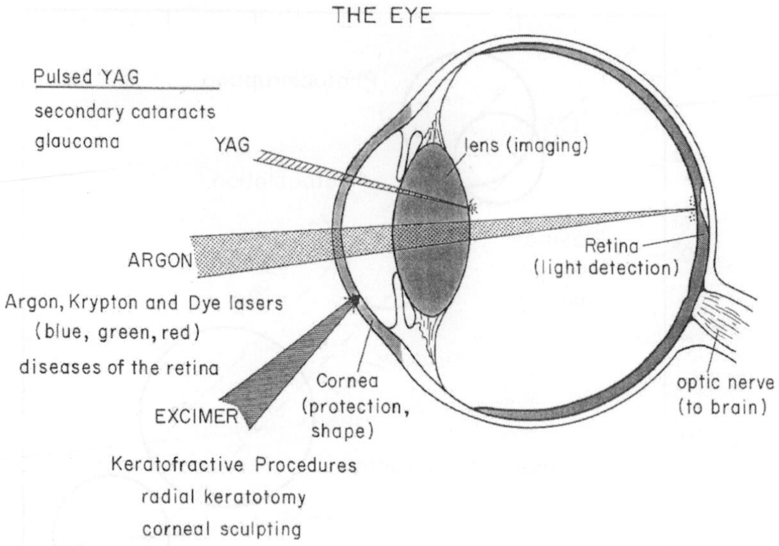

Fig. 3. Clinical laser applications to different structures in the eye.

2. The Eye

The human eye is an organ that is amenable to alteration and study using many different lasers (Fig. 3) and virtually all of the light-induced ablative mechanisms illustrated in Fig. 2.

With respect to the *thermal mechanisms*, the argon and krypton ion gas lasers, and diode lasers are being used for thermal photocoagulation of pigmented retinal tissue and hemorrhagic retinal blood vessels. In this application the brown melanin pigment in the retina and the red hemoglobin pigment in blood vessels absorb the light. The photon energy is transferred to the pigment which is then dissipated as heat resulting in a temperature rise destroying the cells and some of the surrounding tissue. The thermal use of laser light has widespread applications in the eye for such problems as diabetic retinopathy and detached retinas.[11,12] In both of these cases thermal photocoagulation is the main mechanism of light interaction. In the case of the detached retina the laser is used to "spot weld" the separated retina back against the choroid. In the case of diabetic retinopathy the laser is used to destroy the retinal tissue surrounding the proliferating blood vessels to slow progression of the disease.

Another disease of the retina, macular degeneration, is being treated using dye lasers and operating at 620–640 nm. In this application, the laser light is used to induce a *photochemical reaction* by excitation of a porphyrin-based dye molecule that is retained longer by the diseased tissue than the normal tissue, thus providing a window of opportunity for treatment of the diseased tissue. The excited dye molecule transfers its energy to molecular oxygen present in the tissue resulting in the production of highly reactive (and toxic) singlet oxygen. Thus the diseased tissue is selectively destroyed by

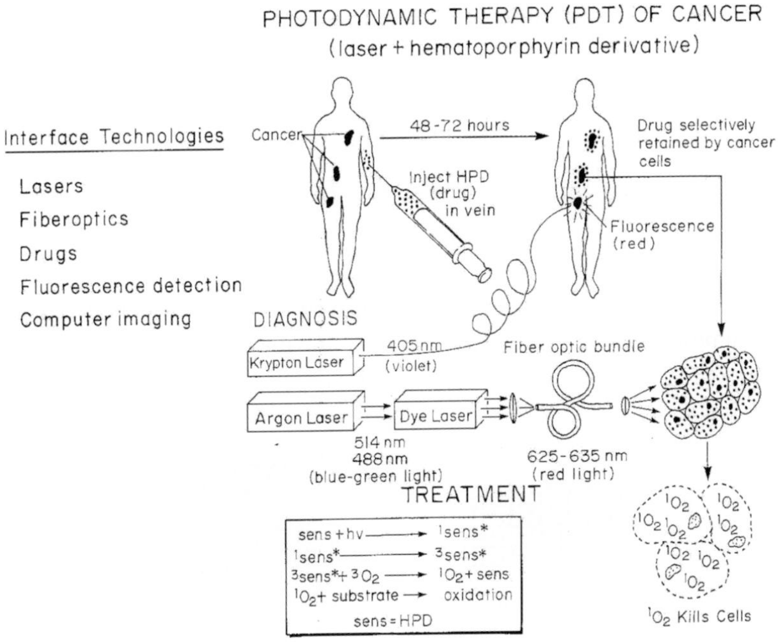

Fig. 4. Laser photodynamic therapy (PDT) and diagnosis of cancer. The use of light-activated dye that localizes in the retina also has been used to effectively treat age-related macular degeneration of the eye.

the photochemical product rather than heat or some other physical mechanism caused by the laser light. This method utilizes long exposure times (seconds to minutes) and low irradiance (mW/cm^2). This method, *photodynamic therapy*, was initially developed for cancer diagnosis and therapy (Fig. 4), but has found a major application in stemming the progression of age-related macular degeneration.[13]

When the pulsed ultraviolet excimer lasers became available in the 1980s, a new application of lasers for tissue ablation became available. The breaking of molecular bonds due to the combination of the high photon energy of 193 nm UV light and the high intensity of the beam was put to use for the controlled precise *photoablation* of small volumes of corneal tissue (Fig. 5). The process of excimer induced *photoablation* has been carefully studied in organic polymers[14] (Fig. 6).

Its application for re-shaping the cornea has become one of the most widespread and publicized uses of the laser in medicine. The ability to remove 0.5 μm of tissue with single laser pulses provides a very precise method to etch and ablate the cornea which is on the order of 500–700 μm in thickness. The procedure termed *LASIK* (***L****aser* ***AS****sisted* ***I****n-situ* ***K****eratomileusis*) is performed on hundreds of thousands of eyes a year. It involves cutting a precision flap in the cornea (usually with a sharp steel blade), and then laser photoablating a very precise amount of the underlying stroma (mostly collagen) so the corneal shape is either steepened or flattened very precisely. Steepening of the cornea is used to correct for far-sightedness (hyperopia) and flattening of the cornea for near-sightedness (*myopia*).

Absorption of high-energy UV photons
⇓
Promotion to repulsive excited states
⇓
Dissociation
⇓
Ejection of fragments (no necrosis).
⇓
Ablation

Fig. 5. Diagram of the time-course of the physical events (in a plastic polymer) that occurs during the ablation process using high energy 193 nm ultraviolet photons from an excimer laser.[14]

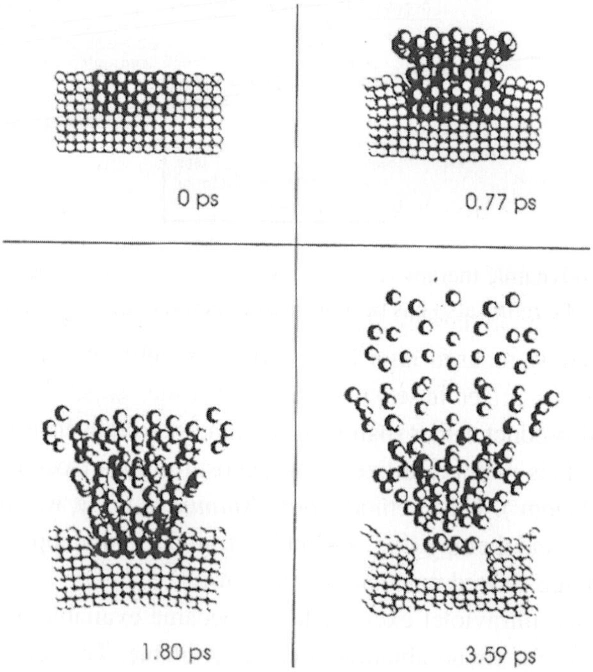

Fig. 6. The steps leading to high energy (193 nm) UV photon ablation of tissue (as applied to the cornea of the eye).

After the laser ablation of the stroma, the flap is replaced down on the corneal surface where it heals. In these procedures the depth of tissue removed from the stroma varies from a few microns (representing only a single cell layer equivalent of tissue) to tens of microns.[15,16] Recently a short-pulsed femtosecond laser has been used to cut the flap as opposed to using a highly honed steel blade that is mechanically moved across the corneal surface in a device called a microkeratome. At the present time the "verdict is still out" as to whether or not the laser flap cutting has any advantage over the mechanical flap-cutting.

When laser pulses of very high irradiance (W/cm^2) are focused to a spot of a few microns or less in diameter, nonlinear ablation events may occur in the focal volume. These are generally termed "optical breakdown" and can involve a variety of physical processes such as plasma formation, shock wave generation, pressure-induced cavitation bubbles, and high electric fields — all of which can result in cellular and tissue disruption.[5] These ablation mechanisms have been applied to the removal of secondary cataracts that form on the rear surface of lens implants following cataract surgery[17] and as a possible method to create channels through the trabecular meshwork allowing for a release of pressure that builds up in the eye of people with glaucoma.[18]

3. Cell Surgery (Laser Scissors)

3.1 *History and background*

The goal of altering either a single cell or a sub-cellular region with radiation had its early roots in the work of the Russian, Tschachotin,[19] who really invented the technique of "partial cell irradiation" (PCI), using focused ultraviolet light. Over the course of much of the 20th century PCI was conducted not only with UV light, but also with X-rays, gamma rays, electron beams, proton beams, and lasers.[20] Lasers were first focused through the microscope by Marcel Bessis in Paris not long after the laser was invented in 1960.[21]

Following the early studies by Bessis and his colleagues, many other researchers began to use and apply focused laser beams to study cells and their organelles. These studies have been summarized in several reviews over the years.[8,22–24] As a result of these studies, several commercial devices and companies have evolved, resulting in these technologies being made accessible to the scientific community (www.palm-microlaser.com; www.cellrobotics.com; www.cyntellect.com). With the current and rapid development of web-based technologies, a robotic laser microscope ("RoboLase") has been developed which can be accessed online using high-speed Internet connections.[25]

3.2 *Selectivity on cell organelles*

3.2.1 *Pre-fusion protein*

In 1969 "laser scissors" were used to damage micron-size regions of chromosomes in live tissue culture cells.[26] Electron microscopy subsequently confirmed that the damage was precisely confined to the beam focal point and matched precisely with the damage observed under live phase contrast microscopy (Fig. 7).

These studies led to the ability to remove selective genetic regions (the ribosomal genes) from specific chromosomes followed by the establishment of genetically deficient cell clones.[24,27] The mechanism of laser alteration of the chromosome was most likely thermal in the early studies because of the use of light-absorbing dyes such as acridine orange[26] in combination with blue-green argon ion lasers as well as the relatively long microsecond pulse-width of the laser. In later studies, nonlinear photoablation and multi-photon

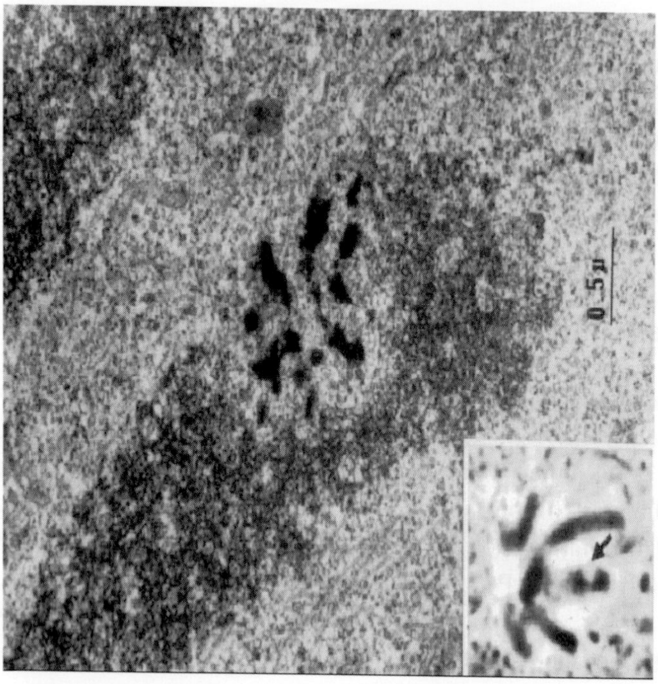

Fig. 7. Phase contrast light microscope picture (in-set) of a group of chromosomes in a live cell from the rat kangaroo (*Potorous tridactylis*) undergoing cell division (mitosis) with a small light spot on a single chromosome indicating the lesion (arrow) produced by a focused pulsed green laser beam. The large image is a transmission electron micrograph (TEM) of the same cell (inset). Note that the lesion is confined to the focal spot and there is no visible damage outside that zone.

absorption were more likely mechanisms.[20,28] In these studies the second harmonic green (532 nm), third harmonic (355 nm), and fourth harmonic (266 nm) from a ten-nanosecond neodymium YAG laser were used. These studies were the first to demonstrate multi-photon absorption in live cells and opened the way for a large number of studies employing multi-photon ablation on a variety of cell organelles: mitotic spindle microtubules, cytoplasmic microtubules,[29] chromosome kinetochores,[30] microfilaments,[31] and mitochondria.[32] Paralleling much of this work, a series of experiments using the laser for cell surgery was being conducted by Greulich and colleagues in Germany.[23]

The advent of femtosecond lasers has made really short laser pulses for ablation available.[33] It has been proposed that short femtosecond lasers are superior to the longer pulse-width lasers because the plasma produced in a small diffraction-limited volume would be confined to the focal volume with little collateral damage.[34,35] However many factors in addition to the laser pulse duration play determinant roles in the production of damage confined to the focal volume. These are: wavelength, focusing optics, beam mode structure, and the physical properties of the target itself. At this time it is unclear if femtosecond lasers really can result in focal spot damage that is smaller and more confined than other laser systems. Current laboratory tutorials on laser scissors can be found in Celis' "Cell Biology: A Laboratory Handbook," and Berns and Greulich, "Laser Manipulation of Cells and Tissues."[36,47]

3.2.2 *Fusion proteins*

The use of fluorescent proteins that are incorporated into specific proteins of interest associated with specific organelles has greatly expanded the applicability of laser scissors (Fig. 8). In 1997, Conly Rieder and colleagues at the Wadsworth Medical Center in Albany, NY, published a paper describing the use of green fluorescent protein (GFP) in combination with a nanosecond second harmonic green YAG laser beam to destroy bundles of spindle microtubules in live cells.[37] The ability to see structures in live cells not previously visible with the light microscope has greatly increased the versatility of laser scissors. Instead of shooting "blind" at organelles or regions of organelles,[29] GFP (and its mutant fluorescing variants) has made these structures visible so they can be altered and studied.[6,34]

Perhaps one of the most elegant studies illustrating the precision of GFP-assisted laser microsurgery was the destruction of one of the two centrioles in the centrosome of a mitotic cell.[38] In this study the fluorescing GFP protein was incorporated into the centrin protein of the centriole. Thus, two single fluorescing dots corresponding to each of the two centrioles (Fig. 9) were visible at each pole of the mitotic cell.

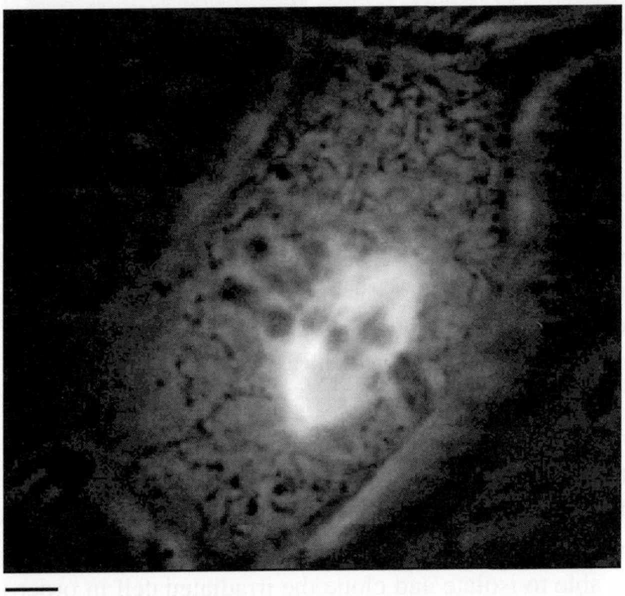

Fig. 8. A combined phase contrast and fluorescence image of a rat kangaroo cell in metaphase of mitosis. Note the dark chromosomes aligned along the metaphase plate, and the bright fluorescing microtubules of the mitotic spindle. The microtubules are fluorescing because of the incorporation of a green fluorescent protein variant of tubulin into the microtubules (see Botvinick et al, 2004, for further explanation).

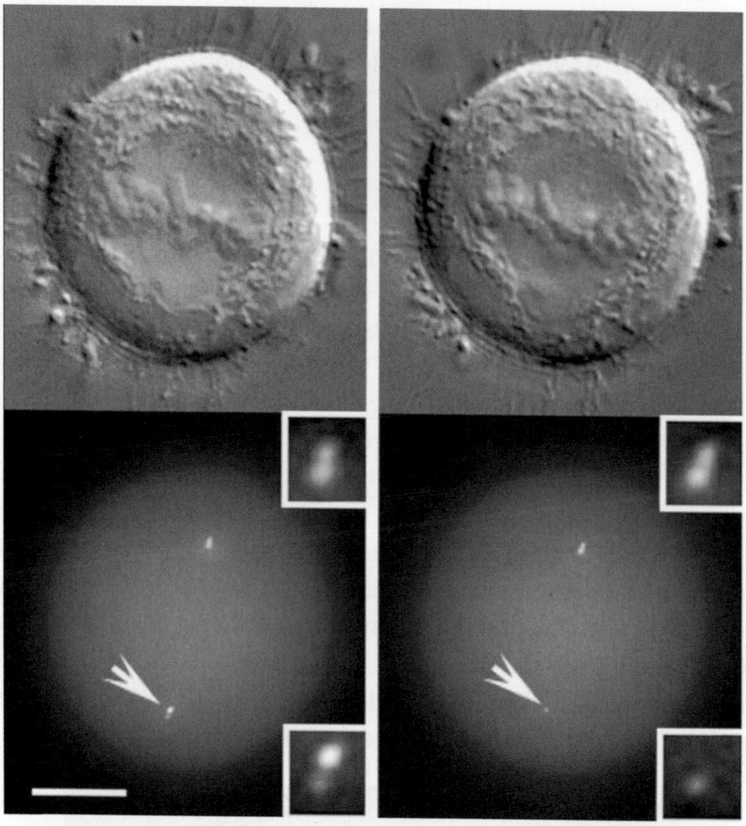

Fig. 9. Cancer cells (HeLa) viewed with differential interference microscopy (top to images) in mitosis with the chromosomes aligned across the middle of the cell. Bottom two pictures show the fluorescent centrioles at each pole of the above cells (insets show higher magnification of the two centrioles at each pole). First lower image shows two centrioles at each pole. Second lower picture shows one pole in which one of the two centrioles has been destroyed by laser irradiation. The other pole has both centrioles. The centrioles are visible because of the presence of a GFP protein (centrin) which fluoresces, thus making the structure visible for targeting by the laser. The laser used was a nanosecond frequency doubled green (532 nm) Nd:YAG. Bar = 5 μm. Each individual centriole is 0.1–0.2 μm in diameter.

A second harmonic green (532 nm) ten-nanosecond laser was focused onto one of the fluorescing dots, which immediately lost its fluorescence. Electron microscopy confirmed that one of the two centrioles at the irradiated mitotic pole had been destroyed. The authors were able to isolate and clone the irradiated cell in order to study the control of centriole replication in subsequent daughter cells. They demonstrated that cells that do not receive any centriole can produce their own centrioles *de novo*. However, the cells produced many more than the two centrioles needed thus suggesting that the control system for centriole production had been compromised. These studies have paved the way for additional studies on the control of centriole formation during the cell cycle. But more importantly, from a technical perspective, these studies have demonstrated the precision with which laser scissors can be used in conjunction with fluorescent gene fusion molecular techniques.

3.3 *Optoinjection and optoporation*

3.3.1 *Optoinjection*

In 1984, Tsukakoshi and colleagues published a paper demonstrating that a pulsed laser could be used to temporarily make either a small hole or create a localized change in permeability in the cell membrane so that foreign DNA could enter the cell and be incorporated into the cell genome.[39] This was a rather significant biotechnical achievement because it provided a potential way to transfect cells with foreign genes — especially cells that are particularly resistant to transfection. The transfection and cloning of transfection-resistant human fibrosarcoma cells was accomplished using a semi-automated laser scissors.[40] In these studies, the third harmonic 355 nm wavelength of a ten-anosecond YAG laser was used.

Perhaps the most intriguing use of optoinjection has been the insertion of genes into an undifferentiated rice cell that subsequently developed into a fully differentiated plant that expressed the inserted genes in every cell in the organism.[41] In this study, genes were inserted into undifferentiated rice callus cells using 355 nm laser light and plasmid DNA. Fully differentiated plantlets were formed that expressed the kanamycin resistance antibiotic gene as well as the gene for beta-glucuronidase. Because the plant cells exist with high internal turgor pressure, initial studies using the laser to puncture the cell wall and cell membrane resulted in the extrusion of all the cell components into the surrounding medium. However, by performing the experiment in media that was hypertonic to the cells, the DNA in the external medium flowed into the cells when the cell wall and cell membrane were punctured. A transfection frequency of 0.5% was achieved.

3.3.2 *Optoporation*

Optoporation is a variation of optoinjection whereby numerous cells can have a transient change made in their membrane permeability due to a nearby laser induced shock wave. In this way, external molecules can enter into a group of cells rather than into a single cell, one cell at a time. The exact nature of the change in the cell membrane that allows molecules to move across it and into the cell is not known. However, it is likely a physical change is brought on by the laser induced shock wave that is transmitted through the surrounding medium to the cells. The shock wave is produced as a result of a small microplasma that is formed either in the medium or on the surface of the vessel in which the cells are growing. Such plasma-generated shock waves may be similar to those formed when IR lasers are focused behind the lens capsule in the eye to destroy cells that have grown over lens implants.[17]

4. The Future

The future for laser scissors is really closely tied to its "companion" technologies. Laser tweezers have not been covered in this chapter, but the combination of both have a significant role in cellular and biophysical studies. As has been demonstrated in early studies, laser scissors and tweezers can be used to manipulate cells and their organelles when used either alone, or in concert with each other.[10,42,43] In addition, the use of non-manipulative optical imaging technologies such as fluorescence recovery after photobleaching[44] (FRAP), fluorescent resonance energy transfer[45] (FRET), and fluorescence correlation spectroscopy[46] (FCS) are powerful analytical tools that may be used alone or in combination with laser tweezers and/or scissors. From a bioengineering perspective, all of these technologies have been combined into one microscope platform that is accessible via the Internet in a system called "RoboLase"[25,47] (Fig. 10). This "robotic"

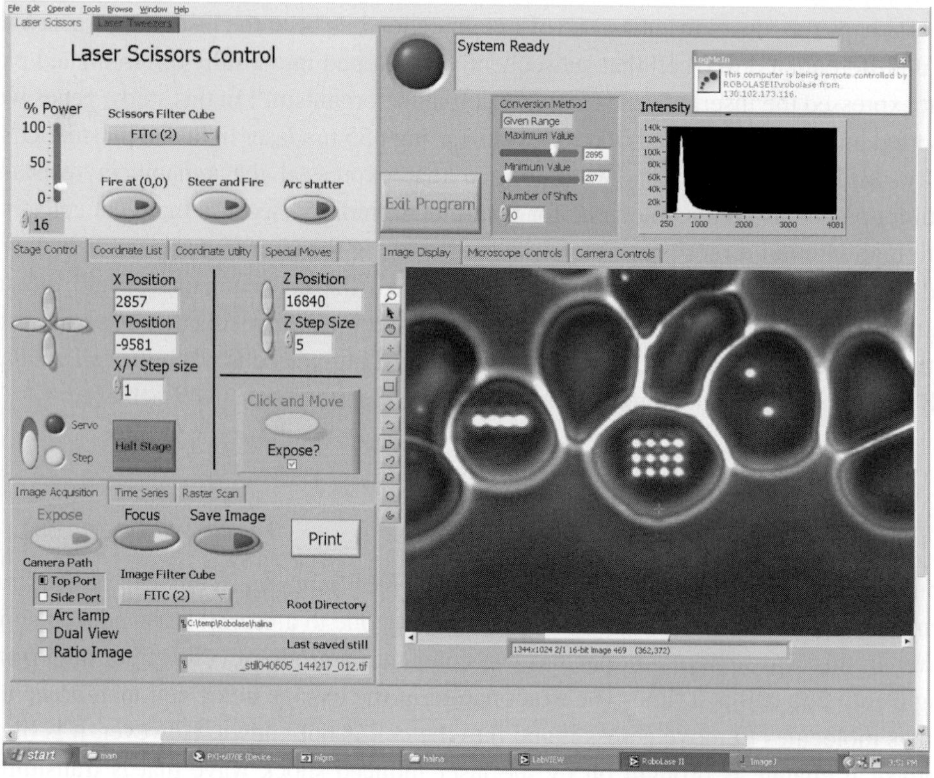

Fig. 10. Control panel for the "RoboLase" microscope system that can be operated via the Internet. Images are of human blood cells with a series of precise sub-micron holes produced by remote operation from Queensland, Australia. The microscope system was in California.[25]

laser microscope, with its Internet accessibility, makes these optical tools available to researchers and students all over the world.

References

1. A. Ashkin, Forces of a single-beam gradient laser trap on a dielectric sphere in the ray optics regime. In: *Laser Tweezers in Cell Biology*, Methods in Cell Biology, Vol. 55, ed. M. P. Sheetz (Academic Press Inc., San Diego, CA, 1998), Chap. 1, pp. 1–27.
2. M. P. Sheetz, ed., *Laser Tweezers in Cell Biology*, Methods in Cell Biology, Vol. 55 (Academic Press Inc., San Diego, CA, 1998).
3. M. W. Berns, Y. Tadir, H. Liang and B. Tromberg, Laser scissors and tweezers. In: *Laser Tweezers in Cell Biology*, Methods in Cell Biology, Vol. 55, ed. M. P. Sheetz (Academic Press Inc., San Diego, CA, 1998), Chap. 5, pp. 71–98.
4. T. Vo-Dinh, *Biomedical Photonics Handbook* (CRC Press, 2003), p. 1872.
5. M. H. Niemz, *Laser-Tissue Interactions: Fundamentals and Applications* (Springer, Berlin, 2004), p. 305.
6. E. L. Botvinick, V. Venugopalan, J. V. Shah, L. H. Liaw and M. W. Berns, Controlled ablation of microtubules using a picosecond laser, *Biophys. J.* **87**: 4203–4212 (2004).
7. A. Vogel and V. Venugopalan, Mechanisms of pulsed laser ablation of biological tissues, *Chem. Rev.* **103**: 577–644 (2003).
8. A. Vogel, J. Noack, G. Huttman and G. Paltauf, Mechanisms of femtosecond laser nanosurgery of cells and tissues *Appl. Phys. B* **81**: 1015–1047 (2005).
9. J.-L. Boulnois, Photophysical processes in recent medical developments, *Lasers Med. Sci.* **1**: 47–66 (1986).
10. M. W. Berns, Laser scissors and tweezers, *Sci. Am.* **278**: 62–67 (1998).
11. M. W. Berns, Laser surgery, *Sci. Am.* **264**: 84–90 (1991).
12. F. A. L'Esperance, *Ophtalmic Lasers* (C. V. Mosby, St. Louis, 1989), p. 104.
13. S. J. Keam, L. J. Scott and M. P. Curran, Verteporfin: a review of its use in the management of subfoveal choroidal neovascularisation, *Drugs* **63**: 2521–2554 (2003).
14. B. J. Garrison and R. Srinivasan, Laser ablation of organic polymers: microscopic models for photochemical and thermal processes, *J. Appl. Phys.* **57**: 2909–2914 (1985).
15. F. E. Fantes and G. O. Waring, 3rd, Effect of excimer laser radiant exposure on uniformity of ablated corneal surface, *Lasers Surg. Med.* **9**: 533–542 (1989).
16. I. G. Pallikaris and D. S. Siganos, Excimer laser *in situ* keratomileusis and photorefractive keratectomy for correction of high myopia, *J. Refract. Corneal Surg.* **10**: 498–510 (1994).
17. D. Aron-Rosa, J. J. Aron, M. Griesemann and R. Thyzel, Use of the neodymium-YAG laser to open the posterior capsule after lens implant surgery: a preliminary report, *J. Am. Intraocul. Implant. Soc.* **6**: 352–354 (1980).
18. R. A. Hill, G. Baerveldt, S. A. Ozler, M. Pickford, G. A. Profeta and M. W. Berns, Laser trabecular ablation (LTA), *Lasers Surg. Med.* **11**: 341–346 (1991).
19. S. Tschachotin, Die mikrokopische strahlenstrich methode, eine zelloperationsmethode, *Biol. Zenttralbl.* **32**: 623 (1912).
20. M. W. Berns, A possible two-photon effect *in vitro* using a focused laser beam, *Biophys. J.* **16**: 973–977 (1976).
21. M. Bessis, F. Gires, G. Mayer and G. Nomarski, Irradiation des organites cellulaires a l'aide laser rubis, *C. R. Acad. Sci.* **255**: 1010–1012 (1962).
22. M. W. Berns and C. Salet, Laser microbeams for partial cell irradiation, *Int. Rev. Cytol.* **33**: 131–156 (1972).

23. K. O. Greulich, *Micromanipulation by Light in Biology and Medicine* (Birkhauser, Berlin, 1999).
24. M. W. Berns, J. Aist, J. Edwards, K. Strahs, J. Girton, P. McNeill, J. B. Rattner, M. Kitzes, M. Hammer-Wilson, L. H. Liaw *et al.*, Laser microsurgery in cell and developmental biology, *Science* **213**: 505–513 (1981).
25. E. L. Botvinick and M. W. Berns, Internet-based robotic laser scissors and tweezers microscopy, *Microsc. Res. Tech.* **68**: 65–74 (2005).
26. M. W. Berns, R. S. Olson and D. E. Rounds, *In vitro* production of chromosomal lesions with an argon laser microbeam, *Nature* **221**: 74–75 (1969).
27. M. W. Berns, L. K. Chong, M. Hammer-Wilson, K. Miller and A. Siemens, Genetic microsurgery by laser: establishment of a clonal population of rat kangaroo cells (PTK2) with a directed deficiency in a chromosomal nucleolar organizer, *Chromosoma* **73**: 1–8 (1979).
28. P. P. Calmettes and M. W. Berns, Laser-induced multiphoton processes in living cells, *Proc. Natl. Acad. Sci. USA* **80**: 7197–7199 (1983).
29. W. Tao, R. J. Walter and M. W. Berns, Laser-transected microtubules exhibit individuality of regrowth, however most free new ends of the microtubules are stable, *J. Cell. Biol.* **107**: 1025–1035 (1988).
30. P. A. McNeill and M. W. Berns, Chromosome behavior after laser microirradiation of a single kinetochore in mitotic PtK2 cells, *J. Cell. Biol.* **88**: 543–553 (1981).
31. K. R. Strahs and M. W. Berns, Laser microirradiation of stress fibers and intermediate filaments in non-muscle cells from cultured rat heart, *Exp. Cell. Res.* **119**: 31–45 (1979).
32. M. Kitzes, G. Twiggs and M. W. Berns, Alteration of membrane electrical activity in rat myocardial cells following selective laser microbeam irradiation, *J. Cell. Physiol.* **93**: 99–104 (1977).
33. K. Konig, I. Riemann, P. Fischer and K. J. Halbhuber, Intracellular nanosurgery with near infrared femtosecond laser pulses, *Cell Mol. Biol. (Noisy-le-grand)* **45**: 195–201 (1999).
34. L. Sacconi, I. M. Tolic-Norrelykke, R. Antolini and F. S. Pavone, Combined intracellular three-dimensional imaging and selective nanosurgery by a nonlinear microscope, *J. Biomed. Opt.* **10**: 14002 (2005).
35. A. Heisterkamp, I. Z. Maxwell, E. Mazur, J. M. Underwood, J. A. Nickerson, S. Kumar and D. E. Ingber, Pulse energy dependence of subcellular dissection by femtosecond laser pulses, *Opt. Express* **13**: 3690–3696 (2005).
36. J. E. Celis, *Cell Biology: A Laboratory Handbook*, 3rd ed., Vol. 2 (Elsevier Acdamic Press, MA, 2005).
37. A. Khodjakov, R. W. Cole and C. L. Rieder, A synergy of technologies: combining laser microsurgery with green fluorescent protein tagging, *Cell Motil. Cytoskeleton* **38**: 311–317 (1997).
38. S. La Terra, C. N. English, P. Hergert, B. F. McEwen, G. Sluder and A. Khodjakov, The *de novo* centriole assembly pathway in HeLa cells: cell cycle progression and centriole assembly/maturation, *J. Cell. Biol.* **168**: 713–722 (2005).
39. M. Tsukakoshi, S. Kurata, Y. Nominya, Y. Ikawa and T. Kasuya, A novel method of DNA transfection by laser microbeam cell surgery, *Appl. Phys. B* **35**: 135–140 (1984).
40. W. Tao, R. J. Walter and M. W. Berns, Laser-transected microtubules exhibit individuality of regrowth, however most free new ends of the microtubules are stable, *J. Cell. Biol.* **107**: 1025–1035 (1988).
41. Y. Guo, H. Liang and M. W. Berns, Laser-mediated gene transfer in rice, *Physiol. Plant.* **93**: 19–24 (1995).
42. R. W. Steubing, S. Cheng, W. H. Wright, Y. Numajiri and M. W. Berns, Laser induced cell fusion in combination with optical tweezers: the laser cell fusion trap, *Cytometry* **12**: 505–510 (1991).
43. H. Liang, W. H. Wright, C. L. Rieder, E. D. Salmon, G. Profeta, J. Andrews, Y. Liu, G. J. Sonek and M. W. Berns, Directed movement of chromosome arms and fragments in mitotic newt lung cells using optical scissors and optical tweezers, *Exp. Cell. Res.* **213**: 308–312 (1994).

44. J. V. Shah, E. Botvinick, Z. Bonday, F. Furnari, M. Berns and D. W. Cleveland, Dynamics of centromere and kinetochore proteins; implications for checkpoint signaling and silencing, *Curr. Biol.* **14**: 942–952 (2004).
45. Y. Wang, E. L. Botvinick, Y. Zhao, M. W. Berns, S. Usami, R. Y. Tsien and S. Chien, Visualizing the mechanical activation of Src, *Nature* **434**: 1040–1045 (2005).
46. Z. Wang, J. V. Shah, Z. Chen, C. H. Sun and M. W. Berns, Fluorescence correlation spectroscopy investigation of a GFP mutant-enhanced cyan fluorescent protein and its tubulin fusion in living cells with two-photon excitation, *J. Biomed. Opt.* **9**: 395–403 (2004).
47. M. W. Berns and K. O. Greulich, eds., *Laser Manipulation of Cells and Tissue*, Methods in Cell Biology, Vol. 82 (Elsevier/Academic Press, San Diego, CA, 2007).

CHAPTER 22

MICROELECTRONIC ARRAYS: APPLICATIONS FROM DNA HYBRIDIZATION DIAGNOSTICS TO DIRECTED SELF-ASSEMBLY NANOFABRICATION

Michael J. Heller and Dietrich Dehlinger

Abstract

A variety of microelectronic array devices have been developed for DNA hybridization analysis and clinical genotyping diagnostics. In addition to these DNA research and diagnostic applications, such devices are now being used to carry out layer-by-layer (LBL) directed self-assembly of a wide variety of molecular and nanoparticle entities into higher order structures. Such microelectronic array devices are able to produce electric fields on their surfaces that allow charged molecules and nanostructures to be rapidly transported and bound to any site on the surface of the array. Such devices can utilize either DC electric fields which affect the electrophoretic transport of the entities, or AC electric fields which allows entities to be selectively positioned by dielectrophoresis (DEP). In the past, microelectronic array devices have been used to carry out the selective transport and binding of DNA, RNA, peptides, proteins, nanoparticles, cells and even micron scale semiconductor components. More recently, these devices have demonstrated the ability to carry out the directed self-assembly of biotin and streptavidin derivatized nanoparticles into multi-layer structures. Nanoparticle addressing can be carried out in about 15 seconds, and more than 40 nanoparticle layers can be completed in less than one hour. Microelectronic array based directed self-assembly represents an example of combining "top-down" and "bottom-up" technologies into viable nanofabrication process for the assembly and integration of nanocomponents into higher order structures.

1. Introduction

Over the past decade microelectronic array devices produced by a top-down photolithography process have been developed for DNA hybridization and genotyping diagnostic applications. In these applications, microelectronic array devices which produce reconfigurable electric fields on their surface are first used to address and bind negatively charged biotinylated DNA molecules to selected test sites on the microarray surface. In the next step, samples containing unknown DNA sequences are then rapidly transported and selectively hybridized to the DNA sequences bound at the specific test sites.[1–7] Thus, these devices are able to direct and accelerate the self-assembly or "bottom-up" process of DNA hybridization occurring on the microarray. Microelectronic arrays have several important features that make them attractive for DNA hybridization and other applications. First, a permeation layer or porous hydrogel is used to cover the microelectrode structures on the array. The permeation layer is usually impregnated with streptavidin which allows biotinylated DNA to be bound at the selected test site. This layer also allows relatively high DC electric field strengths to be used for rapid electrophoretic transport of molecules and nanostructures, while protecting the more sensitive DNA from the adverse effects of the electrolysis products generated at the electrodes.[6,7] A second feature of electronic array devices is that they can be designed in a wide variety of shapes and sizes. To date, arrays have been fabricated in sizes from 2×2 mm to over 2.5×2.5 cm, with 25 to 10,000 electrodes and with electrode structures which range in size from ten microns to several millimeters. A third feature is that sophisticated CMOS control elements can be integrated into the underlying silicon structure of electronic arrays which allows precise control of currents and voltages to each of the individual microelectrodes on the array.[8] Finally, size reduction in the electronic array controller system now provides a relatively compact control unit that can be run with a laptop computer. In addition to the directed transport and hybridization of DNA molecules, the ability of electronic arrays to carry out the rapid patterned deposition of charged nanoparticles was also demonstrated early in the development of the technology.[1] Ultimately, electronic microarrays have now been used to carry out transport, addressing and selective binding of a variety of charged entities that includes DNA, RNA, biotin/streptavidin, antibodies,[9,10] nanoparticles,[9,10] cells,[10–13] and even 20 μm sized light emitting diode (LED) semiconductor devices.[14–17] Overall, these devices have proven their ability to manipulate a surprising large variety of entities that range in size from small molecules to micron scale objects.[9,16]

2. Nanotechnology and Self-Assembly

Advances in nanotechnology and nanoscience are likely to enable a wide range of new applications and improvements for nanoelectronics, nanophotonics, high performance fuel cells, photovoltaics and batteries, nanocomposite materials, and biosensor, drug delivery and other biomedical devices.[18,19] With regard to nanoelectronics, the continued improvement of classical photolithography, a top-down process, is allowing the microelectronics industry to already reduce silicon/CMOS semiconductor feature sizes down into the nanoscale level (<100 nm). Thus, "nanoelectronics" via classical top-down fabrication technology is assured, the only question being the final size limit at which the photolithographic type processes become impractical. On the other hand, while there are numerous examples of novel nanocomponents (metallic nanoparticles, quantum dots, carbon nanotubes, nanowires and various organic electron transfer molecules), their organization into higher order structures by a "bottom-up" or self-assembly process still remains a challenge.[9]

Biology provides some of the best examples of self-assembly or self-organizational processes that may provide guidance for new bottom-up nanofabrication techniques. Living systems have many unique macromolecules, nanostructures and mechanisms which include high information content molecules such as deoxyribonucleic acid (DNA) and ribonucleic acid (RNA), structural components (proteins, fatty acids, etc.), efficient electron transfer and photonic transfer systems, energy conversion systems, and a wide variety of highly efficient chemomechanical or catalytic molecules called enzymes. Of all the biologically based molecules available with intrinsic recognition and self-assembly properties, the nucleic acids represent the most promising material which may prove useful for creating molecular electronic/photonic circuits, organized nanostructures, and even integrated microelectronic and photonic devices. The nucleic acids, DNA and RNA are programmable molecules (via their base sequence) which have intrinsic molecular recognition and self-assembly properties. DNA sequences are easily synthesized (oligonucleotides) and readily modified with a variety of functional groups. Functionalized DNA molecules with fluorescent or charge transfer groups provide a way to incorporate both electronic and photonic properties directly into the molecular structures. Additionally, DNA can be used to functionalize larger molecules, nanostructures (quantum dots, gold nanoparticles, carbon nanotubes, etc.) or even micron size components which can then be selectively attached to silicon or other surfaces. Thus, DNA molecules and their functionalized entities represent a type of "molecular legos" for self-assembly of nanocomponents into higher order two- and three-dimensional structures.

3. Microelectronic Array Technology

3.1 *Arrays and instruments*

A variety of microelectronic array devices have been fabricated by Nanogen (San Diego, CA) that are used primarily for DNA genotyping diagnostic applications. These include an early stage 25 test site prototype device as well as more advanced devices with 100, 400 and 10,000 test sites. The 100 test site array which was first to be commercialized, has an inner set of 80 μm diameter test sites with underlying platinum microelectrodes, and an outside set of auxiliary microelectrodes (Fig. 1A). The outer group of electrodes can be biased negative, which allows the DNA in the sample solution to be concentrated at the specific internal test sites (biased positive). Each microelectrode

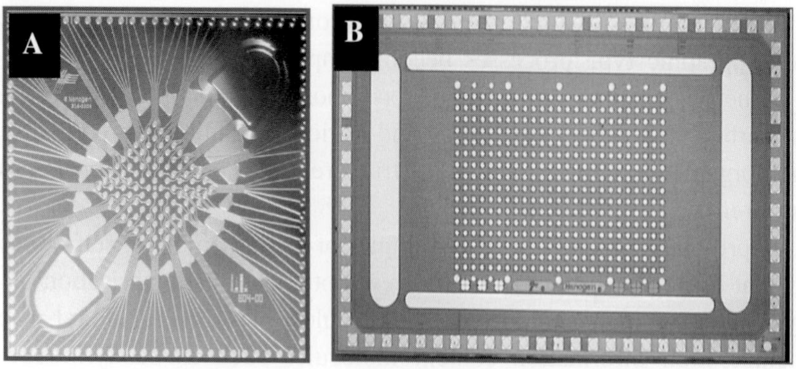

Fig. 1. (**A**) An advanced version of Nanogen's microelectronic DNA chip device with 100 test sites or microlocations. The test sites are approximately 80 μm in diameter, with underlying platinum microelectrodes. The outside ring of 20 microelectrodes can be used as counter electrodes to the test site electrodes (overall chip dimensions about 1 cm^2). (**B**) A photo of a CMOS microarray with 400 54 μm test site microelectrodes and four large perimeter counter electrodes used to produce electric field geometries that encompass the whole microarray surface area (overall dimensions ~4 × 6 mm).

has an individual wire interconnect through which current and voltage are applied and regulated. The 100-site microelectronic arrays are fabricated from silicon wafers, with insulating layers of silicon dioxide, platinum microelectrodes and gold connecting wires. Silicon dioxide/silicon nitride is used to cover and insulate the conducting wires. The whole surface of the array is ultimately covered with several microns of hydrogel (agarose or polyacrylamide) which forms the permeation layer. The permeation layer is impregnated with a coupling agent (streptavidin) which allows attachment of biotinylated DNA probes or other entities.[5,6,20] The ability to use silicon and microlithography for fabrication of these DNA chips allows a wide variety of devices to be designed and tested. The higher density arrays (400 to 10,000) represent more sophisticated devices that have on-chip CMOS control elements for regulating the current and voltages to the microelectrode at each test site.[8] These control elements are located in the underlying silicon structure and are not exposed to the aqueous samples that are applied to the chip surface when carrying out the DNA hybridization reactions. The CMOS 400 test site microarray which has been recently commercialized has 54 μm test sites (Fig. 1B).

The 100 and 400 test site microelectronic array devices have been incorporated into a cartridge package (NanoChip™ cartridge) which provides for the electronic, optical, and fluidic interfacing. The chip itself is mounted (flip chip bonded) onto a ceramic plate and pinned out for the electrical connections. The chip/ceramic plate component is mounted into a plastic cartridge that provides several fluidic input and output ports for addition and removal of DNA samples and reagents. The area over the active test site portion of the array is an enclosed sample chamber covered with a quartz glass window. This window allows for fluorescent detection to be carried out on the DNA

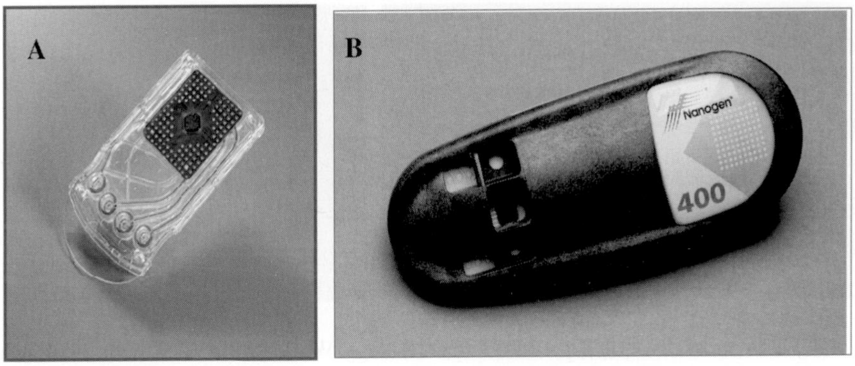

Fig. 2. **(A)** The 100 test site NanoChip™ cartridge. **(B)** The 400 test site NanoChip™ cartridge. The cartridge protects the chips, as well as provides the electronic, optical, and fluidic interfacing to the instrument.

hybridization reactions that occur at the test sites on the array surface. The NanoChip™ cartridge assemblies for both the 100- and 400-site CMOS chips are shown in Fig. 2.

A complete instrument system (NanoChip™ Molecular Biology Workstation) provides a chip loader component, fluorescent detection/reader component, computer control interface and data display screen component. The probe loading component allows DNA probes or target sequences (DNA, RNA, PCR amplicons) to be selectively addressed to the array test sites, and provides a "make your own chip" capability. The automated probe loader system allows NanoChip™ cartridges to be loaded with DNA probes or samples from a 96- or 384-well microtiter plate. Probes or target sequences are usually biotinylated, which allows them to become bound to streptavidin within the permeation layer of the specified test site. In the electronic addressing procedure, the probe loader component deliver the desired biotinylated probe to specified test sites which are biased positive. The electric (electrophoretic) field causes the negatively charged DNA molecules to concentrate onto the positively activated test sites, with subsequent binding via the biotin/streptavidin reaction. Figure 3 shows the instrument systems which have been designed for the original 100 test site arrays (A), and for the newer CMOS 400 test site arrays (B).

3.2 *Electronic DNA hybridization*

DNA hybridization is carried out on the microelectronic array by first using a DC electric field to electrophoretically transport and address biotinylated oligonucleotide "capture" probes or target DNA sequences to the selected test sites on the array surface. In this process the biotinylated DNA molecules become strongly bound to streptavidin molecules which are cross-linked to the hydrogel layer covering the underlying microelectrode. In the next steps, electric fields are used to control and

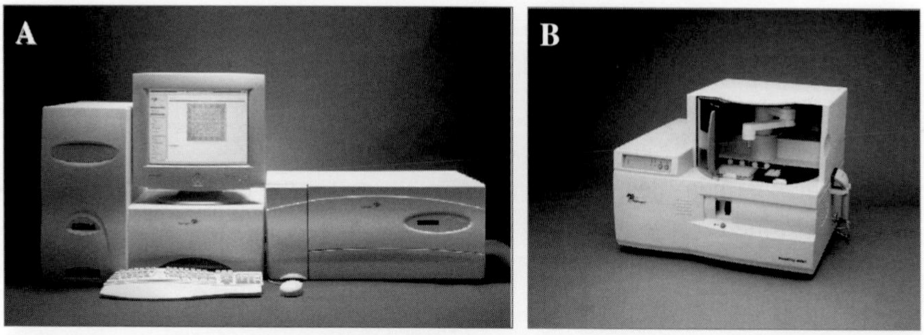

Fig. 3. **(A)** The earlier NanoChip™ Molecular Biology Workstation used with the 100 test site microarray devices for bioresearch and DNA diagnostic applications. This system provides a chip loader component, fluorescent detection/reader component, computer control interface and data display component. The probe loader component is used for automated DNA probe addressing or spotting, and provides the end-user with "Make Your Own Chip" capabilities. **(B)** The newer NanoChip™ Molecular Biology Workstation used with the CMOS 400 test site microarray devices for bioresearch and DNA diagnostic applications. This system provides a chip loader component, fluorescent detection/reader component, computer control interface and data display component.

direct the hybridization of the DNA molecules in the sample to the DNA sequences attached to the selected test sites. The microarrays can be used for a variety of hybridization formats which include reverse dot blot format (capture/identity sequences bound to test sites), sandwich format (capture sequences bound to test sites) and dot blot format (target sequences bound to test sites). DNA hybridization assays involve the use of fluorescent reporter probes or target DNA sequences. The reporter groups are usually organic fluorophores that have either been attached to oligonucleotide probes or to the sample DNA sequences. After electronic addressing and hybridization are carried out, the array is analyzed by the fluorescent detection system. The fluorescent detector monitors for signals at two wavelengths (530 and 630 nm). References 1–7, 21, and 22 provide a number of examples of how single nucleotide polymorphism (SNP), other types of DNA genotyping and gene expression analysis are carried out on the microelectronic array devices. DNA genotyping analysis is used for clinical diagnostics of cancer, genetic diseases and for the detection of infectious bacterial and viral agents.

3.3 *Cell separation by DEP*

Microelectronic arrays have also been used to carry out a variety of cell separation applications by dielectrophoresis (DEP). Disease diagnostics often involves identifying a small number of bacteria or virus in a blood sample, fetal cells in maternal blood, or tumor cells in a background of normal cells. One basic electronic method for cell separation is called dielectrophoresis (DEP). This process involves the application of an asymmetric alternating current (AC) electric field to the cell population. Active microelectronic arrays have been used to achieve the separation of bacteria from whole blood[11] and for the separation of cervical carcinoma cells from blood.[12] In the case of mixed blood/bacteria samples, DEP causes cell separation to occur, where the blood cells can be positioned in the low field regions between the electrodes and the bacteria can be positioned in the high field regions near the microelectrode surface. While maintaining the AC field, the microarray can be washed with a buffer solution that removes the blood cells (low field regions) from the more firmly bound bacteria (high field regions) near the microelectrodes. The bacteria can then be released and collected or electronically lysed to release the genomic DNA or RNA for further manipulation and analysis.[12] DEP represents a particularly useful process that allows difficult cell separation applications to be carried out rapidly and with high selectivity. The DEP process may also be useful for the manipulation of nanoparticles and for nanofabrication applications.

4. Electronic Field Directed Self-Assembly

4.1 *Background*

One of the grand challenges in nanotechnology is the development of fabrication technologies that will lead to cost effective nanomanufacturing processes. In addition to the more classical top-down processes such as photolithography, so-called bottom-up processes are also being developed for carrying out self-assembly of nanostructures into higher order structures, materials and devices.[23–28] To this end, considerable efforts have been directed at both passive and active types of layer-by-layer (LBL) self-assembly processes as a way to make three-dimensional layered structures which can have macroscopic x-y dimensions.[29–46] In cases where patterned structures are desired, the substrate material is generally pre-patterned using masking and a photolithographic process.[47,48] Other approaches to patterning include the use of optically patterned ITO films and active deposition of the nanoparticles.[49,50] Nevertheless, limitations of passive LBL and as well as active assembly processes provide considerable incentive to continue the development of better paradigms for self-assembly based nanofabrication and heterogeneous integration.

As was discussed earlier, over the past decade microelectronic array devices have been developed for DNA diagnostic applications. In these applications, electronic microarray devices which produce reconfigurable electric fields on their surfaces are first used to address and bind negatively charged biotinylated DNA molecules to selected test sites on the microarray surface, followed by the hybridization of target DNA sequences to the specific test sites. Thus, these devices are able to direct and accelerate the self-assembly or "bottom-up" process of DNA hybridization occurring on

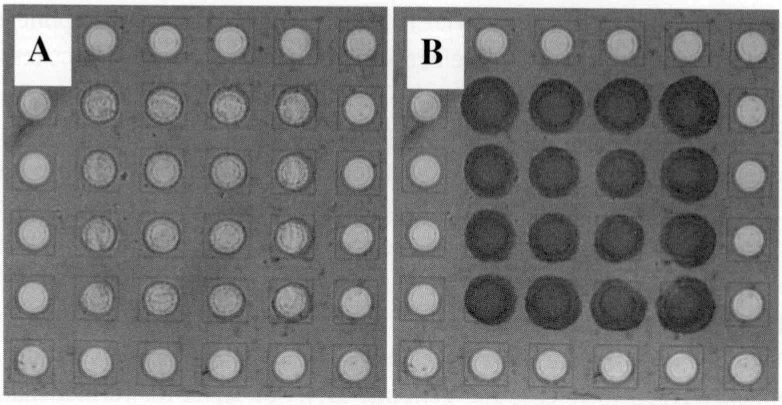

Fig. 4. The sequence for transport and addressing of 250 nm negatively charged polystyrene nanoparticles using the 400 test site CMOS microarray (54 μm microelectrodes). **(A)** The array before the negatively charged 250 nm nanoparticles are selectively transported and concentrated onto six positively charged test sites. **(B)** The nanoparticles now concentrated onto the six test sites of the 400-site CMOS microarray (one minute run time at 3.5 volts DC).

the microarray. Microelectronic arrays have also been used to direct the binding of derivatized nanospheres and microspheres onto selected locations on the microarray surface. In this case, fluorescent and non-fluorescent polystyrene nanospheres and microspheres derivatized with specific DNA oligonucleotides are transported and bound to selected test sites or microlocations derivatized with the specific complementary oligonucleotide sequences. Ultimately, these devices have shown a proven ability to manipulate a surprisingly large variety of entities that range in size from small molecules, to nanoparticles, to one micron scale objects. Nanoparticle transport and addressing has also been demonstrated using the 400 test site CMOS microarray device. Figure 4 shows an example of negatively charged 250 nm fluorescent nanoparticles being transported and concentrated onto the test sites of the 400-site CMOS microarray.

The whole process is carried out at 3.5 volts DC, and requires less than one minute. More recently, the 400-site CMOS microarray device and controller system have been used to carry out the rapid and highly parallel directed self-assembly of biotin and streptavidin derivatized nanoparticles into higher order multilayered structures. Figures 5A to 5C show the basic process by which a microelectronic array can be used to carry out the directed self-assembly of biotin and streptavidin derivatized fluorescent nanoparticles into higher order multilayer three-dimensional structures.

4.2 *Experimental methods for nanoparticles layering*

The 400-site CMOS microarray device (Nanogen) is able to independently output currents to each of the 400 different electrodes (54 μm diameter, in 25 columns by 16 rows)

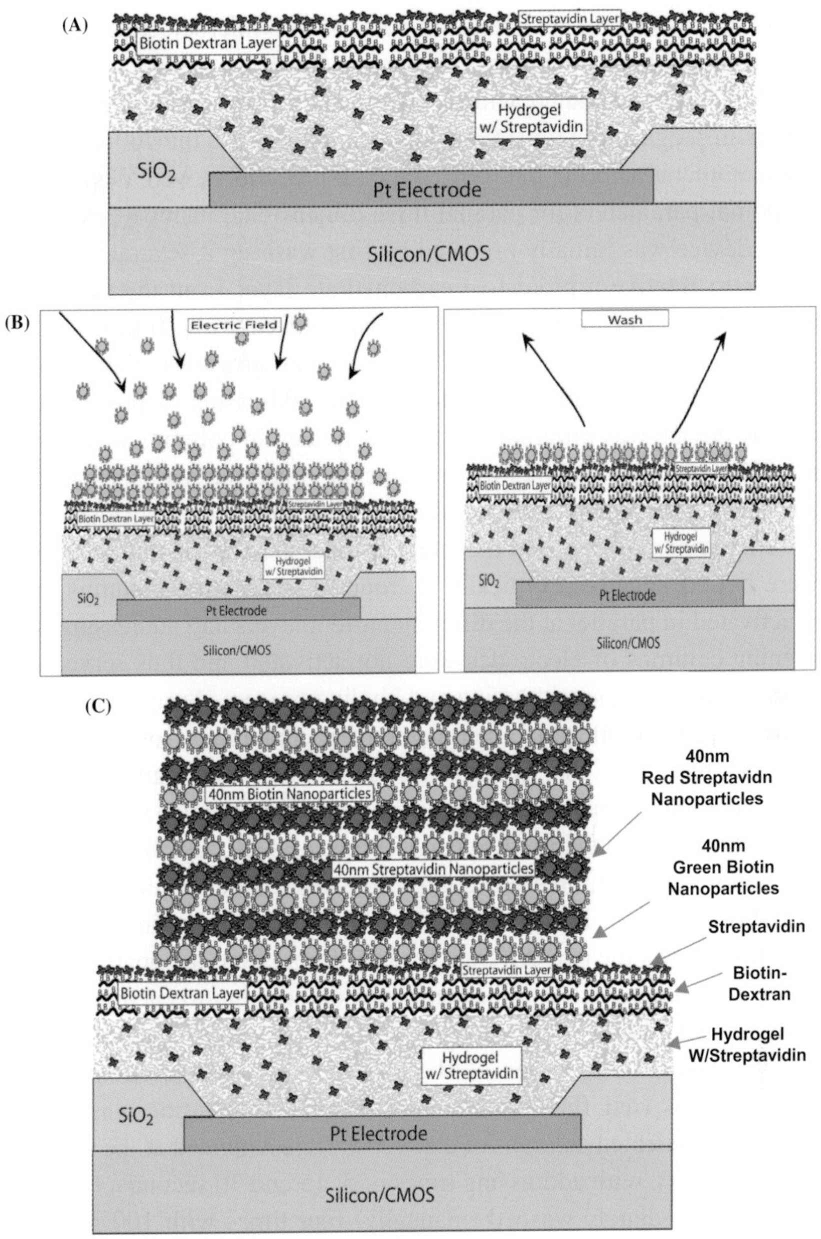

Fig. 5. **(A)** The first step of a generalized scheme for using an electronic microarray to carry out the directed self-assembly process for layering biotin and streptavidin derivatized 40 nm fluorescent nanoparticles. In the first step, the permeation layer covering the platinum microelectrode is over layered with a biotin-dextran and a streptavidin layer before nanoparticle addressing is carried out. **(B)** The next step in the nanoparticle layering process, where the electric field is used to rapidly address and bind the biotin derivatized fluorescent nanoparticles to the streptavidin layer. The excess nanoparticles are removed by washing and the first nanoparticle layer is finished. The process is then repeated with streptavidin derivatized nanoparticles. **(C)** The completed nanoparticle layering process where ten alternate layers of biotin and streptavidin fluorescent nanoparticles have now been deposited onto the original biotin-dextran/strepatavidin layer.

at up to 1 μA at 5 volts. Surrounding the 400 test site microelectrodes are four large counter electrodes which encompass the inner electrode array (see Fig. 1B). The 400-site CMOS array is coated with a 10 μm thick polyacrylamide gel permeation layer which is impregnated with streptavidin. In order to use the 400-site CMOS array device as a nanomanufacturing platform, standard procedures were developed to determine the optimal parameters for parallel three-dimensional nanoparticle layering. The CMOS array device was initially prepared by first washing it several times with ultrapurified water to remove a protective carbohydrate layer from the permeation layer. After washing, the permeation layer surface was reacted with 20 μl of a 2 μM biotin dextran (Sigma B-9264) solution for 30 minutes. The array was then finally washed with a 100 mM L-histidine solution. To test various addressing (deposition) conditions, the microarray device was programmed to be activated in columns with currents varying from 0.025 to 0.4 μA in 0.025 μA increments. Since the nanoparticles used in the experiments have a net negative charge, the electrodes at the desired addressing sites on the array were biased positive, and the larger counter electrodes on the perimeter of the device were biased negative. Typically a group of alternating columns on the array would be activated in parallel at the different current levels and addressing times, while the intervening columns of electrodes were not activated and thus served as negative controls for the nanoparticle addressing and binding process. Array addressing was carried out using 10 μl of 100 mM L-histidine buffer, containing from 1 to 10 nM of the derivatized nanoparticles. The two types of derivatized nanoparticles used in these nanoparticle layering experiments were 40 nm red fluorescent polystyrene nanoparticles derivatized with streptavidin (Molecular Probes F8770, Ex 580 nm – Em 605 nm) and 40 nm yellow-green fluorescent polystyrene nanoparticles derivatized with biotin (Molecular Probes F8766, Ex 505 nm – Em 515 nm). The biotin-streptavidin ligand binding reaction allows the two different types of fluorescent nanoparticles to be bound to each other, but nanoparticles of the same type do not bind to each other.

Nanoparticle addressing, binding and layering experiments were carried out as follows: (1) First Nanoparticles Layer — About 20 μl of a 10 nM solution of 40 nm streptavidin nanoparticles (red fluorescence) in 100 mM L-histidine was placed on the microarray and the selected columns of electrodes were activated at the different current levels (0.025–0.4 μA), with addressing times of 5, 15 and 30 seconds. (2) Wash — The array was then immediately washed (manually) four times with 100 mM L-histidine, which usually took less than one minute. Epifluorescence (Ex 580 nm – Em 605 nm) monitoring of the array was carried out during the process. (3) Second Nanoparticles Layer — About 20 μl of a 10 nM solution of 40 nm biotin nanoparticles (green fluorescence) in 100 mM L-histidine was placed on the array and selected electrodes were activated (currents form 0.025–0.4 μA, with addressing times of 5, 15 and 30 seconds). Epifluorescence (Ex 505 nm – Em 515 nm) monitoring of the array was carried out during the process. (4) Wash — The array was then immediately washed three times with 100 mM L-histidine. (5) Successive Nanoparticles Layers — Steps 1 to 4 were repeated to achieve desired number of nanoparticle layers, and epifluorescence (red and green) monitoring of the array was carried during the whole process. (6) Final Wash — The array was finally washed several times with L-histidine. Using the process described above, a total of 40 alternating addressing and depositions of the 40 nm streptavidin nanoparticles (red fluorescence) and 40 nm biotin nanoparticles (green fluorescence) were carried out in about one hour.

4.3 *Experimental results for nanoparticles layering*

In order to determine the optimal conditions for layering of derivatized nanoparticle, experiments were carried out at addressing times of 5, 15 and 30 seconds. For each of the addressing time experiments, ten columns (16 sites) were activated with DC current levels that ranged from 0.025 to 0.4 μA, in increments of 0.025 μA. The activation of all 160 sites (at different current levels) was carried out in parallel. For each addressing time experiment (5, 15 and 30 seconds), the addressing process was carried out 25 times with alternating 40 nm red fluorescent streptavidin nanoparticles and green fluorescent biotin nanoparticles. In these experiments, the alternate columns of test sites on the array were not activated. Figure 6 shows the fluorescent microscope imaging results for the 25 addressings experiment at 15 seconds. The insert in the photo is an enlargement

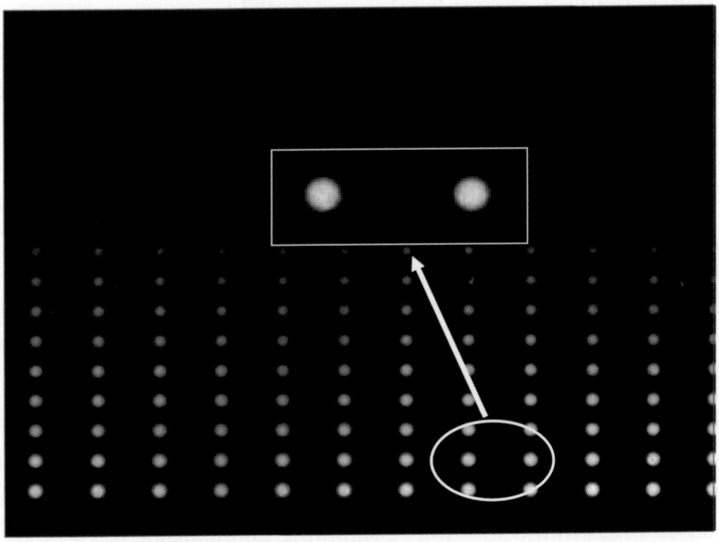

Fig. 6. Fluorescent analysis results for highly parallel electric field 3D layering of 25 alternate layers of 40 nm red fluorescent streptavidin nanoparticles and green fluorescent biotin nanoparticles onto the 54 μm diameter test sites of a CMOS 400-site microarray device. Addressing time was 15 seconds at DC current levels that ranged from 0.025 to 0.40 μA. Inserts are enlargements of a small section of the microarray each showing the results for 0.375 μA on two test sites. Total layering time is one minute or less per layer, including the washing steps.

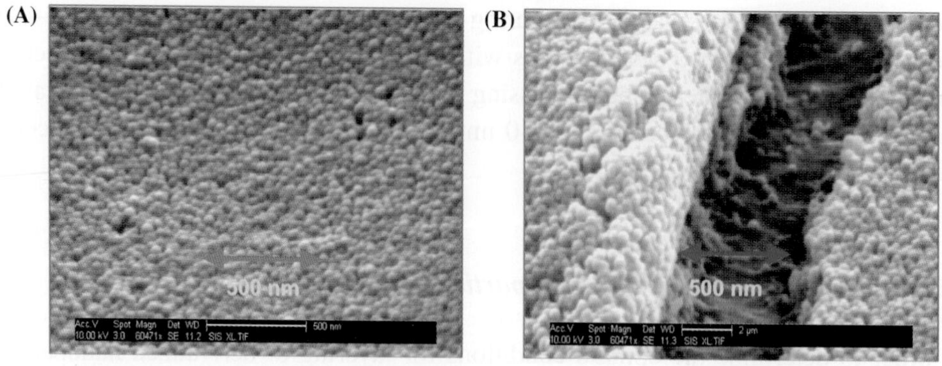

Fig. 7. **(A)** A high resolution SEM image of the top layer of one of the addressed sites after 40 alternate addressings with biotin and streptavidin derivatized 40 nm nanoparticles (scale is 500 nm). **(B)** A fracture through one of the sites reveling the multiple layering of the biotin and streptavidin 40 nm nanoparticles (scale is 500 nm).

showing two of the 54 μm test sites where the fluorescent nanoparticle layers have been deposited at the 0.375 μA DC current level.

The images show no detectable fluorescence is observed on any of the alternate sites which were not activated. By the relative fluorescent intensity of the activated sites, the best conditions for nanoparticle layering at the 5-second addressing time appears to be between the 0.30 to 0.40 μA current level range. At the lower current levels (< 0.30 μA), the overall fluorescent intensity for the layers begins to decrease. At the longer 30-second addressing time, the nanoparticle layers become visibly damaged. Under real-time epifluorescent microscope observation, some of these fractured layers could actually be observed to flap when the sites were activated. In further experiments, up to 40 nanoparticle layers could be deposited. In these experiments scanning electronic microscopy was used to examine the 40-layer nanoparticle structures in more detail. Figure 7A shows high resolution SEM images of the top layer of an addressed site after 40 addressings with nanoparticles. The SEM image of the addressed site shows a very well defined top layer of nanoparticles that appear to about 40 nm in diameter. Figure 7B shows a fracture through one of the sites illustrating the multiple layering of the 40 nm nanoparticles. A number of nanoparticle layers can be seen from the top layer down to what appears to be the lower surface of the permeation layer. The un-addressed site on the array showed little or no nanoparticle accumulation on the permeation gel layer surface, even though it was exposed 40 times to the solutions containing nanoparticles.

5. Conclusions

Electric field directed self-assembly of 40 multiple layer nanoparticle structures could be carried out in a rapid and highly parallel format using a CMOS microelectronic array device. In this process, efficient nanoparticle addressing/deposition was achieved in 15 seconds or less. With a washing step of about 45 seconds, the total time for creating each nanoparticle layer was about one minute. Thus, the 40-layer nanoparticle structures could be completed in less than one hour. The optimal electronic addressing window for creating high quality three-dimensional layered structures appears to be at current levels in the 0.25 to 0.35 μA range. Because the nanoparticles used in these experiments were of the same size (40 nm), it is difficult to determine from the SEM image (Fig. 7) the homogeneity of individual biotin and streptavidin nanoparticle layers within the multilayered structures. Nevertheless, when taken together with fluorescent monitoring of depositions and the experiments showing no layering of like nanoparticles, the SEM images are consistent with a 40-layer alternating nanoparticle structure. Work is now in progress with different sized nanoparticles to better determine true quality of individual layers within the multilayer structures.

The results of the nanoparticles layering experiments have demonstrated a number of advantages relevant to electric field directed self-assembly nanofabrication. First, it was shown that high electric field strengths and electrophoretic transport from the bulk solution could be maintained even through many layers of nanoparticles, allowing the new upper layers of nanoparticles to still be rapidly addressed to the activated sites. Second, the results showed that the integrity of the biotin-streptavidin ligand binding was maintained even after 40 layers of nanoparticles were addressed. This is extremely important for any three-dimensional self-assembly process that is based on biomolecular binding using highly specific protein or DNA ligands. This ability to use high electric field strength with minimal side effects is most certainly due to the permeation layer which separates the electrode from the nanostructures, as well as to the relatively short addressing times (5 to 15 seconds) needed for nanoparticle addressing and deposition. Third, because the high electric field strengths produce enormous concentration effects at the activated sites, only minimal concentrations of nanoparticles were needed for addressing and deposition. In fact, the concentration of nanoparticles used was so low that no significant non-specific binding of nanoparticles occurred during the whole process. Fourth, CMOS electronic array devices appear to allow nanoparticle layering to be carried out much faster (one to two orders of magnitude) than comparable passive layer-by-layer (LBL) self-assembly nanofabrication processes. Finally, with regard to x-y patterning during the fabrication process, the re-configurable CMOS array devices may also have advantages over lithographic and ITO patterning processes. Any lithographic patterning of the substrate and subsequent layers would be a relatively time consuming process,[25,26] while electronic microarray patterning could be carried out almost instantaneously. In the case of ITO (optical) patterning, the patterning step could be rapid, but ITO substrates are not stable at the high DC electric fields needed for rapid addressing and deposition.[27,28]

Overall, the use of a microelectronic array device for directed self-assembly represents a unique example of combining the best aspects of "top-down" and "bottom-up" technologies into a process not only viable for carrying out DNA hybridization diagnostics, but also potentially useful as a nanomanufacturing process. This electric field process allows the heterogeneous integration of many different molecular and nanoscale entities, and their controlled hierarchical self-assembly from the molecular level, to the nanoscale level, to the microscale level and ultimately to the macroscale level. Such a process may be useful for the directed assembly of integrated nano/micro/macrostructures for a variety of nanoelectronic, nanophotonic, nanomaterials, nanobiosensor and many other applications.

Acknowledgments

Funds in support of this study were received from NSF award No. DMI-0327077 and Nanogen CMOS supplied the microarrays and the array controller system. The authors thank Ben Sullivan (UCSD), Professor Sadik Esener, Dr. Dalibor Hodko (Nanogen), and Dr. Paul Swanson (Nanogen) for their advice and support.

References

1. M. J. Heller, An active microelectronics device for multiplex DNA analysis, *IEEE Eng. Med. Biol. Mag.* **15**: 100–103 (1996).
2. R. G. Sosnowski, E. Tu, W. F. Butler, J. P. O'Connell and M. J. Heller, Rapid determination of single base mismatch in DNA hybrids by direct electric field control, *Proc. Natl. Acad. Sci. USA* **94**: 1119–1123 (1997).
3. C. F. Edman, D. E. Raymond, D. J. Wu, E. Tu, R. G. Sosnowski, W. F. Butler, M. Nerenberg and M. J. Heller, Electric field directed nucleic acid hybridization on microchips, *Nucleic Acids Res.* **25**: 4907–4914 (1997).
4. M. J. Heller *et al.*, An integrated microelectronic hybridization system for genomic research and diagnostic applications. In: *Micro Total Analysis Systems* '98, eds. D. J. Harrison and A. van den Berg (Kluwer Academic Publishers, 1998), pp. 221–224.
5. M. J. Heller, E. Tu, A. Holmsen, R. G. Sosnowski and J. P. O'Connell, Active microelectronic arrays for DNA hybridization analysis. In: *DNA Microarrays: A Practical Approach*, ed. M. Schena (Oxford University Press, Oxford, UK, 1999), pp. 167–185.
6. M. J. Heller, A. H. Forster and E. Tu, Active microelectronic chip devices which utilize controlled electrophoretic fields for multiplex DNA hybridization and genomic applications, *Electrophoresis* **21**: 157–164 (2000).
7. C. Gurtner, E. Tu, N. Jamshidi, R. Haigis, T. Onofrey, C. F. Edman, R. Sosnowski, B. Wallace and M. J. Heller, Microelectronic array devices and techniques for electric field enhanced DNA hybridization in low-conductance buffers, *Electrophoresis* **23**: 1543–1550 (2002).
8. P. Swanson, R. Gelbart, E. Atlas, L. Yang, T. Grogan, W. F. Butler, D. E. Ackley and E. Sheldon, A fully multiplexed CMOS biochip for DNA analysis, *Sens. Actuators B Chem.* **64**: 22–30 (2000).
9. M. J. Heller, E. Tu, R. Martinsons, R. R. Anderson, C. Gurtner, A. Forster and R. Sosnowski, Active microelectronic array systems for DNA hybridization, genotyping, pharmacogenomic, and nanofabrication applications. In: *Integrated Microfabricated Biodevices*, eds. M. J. Heller and A. Guttman (Marcel Dekker, New York, NY, 2002), pp. 223–270.

10. Y. Huang, K. L. Ewalt, M. Tirado, R. Haigis, A. Forster, D. Ackley, M. J. Heller, J. P. O'Connell and M. Krihak, Electric manipulation of bioparticles and macromolecules on microfabricated electrodes, *Anal. Chem.* **73**: 1549–1559 (2001).
11. J. Cheng, E. L. Sheldon, L. Wu, A. Uribe, L. O. Gerrue, J. Carrino, M. Heller and J. O'Connell, Electric field controlled preparation and hybridization analysis of DNA/RNA from *E. coli* on micro-fabricated bioelectronic chips, *Nat. Biotechnol.* **16**: 541–546 (1998).
12. J. Cheng, E. L. Sheldon, L. Wu, M. J. Heller and J. O'Connell, Isolation of cultured cervical carcinoma cells mixed with peripheral blood cells on a bioelectronic chip, *Anal. Chem.* **70**: 2321–2326 (1998).
13. Y. Huang, J. Sunghae, M. Duhon, M. J. Heller, B. Wallace and X. Xu, Dielectrophoretic separation and gene expression profiling on microelectronic chip arrays, *Anal. Chem.* **74**: 3362–3371 (2002).
14. C. F. Edman, C. Gurtner, R. E. Formosa, J. J. Coleman and M. J. Heller, Electric-field-directed pick-and-place assembly, *HDI* **10**: 30–35 (2000).
15. C. F. Edman, R. B. Swint, C. Gurthner, R. E. Formosa, S. D. Roh, K. E. Lee, P. D. Swanson, D. E. Ackley, J. J. Colman and M. J. Heller, Electric field directed assembly of an InGaAs LED onto silicon circuitry, *IEEE Photonics Tech. Lett.* **12**: 1198–1200 (2000).
16. S. C. Esener, D. Hartmann, M. J. Heller and J. M. Cable, DNA assisted micro-assembly: a heterogeneous integration technology for optoelectronics, Vol. 70. In: *Critical Reviews of Optical Science and Technology*, eds. A. Hussain and F. Mahmoud (Society of Photo Optical Instrumentation Engineers, Bellingham, WA, 1998).
17. D. M. Hartmann, D. Schwartz, G. Tu, M. Heller, C. Sadik and S. C. Esener, Selective DNA attachment of particles to substrates, *J. Mater. Res.* **17**: 473–478 (2002).
18. Small Wonders, *Endless Frontiers: Review of the National Nanotechnology Initiative* (National Research Council, 2002).
19. The National Nanotechnology Initiative — Strategic Plan (National Science and Technology Council, December 2004).
20. S. K. Kassengne, H. Reese, D. Hodko, J. M. Yang, K. Sarkar, P. Swanson, D. E. Raymond, M. J. Heller and M. J. Madou, Numerical modeling of transport and accumulation of DNA on electronically active biochips, *Sens. Actuators B. Chem.* **94**: 81–98 (2003).
21. C. Gurtner, C. F. Edman, R. E. Formosa and M. J. Heller, Photoelectrophoretic transport and hybridization of DNA on unpatterned silicon substrates, *J. Am. Chem. Soc.* **122**: 8589–8594 (2000).
22. P. N. Gilles, D. J. Wu, C. B. Foster, P. J. Dillion and S. J. Chanock, Single nucleotide polymorphic discrimination by an electronic dot blot assay on semiconductor microchips, *Nat. Biotechnol.* **17**: 365–370 (1999).
23. K. L. Prime and G. M. Whitesides, Self-assembled organic monolayers: model systems for studying adsorption of proteins at surfaces, *Science* **252**: 1164–1167 (1991).
24. B. Kim, S. L. Tripp and A. Wei, Self-organization of large gold nanoparticle arrays, *J. Am. Chem. Soc.* **123**: 7955–7956 (2001).
25. N. Bowden, A. Terfort, J. Carbeck and G. M. Whitesides, Self-assembly of mesoscale objects into ordered two-dimensional arrays, *Science* **276**: 233–235 (1997).
26. J. Fink, C. J. Kiely, D. Bethell and D. J. Schiffrin, Self-organization of nanosized gold particles, *Chem. Mater.* **10**: 922–926 (1998).
27. S. W. Lee, C. Mao, C. E. Flynn and A. M. Belcher, Ordering of quantum dots using genetically engineered viruses, *Science* **296**: 892–895 (2002).
28. C. A. Mirkin, R. L. Letsinger, R. C. Mucic, J. J. Storhoff, A DNA-based method for rationally assembling nanoparticles into macroscopic materials, *Nature* **382**: 607–609 (1996).
29. D. Decher and J. B. Schlenoff, eds., *Multilayer Thin Films-Sequential Assembly of Nanocomposite Materials* (Wiley-VCH Verlag, Weinheim, Germany, 2003).
30. M. P. Hughes, ed., *Nanoelectromechanics in Engineering and Biology* (CRC Press, Boca Raton, FL, 2003).

31. W. A. Goddard, D. W. Brenner, S. E. Lyashevski and G. J. Lafrate, eds., *Handbook of Nanoscience, Engineering and Technology* (CRC Press, Boca Raton, FL, 2003).
32. A. B. Artyukhin, O. Bakajin, P. Stroeve and A. Noy, Layer-by-layer electrostatic self-assembly of polyelectrolyte nanoshells on individual carbon nanotube templates, *Langmuir* **20**: 1442–1448 (2004).
33. J. J. Yuan, S. X. Zhou, B. You and L. M. Wu, Organic pigment particles coated with colloidal nano-silica particles via layer-by-layer assembly, *Chem. Mater.* **17**(14): 3587–3594 (2005).
34. N. Ma, H. Y. Zhang, B. Song, Z. Q. Wang and X. Zhang, Polymer micelles as building blocks for layer-by-layer assembly: an approach for incorporation and controlled release of water-insoluble dyes, *Chem. Mater.* **17**: 5065–5069 (2005).
35. S. Zapotoczny, M. Golonka and M. Nowakowska, Novel photoactive polymeric multilayer films formed via electrostatic self-assembly, *Macromol. Rapid Commun.* **26**: 1049–1054 (2005).
36. M. Wanunu, R. Popovitz-Biro, H. Cohen, A.Vaskevich and I. Rubinstein, Coordination-based gold nanoparticle layers, *J. Am. Chem. Soc.* **127**: 9207–9215 (2005).
37. P. T. Hammond, Form and function in multilayer assembly: new applications at the nanoscale, *Adv. Mater.* **16**: 1271–1293 (2004).
38. H. O. Jacobs, S. A. Campbell and M. G. Steward, Approaching nanoxerography: the use of electrostatic forces to position nanoparticles with 100 nm scale resolution, *Adv. Mater.* **14**: 1553–1557 (2002).
39. C. R. Barry, J. Gu and H. O. Jacobs, Charging process and coulomb-force-directed printing of nanoparticles with sub-100-nm lateral resolution, *Nano Lett.* **5**: 2078–2084 (2005).
40. P. Mardilovich and P. Kornilovitch, Electrochemical fabrication of nanodimensional multilayer films, *Nano Lett.* **5**: 1899–1904 (2005).
41. D. B. Allred, M. Sarikaya, F. Baneyx and D. T. Schwartz, Electrochemical nanofabrication using crystalline protein masks, *Nano Lett.* **5**: 609–613 (2005).
42. D. H. Tsai, S. H. Kim, T. D. Corrigan, R. J. Phaneuf and M. R. Zachariah, Electrostatic-directed deposition of nanoparticles on a field generating substrate, *Nanotechnology* **16**: 1856–1862 (2005).
43. S. O. Lumsdon, E. W. Kaler and O. D. Velev, Two-dimensional crystallization of microspheres by a coplanar AC electric field, *Langmuir* **20**: 2108–2116 (2004).
44. J. C. Lin, M. Z. Yates, A. T. Petkoska and S. Jacobs, Electric-field-driven assembly of oriented molecular-sieve films, *Adv. Mater.* **496**: 649–652 (2006).
45. N. Wang, H. Lin, J. Li, X. Yang and B. Chi, Electrophoretic deposition and optical properties of titania nanotube films, *Thin Solid Films* **496**: 649–652 (2006).
46. B. Ferrari, S. González, R. Moreno C., Baudín, Multilayer coatings with improved reliability produced by aqueous electrophoretic deposition, *J. Eur. Ceram. Soc.* **26**: 27–36 (2006).
47. F. Hua, J. Shi, Y. Lvov and T. Cui, Patterning of layer-by-layer self-assembled multiple types of nanoparticle thin films by lithographic technique, *Nano Lett.* **2**: 1219–1222 (2002).
48. H. H. Ko, C. Y. Jiang and V. V. Tsukruk, Encapsulating nanoparticle arrays into layer-by-layer multilayers by capillary transfer lithography, *Chem. Mater.* **17**: 5489–5497 (2005).
49. M. Gao, J. Sun, E. Dulkeith, N. Gaponik, U. Lemmer and J. Feldmann, Lateral patterning of CdTe nanocrystal films by the electric field directed layer-by-layer assembly method, *Langmuir* **18**: 4098–4102 (2002).
50. R. C. Hayward, D. A. Saville and I. A. Aksay, Electrophoretic assembly of colloidal crystals with optically tunable micropatterns, *Nature* **404**: 56–59 (2000).

SECTION VII

GENOMIC ENGINEERING AND SYSTEMS BIOLOGY

CHAPTER 23

SYSTEMS BIOLOGY: A FOUR-STEP PROCESS

Jennifer L. Reed and Bernhard O. Palsson

Abstract

Systems biology focuses on the study of biological networks through the processes of network reconstruction, computer model formulation, hypothesis generation and experimentation. This chapter sets out to define systems biology and the technological driving forces that have enabled the field to emerge. In addition, the four core steps in systems biology are presented. The first two steps involve identifying components in biological networks and interactions between these components. The result of the first two steps is a network reconstruction: an accounting of all components and their interactions compromising the network. From this network reconstruction, an *in silico* model can be generated (step 3) which can be used to predict and analyze the behavior of biological systems (step 4). The chapter concludes with a discussion of systems biology applications, to address specific biological and industrial questions.

1. Emergence of Systems Biology

Systems biology is becoming more prominent as evidenced by a growth in the number of publications, institutes, and departments devoted to studying this area. Enabled by whole genome sequencing projects, the field emerged in the 1990s as high-throughput technologies made it possible to measure and quantify cellular components and their functional interactions. Advances in genome sequencing and annotation, for example, have generated complete sequences for prokaryotic and eukaryotic genomes, providing us with a "parts list" for a cell. Furthermore, gene expression data provides a quantitative measurement on the abundance of mRNA transcripts within the cell, allowing us to see how regulation affects the expression of genes in response to a stimulus. These technologies, and others, have played an important role in the development of systems biology.

Systems biology captures the information from these high-throughput technologies by developing mathematical models to study biological networks and focuses on the study of whole systems, rather than their isolated parts.[1] As such, knowledge about the components and the interactions between components in a network is critical. This information is captured in what is called a *network reconstruction*. One goal of systems biology is to generate models which can explain or predict cellular behavior and can lead to the generation of new biological hypotheses. The resulting models can also be used to provide biological context for the vast number of large experimental datasets being generated.

2. Historical Roots

Systems biology has emerged at the interface between molecular biology and systems science. In the 20th century, molecular biologists typically took a reductionist approach to studying the cell.[2] These scientists were interested in figuring out what made up a cell and what each of the parts did, effectively taking a reverse engineering approach. Biologists developed methods to detect and quantify cellular components and to determine their roles within cellular processes. During the same time, mathematicians and physicists were interested in developing methods to model biological processes, and created some detailed models for which enough data were available. At that time, the generation of models was hindered by the availability of experimental data, limiting modeling efforts to small systems such as the *lac* operon in *Escherichia coli*[3] and human red blood cell metabolism.[4]

In the 1990s a number of high-throughput technologies were developed, including whole genome sequencing and gene expression *microarrays*, which pushed biological research into a data-rich environment by providing whole cell, system-level, measurements. These scientific and technological achievements were critical in that they enabled the generation of genome-scale models and hence, the emergence of systems biology. There is currently enough information about the components and interactions in biological networks to generate large-scale (often genome-scale) models of cellular networks.

3. Principal Steps in Systems Biology

As shown in Fig. 1, there are four principal steps in a systems biology approach.[1]

First, we must identify all of the components that participate in the network. Second, we have to identify the interactions between these components. These first two steps result in a network reconstruction, where components and interactions are delineated. From such a reconstruction, a model can then be generated. This stage is where variables and equations are introduced to formulate a model, which is then solved either numerically or analytically. Once the network is successfully represented by a system of mathematical equations, the resulting model can be used to analyze and predict

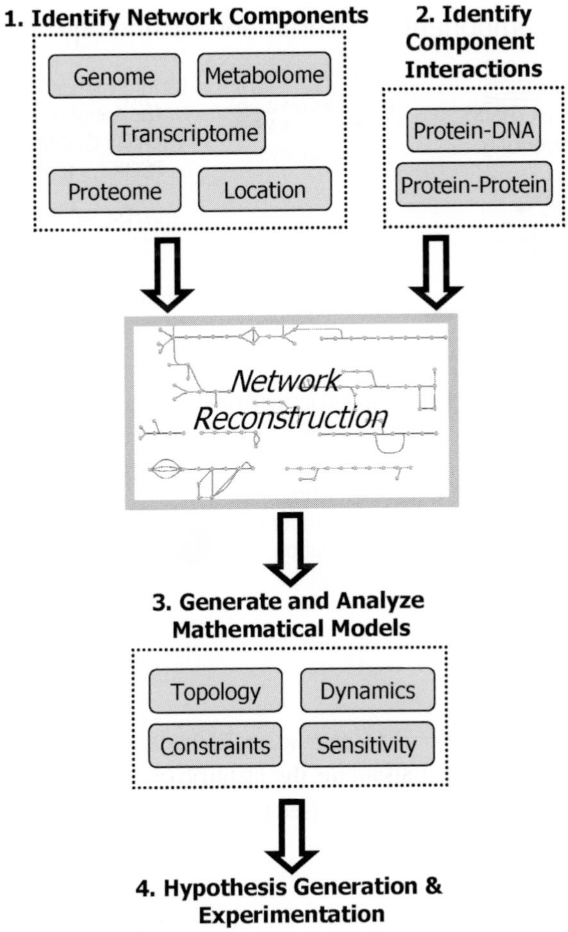

Fig. 1. Four Steps in Systems Biology. The first two steps use available data to identify the network components and component interactions, essentially specifying the nodes and links in a network. This information is used to generate a reconstruction, which is a representation of the components and interactions and serves as a foundation for development of mathematical models. The last two steps involve generating mathematical models that predict the behavior of the network and using the models to generate hypotheses and analyze experimental data.

cellular behaviors. This entire process requires a strong background in a number of fields, including chemistry, engineering, mathematics and biology.

3.1 *Delineation and quantitation of components*

As mentioned previously, the first step in a systems biology approach involves the identification of the components that make up the network. This includes all metabolites, proteins, and other macromolecules that constitute a cell. A number of processes operate

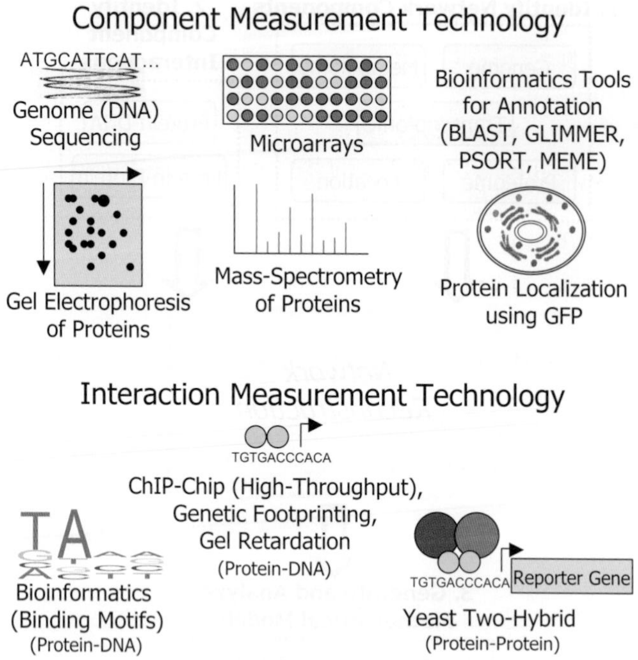

Fig. 2. Component and Interaction Measurement Technologies. A variety of different technologies have been developed that allow for the identification of components and interactions between components in biological networks (see text for more details).

together inside a cell, including metabolism, cell signaling, regulation, transcription, translation, DNA replication, and DNA repair. Each of these networks can be reconstructed independently, or in conjunction with one another. A variety of data types are available that detail, and sometimes quantify, the components of these biological networks (Fig. 2).

Genome sequencing methods have led to the sequencing of hundreds of single- and multicellular organisms. This technology, coupled with bioinformatic analyses, ultimately provides a *genome annotation*. This genome annotation gives the precise location of genes within the genome and characterizes the known or putative function of the resulting gene products. Bioinformatics has played a key role in the delineation of cellular components from DNA sequences. The computational methods developed can predict coding regions within a DNA sequence and predict gene functions based on sequence homology. Other bioinformatics tools can be used to predict transcription factor binding sites, promoter regions, and the sub-cellular localization of gene products.

Gene expression microarrays, another useful technology, have readily been applied to study cellular states. This technology quantitatively measures the expression state of a cell and determines which genes are being transcribed under a particular environmental condition or in a specific tissue. These measurements can be made on a genome-scale, where the mRNA levels for all genes in the genome can be quantified. In some cases, isolated mRNA can be used to form a library of expressed sequence tags (ESTs), which are then used to identify and localize genes in a genome.

In addition to genome sequences and mRNA expression profiles, proteomic measurements using liquid chromatography, *mass spectrometry*, and gel electrophoresis can be used to identify cellular proteins and their modification states. Proteomic analysis of the mitochondria from human cardiac myocytes,[5] for example, has allowed scientists to identify the proteins that are localized within this important organelle. Mass spectrometry can also be used to identify the cellular metabolites found inside a cell, information that is especially valuable for specifying pathway intermediates in metabolic networks. Imaging technology can be used to identify subcellular localization of proteins within the cell; green fluorescent protein (GFP) tagging of individual proteins is one way in which proteins can be visualized locally within the cell.

3.2 *Identify interactions between components*

Once all of the network components are specified, interactions between components that determine the functional state of the network are identified. In general, transcription of a gene begins with transcription factors binding the DNA in the upstream region of a gene and the subunits of the RNA polymerase assembling and binding to the promoter region. In this case, the links include both DNA-protein and protein-protein interactions. For a metabolic network, specifying the interactions can include identifying interacting protein subunits, enzyme complexes, allosteric regulators, enzyme cofactors, and metabolic reactions.

A great deal of information about the links between network components comes from the primary literature. A bottom-up approach attempts to piece together links from a large number of data sources relating to the specific and detailed measurements of different individual interactions. In contrast, a top-down approach uses measurements pertaining to the whole network to extract information about the component interactions. With advances in high-throughput technologies, there is an abundance of large-scale datasets relating to component interactions,[6] some of which are discussed below (Fig. 2).

DNA-protein interactions have traditionally been identified by performing gel retardation and DNA footprinting experiments in order to determine where proteins bind on a DNA molecule. These experiments focus on the binding events of a protein to a small segment of the genome, such as the upstream regulatory region of a single gene. But what if we wanted to measure every binding site for a protein on the entire genome? This can be accomplished by performing chromatin immunoprotection (ChIP) chip experiments.[7] This approach uses antibodies specific to a particular DNA binding protein to isolate bound DNA sequences, which are then detected by hybridization to microarrays. Such experiments have been performed to identify transcription factor binding sites in yeast[8,9] and *E. coli*,[10,11] as well as RNA polymerase binding sites in *E. coli*[12] and human,[13] as a few examples.

Protein-protein interactions can also be measured in a high-throughput fashion. Immunoprecipitation (IP) experiments provide one way to identify interactions between proteins. In this case, an antibody for one protein is used to isolate the proteins it interacts with, which are then identified using mass spectrometry (instead of a microarray). Yeast two-hybrid experiments can also be used to detect protein interactions. In these experiments, two proteins of interest are fused to two different transcription factor domains, if the two proteins interact then the transcription factor domains come together and allow for the expression of a reporter gene which can be measured. These approaches can be useful for detecting interactions between proteins in signaling cascades, enzymatic protein complexes, and regulatory interactions.

3.3 *Mathematical representation of the components and interactions*

These previously described first two steps of systems biology result in a network reconstruction, which is a representation of the network components and their interactions.[14] This reconstruction can be thought of as a biochemically, genetically, and genomically structured database and serves as a foundation from which mathematical models can be built. A reconstruction can be represented as a list or a diagram, but for modeling purposes is more useful if represented mathematically. For example, in a metabolic network the components are the enzymes, their genes, and the metabolites they act on. The component interactions are the associations between genes, enzymes, and reactions, as well as the metabolic reactions themselves. The network of reactions can be represented mathematically as a stoichiometric matrix (**S**), where each row corresponds to an individual metabolite (component) and each column to an individual reaction (interaction between components). The elements of the resulting **S** matrix are the stoichiometric coefficients of the individual reactions. Gene to protein to reaction associations (GPRs) can also be represented mathematically as a set of *Boolean logic statements*, where "and" and "or" statements are used to describe protein complexes and *isozymes*.[14]

With a reconstruction as a starting point, one can build different types of mathematical models to simulate or predict the behavior of a system. These models can be stochastic, deterministic, dynamic, or *steady-state* (see Table 1). The equations that are used to represent the relationship between the variables in the system are all governed by physical and chemical laws. Reaction kinetics, diffusion equations, and mass balance equations are examples of how system variables (such as concentrations, fluxes, and network states) can be interrelated. All cells are limited by constraints, and cells must abide by physico-chemical laws such as conservation of mass, energy, and momentum.[15] Cells are restricted by physical limitations such as

Table 1. Examples of *in silico* analysis methods.

Method	General applications and uses
Topology	Assess distribution of connections
Constraint-based	Genome-scale analysis of phenotypic traits
Kinetic simulations	Evaluate time dependencies for small networks
Stochastic	Situations describing a small number of molecules

diffusion coefficients, thermodynamic properties, and reaction rates; to generate mathematically accurate models of a cellular biochemical network, a number of such parameters are needed.

One approach to simulating genome-scale biochemical networks (e.g. metabolic, regulatory, and signaling networks) is constraint-based modeling.[15] This approach assumes a quasi steady-state and uses constraints (mass balance, enzyme capacity, and thermodynamic) to limit the behavior of the cell. In most cases, the number of constraints (i.e. equations) is less than the number of variables so a unique solution does not exist; instead, a range of allowable solutions that the cell could utilize is identified. A variety of constraint-based modeling methods can then be applied to determine the likely cellular behavior or to determine the range of all allowable behaviors.[15] The advantage to a constraint-based approach is that the needed model parameters are known; knowledge of all the enzyme and metabolite concentrations, diffusion coefficients and reaction kinetics for the network is not necessary. As such, the approach is readily scalable and can be used to simulate large genome-scale networks.

3.4 *Hypothesis generation and testing*

The purpose of a model, no matter how it is formulated, is to predict cellular behavior. For example, models have been used to predict things like gene expression (i.e. the transcriptional state of the cell), intracellular flux distributions, regulatory states (i.e. the location of bound transcription factors on the genome), nutrient requirements, nutrient uptake and secretion rates, cellular growth rates, and essential genes. Model predictions can be used to help analyze and interpret experimental data, as well as to generate new hypotheses. As diagrammed in Fig. 3, two separate but complementary approaches, a model-based and an experimental approach, are used to predict and measure cellular behavior.

Either the model and data agree, which allows us to use the model to understand how the experimental observations are achieved by the cell, or the model and data disagree. Both situations are desirable, the former indicates that the model is a good representation of the system and can be used to interpret experimental data and to

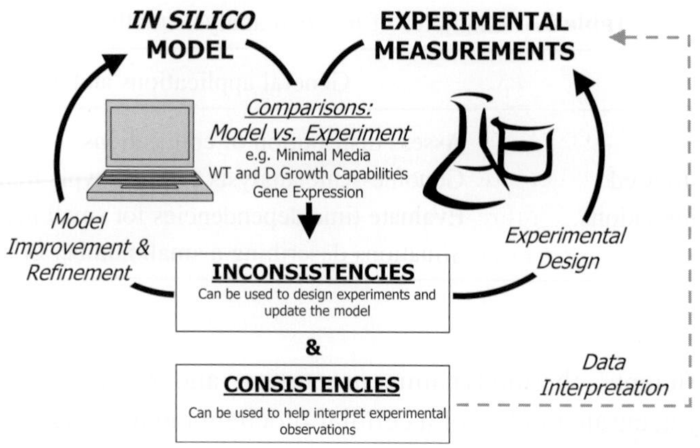

Fig. 3. Model and Data Comparisons. By comparing model predictions with experimental data consistencies and inconsistencies can be identified. The model can be used to interpret the data in the case of consistencies. With inconsistencies, the model can be used to generate hypotheses about network components and interactions, which can then be tested experimentally. Modified from Reed *et al.*[14]

provide biological insights into the behavior of the organism. The latter indicates that there is something about the system or network that is unknown, and that the model can be used to generate new hypotheses about new components of the system or new interactions between components. Either scenario, whether consistencies or inconsistencies, ultimately results in a better understanding of the biological system being studied.

Take a model of a metabolic network as an example, where the model is used to predict whether a cell can grow on a particular carbon source (such as glucose) and where experimental data is generated regarding growth of this organism on glucose. If the model and data agree, then the model can be used to describe either which metabolic pathways are used in the catabolism of glucose or why the cell cannot catabolize glucose, whichever the case may be. If the model and data disagree and the cell can grow experimentally, then the model can be used to generate hypotheses about which pathways are missing from the network reconstruction that are being used by the cell. Conversely, if the model and data disagree and the cell cannot grow experimentally, then the model can be used to identify enzymes or pathways in the model that might not be available to the cell because the required genes are not being transcribed.

As model predictions are compared to more and more experimental data, the models will evolve in an iterative fashion.[2] The models will be used to generate hypotheses and the results from the experimental testing of these hypotheses will be used to update and improve the model. The result of this cyclic and iterative process is that more accurate models will arise, but more importantly, we will simultaneously be increasing our knowledge about the chosen biological system.

4. Contemporary Examples

Models have been used to analyze and predict experimental results for a number of different applications (Table 2), including minimal medium requirements, gene essentiality, regulatory network elucidation, and metabolic engineering.

Different organisms require different nutrients in order to survive; humans for example, need to acquire nine amino acids from their diet (histidine, isoleucine, leucine, lysine, methionine, phenylalanine, threonine, tryptophan, and valine), whereas the remaining 11 can be synthesized from other nutrients. Models of metabolism can

Table 2. Using genome-scale models to drive discovery.

Genome-scale model applications
Evaluating consequences of gene knock-outs
• J. S. Edwards and B. O. Palsson, *PNAS* **97**: 5528–5533 (2000)
• D. Segre, *et al.*, *PNAS* **99**: 15112–15117 (2002).
• T. Shlomi, *et al.*, *PNAS* **102**: 7695–7700 (2005).
Evaluating synthetic lethality
• J. S. Edwards and B. O. Palsson, *J. Biol. Chem.* **274**: 17410–17416 (1999).
• C. M. Ghim, *et al.*, *J. Theor. Biol.* **237**: 401–411 (2005).
Optimal growth rates
• J. S. Edwards, *et al.*, *Nat. Biotechnol.* **19**: 125–130 (2001).
Outcomes of adaptive evolution
• J. S. Edwards, *et al.*, *Nat. Biotechnol.* **19**: 125–130 (2001)
• S. S. Fong, *et al.*, *J. Bacteriol.* **185**: 6400–6408 (2003).
• S. S. Fong and B.O. Palsson, *Nat. Genet.* **36**: 1056–1058 (2004).
• S. S. Fong, *et al.*, *Biotechnol. Bioeng.* **91**: 643–648 (2005).
Horizontal gene transfer
• C. Pal, *et al.*, *Bioinformatics* **21**(Suppl. 2): ii222–ii223 (2005).
• C. Pal, *et al.*, *Nat. Genet.* **37**: 1372–1375 (2005).
Evolution to minimal genomes
• C. Pal, *et al.*, *Nature* **440**: 667–670 (2006).
Metabolic engineering
• A. P. Burgard, *et al.*, *Biotechnol. Bioeng.* **84**: 647–657 (2003).
• P. Pharkya, *et al.*, *Biotechnol. Bioeng.* **84**: 887–899 (2003).
• P. Pharkya, *et al.*, *Genome Res.* **14**: 2367–2376 (2004).
• S. S. Fong, *et al.*, *Biotechnol. Bioeng.* **91**: 643–648 (2005).
• H. Alper, *et al.*, *Metab. Eng.* **7**: 155–164 (2005).
Gap filling and gene discovery
• J. L. Reed, *et al.*, *Genome Biol.* **4**: R54.1–R54.12 (2003).

be used to determine which biomass components (amino acids, nucleic acids, lipids, and vitamins) can be synthesized from specified starting compounds, thereby identifying which components (or metabolic precursors) have to be supplied from the environment. This approach has been used to calculate the minimal medium growth requirements for human pathogens such as *Haemophilus influenzae*[16] and *Helicobacter pylori*.[17]

Metabolic models have also been extensively used to predict the phenotypic effect of deleting a gene from the genome, with the goal of answering the question, "will the cell be able to survive without a particular gene?" In a matter of minutes one can predict, for an entire genome, which metabolic gene deletions would be lethal under a specific growth environment. Gene deletion studies have been conducted for a number of organisms including *E. coli*,[18–21] *H. pylori*,[17,22] *H. influenzae*,[16,23] *Saccharomyces cerevisiae*,[24–26] and *Staphylococcus aureus*[27] with good agreement between model predictions and experimental data. For human pathogens, such gene essentiality studies can be used to identify potential drug targets for antibiotic development.

Models of transcriptional regulation have been used to predict and analyze changes in gene expression. Since regulatory networks are generally less characterized than metabolic networks, there is an interest in the field for using high-throughput data (such as gene expression and ChIP-chip data) to back-calculate or identify the interactions in regulatory networks. Researchers are interested in what genes a transcription factor regulates and how it controls expression of these genes (i.e. by activation or repression). Analysis of gene expression data for wildtype and transcription factor knockout strains by using models based on initial regulatory reconstructions has identified new regulatory interactions in *S. cerevisiae*[28] and *E. coli*.[29] These newly identified regulatory interactions can then be incorporated into an updated regulatory model, and in an iterative fashion the regulatory model can be expanded and improved.

Since chemical synthesis of molecules can be costly, chemical and pharmaceutical companies are looking into ways of using cells to produce molecules of interest (such as ethanol) from inexpensive substrates (such as glucose). Models are currently being used to improve these processes by designing microbial strains for industrial purposes. Computational methods have been developed which can calculate the best genes to knock-out or add-in to an organism to improve production yields.[30–33] Such methods have been applied to engineer strains for the production of lactic acid[30,34] and lycopene,[33] for example.

These are just a few examples of how models have been used both to drive biological discovery and to fulfill industrial and commercial applications. Future advances in the field of systems biology will involve generating new technologies to identify and quantify network components and interactions, as well as generating new modeling methods and computational tools to analyze and predict cellular behavior.

Acknowledgments

The authors wish to thank Andrew Joyce and Ken Gratz for critical review of the text and figures in this chapter. Jennifer Reed received financial support as a Faculty Fellow at the University of California, San Diego.

References

1. B. O. Palsson, *Systems Biology: Properties of Reconstructed Networks* (Cambridge University Press, New York, 2006).
2. B. O. Palsson, The challenges of *in silico* biology, *Nat. Biotechnol.* **18**: 1147–1150 (2000).
3. P. Wong, S. Gladney and J. D. Keasling, Mathematical model of the lac operon: inducer exclusion, catabolite repression, and diauxic growth on glucose and lactose, *Biotechnol. Prog.* **13**: 132–143 (1997).
4. A. Joshi and B. O. Palsson, Metabolic dynamics in the human red cell. Part I — A comprehensive kinetic model, *J. Theor. Biol.* **141**: 515–528 (1989).
5. S. W. Taylor, E. Fahy, B. Zhang, G. M. Glenn, D. E. Warnock, S. Wiley, A. N. Murphy, S. P. Gaucher, R. A. Capaldi, B. W. Gibson and S. S. Ghosh, Characterization of the human heart mitochondrial proteome, *Nat. Biotechnol.* **21**: 281–286 (2003).
6. A. R. Joyce and B. O. Palsson, The model organism as a system: integrating "omics" data sets, *Nat. Rev. Mol. Cell Biol.* **7**: 198–210 (2006).
7. D. Sikder and T. Kodadek, Genomic studies of transcription factor-DNA interactions, *Curr. Opin. Chem. Biol.* **9**: 38–45 (2005).
8. T. I. Lee, N. J. Rinaldi, F. Robert, D. T. Odom, Z. Bar-Joseph, G. K. Gerber, N. M. Hannett, C. T. Harbison, C. M. Thompson, I. Simon, J. Zeitlinger, E. G. Jennings, H. L. Murray, D. B. Gordon, B. Ren, J. J. Wyrick, J. B. Tagne, T. L. Volkert, E. Fraenkel, D. K. Gifford and R. A. Young, Transcriptional regulatory networks in *Saccharomyces cerevisiae*, *Science* **298**: 799–804 (2002).
9. C. T. Harbison, D. B. Gordon, T. I. Lee, N. J. Rinaldi, K. D. Macisaac, T. W. Danford, N. M. Hannett, J. B. Tagne, D. B. Reynolds, J. Yoo, E. G. Jennings, J. Zeitlinger, D. K. Pokholok, M. Kellis, P. A. Rolfe, K. T. Takusagawa, E. S. Lander, D. K. Gifford, E. Fraenkel and R. A. Young, Transcriptional regulatory code of a eukaryotic genome, *Nature* **431**: 99–104 (2004).
10. D. C. Grainger, D. Hurd, M. Harrison, J. Holdstock and S. J. Busby, Studies of the distribution of *Escherichia coli* cAMP-receptor protein and RNA polymerase along the *E. coli* chromosome, *Proc. Natl. Acad. Sci. USA* **102**: 17693–17698 (2005).
11. D. C. Grainger, T. W. Overton, N. Reppas, J. T. Wade, E. Tamai, J. L. Hobman, C. Constantinidou, K. Struhl, G. Church and S. J. Busby, Genomic studies with *Escherichia coli* MelR protein: applications of chromatin immunoprecipitation and microarrays, *J. Bacteriol.* **186**: 6938–6943 (2004).
12. C. D. Herring, M. Raffaelle, T. E. Allen, E. I. Kanin, R. Landick, A. Z. Ansari and B. O. Palsson, Immobilization of *Escherichia coli* RNA polymerase and location of binding sites by use of chromatin immunoprecipitation and microarrays, *J. Bacteriol.* **187**: 6166–6174 (2005).
13. T. H. Kim, L. O. Barrera, M. Zheng, C. Qu, M. A. Singer, T. A. Richmond, Y. Wu, R. D. Green and B. Ren, A high-resolution map of active promoters in the human genome, *Nature* **436**: 876–880 (2005).
14. J. L. Reed, I. Famili, I. Thiele and B. O. Palsson, Towards multidimensional genome annotation, *Nat. Rev. Genet.* **7**: 130–141 (2006).
15. N. D. Price, J. L. Reed and B. O. Palsson, Genome-scale models of microbial cells: evaluating the consequences of constraints, *Nat. Rev. Microbiol.* **2**: 886–897 (2004).

16. C. H. Schilling and B. O. Palsson, Assessment of the metabolic capabilities of *Haemophilus influenzae* Rd through a genome-scale pathway analysis, *J. Theor. Biol.* **203**: 249–283 (2000).
17. C. H. Schilling, M. W. Covert, I. Famili, G. M. Church, J. S. Edwards and B. O. Palsson, Genome-scale metabolic model of *Helicobacter pylori* 26695, *J. Bacteriol.* **184**: 4582–4593 (2002).
18. J. S. Edwards and B. O. Palsson, The *Escherichia coli* MG1655 *in silico* metabolic genotype: its definition, characteristics, and capabilities, *Proc. Natl. Acad. Sci. USA*, **97**: 5528–5533 (2000).
19. M. Covert and B. O. Palsson, Constraints-based models: regulation of gene expression reduces the steady-state solution space, *J. Theor. Biol.* **221**: 309–325 (2003).
20. M. W. Covert and B. O. Palsson, Transcriptional regulation in constraints-based metabolic models of *Escherichia coli*, *J. Biol. Chem.* **277**: 28058–28064 (2002).
21. J. Stelling, S. Klamt, K. Bettenbrock, S. Schuster and E. D. Gilles, Metabolic network structure determines key aspects of functionality and regulation, *Nature* **420**: 190–193 (2002).
22. I. Thiele, T. D. Vo, N. D. Price and B. Palsson, An expanded metabolic reconstruction of *Helicobacter pylori* (*i*IT341 GSM/GPR): an *in silico* genome-scale characterization of single and double deletion mutants, *J. Bacteriol.* **187**: 5818–5830 (2005).
23. J. S. Edwards and B. O. Palsson, Systems properties of the *Haemophilus influenzae* Rd metabolic genotype, *J. Biol. Chem.* **274**: 17410–17416 (1999).
24. N. C. Duarte, M. J. Herrgard and B. Palsson, Reconstruction and validation of *Saccharomyces cerevisiae* iND750, a fully compartmentalized genome-scale metabolic model, *Genome Res.* **14**: 1298–1309 (2004).
25. J. Forster, I. Famili, B. O. Palsson and J. Nielsen, Large-scale evaluation of *in silico* gene knockouts in *Saccharomyces cerevisiae*, *Omics* **7**: 193–202 (2003).
26. L. Kuepfer, U. Sauer and L. M. Blank, Metabolic functions of duplicate genes in *Saccharomyces cerevisiae*, *Genome Res.* **15**: 1421–1430 (2005).
27. S. A. Becker and B. O. Palsson, Genome-scale reconstruction of the metabolic network in *Staphylococcus aureus* N315: an initial draft to the two-dimensional annotation, *BMC Microbiol.* **5**: 8 (2005).
28. T. Ideker, V. Thorsson, J. A. Ranish, R. Christmas, J. Buhler, J. K. Eng, R. Bumgarner, D. R. Goodlett, R. Aebersold and L. Hood, Integrated genomic and proteomic analyses of a systematically perturbed metabolic network, *Science* **292**: 929–934 (2001).
29. M. W. Covert, E. M. Knight, J. L. Reed, M. J. Herrgard and B. O. Palsson, Integrating high-throughput and computational data elucidates bacterial networks, *Nature* **429**: 92–96 (2004).
30. A. P. Burgard, P. Pharkya and C. D. Maranas, Optknock: a bilevel programming framework for identifying gene knockout strategies for microbial strain optimization, *Biotechnol. Bioeng.* **84**: 647–657 (2003).
31. P. Pharkya, A. P. Burgard and C. D. Maranas, OptStrain: a computational framework for redesign of microbial production systems, *Genome Res.* **14**: 2367–2376 (2004).
32. D. Segre, D. Vitkup and G. M. Church, Analysis of optimality in natural and perturbed metabolic networks, *Proc. Natl. Acad. Sci. USA* **99**: 15112–15117 (2002).
33. H. Alper, Y. S. Jin, J. F. Moxley and G. Stephanopoulos, Identifying gene targets for the metabolic engineering of lycopene biosynthesis in *Escherichia coli*, *Metab. Eng.* **7**: 155–164 (2005).
34. S. S. Fong, A. P. Burgard, C. D. Herring, E. M. Knight, F. R. Blattner, C. D. Maranas and B. O. Palsson, *In silico* design and adaptive evolution of *Escherichia coli* for production of lactic acid, *Biotechnol. Bioeng.* **91**: 643–648 (2005).

Additional Recommended Reading

A. R. Joyce and B. O. Palsson, The model organism as a system: integrating "omics" data sets, *Nat. Rev. Mol. Cell. Biol.* **7**: 198–210 (2006).

B. O. Palsson, *Systems Biology: Properties of Reconstructed Networks* (Cambridge University Press, New York, 2006), p. 322.

N. D. Price, J. L. Reed and B. O. Palsson, Genome-scale models of microbial cells: evaluating the consequences of constraints, *Nat. Rev. Microbiol.* **2**: 886–897 (2004).

J. L. Reed, I. Famili, I. Thiele and B. O. Palsson, Towards multidimensional genome annotation, *Nat. Rev. Genet.* **7**: 130–141 (2006).

Definitions

Boolean logic statements: Logic statements using "and" and "or" to describe the relationship between components. For example, Gene A is expressed if Transcription Factor B is active.

Genome annotation: A description of genes in a genome, including their genome location (start and stop codon), function, and cellular location.

Isozymes: Two or more enzymes that catalyze the same reaction(s).

Mass spectrometry: Separates molecules based on their mass-to-charge ratio and can be used to quantify the relative abundances of the different molecules. The method can also be used for identifying compounds.

Microarray: Typically used to measure mRNA concentrations, it contains an array of spots with short nucleotide sequences which can hybridize to complementary strands of mRNA molecules.

Model: A set of mathematical equations used to predict the behavior of a system and which account for the information in a network reconstruction.

Network reconstruction: A detailed, summary or accounting of the components and component interactions in a network.

Steady-state: Assumes that variables do not change with time, for a metabolic network this means concentrations and fluxes are independent of time and will stay fixed.

Stoichiometric coefficient: The coefficient in a metabolic reaction for a particular metabolite. For example, if two molecules of A are consumed to make one molecule of B, the stoichiometric coefficients for A and B would be –2 and 1, respectively.

CHAPTER 24

BIOINFORMATICS AND SYSTEMS BIOLOGY: OBTAINING THE DESIGN PRINCIPLES OF LIVING SYSTEMS

Shankar Subramaniam

Abstract

All living systems obey the principal laws of nature, including the laws of thermodynamics. However, an important feature characterizes living systems. Because of the need for living systems to self-organize and self-evolve, the coupling between multiple time scales or multiple length scales are non-hierarchical. For instance, the absorption of a photon with a primary event happening in femptosecond time scales leads to long time-scale processes to result in vision. Events that happen in seconds and minute time scales give rise to developmental processes that happen in days to weeks. Also processes that can be traced to events in a single cell can result in a systemic response spanning entire physiology. The blueprint for continuous adaptation, error-checking and optimization of the system is built into living systems such that they sample infinite number of states with no two states being identical. Yet the end-point physiology is often similar or in some cases identical. This implies that multiple solutions lead to nearly the same optimality and behavior of the system. In this chapter we will explore the features of living systems from an engineering and design perspective and attempt to identify methods we can use to decipher the rules that govern living systems.

1. Introduction

Self-reproduction, self-evolution and self-organization are hallmarks of living systems. Utilizing or converting available resources living systems use blueprints to reproduce themselves but continue to slightly change the blueprint to adapt to the pressures exerted by the environment. This entire process also involves self-reorganization to accommodate the changes in both the environment and the blueprint for reproduction. While the only apparent "intelligence" of this design has been proven time and again to be merely a consequence of natural selection according to Darwinian principles, there is nothing preventing us from reverse engineering living systems in order to understand the design principles that govern living systems in defined environments.

All living systems obey the principal laws of nature including the laws of thermodynamics. However, an important feature characterizes living systems. Because of the need for living systems to self-organize and self-evolve, the coupling between multiple time scales or multiple length scales are non-hierarchical and cannot be scaled up or down like in physical systems. For instance, the absorption of a photon with a primary event happening in femptosecond time scales gives couples to long time processes leading to vision. Events that happen in seconds and minute time scales give rise to developmental processes that happen in days to weeks. Also processes that can be traced to events in a single cell can result in a systemic response spanning entire physiology. The blueprint for continuous adaptation, error-checking and optimization of the system is built into living systems such that they sample infinite number of states with no two states being identical. Yet the end-point physiology is often similar or in some cases identical. This implies that multiple solutions lead to nearly the same optimality and behavior of the system. In this chapter we will explore from an engineering and design perspective features of living systems and attempt to identify methods we can use.

2. The Coding Problem — A Blueprint — How do we know?

For a species to consistently reproduce and function, there has to be a well-defined blueprint. Early in the 20th century, Alan Turing regarded as the father of the computer postulated that: *"There exists an abstract universal computer whose repertoire includes any computation that any physically possible object can perform."* This notion of a computer was based on binary arithmetic where any discrete problem could be mapped using truth tables involving 0's and 1's into a computable algorithm. The code of life is not very different from this principle. The essential blueprint for life is contained in the genome that from an information point of view is a long string with four alphabets (analogous to mapping to quarternary code in Turing machines involving 0, 1, 2 and 3). The human genome contains three billion elements where each element has one of four alphabets A, T, G or C (actually these alphabets stand for the base pairs that are the distinguishing part of the sugar phosphate-nucleic acid polymer) and the specific combination of these alphabets provide the code for all of physiology from reproduction to function to health and disease (see Box 1 for some basic biology details). The genomes of living systems have higher order structure. For instance, at the simplest level the DNA that constitutes the genome is double-stranded. This enables one strand to serve as a template in replication (making a copy) from individual components (nucleotides). As you may have guessed, replication is essentially a template-driven process. The pairing of the strands is driven by a very

Box 1: Mapping a Genome Sequence

The location of a gene in a chromosome is designated as it locus. A genetic linkage map of a chromosome is a picture showing the order and relative distance among genes (loci). This map does not provide the actual distance in terms of base pairs.

A physical map can tell the location of certain markers, which are precisely known small sequences, within 10^4 base pairs or so. This map provides the actual distance in base pairs between genes or markers.

Small segments of DNA up to 10^3 base pairs can be sequenced directly by modern sequencing techniques and is often referred to as a sequence (or a sequence map).

Chromosome — Genetic linkage map (works on 10^7–10^8 bp range)
Clones — Physical map (works on 10^5–10^6 bp range)
Sequence — Sequence maps (works on 10^3–10^4 bp range)

specific coupling of the alphabets where A and T are paired and where G and C are paired. This pairing specificity, which is a result of optimal chemical interactions, produces unique and interesting consequences.

The sequencing of the human genome was a major scientific endeavor of the laterpart of the 20th century (Box 1). Engineering and information technology played a major role in the sequencing and decoding of the genome. Once the DNA from human cells was isolated the task was to cut, copy, read and assemble large portions of the genome (see Box 2 for details of genome sequencing and Fig. 1 for the high throughput sequencing assembly at the Whitehead Institute engaged in human genome sequencing).

The problem of genome assemble is a fundamental problem in bioinformatics. How can you take strings of alphabets and decipher how to join these smaller strings to make larger strings? This is a very hard problem and comes under the category of what computer scientists call the NP (Non Polynomial order) hard problem. This means that for computing the problem does not scale as some power of N, which is the dimension of the problem. The heuristic way to do this is to use some bootstrapping strategy to associate some number of strings with each other to form larger strings and keep doing this process until very large assemblies are formed. How do you associate two strings as being similar? How similar should they be before they can be considered to be part of the same large string? Note that if the experimental process of sequencing introduces errors, because of which the same sequence can display some minor variations! The method of string comparisons or string alignment is

Box 2: The Mapping Procedure

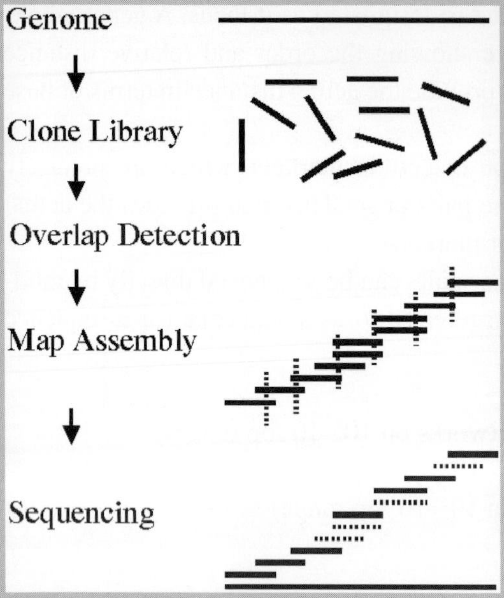

The goal of the Human Genome Project is the determination of the location of the human genes on the DNA and the sequencing (the determination of the sequence of nucleotides A, C, G and T) of the entire human genome. In total, the human genome contains about three billion letters distributed over 23 chromosomes. Maps that reflect the actual distance in base pairs are called *physical maps*. Physical maps are powerful tools for localization and isolation of genes, studying the organization and evolution of genomes and as a preparatory step for efficient sequencing.

The procedure of physical mapping coarsely divides into two steps (see figure): First large pieces of DNA (contigs), a library of cloned fragments are ordered according to their position in the genome. Different experimental techniques are used to do that. Roughly, these are clone-probe hybridization mapping, restriction mapping, radiation-hybrid mapping and optical mapping. On the right you see mapping using hybridization data. Second the cloned fragments are cut by restriction enzymes, smaller DNA fragments are obtained which are sequenced in detail (shotgun-sequencing), and the overall sequence in detail is obtained by sequence assembly.

an old problem very similar to the problem of a traveling salesman. Given a number of cities and the costs of traveling from any city to any other city, what is the cheapest round-trip route that visits each city once and then returns to the starting city? This comes under a class of problems involving discrete or combinatorial optimization. The most rigorous method for solving this is using dynamic programming. Box 3 provides an introduction to sequence alignment.

It is usually used to study the evolution of the sequences from a common ancestor, especially biological sequences such as protein sequences or DNA sequences. Mismatches in the alignment correspond to mutations, and gaps correspond to insertions or deletions. Sequence alignment can also be used to study things like the

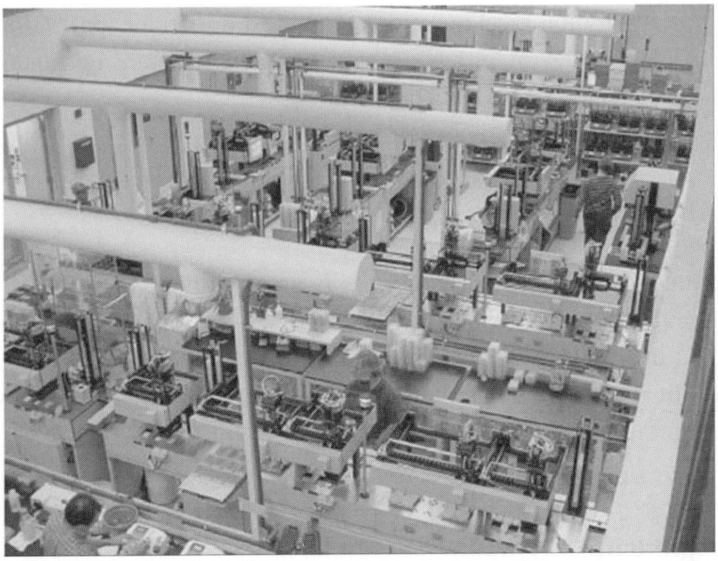

Fig. 1. Automated sequencing machines at the Center for Genome Research in the Whitehead Institute.

Box 3: Sequence Alignment

From Wikipedia, the free encyclopedia:

Sequence alignment is an arrangement of two or more sequences, highlighting their similarity. The sequences are padded with gaps (usually denoted by dashes) so that wherever possible, columns contain identical or similar characters from the sequences involved:

```
tcctctgcctctgccatcat---caaccccaaagt
|||| ||| ||||| |||||   ||||||||||||
tcctgtgcatctgcaatcatgggcaaccccaaagt
```

evolution of languages and the similarity between texts. The term sequence alignment may also refer to the process of constructing such an alignment or finding significant alignments in a database of potentially unrelated sequences.

Pairwise sequence alignment methods are concerned with finding the best-matching piecewise (local) or global alignments of protein (amino acid) or DNA (nucleic acid) sequences. Typically, the purpose of this is to find homologues (relatives) of a gene or gene-product in a database of known examples. This information is useful for answering a variety of biological questions. The most important application of pairwise alignment is identification of sequences of unknown structure or function. Another important use is the study of molecular evolution.

3. Global Alignment

A global alignment between two sequences is an alignment in which all the characters in both sequences participate in the alignment. Global alignments are useful mostly for finding closely related sequences. As these sequences are also easily identified by local alignment methods, global alignment is now somewhat deprecated as a technique.

4. Local Alignment

Local alignment methods find related regions within sequences — in other words they can consist of a subset of the characters within each sequence. For example, positions 20 – 40 of sequence A might be aligned with positions 50 – 70 of sequence B. This is a more flexible technique than global alignment and has the advantage that related regions, which appear in a different order in the two proteins (which is known as domain shuffling) can be identified as being related. This is not possible with global alignment methods.

5. Significance of Alignment

The typical assumption in the use of alignment is the mechanism of molecular evolution. DNA carries over genetic material from generation to generation, by virtue of its semi-conservative duplication mechanism. Changes in the material are introduced by occasional errors and mutations in the duplication, and by viruses and other mechanisms, which sometimes move sub-sequences within the chromosome and between individuals. Consequently, an alignment between sequences indicates that the sequences evolved from a common ancestor, which contained the matching subsequences. In the case of genetic sequences, it implies that their carriers evolved from a common ancestor. The actual biological meaning of any alignment can never be absolutely guaranteed. However, statistical methods can be used to assess the likelihood of finding an alignment between two regions (or sequences) by chance, given the size of the database and its composition.

Two important related issues for sequence alignment are:

1. How should we choose the best alignment between two sequences (or regions of sequences)?

2. How do we rank the alignments between a query and the many sequences in the database, according to their significance in the domain (such as biological significance)?

In studying evolution, the first question can be addressed by developing a model of how likely are certain changes between characters in the sequence. There are many ways to do this, none of which is generally superior. The models are derived empirically using related sequences, and are expressed as substitution matrices. These matrices, along with gap penalties, are used by the algorithms named below to evaluate alternative alignments between two sequences. The actual biological quality of the alignments then depends upon the evolutionary model used to generate the score. The second question is purely statistical. Research has determined a few hard theoretical rules and many approximations. It is now generally accepted that the scores of alignments between random sequences follow the extreme value distribution.

6. Multiple Alignment

Multiple alignment is an extension of pairwise alignment to incorporate more than two sequences into an alignment. Multiple alignment methods try to align all of the sequences in a specified set. Alignments help in the identification of common regions between the sequences. There are several approaches to creating multiple sequence alignments, one of the most popular being the progressive alignment strategy used by the Clustal family of programs. Clustal is used in cladistics to build phylogenetic trees, and to build sequence profiles which are used by PSI-BLAST and Hidden Markov model (HMM) methods to search sequence databases for more distant relatives. Multiple sequence alignment is computationally difficult and is classified as an NP-Hard problem.

In the majority of alignments methods like dynamic programming are computationally very intensive. Often simple heuristic methods based on a hashing scheme are employed. Box 4 below provides an insight into the popular heuristic, BLAST, algorithm.

7. Algorithm

To run, BLAST requires two sequences as input: a query sequence (also called the target sequence) and a sequence database. BLAST will find subsequences in the query that are similar to subsequences in the database. In typical usage, the query sequence is much smaller than the database, e.g. the query may be one thousand nucleotides while the database is several billion nucleotides.

Box 4: BLAST

From Wikipedia, the free encyclopedia:

In bioinformatics, **B**asic **L**ocal **A**lignment **S**earch **T**ool, or **BLAST**, is an algorithm for comparing biological sequences, such as the amino-acid sequences of different proteins or the DNA sequences. Given a library or database of sequences, a *BLAST search* enables a researcher to look for sequences that resemble a given sequence of interest. For example, following the discovery of a previously unknown gene in the mouse, a scientist will typically perform a BLAST search of the human genome to see if human beings carry a similar gene; BLAST will identify sequences in the human genome that resemble the mouse gene based on similarity of sequence.

BLAST is one of the most widely used bioinformatics programs, probably because it addresses a fundamental problem and the algorithm emphasizes speed over sensitivity. This emphasis on speed is vital to making the algorithm practical on the huge genome databases currently available, although subsequent algorithms can be even faster.

Examples of other questions that researchers use BLAST to answer are:

- Which bacterial species have a protein that is related in lineage to a certain protein whose amino-acid sequence I know?
- Where does the DNA that I've just sequenced come from?
- What other genes encode proteins that exhibit structures or motifs such as the one I've just determined?

The original paper "S. F. Altschul, W. Gish, W. Miller, E. W. Myers and D. J. Lipman, Basic local alignment search tool, *J. Mol. Biol.* **215**(3): 403–410 (1990)" was the most highly cited paper published in the 1990s.

BLAST searches for high scoring sequence alignments between the query sequence and sequences in the database using a heuristic approach that approximates the Smith-Waterman algorithm. The exhaustive Smith-Waterman approach is too slow for searching large genomic databases such as GenBank. Therefore, the BLAST algorithm uses a heuristic approach that is slightly less accurate than Smith-Waterman but over 50 times faster. The speed and relatively good accuracy of BLAST are the key technical innovations of the BLAST programs and arguably why the tool is the most popular bioinformatics search tool.

The BLAST algorithm can be conceptually divided into three stages. In the first stage, BLAST searches for short matches of a fixed length W between the query and sequences in the database. For example, given the sequences AGTTAC and ACTTAG and a word length $W = 3$, BLAST would identify the matching substring TTA that is common to both sequences. In the second stage, BLAST performs an ungapped alignment between the query and database sequence if they share a common word. The ungapped alignment process extends the initial match of length W in each direction in an attempt to boost the alignment score. Insertions and deletions are not considered during this stage. For our example, the ungapped alignment between the sequences AGTTAC and ACTTAG centered around the common word TTA would be:

```
AGTTAC
| |||

ACTTAG
```

If a high-scoring ungapped alignment is found, the database sequence is passed on to the third stage. In the third stage, BLAST performs a gapped alignment between the query sequence and the database sequence using a variation of the Smith-Waterman algorithm. Statistically significant alignments are then displayed to the user.

The complete genomes of a few hundred microorganisms and several eukaryotic species have now been sequenced and the information on the sequences is now available in databases. The genomes only contain the blueprints for living systems. To decode the structure and function of living systems it is essential to map all the macromolecules, their organization and dynamics. The genomes contain genes that code for specific proteins or other nucleic acid molecules that are functional units in living systems. To a large extent the proteins are the workhorses of living systems. The central dogma of biology which was based on painstaking experimental work on living systems states that DNA codes for RNA (another template) which in turn codes for proteins (the macromolecules that are a polymer of amino acid units). The making of RNA using the DNA template is called the transcription process while the synthesis of proteins using the RNA code is called the translation process. Unlike DNA and RNA that have four alphabets, proteins contain 20 alphabets which provide the rich repertoire of proteins that can carry out myriad functions. Again, in order for consistent replication and modular organization of life, the code has to be unique for each amino acid. In the RNA and hence in the DNA a triplet of nucleotides form a unit called the codon and each codon codes for a specific amino acid. You can now use simple arithmetic to figure out why a triplet is needed for a template constituted with a combination of only four alphabets and specified pairing between two pairs of alphabets is needed for creating a polymer constituted by a combination of 20 possible alphabets. The codon table is called universal, even though some organisms can have slight differences in the code. Figure 2 provides the codon table for eukaryotic species. Our knowledge of genes and proteins comes largely from comparative analysis of the respective sequences between species.

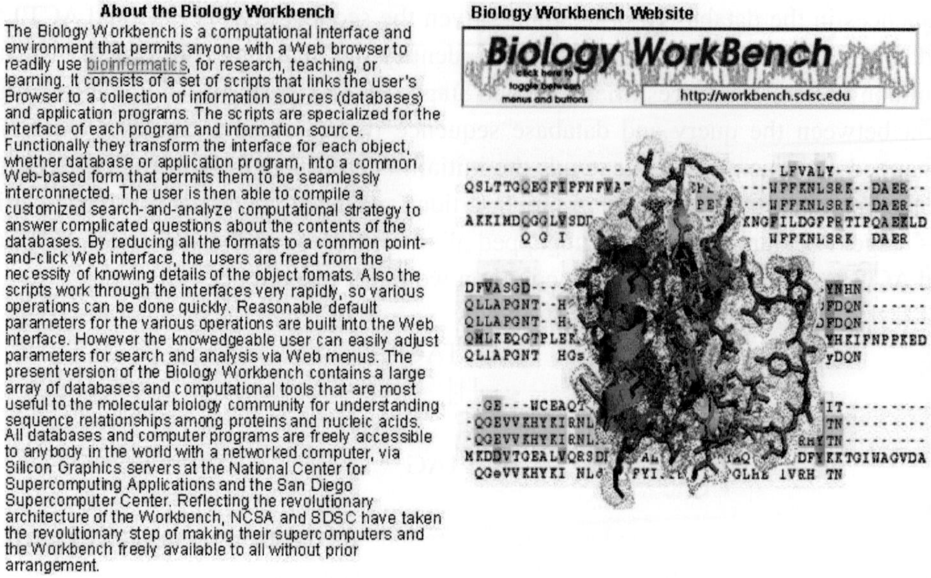
About the Biology Workbench
The Biology Workbench is a computational interface and environment that permits anyone with a Web browser to readily use bioinformatics, for research, teaching, or learning. It consists of a set of scripts that links the user's Browser to a collection of information sources (databases) and application programs. The scripts are specialized for the interface of each program and information source. Functionally they transform the interface for each object, whether database or application program, into a common Web-based form that permits them to be seamlessly interconnected. The user is then able to compile a customized search-and-analyze computational strategy to answer complicated questions about the contents of the databases. By reducing all the formats to a common point-and-click Web interface, the users are freed from the necessity of knowing details of the object fomats. Also the scripts work through the interfaces very rapidly, so various operations can be done quickly. Reasonable default parameters for the various operations are built into the Web interface. However the knowedgeable user can easily adjust parameters for search and analysis via Web menus. The present version of the Biology Workbench contains a large array of databases and computational tools that are most useful to the molecular biology community for understanding sequence relationships among proteins and nucleic acids. All databases and computer programs are freely accessible to anybody in the world with a networked computer, via Silicon Graphics servers at the National Center for Supercomputing Applications and the San Diego Supercomputer Center. Reflecting the revolutionary architecture of the Workbench, NCSA and SDSC have taken the revolutionary step of making their supercomputers and the Workbench freely available to all without prior arrangement.
Biology Workbench Website
Biology WorkBench
click here to toggle between menus and buttons
http://workbench.sdsc.edu

Fig. 2. Biology workbench.

Once again bioinformatics methods such as BLAST enable us annotate the similarities or otherwise amongst proteins.

With the sequencing of complete genomes of hundreds of living systems, researchers need complex informatics infrastructures that can seamlessly integrate data from multiple databases, perform comparative analysis and present the results in a biologist-friendly manner. This was an early challenge in bioinformatics. With the advent of the World Wide Web, user-friendly architectures began to emerge. The Biology Workbench is the harbinger of such infrastructures. Figure 2 provides information on the biology workbench and its use: (http://workbench.sdsc.edu; http://www.bioquest.org/bioinformatics/module/tutorials/How3.2/).

8. Energy Needs in Design and Life Forms — Harnessing Energy and Computing Energy Needs

The very basic principle that governs living systems follows the same principles that govern the universe at large, i.e. conversion of one form of energy into another. The two grand strategies of life are photosynthesis and metabolism (Fig. 3).

In living plants, the energy flow through the system is supplied principally by solar radiation. In fact, leaves provide relatively large surface areas per unit volume for most plants, allowing them to "capture" the necessary solar energy to maintain themselves far from equilibrium. This solar energy is converted into the necessary useful work to maintain the plant in its complex, high-energy configuration by a complicated process

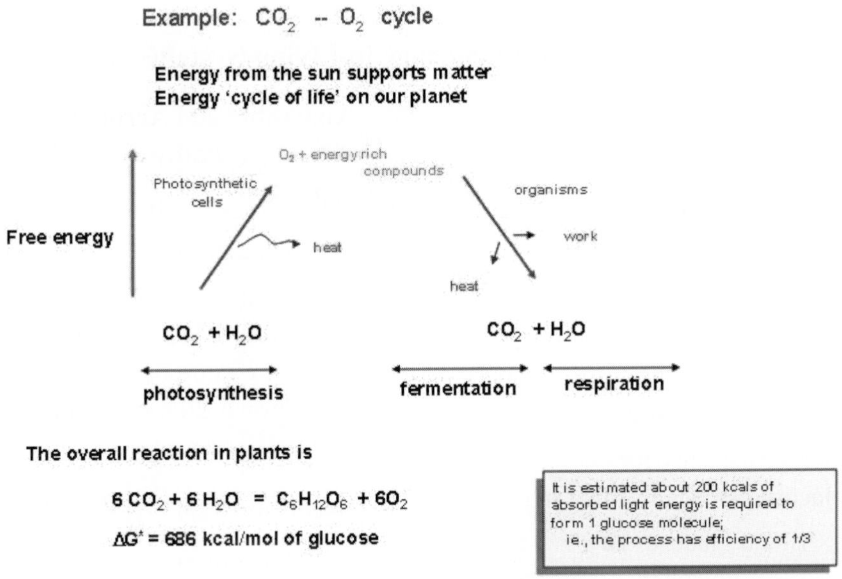

Fig. 3. Grand strategies of living systems.

called photosynthesis. Mass, such as water and carbon dioxide, also flows through plants, providing necessary raw materials, but not energy. In collecting and storing useful energy, plants serve the entire biological world (see Box 5 for details on Energy Storage in Animals).

For animals, energy flow through the system is provided by eating high energy biomass, either plant or animal. The breaking down of this energy-rich biomass, and the subsequent oxidation of part of it (e.g. carbohydrates), provides a continuous source of energy as well as raw materials. If plants are deprived of sunlight or animals of food, dissipation within the system will result in death. Maintenance of the complex, high-energy condition associated with life is not possible apart from a continuous source of energy. A source of energy alone is not sufficient, however, to explain the origin or maintenance of living systems. The additional crucial factor is *a means of converting this energy* into the necessary useful work to build and maintain complex living systems from the simple biomonomers that constitute their molecular building blocks.

With the knowledge of the genome and the use of bioinformatics methods in deciphering steps in biochemical pathways that lead to a function such as energy metabolism, we are now in a position to calculate more precisely the energetics associated with cellular function. Such approaches form the basis for systems biology. Let us consider the energetics associated with mitochondrial metabolism. Winslow and his co-workers created a detailed reaction model of cardiac mitochondrial bioenergetics represented below in Fig. 4, which can be translated into a mathematical model for calculation of energy changes associated with metabolism. Since there are several coupled reactions, we can use standard tools of thermodynamics (recall free energy) and equations involving equilibrium calculations (e.g. Nernst equation).

Box 5: Energy Storage in Living Systems

Energy Sources

A. Carbohydrate

1. Energy yield – 4 kcal/gram
2. Types of carbohydrate:
   - Monosaccharide (glucose) vs. disaccharide (sucrose) vs. polysaccharides (glycogen, starch, cellulose)
3. Storage capacity for glucose/glycogen (70 kg male)
   - Blood glucose: 10 g (40 kcal)
   - Liver glycogen: 60 g (240 kcal)
   - Muscle glycogen: 350 g (1400 kcal)

B. Fat: C_xH_x-COOH

1. Energy yield – 9 kcal/gram
2. Types of fat:
   - Fatty acids vs. triglycerides
   - Phospholipids and steroids
3. Storage capacity of fat:
   - Adipocytes (fat cells): 14 kg (107,800 kcal)
   - Muscle: 0.5 kg (3850 kcal)

C. Protein: R-CH-COOH-NH_2

1. Energy yield – 4 kcal/gram
2. Components of proteins: amino acids (20): valine, leucine, alanine, etc.
3. Usually less than 15% of the energy required for exercise comes from protein

Anaerobic and Aerobic Energy Pathways

A. ATP-PC System: Immediate Energy

1. Breakdown of phosphocreatine by creatine kinase (enzyme)
2. Formation of ATP from ADP
3. Does not require oxygen
4. Energy for 1–20 seconds of maximal exercise

B. Glycolysis: Rapid ATP Production

1. Breakdown of glucose or glycogen to yield molecules of pyruvate or lactate
2. Yields two ATP per glucose molecule
3. Can occur without oxygen
4. Takes place in the cytoplasm
5. When lactate is produced, muscle and blood pH falls (acidosis) due to lactic acid

C. Aerobic ATP Production

1. Conversion of pyruvate to acetyl CoA, ultimately into CO_2 and H_2O. Yields 36 ATP per glucose molecule
2. Occurs within the mitochondria
3. Occurs only in the presence of O_2
4. Involves the interaction of:
   - Krebs (citric acid) cycle — forms carbon dioxide and electrons
   - Electron transport chain — Oxidative phosphorylation — uses oxygen and forms water

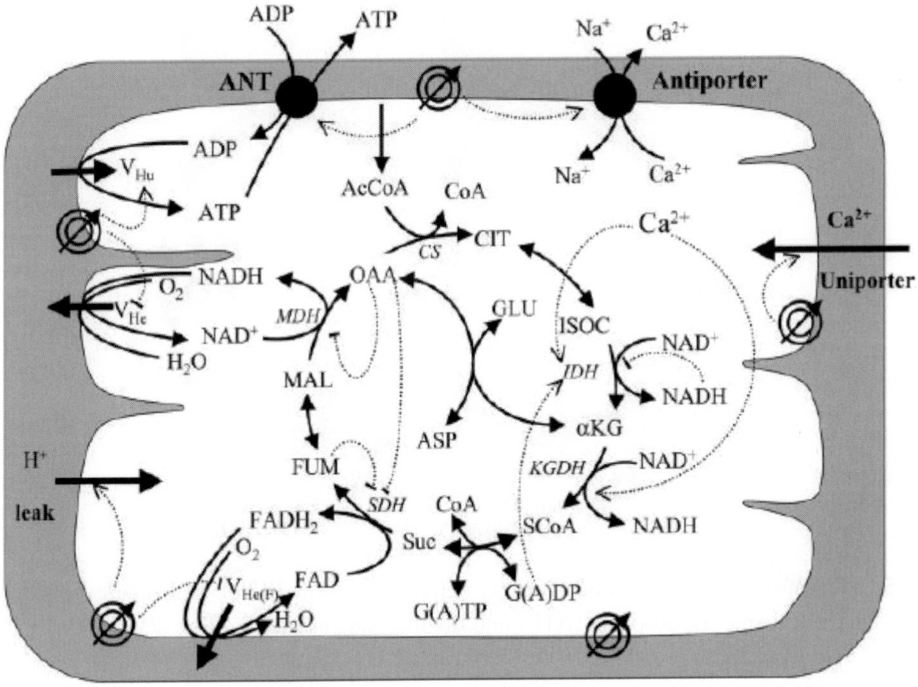

Fig. 4. Cardiac mitochondrial energy metabolism model from Winslow.[4]

Most metabolic processes in the body use **chemical energy**, which is held in the bonds between the atoms of molecules and is released when these bonds are broken. The muscles are no different. When muscles work, they require energy so that they can contract. The unique feature about muscular contraction is that the chemical energy is transformed into mechanical energy — movement. Although extended muscle activity depends on the provision of important nutrients such as carbohydrates, fats and even protein, the basic source of chemical energy for muscle contraction is **adenosine triphosphate** or **ATP**, which has the following basic chemical formula:

$$\text{Adenosine} - PO_4 \sim PO_3 \sim PO_3$$

The bonds attaching the last two phosphate radicals to the molecule, designated by the symbol ~, are so-called high energy phosphate bonds. When these bonds are broken (and new bonds are formed), a large amount of chemical energy is released (Fig. 5). In fact, each of these bonds stores about 11,000 calories of energy (per mole of ATP) in the body. (A **mole** is equivalent to Avagadro's number — 6.02 × 10 — of molecules.) Therefore, when one phosphate radical is removed from one mole (6.02 ×10 molecules), 11,000 calories of energy that can be used to energize the muscle contraction process are released. Then, when the second phosphate radical is removed, still another 11,000 calories become available. Removal of the first phosphate converts the ATP into **adenosine diphosphate**

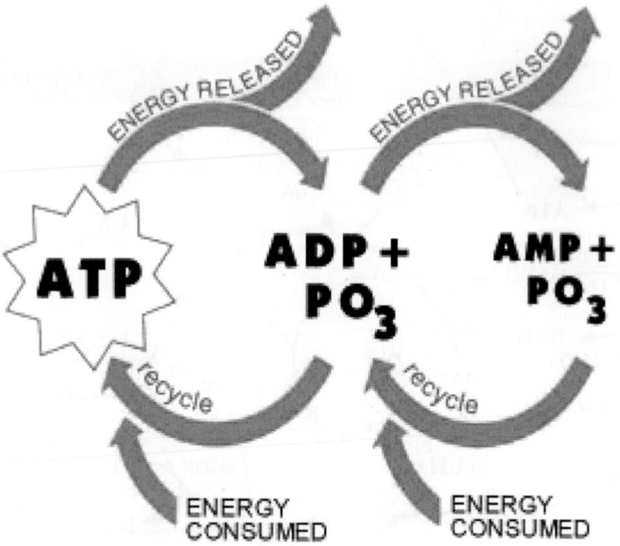

Fig. 5. Transformation of chemical energy released by removal of high energy phosphate bonds from adenosine triphsophate (ATP) into the mechanical energy of muscle contraction.

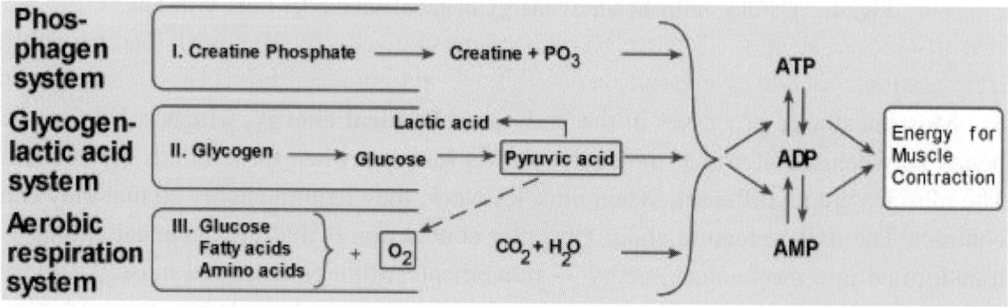

Fig. 6. The three important metabolic systems that supply energy for muscle contraction: (I) the phosphagen energy system, (II) the glycogen-lactic acid system, and (III) the aerobic system.

or **ADP** and removal of the second converts this ADP into **adenosine monophosphate** or **AMP**.

The left-hand side of Fig. 6 shows the three different metabolic mechanisms that are responsible for recycling AMP and ADP back into ATP in order to provide a continuous supply of ATP in the muscle fibers. Why do we have three different metabolic systems in our bodies to produce the ATP? Well, each one serves a different metabolic need that we may have depending on the level of movement or activity that we participate in. The longer and more intense the muscular activity, the greater is our need to supply ATP more rapidly to those muscles. Let us examine how our bodies provide this important fuel.

Unfortunately, the amount of ATP that is present in the muscle cells, even in the well-trained athlete, is only sufficient to sustain maximal muscle power for five or six seconds, maybe enough for a 50-meter dash. Therefore, except for a few seconds at a time, it is essential that **new ATP** be formed continuously, particularly during the performance of athletic events. However, different kinds of athletic events require different amounts of energy. Some athletic events such as a 100-meter dash, weight lifting or certain football plays require a quick "burst" of energy. In this case, the amount of ATP present in the muscle cells is not enough to sustain the muscle power that is needed. Therefore, the body must rely on the **phosphagen system**, which utilizes a substance called **creatine phosphate** (which also contains a high energy phosphate bond) to recycle ADP back into ATP to provide fuel for our muscles (Fig. 6). This is accomplished when the high energy phosphate bond in creatine phosphate breaks down and releases phosphate and energy. The phosphate molecule and the energy then combine with ADP to form ATP. In simple terms, the reaction goes like this:

$$\text{Creatine} \sim \text{PO}_4 \rightarrow \text{Creatine} + \text{PO}_4 + \text{Energy}$$
$$\text{PO}_4 + \text{Energy} + \text{ADP} \rightarrow \text{ATP}$$

Therefore, ATP can be produced using the phosphagen system so that our bodies can sustain about 10 to 15 more seconds of muscle activity. This is all fine for short, quick, intense bursts of energy. But the significant energy producing mechanisms in the body, those that allow us to use our muscles for longer and more intense periods, require the breakdown of the sugars in our bodies, glucose. The complete breakdown of glucose, which is needed to supply ATP during heavy muscular activity, occurs in two steps. The first stage of glucose utilization is called anaerobic because oxygen is not used. The second step, when physical activity becomes even more vigorous, is called aerobic respiration (aerobic = occurring in the presence of oxygen) because oxygen is used to produce even more energy. Certain athletic events, such as longer track events (200- or 400-meter dash events and longer), require extra ATP to fuel the muscles. During these times of heavy physical activity, glucose molecules come to the rescue. The extra ATP that is needed by the muscles is provided by a system that breaks down glucose in the absence of oxygen. This system — the glycogen-lactic acid system — is the anaerobic step of glucose breakdown (Fig. 7). In this step, each glucose molecule is split into two pyruvic acid molecules, and energy is released to form several ATP molecules providing about 30 to 40 seconds of maximal muscle activity in addition to the 10 to 15 seconds provided by the phosphagen system. The pyruvic acid will then partly break down further to produce lactic acid. If the lactic acid is allowed to accumulate in the muscle, you will experience muscle fatigue (often characterized by pain such as cramps).

The aerobic system in the body is utilized for those sports that require an extensive expenditure of energy, such as a marathon race or cross-country skiing. Lots of

ATP must be provided to your muscles in order to sustain the muscle power that you need to perform such events without excessive production of lactic acid. Therefore, in the presence of oxygen, pyruvic acid breaks down into carbon dioxide (CO_2), water, and energy by way of a very complex series of reactions known as the citric acid cycle, providing essentially unlimited time (as long as nutrients in the body last) to continue muscle utilization (Fig. 7). As you are aware, oxygen travels around the body in the blood vessels by attaching to the hemoglobin of the red blood cells. However, in the muscles, some oxygen can be stored in a special chemical substance found within the muscle fibers called myoglobin (remember, myo = refers to muscle). The myoglobin is similar to hemoglobin in its oxygen-binding capability but it can only store small amounts of oxygen. Therefore, for very heavy, sustained exercise, new oxygen must be provided from outside the body if you expect to keep working. For more details on

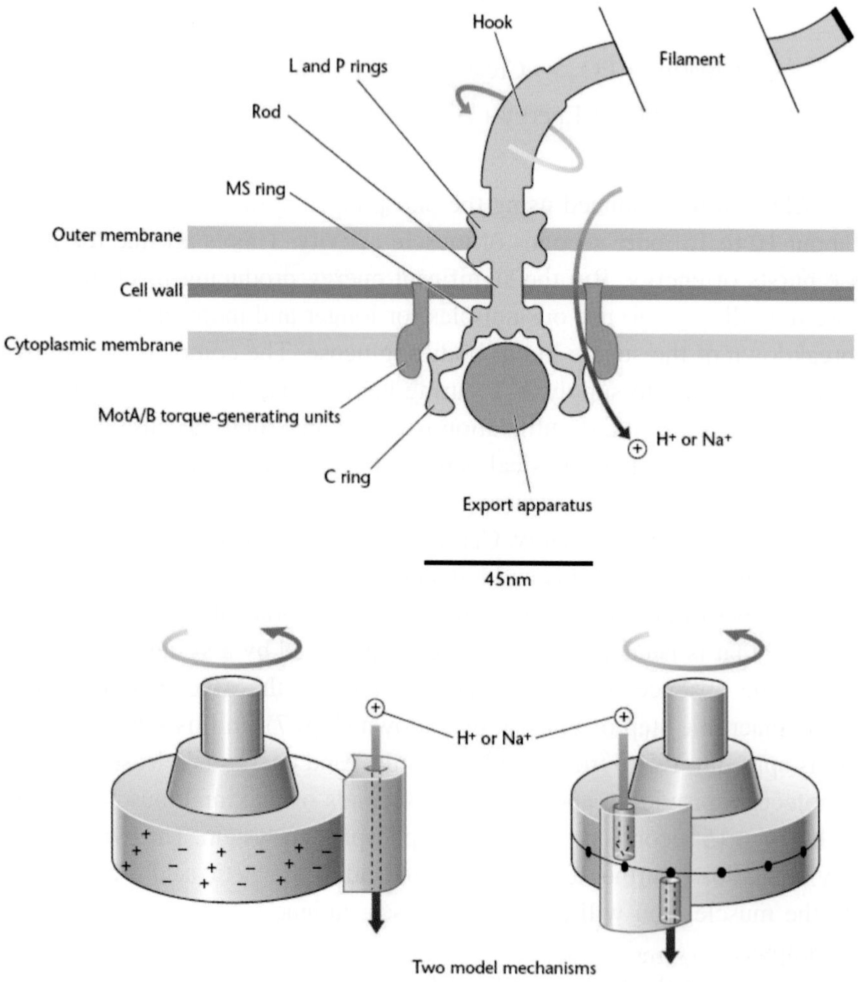

Fig. 7. Bacterial flagella: flagella motor.[5,6]

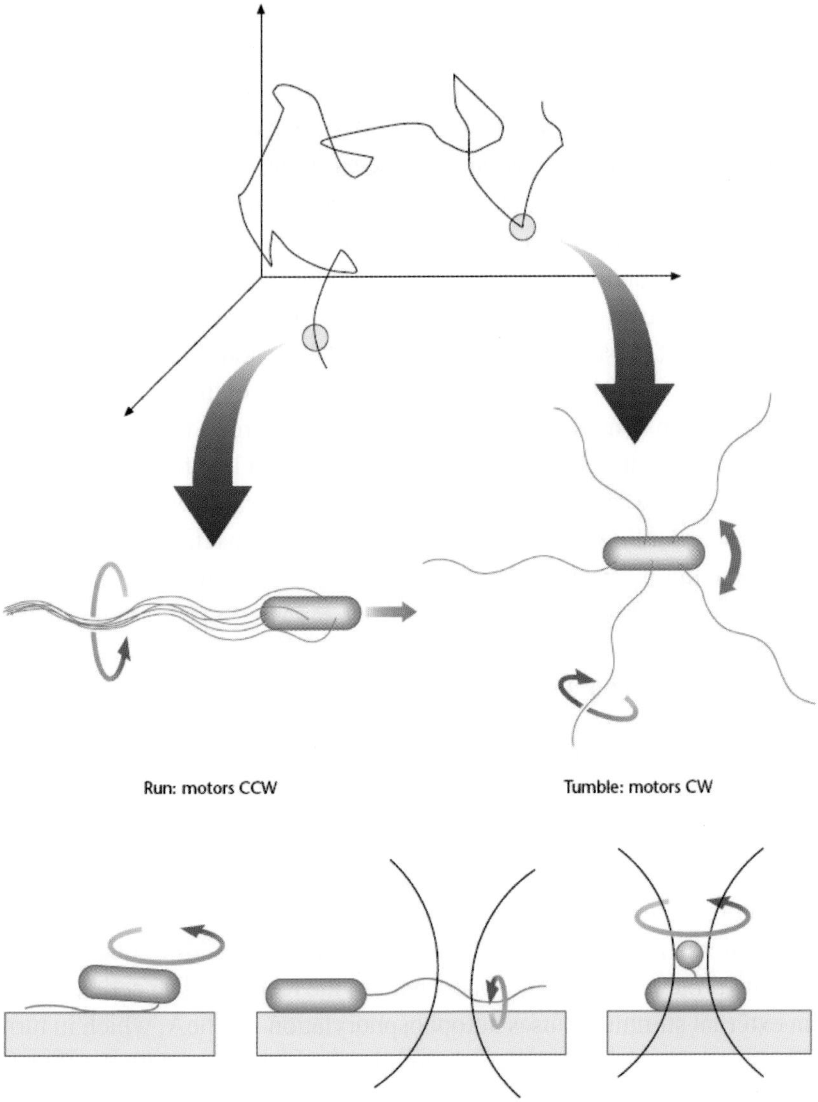

Fig. 7. (*Continued*).

energetics associated with muscle activity, you can refer to, http://www.nsbri.org/HumanPhysSpace/.

9. Sensing and Response — Designed for Adaptation?

A hallmark of living organisms is adaptation to the environment and a long-term evolution that is a result of selection pressures. Within the lifetime, cells composing a living organism sense their surroundings and respond to stimuli. This sensing and response is handled by the molecules of life through a combination of interactions, changes and transport. The simplest example of this phenomenon is bacteria swimming towards the highest concentration of sugar molecules in a solution or away from a toxic substance like phenol. The sensing and response in this case is orchestrated by a number of molecules on the surface and inside the bacterium. The overall movement of a bacterium is the result of alternating tumble and swim phases. If one watches a bacterium swimming in a uniform environment, its movement will look like a random walk with relatively straight swims interrupted by random tumbles that reorient the bacterium. Bacteria such as *E. coli* are unable to choose which direction they swim and are unable to swim in a straight line for more than a few seconds due to rotational diffusion. In the presence of a chemical gradient bacteria will chemotax or will direct their overall motion based on the gradient. If the bacterium senses that it is moving in the correct direction (toward attractant/away from repellent) it will keep swimming in a straight line for longer before tumbling. If it is moving in the incorrect direction it will tumble sooner and try a new direction at random. The tumbling is accomplished by motion of bacterial flagella and the chemotaxis phenomenon is an exemplar of design and engineering of a molecular motor in living systems. See Fig. 7 below taken from R. M. Berry's article in the *Encyclopedia of Life Sciences*.[1]

Bacteria like *E. coli* use temporal sensing to decide whether life is getting better or worse. This way, it finds the location with the highest concentration of attractant (usually the source) quite well. Even under very high concentrations, it can still distinguish very small differences in concentration. This initial sensing is accomplished by transmembrane receptors on the surface of the bacterium. A bacterium has three types of transmembrane receptors, for attractants, repellents and periplasmic proteins. The signals from these receptors are transmitted across the plasma membrane into the cytosol, where "che" proteins are activated. The che proteins alter the tumbling frequency, and alter the receptors. The proteins CheW and CheA bind to the receptor. The activation of the receptor by an external stimulus causes autophosphorylation in CheA, which in turn phosphorylates CheB and CheY. CheY induced tumbling by interacting with the flagellum protein FliM. CheB, which was activated by CheA, is a methylesterase, removes methyl residues from glutamate residues on the cytosolic side of the receptor. It works against CheR, a methyltransferase, which adds methyl residues to the glutamate residues. The more methyl residues are attached to the receptor, the more sensitive the receptor. As the signal from the receptor induces demethylation of the receptor in a feedback loop, the system is continuously adjusted to environmental chemical levels, remaining sensitive for small changes even under extreme chemical concentrations. This regulation allows the bacterium to "remember" chemical concentrations from the recent past and compare them to those it is currently experiencing, thus sense whether it is traveling up or down a gradient. A model of chemotaxis that you can compute an input-response relationship is presented in Fig. 8. For the most part, eukaryotic cells sense the presence of chemotactic stimuli though the use of 7-transmembrane (or serpentine) heterotrimeric G-protein coupled receptors. This class of receptors is huge,

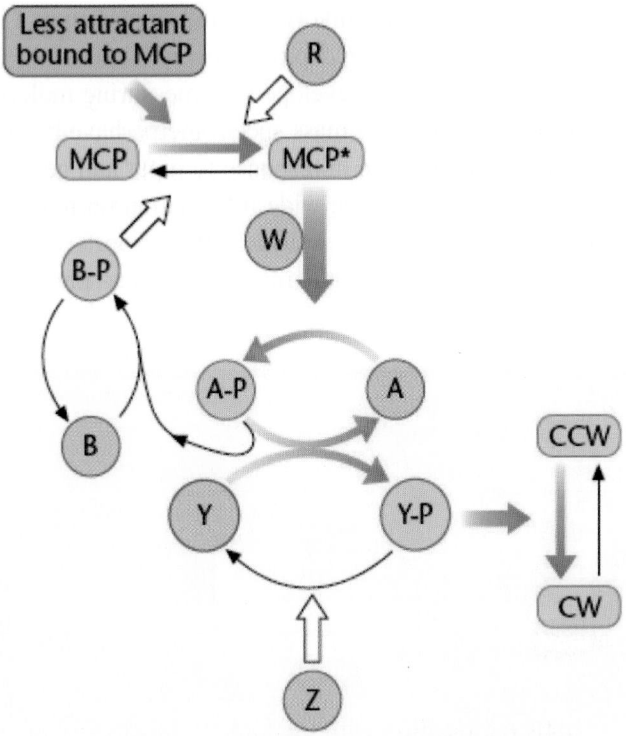

Fig. 8. The bacterial chemataxis pathway.[7]

representing a significant portion of the genome. Some members of this gene superfamily are used in vision (rhodopsins) as well as in olfaction (smelling).

A mathematical model of bacterial chemotaxis and the solutions to the model can be found in the paper by Spiro *et al.*[2]

How does the animal or insect host sense pathogens? Our present concepts grew directly from longstanding efforts to understand infectious disease: how microbes harm the host, what molecules are sensed and, ultimately, the nature of the receptors that the host uses. The discovery of the host sensors, the Toll-like receptors, was rooted in chemical, biological, and genetic analyses that were centered on a bacterial poison, termed endotoxin. Toll-like receptors (TLRs) are a family of pattern recognition receptors that are activated by specific components of microbes and certain host molecules. They constitute the first line of defense against many pathogens and play a crucial role in the function of the innate immune system. Recently, TLRs were observed to influence the development of adaptive immune responses, presumably by activating antigen-presenting cells, which has important implications for our understanding of how the host tailors its immune response as a function of specific pathogen recognition.

A major response of the Toll-like receptors sensing of infection is the stimulation of gene expression of molecules called cytokines. Cytokines lead to inflammatory and innate immune responses. From a systems biology point of view, we now have an intimate understanding of the specific pathways involved in inflammation and innate immunity. Figure 9 provides a cartoon of the Toll-receptor pathway.

Several new techniques have been developed for measuring molecular responses to input in cells. Several methods, notably mass spectrometry, have been used to measure lipids and proteins as a function of time in response to input and gene expression measurements have provided valuable genome wide information on transcriptional changes following input into cells. For instance, a large study reports data from large-scale measurements of diverse responses to input to macrophages and provides the rich data set

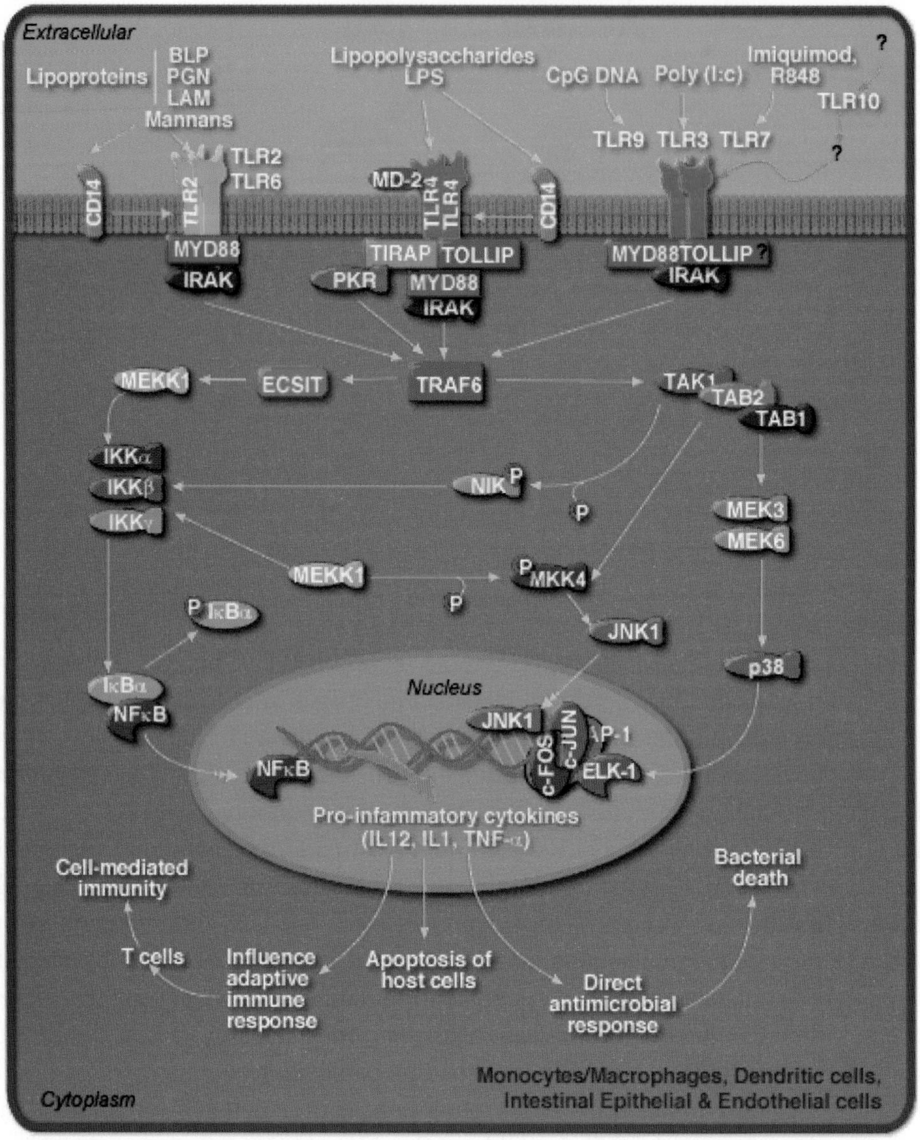

Fig. 9. Toll-receptor pathway from BioCarta (http://www.biocarta.com).

needed for a systems biology approach towards understanding cellular signaling (http://www.cellular-signaling.org/). A challenge for engineering is the reconstruction of the signaling networks from the data using a variety of statistical and computational methods.

DNA microarrays are a revolutionary technology that provides us with a genome-wide understanding of responses of cells. A **DNA microarray** is a collection of microscopic DNA spots attached to a solid surface, such as glass, plastic or silicon chip forming an array. Scientists use DNA microarrays to measure the expression levels of large numbers of genes simultaneously. The affixed DNA segments are known as *reporters*, thousands of which can be used in a single DNA microarray. Microarray technology evolved from Southern blotting, where fragmented DNA is attached to a substrate and then probed with a known gene or fragment. Measuring gene expression using microarrays is relevant to many areas of biology and medicine, such as studying treatments, disease and developmental stages. For instance, the gene expression response of a macrophage cell to bacterial infection can be followed by significant genes and thereby pathways enriched in the response. This requires sophisticated statistical methods. See Fig. 10 for one such gene expression experiment.

Sensing and recognition first occurs at a molecular level. The physical principle governing recognition between molecules is both intermolecular and intramolecular dynamics. The dynamics of molecules in a cellular process spans multiple time and length scales. Figure 11 provides a view of the scales spanned in molecular motion.

At the molecular level, sensing and specificity in recognition has its origin in intermolecular forces. For instance, two proteins that share a complementary electrostatic profile (complementary charges on their surfaces) bind specifically and play a role in a specific step of the cellular pathway. Further, these simulations lead to understanding of rates of reactions and hence to experimentally measurable temporal phenotypes in cells (see Box 6 for more details).

Box 6: Diffusional Encounters Between Molecules (From Livesay *et al.*[3])

The combination probability of a pair of reactants diffusing from some given initial separation obtained from Brownian dynamics trajectories can be used to calculate the diffusion-controlled rate constant. The basic principle that governs this algorithm is that given the positions of particles at time t, the Langevin equation can be used to predict a multivariate Gaussian distribution function from which particle displacements can be randomly selected to provide a new configuration. The position $\mathbf{r}$ after a time step Dt, is given by,

$$\mathbf{r} = \mathbf{r}_0 + (k_B T)^{-1} D\, \mathbf{F}(\mathbf{r}_0) Dt + \mathbf{R} \tag{1}$$

(*Continued*)

Box 6: (*Continued*)

where $\mathbf{r}_0$ is the position before a step is taken, k_B is the Boltzmann constant, T is the temperature, D is the relative diffusion constant, $F(\mathbf{r}_0)$ is the force on the substrate at $\mathbf{r}_0$, Dt the time step, and $\mathbf{R}$ is a random vector with moments

$$<R>=0 \text{ and } <R_i R_j>=2Dd_{ij}Dt \tag{2}$$

where i and j indicate the Cartesian components of $\mathbf{R}$. The time step, Dt, in any region should be small enough that the force on the particle only changes by a few percent during any step. Statistics from a large number of trajectories are utilized to calculate the bimolecular rate constants. An assumption of dilute solution is made in that only a pair of reactants is considered in the calculation. Trajectories of the mobile partner are initiated from a sphere of radius b about the stationary target molecule and are terminated when the reactants either react or reach a termination sphere of radius q. Several thousand trajectories are run to yield a reaction probability, b. The diffusion-controlled rate constant, which includes the correction for the probability of the mobile partner reaching q and returning to react, is given by

$$k = k_D(b)b[1-(1-b)k_D(b)/k_D(q)]^{-1}. \tag{3}$$

Here k_D is the steady state rate at which mobile reactants with $r > b$ will first strike the spherical surface at $r = b$.

The forces of interaction (the F term in the equation above) can be calculated for interactions between the molecules. In the case of charged molecules, the interaction is predominantly electrostatic. Continuum models provide a computationally efficient means of incorporating electrostatic effects of an aqueous solvent containing ions in biomolecular simulations. The nonlinear Poisson-Boltzmann equation which provides the continuum description of the electrostatic field around a macromolecule in an ionic medium is given by:

$$-\nabla.[\varepsilon(r)\nabla u(r)] + \varepsilon(r)\kappa^2(r)\sin h[u(r)] = \frac{e_c^2}{k_B T}\sum_{i=1}^{N_m} z_i \delta(r - r_i) \quad in\ \Omega \subset \Re^3 \tag{4}$$

$$u(r) = g(r) \text{ on the boundary } \partial\Omega.$$

The constants e_c, k_B and T in equation (1) represent the elementary unit of charge, Boltzmann's constant, and the absolute temperature, respectively. The piecewise constant dielectric is $e(r)$ and the inverse Debye-Huckel length $k^2(r)$ is given by $k^2(r) = 2Ie_c^2/ek_BT$, where I is the solvent ionic strength.

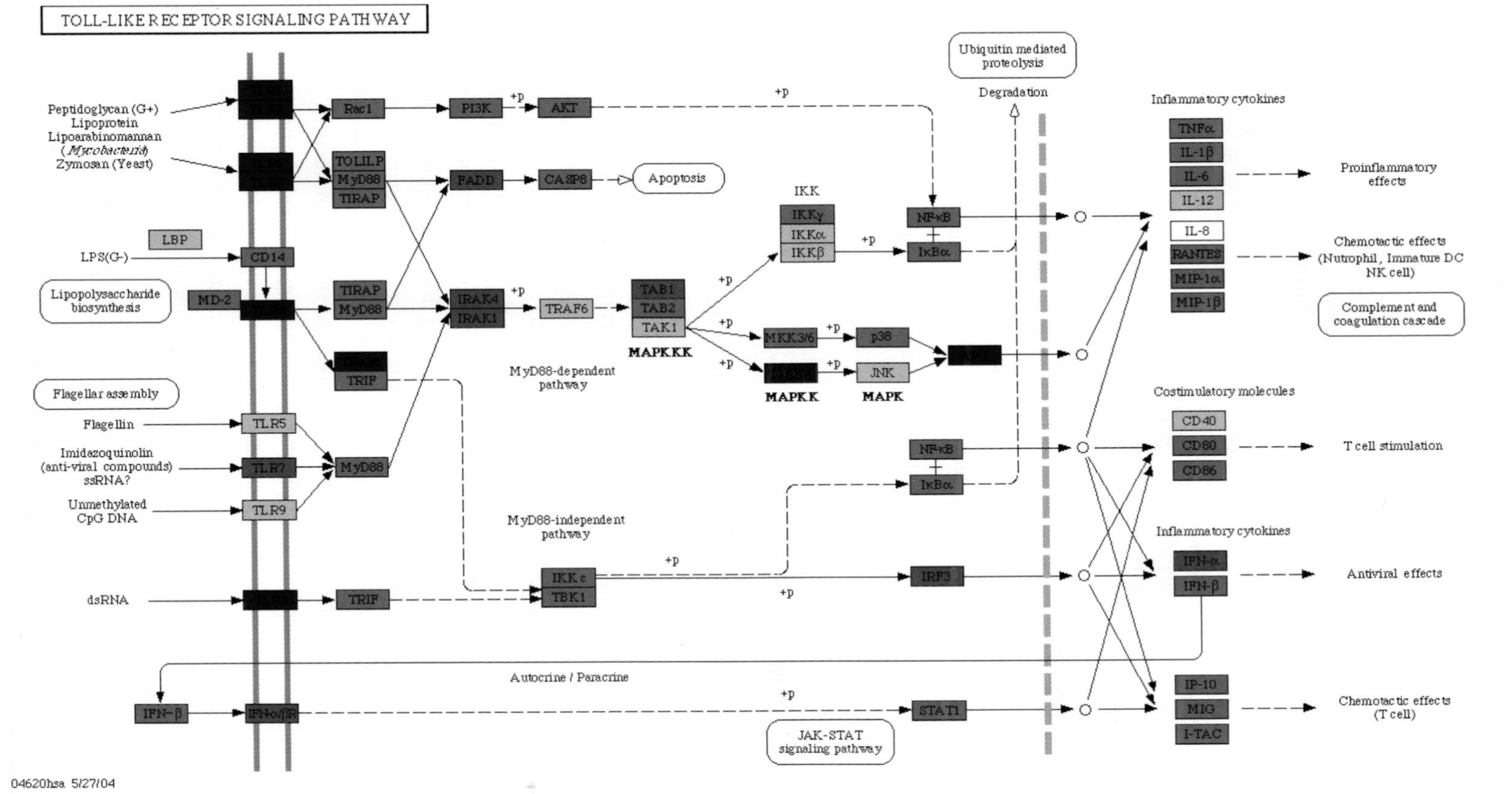

Fig. 10. The coloring of each gene product in the map indicates the expression profile over time. Red codes for genes that were up-regulated at all the time points that were significant. Green codes for gene that were down-regulated at all the time points that were significant. Blue is for those genes that have mixed expression over time. Gray color indicates the gene was not significantly regulated, while white color indicates that the gene was not on the chip.

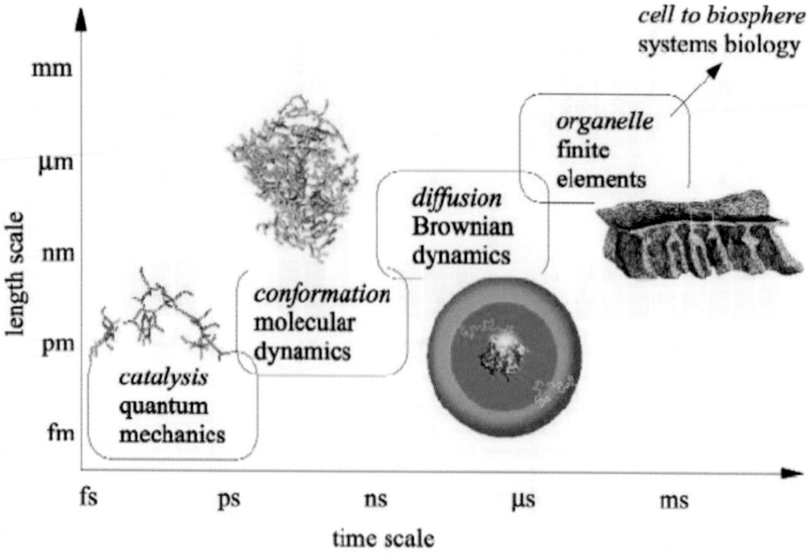

Fig. 11. Multiscale methods applied to biomolecular systems. Each scale defines a different mathematical approach to the dynamics.

10. Modeling of Cellular Networks

The data from high throughput measurements in living systems poses two challenges to quantitative biology and engineering. First, can we use this data in combination with biology to reconstruct networks in living systems that can help map qualitatively and quantitatively the response to stimulus? Second, can we decipher the rules by which living systems operate so that we can carry out defined perturbations that yield desired phenotypes or in other words redesign cellular networks to produce desired outcomes? A fundamental assumption we have to make is that, given the enormous dynamic complexity in the functioning of living systems, we will never know every precise molecular detail. However, such precision may not be necessary for both understanding and designing specific phenotypic outcomes. This is typically an engineering approach to biology and has already given rise to significant successes. For instance, from metabolic networks, we can now engineer microbes to make specific organic molecules. We can perturb cells precisely to produce desired phenotypes in living species; for instance, we can introduce specific mutations in mice that produce a given type of tumor or a defined disease condition. We can introduce specific signals into cellular systems to cause cells to behave in particular ways. For instance, we can cause bone marrow cells to differentiate into macrophages with the input of specific factors.

Much of these accomplishments have however arisen from qualitative considerations. Only recently have quantitative methods begun to be used. We can use techniques of flux balance analysis, used routinely by chemical engineers, to model microbial metabolism and understand the stability of microbes in specified growth media. We can

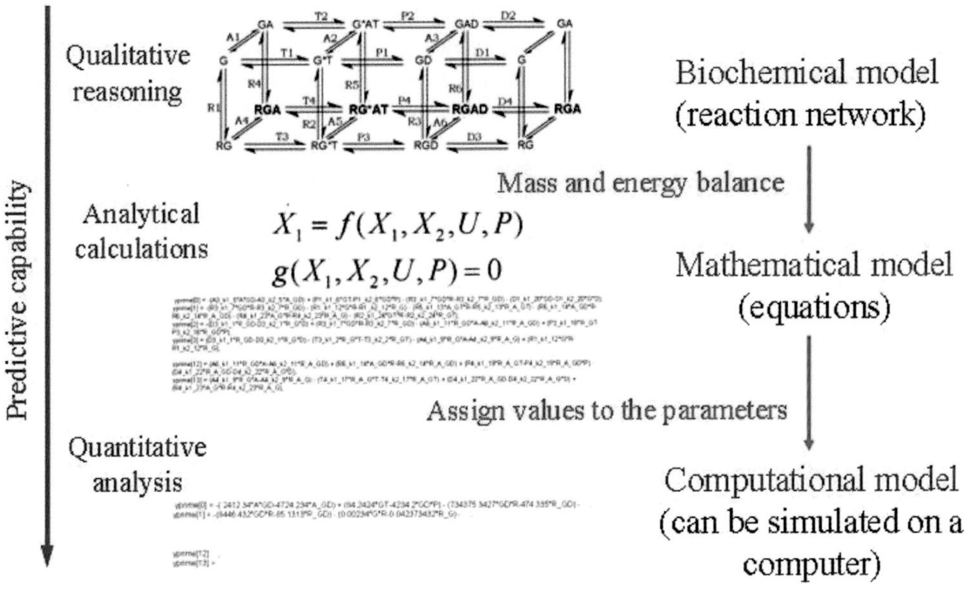

Fig. 12. Models for biological systems: three levels.

use dynamical and kinetic models to understand temporal behavior of cellular signaling networks. The timing and amplitude of signaling in cellular signaling networks depends on the local concentrations and kinetics of component proteins. The complexity of these networks has precluded comprehensive quantitative analysis of signaling. However, the networks can be divided into discrete signaling modules that can be quantitatively modeled and analyzed independently. Figure 12 provides a cartoon of the approaches taken in modeling of reaction networks.

An example of such models for cellular division can be seen in the following: The Cell Cycle and Mitosis tutorial http://www.biology.arizona.edu/cell_bio/tutorials/cell_cycle/cells1.html.

References

1. R. M. Berry, Bacterial flagella: flagellar motor. In: *Encyclopedia of Life Sciences* (Nature Publishing Group, 2005).
2. P. A. Spiro, J. S. Parkinson and H. G. Othmer, A model of excitation and adaptation in bacterial chemotaxis. *Proc. Natl. Acad. Sci. USA* **94**: 7263–7268 (1997).
3. D. R. Livesay, P. Jambeck, A, Rojnuckarin and S. Subramaniam, Conservation of electrostatic properties within enzyme families and superfamilies, *Biochemistry* **42**: 3464–3473 (2003).
4. S. Cortassa, M. A. Aon, E. Marban, R. L. Winslow and B. O'Rourke, Schematic representation of the mitochondrial electrophysiological and metabolic processes and their interaction from an integrated model of cardiac mitochondrial energy metabolism and calcium dynamics, *Biophys. J.* **84**: 2734–2755 (2003).

5. R. M. Berry, The bacterial flagellum from bacterial flagella: flagellar motor. In: *Encyclopedia of Life Sciences* (John Wiley & Sons Ltd., 2005), http://www.els.net/ [DOI: 10.1038/npg.els.0003931].
6. R. M. Berry, Swimming and flagellar rotation in *Escherichia coli* from bacterial flagella: flagellar motor. In: *Encyclopedia of Life Sciences* (John Wiley & Sons Ltd., 2005), http://www.els.net/ [DOI: 10.1038/npg.els.0003931].
7. R. M. Berry, The chemotaxis pathway in *Escherichia coli* from bacterial flagella: flagellar motor. In: *Encyclopedia of Life Sciences* (John Wiley & Sons Ltd., 2005), http://www.els.net/ [DOI: 10.1038/npg.els.0003931].

CHAPTER 25

SYNTHETIC BIOLOGY: BIOENGINEERING AT THE GENOMIC LEVEL

Natalie Ostroff, Mike Ferry, Scott Cookson, Tracy Johnson and Jeff Hasty

Abstract

The developing discipline of synthetic biology attempts to recreate in artificial systems the emergent properties found in natural biology. Because the genetic networks found in cells are often highly integrated and quite complex, redesigning simpler synthetic systems for study is a valuable approach not only at the genome level but also at the gene network level. Recently, there has been significant activity directed towards designing synthetic gene networks that mimic the functionality of natural systems. In addition to being easier to construct, the reduced complexity and increased isolation of these networks makes them more amenable to both tractable experimentation and mathematical modeling. The process of constructing and testing artificial systems resembling naturally occurring systems promises to advance our understanding of how biological systems function by providing information about cellular processes that cannot be obtained by studying intact native systems.

1. Introduction

The emergence and growing prominence of synthetic biology is largely due to vast improvements in molecular biology techniques, advances in the microfabrication of microscopy devices, and the increased application of mathematical modeling to biological systems over the past few decades (Fig. 1). Therefore, it is important to understand the significant contribution that each of these aspects makes to the overall goal of systems biology. First, the tools of molecular biology allow scientists to easily manipulate and clone DNA, enabling the rapid design and construction of simple genetic circuits. Second, microfabricated microfluidic devices, tailored to the acquisition

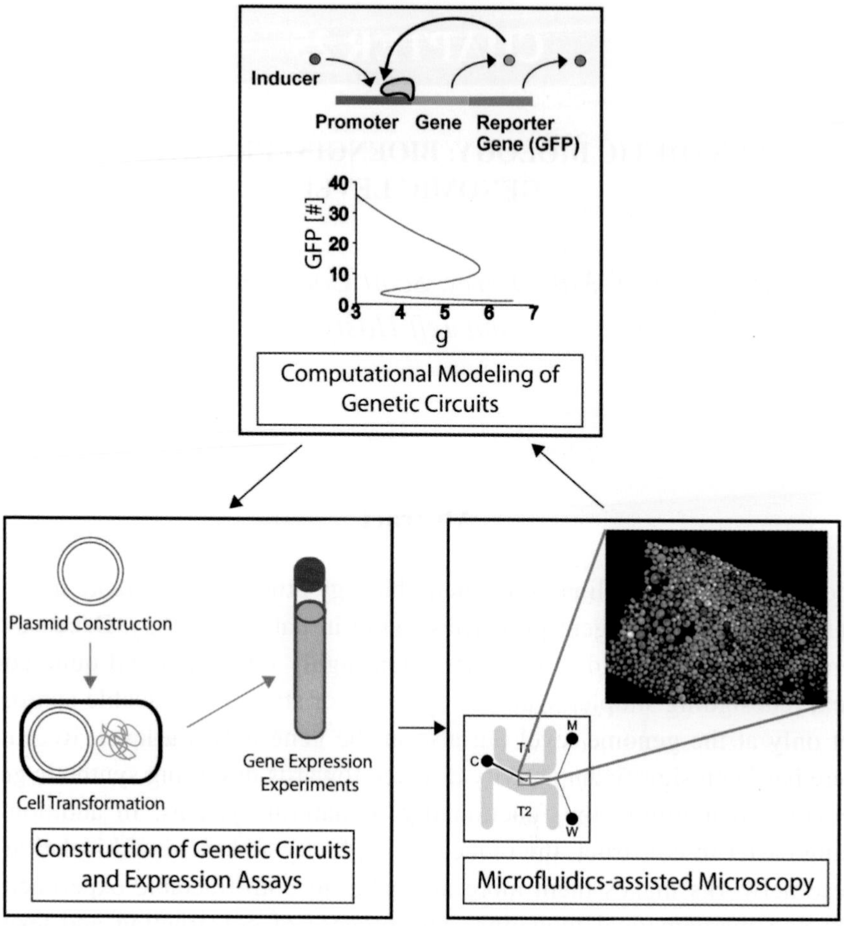

Fig. 1. Synthetic biology applies a combination of quantitative modeling, molecular biology, and microfabricated microscopy devices to the study of synthetic biological systems. A computational model of a theoretical circuit (top) provides guidance for the wet-lab realization of the genetic module (bottom left). Once cells have been transformed with the circuit, gene expression data from single-cell microscopy (bottom right) can be used to refine the original model, leading to a more accurate quantitative understanding of gene regulation.

of long-term data from large populations of individual cells, provide scientists with a means to monitor the dynamic behavior of a novel biological circuit. And finally, the application of engineering-based modeling techniques to the study of biological systems provides insight into the design of new circuits, as well-formulated models can predict the behavior of a circuit before it is even constructed. The growing excitement and recent successes in synthetic biology illustrate the great impact that the union of biology, engineering, and technology can have on our understanding of biological systems.

2. Molecular Biology

The increasing ease with which scientists are able to modify DNA and genetically engineer organisms has led to a recent burst of interest in synthetic biology. Over the past few decades, several breakthroughs in molecular biology techniques have led to a more detailed understanding of basic biological functions and a greater ability to mimic these functions for the purpose of modifying natural biological systems. Therefore, in order to understand the purpose and potential of synthetic biology, it is important to first understand the fundamental processes of life and how scientists are able to utilize these processes to manipulate and probe biological organisms.

2.1 *The central dogma of molecular biology*

DNA is a chemical "code" that stores the hereditary information of a cell. The entire collection of a cell's DNA is called the genome, which contains the code for thousands of proteins that form the structure and perform the vital functions of a cell. Each segment of DNA that codes for a separate protein is called a gene. DNA serves as a template both for its own replication, occurring once each cell cycle, and for the transcription of mRNA, a messenger molecule which serves as an intermediate between a gene and the protein that it encodes. Many mRNA transcripts can be synthesized from a single gene, and these molecules are then translated into proteins by enzymes called ribosomes. The flow of information from DNA to RNA to protein is referred to as the "central dogma" of molecular biology (Fig. 2).

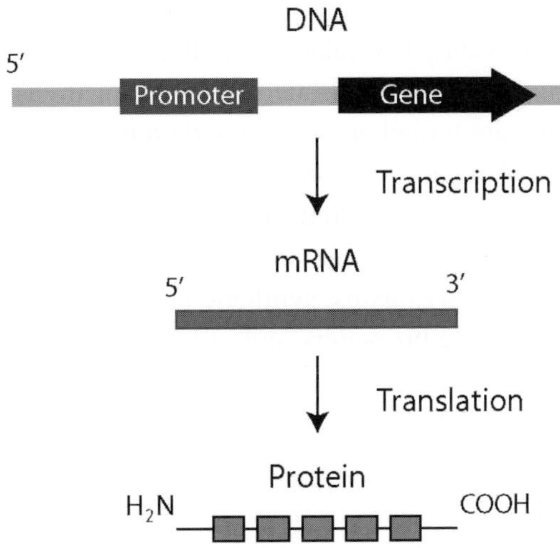

Fig. 2. The central dogma of molecular biology states that mRNA is produced from a DNA template by transcription and proteins are produced from an mRNA template by translation.

2.2 *Gene regulation*

The act of producing a protein from a DNA template is called gene expression, and this process is tightly regulated in order to minimize the resources and energy utilized by a cell. Gene expression is typically controlled at the transcriptional or translational level. Protein synthesis is a metabolically expensive process, and therefore it is most economical to regulate the process at the earliest stage possible. While a cell may contain thousands of genes, it is wasteful and can even be harmful to express a gene at an inappropriate time, and therefore cells have evolved very elaborate and effective mechanisms of transcriptional gene regulation.

Transcription of an mRNA molecule from a gene is catalyzed by an enzyme called RNA polymerase. Regulation of this process is achieved through the action of regulatory proteins which can either interfere with RNA polymerase (repressors), thereby shutting down transcription, or enhance the action of RNA polymerase (activators), thereby increasing the rate of transcription. Stretches of regulatory DNA are interspersed among genes, and these non-coding regions bind to regulatory proteins that control the local rate of transcription. Specifically, a promoter is the DNA region upstream of a gene where the RNA polymerase first binds before initiating transcription, and by interacting with the DNA near a gene's promoter, regulatory proteins are able to interact with RNA polymerase and control the rate of transcription. In this way, the rate of transcription of different genes is controlled independently, according to need. Negative control of gene expression is achieved when a repressor protein binds to DNA at a point that interferes with the action of RNA polymerase on a specific gene, blocking its transcription. In contrast, positive control occurs when an activator binds to a gene's promoter and enhances the ability for RNA polymerase to bind the DNA and proceed with transcription.

Some genes are expressed under all circumstances, such as the "housekeeping" genes required to maintain essential cellular functions. These constitutively expressed genes will be produced regardless of the state of the cell or the environmental conditions. However, other genes are turned on or off based on need. For example, a cell may "turn on" expression of DNA repair genes upon exposure to damaging UV radiation or "turn off" expression of an enzyme required to metabolize a nutrient that is lacking in the current environment. Using knowledge about well-understood regulatory proteins and promoters in simple model organisms, synthetic biologists can mimic various constitutive, inducible, or repressible promoters and utilize them to design synthetic gene networks.

2.3 *Working with DNA*

The manipulation of DNA has become increasingly easy over the past few decades. A scientist can isolate a specific gene, make unlimited copies of it, and determine its exact sequence all in a single day. Not only can we copy a gene, but we can also alter it and transfer it back into a cell to be incorporated into the organism's genome. The ease with which we can genetically engineer organisms has already served many purposes, such as aiding in the discovery of the function of new genes, advancing the understanding of evolutionary links between organisms, and enabling the mass production of proteins for use as hormones or vaccines. In addition, now that we have access to the genomic sequences of many organisms, we can begin to unravel complex regulatory networks, develop a strong understanding of native biological systems, and even piece network components back together to create new functionality. Ultimately, we can envision the ability to construct simple gene modules that could be integrated into a diseased cell to perform the function of a damaged or missing cellular component. Progress in technology for analyzing proteins, DNA, and RNA has led to the possibility of obtaining a complete understanding of how a cell's genetic network regulates its behavior and its response to the extracellular environment.

Recombinant DNA technology allows for the artificial rearrangement and cloning of DNA, making many tasks that were previously tedious or even impossible efficient and cost-effective. There are several techniques that are essential to the field of synthetic biology. First is the ability to cleave DNA, using enzymes called **restriction nucleases** that cut DNA at a specific target sequence. Second is **DNA cloning**, which produces billions of identical copies of a DNA molecule of interest. Third is rapid and reliable sequencing of a purified DNA fragment.

The ability to clone DNA has revolutionized the field of molecular biology. While the term cloning carries with it many connotations, in cell and molecular biology it generally refers to the act of making many copies of a specific DNA molecule. There are two standard ways to clone a DNA molecule. One involves inserting the fragment into a self-replicating genetic element, inserting this cloning vector into a bacterial cell, and using the normal replication mechanisms of the organism to make billions of copies of the vector. The most common type of vector is called a plasmid, which is a small, circular DNA molecule that can be maintained and replicated to create many copies within a single bacterial cell. A plasmid can easily be used as a cloning vector with the help of restriction nucleases to "cut" both the plasmid and the fragment to be cloned, and a ligase to "glue" the two pieces back together, creating a new molecule of recombinant plasmid DNA. This plasmid can then be reintroduced into bacterial cells, and as the cells grow and divide, they replicate the recombinant plasmid along with their own genome. Finally, the cells can be broken apart (lysed) and the plasmids can be easily separated away from genomic DNA due to their small size. The end result is a sample of purified DNA containing millions of copies of the DNA of interest (Fig. 3).

Another method of cloning, which can be performed in a test tube with no need for a cell culture is called the **polymerase chain reaction (PCR)**. The development of PCR was one of the most significant breakthroughs of recombinant DNA technology, as it allows the rapid and specific amplification of any region of DNA from a supplied

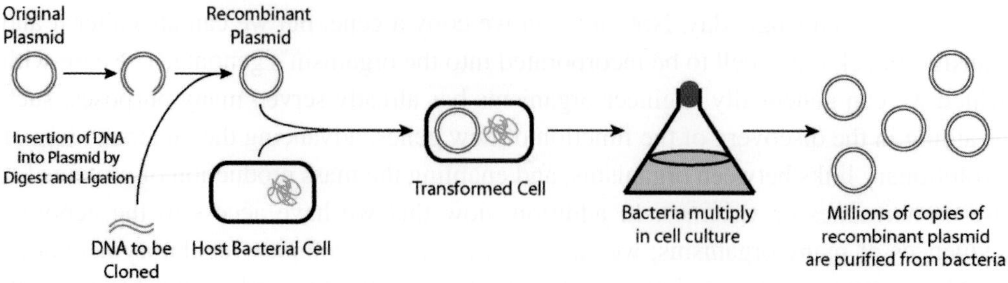

Fig. 3. A foreign fragment of DNA can easily be inserted into a bacterial plasmid. After the plasmid is "cut" using a restriction enzyme, the insert can be "pasted" in using a DNA ligase to bond the two fragments together. The new recombinant plasmid can then be purified and amplified by DNA cloning in an *E. coli* bacterium.

template. The method is extremely sensitive, needing very little template to produce billions of copies of the desired region, and it yields a sample of essentially purified fragments of the desired DNA sequence.

Finally, as robust as these techniques are, it is always a good idea to double-check the resulting DNA sample with a sequencing reaction. The dideoxy method was developed as a simple and quick means to sequence any purified DNA fragment, and with recent improvements it has evolved into a completely automated technology. A molecular biology lab can send out a newly created recombinant DNA fragment one afternoon and receive the sequence the next morning.

The tools of recombinant DNA technology, including the techniques to "cut and paste," clone, and rapidly sequence DNA, have become indispensable to bioscientists interested in probing natural systems. They have also made possible the birth of a new era of synthetic biology, in which we will use these techniques to genetically engineer novel genetic circuits as we strive not only to understand native systems but also to design and build novel biological functions and systems. In the next section we describe the findings of some recent successful studies in this new and exciting field.

2.4 *Successes in synthetic biology*

The field of synthetic biology emerged in the wake of the systems biology boom, which has produced huge data sets revealing the enormous complexity of biological networks. Systems biology aims to reconstruct these complex networks and develop quantitative models to describe how complex cellular functions arise from their connectivity. While originally systems biology involved the use of high-throughput technologies to develop static genome-scale maps of biological networks (the "top-down" approach), in recent years synthetic biologists have joined the field, striving to reveal

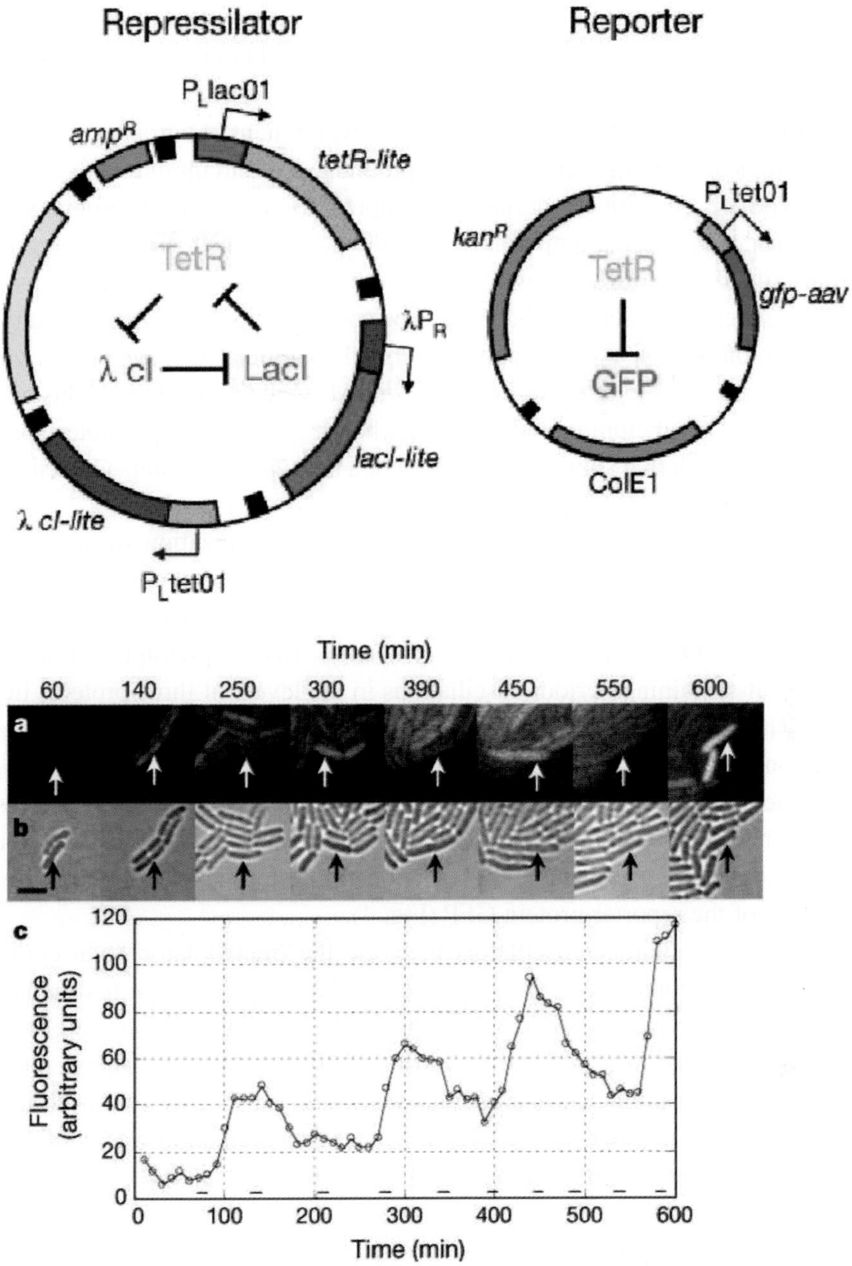

Fig. 4. Top: The repressilator network. The repressilator is a cyclic negative-feedback loop composed of three repressor genes and their corresponding promoters, as shown schematically in the center of the left-hand plasmid. The compatible reporter plasmid (right) expresses an intermediately stable GFP variant (gfp-aav). Bottom, **(a)** and **(b)**: The growth and timecourse of GFP expression for a single cell of the *E. coli* host strain containing the repressilator plasmids. Snapshots of a growing microcolony were taken periodically both in fluorescence (a) and bright-field (b). **(c)** The pictures in (a) and (b) correspond to peaks and troughs in the timecourse of GFP fluorescence of the selected cell.

the dynamic properties of biological networks by mimicking their function with simpler gene "circuits" and studying their behavior in a well controlled environment (the "bottom-up" approach).

One result of systems biology has been the development of a graphical language to describe biological systems that is analogous to the use of circuit diagrams in electronics. This type of description defines the inputs to a regulated promoter as the protein/inducer pair, and the output as "on" if the gene downstream of the promoter is being transcribed and "off" otherwise. Such complexes can be represented as simple biological logic gates, which can themselves be combined to yield circuits of any given complexity. In one study, a novel approach termed "combinatorial synthesis" was used to generate a myriad of logical gene circuits with varying connectivity.[1] This approach involved the use of subcloning and ligation, whereby 15 distinct promoter-gene units were constructed such that subsequent ligation of a mixture of the units yielded a library of three-gene networks.

Cellular rhythms occur at all levels of biological organization from genetic networks to animal populations.[2] Several synthetic clocks have been developed to help us understand the molecular design principles responsible for generating oscillations in natural systems.[3–5] One synthetic oscillatory network (the "repressilator")[3] was shown to generate self-sustaining periodic oscillations in the levels of three proteins in a bacterial cell. The design operates on the same general principle as a ring oscillator in microelectronics. Accordingly, the repressilator network architecture is cyclic, in which the LacI protein represses the promoter for the *TetR* gene, the TetR protein represses the promoter for the *cI* gene, and the cI protein represses the promoter for the *lacI* gene. The network produced roughly sinusoidal oscillations in protein levels, observed by parallel expression of the reporter protein GFP (Fig. 4).

More recently, a synthetic oscillatory network (the "metabolator") was constructed by linking transcriptional regulation with metabolism.[4] The network is designed such that two metabolite pools (acetyl-CoA versus acetyl phosphate and acetate) are interconverted by two enzymes that are transcriptionally regulated by acetyl phosphate. Theoretical modeling predicted that oscillations would only occur for specific combinations of gene copy numbers of the two enzymes. This prediction was verified experimentally by showing that oscillations disappear when one of the enzymes (Acs) is removed from the network. The construction of a synthetic genetic oscillator that is linked to metabolism represents an important step towards understanding the function of circadian clocks which are also integrated with cellular metabolism.

In an autorepressive gene network study, both a negatively controlled and an unregulated promoter were utilized to study the effect of regulation on variations in cellular protein concentration.[6] The central result is that negative feedback decreases the cell-to-cell fluctuations in protein concentration measurements. This study empirically demonstrated network-induced stability through the measurement of protein fluorescence distributions over a population of cells. The findings show that, for a repressive network, the fluorescence distribution is significantly tightened and such tightening is proportional to the degree to which the promoter is negatively controlled. These results suggest that negative feedback is utilized in cellular design as a means for mitigating variations in cellular protein concentrations.

The notion that noise could be important in the choice of a developmental pathway for an organism has induced a flurry of modeling research devoted to the role of fluctuations in gene regulation.[7–11] Theoretical models have been combined with engineered gene networks to elucidate the dominant source of internal noise in a single-gene network.[12,13] Given the two-step process of transcription and translation, the specific goal of this work was to determine their relative contribution to the fluctuations observed in the expressed protein levels within a cell. Point mutations were utilized to independently vary the transcriptional and translational rates, and the fluctuations in the expressed protein levels were observed to increase linearly with the translational efficiency while demonstrating only a mild increase with the transcriptional efficiency.[13]

A method utilizing two different fluorescent reporter proteins expressed from identical promoters was developed to study noise in gene expression in *E. coli.*[14] This study demonstrated that noise in gene expression results in fluctuations in protein levels in a clonal population. Both intrinsic and extrinsic noises were found to contribute to total noise in gene expression. The two-reporter method was also utilized to investigate noise in gene expression in the yeast *S. cerevisiae.*[15] This study revealed that extrinsic noise makes the most significant contribution to the total noise in gene expression. Additional studies have confirmed that gene expression variability is dominated by extrinsic noise.[16–18]

These studies represent important advances in the methodology of designing and constructing synthetic gene networks that are much simpler than their natural counterparts. The process of constructing synthetic gene regulatory circuits and analyzing the behavior of these simplified circuits has the potential to advance our understanding of natural gene expression behavior.

3. Microfluidics and Microscopy

3.1 *Motivation*

Recent progress in reconstructing gene regulatory networks has established a framework for a quantitative description of the dynamics of many important cellular processes. Such a description will require novel experimental techniques that enable the generation of time series data for the governing regulatory proteins in a large number of individual living cells. An ideal data acquisition system would allow for the growth of a large population of cells in a defined environment which can be monitored by high resolution microscopy for a lengthy period of time. With such a setup, the gene expression state of each cell could be monitored for the length of the experiment, giving the experimentalist accurate data about the temporal progression of each individual cell within the larger population. To this end, bioengineers have increasingly used devices with fluid channels on the micron scale known as microfluidic devices.

Microtechnology, and microfluidics in particular, can facilitate the accurate study of cellular behavior *in vitro* because it provides the necessary tools for recreating *in vivo*-like cellular microenvironments. Microfluidics involves the handling and manipulation of very small fluid volumes, enabling the creation and control of micro liter-volume reactors while drawing advantages from low thermal mass, efficient mass transport, and large surface area-to-volume ratios. Because fluid viscosity, not inertia, dominates fluid behavior at this scale, microfluidic flow is laminar, ensuring that the system does not include turbulent flows which would be detrimental for observing cellular behavior under high magnification. Lately, microfluidic "lab-on-a-chip" devices have become increasingly valuable as the known complexity of gene networks grows, driving the need for reduced-scale assays in probing entire parameter spaces of genetic circuits. The result has been the development of integrated microfluidic circuits analogous to their electrical counterparts, which aim to support large-scale multi-parameter analysis in parallel. Recent applications of microfluidics in biotechnology include DNA amplification, purification, separation,[19] DNA sequencing,[20] large-scale proteomic analysis,[21] development of memory storage devices,[22] cell sorting,[23] and single-cell gene expression profiling.

3.2 *Device fabrication*

In recent years, soft lithography has become the preferred method for fabricating microfluidic devices for biology. Soft lithography includes a suite of methods for replicating a pattern using blastomeric polymers (Fig. 5). Soft lithography can be represented as a three-step process comprised of concept developing, rapid prototyping, and replica molding.

The first step, concept developing, involves drafting a device design in a computer-aided design (CAD) program. Here, a general idea for a device that serves some purpose is fleshed out using engineering approaches. Using the laws of fluid dynamics under the condition of low Reynolds number for microvolume flow, fluid channel resistances are calculated and modified to satisfy desired driving pressures and flow rates. Following fine-tuning of the entire channel architecture, the device design is broken up into multiple layers, where all features of a given height are placed on a single layer for photolithographic purposes. Finally, all device layers are printed at high resolution onto transparency film. These are then fastened to ultra-transmissive borosilicate glass for use as a photomask set in the following contact lithography step.

For rapid prototyping, a positive or negative photoresist is spin coated onto a clean silicon wafer at a specified thickness and then exposed to UV light through the photomask to selectively crosslink the features represented by the mask. Since each exposure iteration creates all device features of a given height (defined by the depth of the

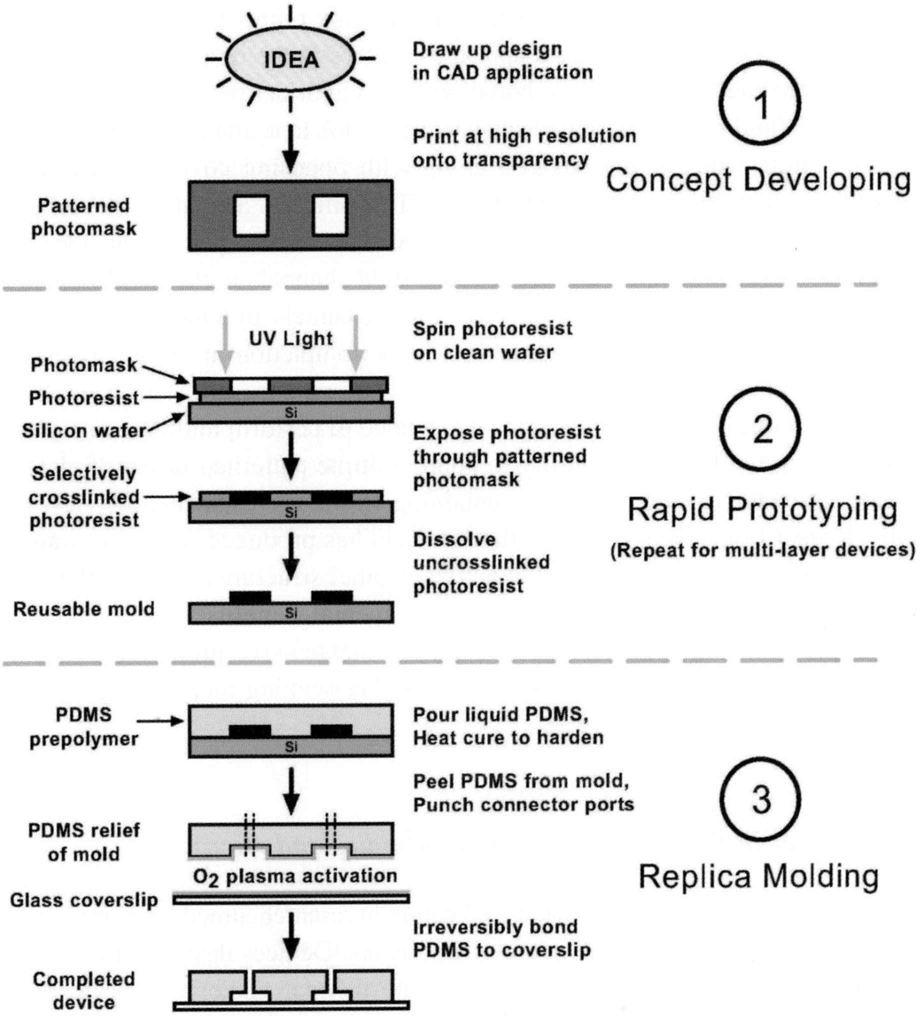

Fig. 5. Schematic of microfluidic device fabrication using soft lithography (adapted from Ng *et al.*[24])

photoresist layer), this process can be repeated to pattern the wafer for multi-layer device features. The final result is a positive relief of photoresist on the silicon wafer, known as a "master mold," whose topology precisely reflects the desired device channel and feature structures and can be used repeatedly to form successive batches of devices. Fabrication of this mold mold completes the rapid prototyping step of soft lithography.

The final step, called replica molding, involves the casting of a transparent, silicone-based liquid prepolymer (usually PDMS) against the master mold to generate a negative replica of the mold. The prepolymer is first poured onto the wafer and heat-cured in place to form a rubbery silicone solid. This silicone monolith is then peeled from the mold to reveal the inverted feature topology represented by the mold. For example, ridges on the master mold appear as valleys in the replica. This monolith is then diced into individual devices, bored with a cylindrical punch to form holes for connection to fluid reservoirs, and cleaned using Scotch tape and methanol. In the final step, the feature sides of the devices, along with opposing coverslip surfaces, are briefly treated with low power oxygen plasma. This process "activates" the surfaces of the PDMS devices and glass coverslips so that they form a permanent bond when placed in contact. In bonding the two objects, fluid channels in the PDMS are sealed against the flat coverslip surface to form microchannels internally connecting the device fluidic ports. These finished devices mark completion of the replica molding step of soft lithography.

The techniques described here can be extended to perform multilayer soft lithography, which provides the capability to bond multiple patterned layers of elastomer to create active microfluidic systems containing on-off valves, switching valves, and pumps. Recent research in the microfluidics field has produced several examples of complex devices with hugely parallel active channel structures for high-throughput cell analysis. In approaching years, the fundamental benefits of soft lithography for biology, which include ease of fabrication, inexpensive production, and rapid device turnover, will continue to aid the researcher seeking increasingly functional cell assays.

3.3 *Successes in the integration of microfluidics and microscopy*

Microfluidics has recently found wide application in research aimed at observing cellular development within dynamic microenvironments. Devices designed for these purposes frequently possess the ability to generate thermal and/or chemical gradients across the cell development volume.[25–27] Another recently demonstrated strength of microfluidics is the ability to generate large-scale and highly parallel integrated circuits of fluidic channels for high-throughput cellular analysis.[28–30] However, for researchers interested in studying the behavior of synthetic gene circuits, the most challenging goal of microfluidics has been in supporting long-term single-cell analysis for large sample populations. Therefore, much recent research has focused on this goal using various design strategies.

One group approached the difficulties in single-cell analysis by developing a microfluidic network enabling the passive and gentle separation of a single cell from bulk suspension.[31] This individual cell is focused by hydrostatic pressure and laminar flow streams to a trapping region, where integrated valves and pumps enable the precise delivery of nanoliter volumes of reagents to that cell. Whereas this research focused on individual cells over a relatively short time span, another group developed a microfluidic platform for long-term cell culture studies spanning the entire differentiation process of mammalian cells.[32] They demonstrated operation of this device by observing a culture of muscle cells differentiating from myoblasts to myotubes over the course of two weeks.

To researchers interested in long-term gene expression variability within single-celled prokaryotic and eukaryotic populations, a chemostat likely represents the ideal cell assay. In recent years, the many challenges involved in operating continuous macroscale bioreactors (such as the need for large quantities of reagents) have driven the miniaturization of these devices into microfluidic chip-based formats. By continually providing fresh nutrients and removing cellular waste to support exponential growth, the microfluidic chemostat presents a nearly constant environment that is ideal for long-term cell culture monitoring with single-cell resolution. Recently, one group presented a microfluidic chemostat for culturing bacterial and yeast cells in an array of shallow microscopic chambers with support for dynamically-defined media.[33] Similarly, a recent implementation of a microfluidic bioreactor has enabled long-term culturing and monitoring of small populations of bacteria with single-cell resolution.[34] This microchemostat contained an integrated peristaltic pump and a series of micromechanical valves to add medium, remove waste, and recover cells. The device was used to observe the dynamics of an *E. coli* strain carrying a synthetic "population control" circuit that regulates cell density through a feedback mechanism based on quorum sensing.

A final implementation of the chemostat design was utilized to precisely control and constrain exponential growth of the yeast *S. cerevisiae* and *E. coli* to a monolayer.[35] Here, dimensions of the chemostat device were precisely controlled to constrain exponential growth of yeast and *E. coli* cells to a monolayer. The device, termed the Tesla microchemostat, was based on an implementation of the classic Tesla diode loop,[36–38] modified for imaging a culture of cells growing in exponential phase for many generations. The construction was such that the side-arm of the diode formed a shallow trapping region which constrained a population of cells to the same focal plane (Fig. 6). The significant advantage of monolayer growth in a height-constrained chamber was demonstrated by visualization of a group of cells residing at the trapping region boundary (Fig. 7). Through directed planar growth, the researchers were able to resolve the temporal evolution of single-cell gene expression levels with the aid of segmentation and tracking software. Advantages of this device design and software package included simple operation and automated single-cell fluorescence trajectory extraction. Such novel data should prove useful in investigating the timing and variability of gene expression within various synthetic gene regulatory network architectures on the time scale of many cellular generations.

The ability to perform long duration experiments which yield consistent single-cell data is critical for the generation of computational models with predictive abilities, a hallmark of synthetic biology. To this end, microfluidics offers new opportunities to automate and improve many experimental techniques and to obtain data that was

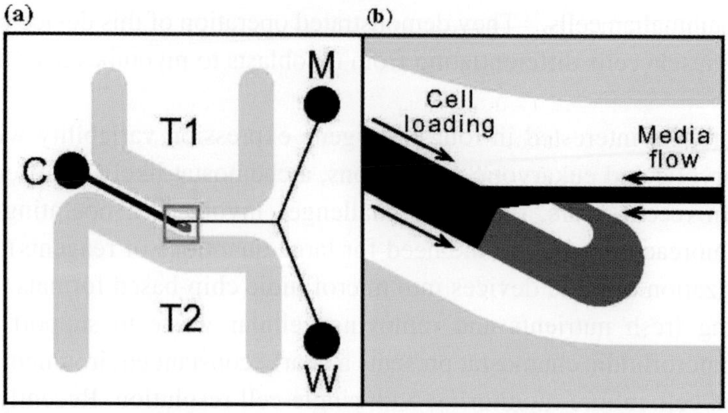

Fig. 6. The Tesla microchemostat design was optimized to allow for long-term growth of cells in a monolayer. **(a)** Three separate ports for cell loading (C), media supply (M), and waste (W) minimize potential clogging of media supply lines. **(b)** Zoom of the diode loop. The height of the trapping region (dark grey) is customized based on species. The flow channels (black) are two to three times higher than the trapping region. An open trapping region (black/grey interface) allows for peripheral cells to be pushed from the observation region as the colony grows.

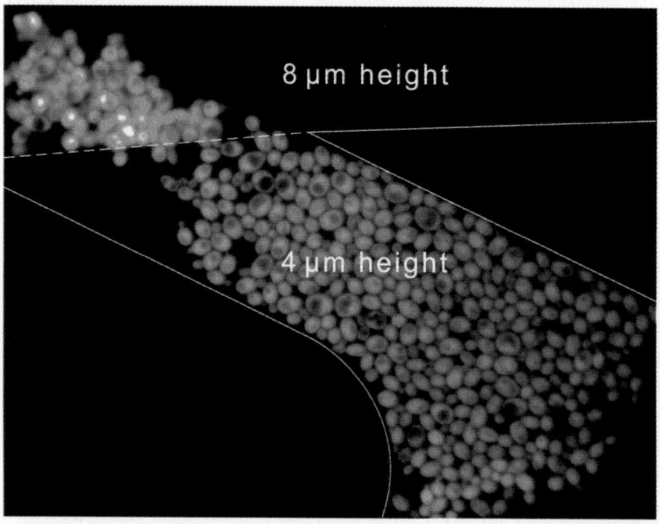

Fig. 7. Shallow trapping regions confine cells to a monolayer. Cells residing at the trapping region entrance highlight the benefit of a height-constrained growth environment, as cells just outside the chamber grow into a blurry aggregate.

previously unavailable. The unique capability of microfluidic chemostats for studying both the dynamics and the variability of biological processes within a population of living cells represents an important step toward bringing quantitative single-cell data to the field of synthetic biology.

4. Quantitative Modeling of Genetic Circuit Behavior

Several levels of detail are typically employed in modeling gene regulation.[39] In the "logical" or "binary" approach,[40] each gene is treated as having two discrete states, ON or OFF, and the dynamics describe how groups of genes act to change one another's states over time. Such models are relatively easy to implement, simplifying the examination of large sets of genes. A disadvantage of the logical approach is that the abstraction of genes to ON/OFF switches makes it difficult or impossible to include many of the details of cellular biology.

A more detailed level of description is used in the "chemical kinetics" or "rate equation" approach, where the variables of interest are the concentrations of individual proteins within the cell, and the dynamics describe the rates of production and decay of these proteins.[41–45] Such a model is a system of ordinary differential equations, permitting the modeler to apply the analytical techniques of nonlinear dynamics. These techniques have been developed considerably in recent decades, making the rate equation approach a promising avenue for combining mathematical analysis and computational simulation.

Though the basic rate equation approach is completely deterministic (no random component exists in the dynamical equations), the equations may be augmented with noise terms to account for fluctuations in concentration within the cell.[9,46,47] The "stochastic kinetics" modeling approach[8,48] provides the most detailed level of description; techniques for simulating the behavior of chemical reactions involving small numbers of molecules[48] are applied to the reactions involved in protein-DNA binding, transcription, and translation. This approach is impressively complete and yields a detailed picture of the behavior of the system modeled. However, such completeness comes at a high computational cost and sacrifices any immediate prospect of analytical treatment. Alternatively, the effects of internal noise can be treated by the master equation approach.[9,47] The advantage of this formulation is that analytic tools can be utilized.

4.1 *Modeling examples*

In this section, we utilize two examples to sketch the kinetic approach to modeling gene regulation. We first explore an unregulated (constitutive) example, where transcription and translation of a gene lead to the production of protein with no feedback. We will use this example to highlight how an inherent separation of time scales necessitates the use of multiple time scale analysis. As a second example, we will discuss a recent study that tightly coupled model with experiment in the context of an autoregulatory network. Here we will demonstrate how one utilizes nonlinear computational tools to analyze the model equations.

Constitutive Expression. Consider a simple example where a gene is transcribed and translated into a protein, and the protein can dimerize or degrade. We simplify even further and lump transcription and translation into a single step, denoting this as the production rate. The reactions describing the processes are:

$$\begin{aligned} &2X \underset{k_b}{\overset{k_f}{\Leftrightarrow}} X_2 \\ &D \overset{k_t}{\rightarrow} D + X \\ &X \overset{k_d}{\rightarrow}. \end{aligned} \tag{1}$$

The kinetic rate equations for the reactions are given by

$$\dot{x} = 2k_b y - 2k_f x^2 + k_t d - k_d x \tag{2}$$

$$\dot{y} = k_f x^2 - k_b y. \tag{3}$$

Kinetic rates are often difficult to ascertain, but it is generally thought that protein multimerization events occur on time scales that are many orders of magnitude faster than production and loss. In the context of our model, this implies that $k_f, k_b >> k_t, k_d$, and this fact can be used to reduce the number of equations from two to one. This is perhaps not so essential with only two equations, but in gene regulation there are typically a large number of "fast" reactions, and it is thus important to understand how to correctly reduce the system.

It may be tempting to assume that the dimerization reaction is in equilibrium with respect to the variable x, set $\dot{y} = 0$, and obtain

$$\dot{x} = k_t d - k_d x. \tag{4}$$

This equation is incorrect, though the technique for obtaining it is in widespread use. The problem with the reasoning is that the dimerization reaction is not in equilibrium with respect to x, since x is one of the players in the dimerization reaction. One correct approach is to first write the equations in terms of a natural small parameter. We are interested in phenomena which occur on time scales relevant to the production rate, so scale time by this rate by defining a dimensionless time $\tilde{t} = k_t t$. The small parameter is given by the quotient of the rates $\varepsilon = k_t/k_b$, and we also define an equilibrium constant $c = k_f/k_b$ which we assume is large compared with ε. Substituting these definitions into equations (2) and (3), we obtain

$$\dot{x} = \left(\frac{2}{\varepsilon}\right) y - \left(\frac{2c}{\varepsilon}\right) x^2 + d - \gamma x \tag{5}$$

$$\varepsilon \dot{y} = cx^2 - y \tag{6}$$

where $\gamma \equiv k_d/k_t$.

In this form, we can explicitly see the problem with assuming equilibrium and setting $\dot{y} = 0$, since sending $\varepsilon \to 0$ creates a divergent term in the $\dot{x}$ equation. The trick is to change variables; let $z = x + 2y$, and then in terms of z and x we have

$$\dot{z} = d - \gamma(z - 2y) \tag{7}$$

$$\varepsilon\dot{y} = c(z - 2y)^2 - y. \tag{8}$$

We may now set $\varepsilon = 0$, since this causes no divergence in the set of equations. This is equivalent to saying that the dimerization reaction is in equilibrium with respect to the *total* number of proteins given by the variable z. After setting $\varepsilon = 0$ and changing variables back to x, we obtain

$$\dot{x} = \left(\frac{1}{1+4cx}\right)(d - \gamma x). \tag{9}$$

Comparing with equation (4) we see that the correct equation carries an extra prefactor which affects the dynamics (it does not affect the steady state). Since the prefactor term is driven by the dimerization process, we make the general observation that multimerization slows the dynamical response of the gene regulatory system.

An Autoregulatory Network. We next consider a detailed example of regulation. We focus on recent work which utilized a theoretical model[46,49] to motivate the construction of an autoregulatory network in *E. coli.*[50] The approach of utilizing computational modeling is illustrated in Fig. 8. The design of the network was inspired by our theoretical results[46,49] predicting bistability in a single-gene system constructed from a promoter controlling the lysogenic state of the bacteriophage λ. Autoregulatory activation and protein multimerization yield nonlinearities which can produce a multistable regime in the steady-state protein concentration. These considerations lead to two critical design features for an engineered bistable single-gene network: (1) activation of sufficient strength as to induce a significant bistable regime and (2) the ability to probe for the bistable regime by tuning the degradation rate of the activator protein.

We briefly outline the derivation of the governing model equations for the autoregulatory system. For complete details, the full derivation can be downloaded from the web (http://biodynamics.ucsd.edu/publications/Isaacs-supplement.pdf). The chemical reactions describing the network of Fig. 8 are naturally divided into two categories — fast and slow. If we let X, X_2, and D_i denote the cI, cI dimer, and DNA promoter sites with i dimers bound, then we may write the fast reactions as

$$\begin{aligned} X + X &\underset{k_{-1}}{\overset{k_1/V}{\Leftrightarrow}} X_2 \\ D_i + X_2 &\underset{k_{-(i+2)}}{\overset{k_{(i+2)}/V}{\Leftrightarrow}} D_{(i+1)} \end{aligned} \tag{10}$$

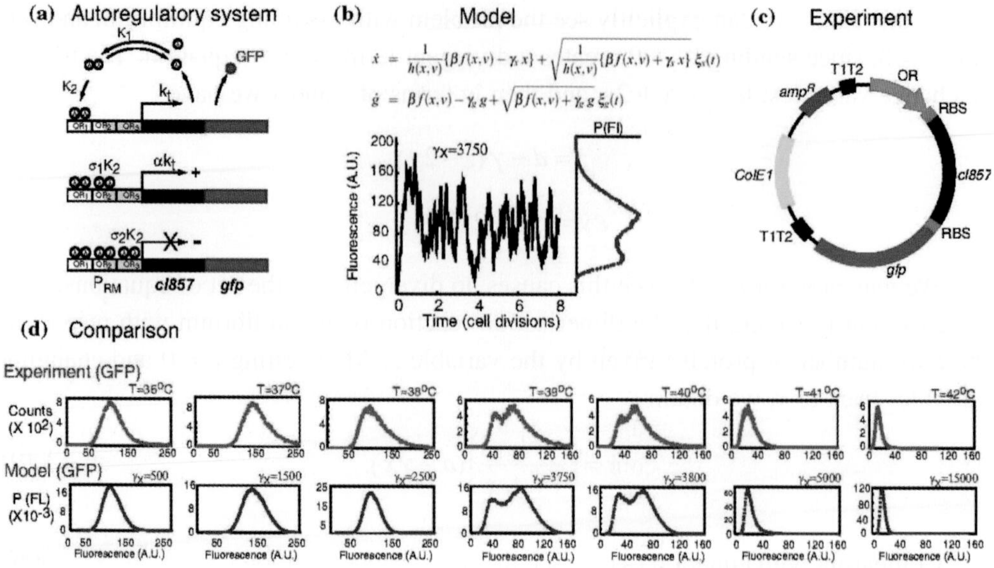

Fig. 8. Design process for an autoregulatory gene circuit. **(a)** The cI857 and gfp genes code for the cI857 protein (denoted by λ) and the green fluorescent protein (GFP), respectively. When two λ dimers are bound to the PRM promoter, transcription is enhanced by a factor α, whereas three dimers repress transcription. The parameters σi denote the strength of binding to the second (σ_1) and third (σ_2) operator sites relative to the first operator site. **(b)** The model is defined by Langevin equations which were systematically derived from the underlying biochemistry. These equations were simulated to produce time series and probability distribution data for cI857 and GFP at differing values of the cI857 degradation parameter γx. **(c)** Plasmids containing the cI857 and gfp genes were inserted into live cells. Since the cI857 protein denatures as the temperature is increased, the effective degradation of cI857 can be experimentally tuned. **(d)** Comparison of experimental and model results. In the experiment, cI857 was effectively degraded by increasing the temperature, while in the simulations cI857 was degraded by increasing the model parameter corresponding to degradation (γx).

where the indices in the second reaction denote the three protein-DNA binding reactions (i = 0, 1, 2), and the volume dependence is explicit in the reaction rates since it is a slowly evolving function of time due to the growth of the host cell.

The slow irreversible reactions are transcription, translation and protein denaturation or destabilization. If no cI is bound to the operator region, or if a single cI dimer is bound to OR1, transcription proceeds at a basal rate. If, however, a cI dimer is bound to OR2, the binding affinity of RNA polymerase to the promoter region is enhanced by a factor α_2, leading to an amplification of transcription. Denaturation is due to the temperature-induced destabilization of cI monomers. We write the reactions governing these processes as

$$\begin{aligned} D_i &\xrightarrow{\alpha_i k_t} D_i + X \\ X &\xrightarrow{k_d} \end{aligned} \tag{11}$$

where the subscripts in the first reaction denote the three states of the operator region ($i = 0, 1, 2$), and transcription and translation are modeled as a single reaction with rate constant $a_i k_t$. In the λ example, this production rate is unaffected when the operator region is vacant or bound by a single dimer ($\alpha_0 = \alpha_1 = 1$), and is amplified by a factor of approximately 11[51] when two dimers are bound ($\alpha_2 = 11$).

The cellular volume increase and division are modeled as follows. For time just after cell division to just before, we let the volume increase as $V = V_0 e^{\ln(2)\, T/\tau 0}$, where V_0 is the volume of the host cell at the beginning of its growth phase, and τ_0 is the cell division time. At times $T = q\tau_0$, where q is an integer, we let $V \rightarrow V/2$ and $n \rightarrow n/2$ for each of the protein species. This operation represents the halving of the volume at division, along with the redistribution of molecules to one daughter cell. Defining the dimensionless variables $v = V/V_0$ and $t = T/\tau_0$, we have $v = e^{\ln(2)t}$, so that, in these units, time is measured in terms of the cell division time ($t \in [0, 1]$) and the volume oscillates between 1 and 2.

Using the biochemical reactions described above, one can write dynamical equations for the number of cI and GFP molecules. The protein multimers and the complexes can be eliminated by utilizing the inherent separation of time scales; the multimerization processes are known to be governed by rate constants that are extremely fast with respect to cellular growth and transcription. This allows for algebraic substitution[49] and leads to the following set of equations governing the temporal evolution of the number of cI (x) and GFP (g) monomers,

$$\begin{aligned} \dot{x} &= \frac{1}{h(x,v)}(\mathrm{B}f(x,v) - \gamma_x x) \\ \dot{g} &= \eta \mathrm{B} f(x,v) - \gamma_g g \\ v &= e^{\ln(2)t} \end{aligned} \tag{12}$$

where the "synthesis function" $f(x,v)$ represents the net effect of transcription and translation, while $h(x,v)$ arises from the vast separation of time scales set by the transcription and protein dimerization rates.[9] A detailed analysis yields their functional form,

$$\begin{aligned} f &= \frac{m(1 + c\tilde{x}^2 + \alpha_2 \sigma_1 c^2 \tilde{x}^4)}{1 + c\tilde{x}^2 + \sigma_1 c^2 \tilde{x}^4 + \sigma_1 \sigma_2 c^3 \tilde{x}^6} \\ h &= 1 + 4c_1 \tilde{x} + 4\frac{cd\tilde{x}}{v} + 16\frac{\sigma_1 c^2 d\tilde{x}^3}{v} + 36\frac{\sigma_1 \sigma_2 c^3 d\tilde{x}^5}{v} \\ d &= \frac{m}{1 + c\tilde{x}^2 + \sigma_1 c^2 \tilde{x}^4 + \sigma_1 \sigma_2 c^3 \tilde{x}^6} \end{aligned} \tag{13}$$

where $\tilde{x} = x/v$ and $c \equiv c_1 c_2$. The form of the synthesis term $f(x,v)$ dictates the equilibrium number of repressor monomers, and its functional dependence on x along with the coefficients can be understood as follows. The even polynomials in x occur due to dimerization and subsequent binding to the promoter region. As depicted in Fig. 8, the σ_i prefactors denote the relative affinities for dimer binding to OR1 versus that of binding to OR2 ($\sigma 1$) and OR3 (σ_2). The prefactor $\alpha_2 > 1$ on the x^4 term is present because transcription is enhanced when the two operator sites OR1 and OR2 are occupied (x^2x^2). The x^6 term represents the occupation of all three operator sites, and arises in the denominator because dimer occupation of OR3 inhibits polymerase binding and shuts off transcription.

The results of our measurements of fluorescent protein expression from single cells agreed well with the simulation results from our model (Fig. 8). This experimental confirmation of our theoretical predictions has validated our mathematical modeling approach, and supported the notion of a forward-engineered genetic circuit. Importantly, the autoregulatory network forms a basic module that can be used to construct higher-order regulatory networks. For example, we have shown how the positive feedback nature of the module can be utilized as a basic element for a genetic relaxation oscillator.[39,49]

5. Conclusion: Towards the Future

Eukaryotic cells must respond rapidly to changes in environmental conditions. In microbial eukaryotes this is often reflected in the cell's ability to metabolize a wide variety of nutrient sources. For example, the yeast *S. cerevisiae* utilizes a sophisticated molecular network to allow the differential metabolism of galactose versus glucose as its primary carbon source. Once galactose is imported into the cytoplasm by the activity of GAL2, galactose-dependent activation of a well-characterized cascade of transcriptional events leads to the expression of essential metabolic proteins (Fig. 9).

While the specific signals may differ, higher eukaryotic cells, like yeast, must also respond rapidly to changes in environmental conditions. The cell's ability to mount the appropriate transcriptional response upon exposure to specific cues is crucial for every cellular function — from controlled cell growth and proliferation to metabolism of vital nutrients in the cellular environment. Hence, understanding how cells respond to dynamic environments will provide important insights into similar mechanisms in higher eukaryotes.

In addition, microbial studies of the GAL regulatory network may yield specific insights into the galactose regulatory network in humans that are relevant to human pathology. Human galactosemia is an inherited defect caused by genetic mutations that render the cell incapable of properly metabolizing galactose, and it is usually diagnosed within the first week of life when infants show an adverse response to lactose in milk. If milk is not removed from the diet, infants with galactosemia die from the toxic effects of one of the metabolic intermediates of lactose metabolism, galactose-1-phosphate,

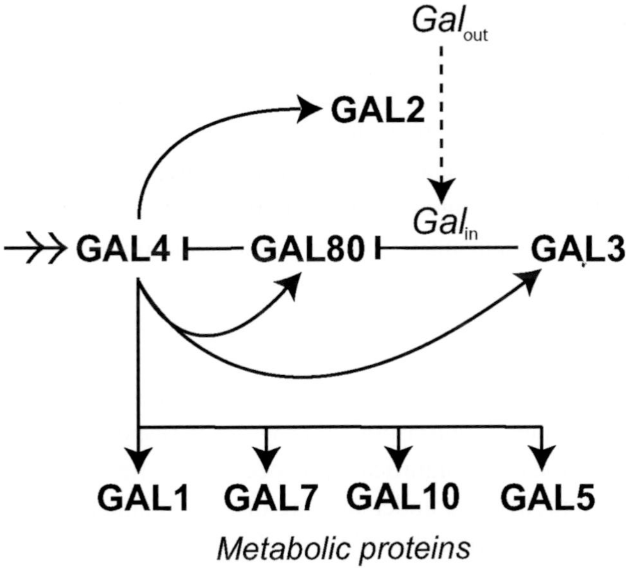

Fig. 9. The galactose utilization network. The GAL4 protein is constitutively produced (double arrow) and is a global transcriptional activator (solid arrows) for the galactose genes. Negative feedback: GAL4 activates GAL80, which in turn binds to GAL4 and renders it unable to activate the galactose genes. Positive feedback: galactose is transported (dashed line) across the cell membrane via GAL2 and forms a complex with GAL3. Positive feedback results from the binding of the GAL3/galactose complex to GAL80, which renders GAL80 unable to bind and inhibit GAL4 activation.

which, in healthy cells, is converted to glucose-1-phosphate. In some cases, and for reasons that are still unclear, even if lactose is rapidly withdrawn from the diet, the patient will continue to suffer long-term complications.

Eukaryotic cells take up galactose by transporters such as GAL2 in yeast and its well-conserved mammalian orthologs.[52,53] Productive utilization of galactose in yeast requires the proper expression of a number of metabolic proteins with human orthologs that have been implicated in galactosemia: GAL1 and GAL31 (galactokinases), Gal7 (a uridyltranserase), and GAL10 (an epimerase). Similar metabolic functions have been attributed to the human homologs of each of these proteins — GALK, GALT, and GALE, respectively, and defects in the functions of these proteins have important pathological effects. Patients with insufficient levels of GALK exhibit neonatal symptoms, most notably neonatal cataracts. Patients with transferase or epimerase deficiencies, on the other hand, exhibit a broad array of potentially deadly symptoms including diarrhea, vomiting, hapatosplenomegaly, and *E. coli* sepsis. Long-term complications include mental and growth retardation, ataxia, and premature ovarian failure in girls.[54]

As with other diseases, *S. cerevisiae* has proven to be a valuable genetic and biochemical model for studies of galactose metabolism in eukaryotes. For example, yeast and human cells that are defective in the transferase (GAL7 null mutants in yeast) experience severe toxicity in the presence of galactose and accumulate dangerously toxic levels of galactose-1-phosphate.[55] Overexpression of the protein yUGP1 in yeast cells or, remarkably, the mammalian protein UGP2, reverses the effect of the GAL7 null phenotype by lowering galactose-1–phosphate levels. This study highlights the important similarities between the yeast galactose metabolic pathway in humans and in yeast, and it is consistent with the growing understanding that studies in yeast can lead to important insights into the mechanisms of similar pathways in humans. In fact, a significant number of genes that have been implicated in human disease have orthologs in yeast, making the biochemically and genetically tractable organism a particularly attractive model system.[56,57]

One of the challenges to understanding human metabolic diseases such as galactosemia is that cells mount a transcriptional program in response to dynamic changes in environmental conditions. The significant prospect for the field of synthetic biology lies in the fact that a comprehensive understanding of physiological disorders requires new approaches aimed at evaluating the effect of dynamic environmental changes at the gene regulatory level. The utilization of synthetic biology for constructing the underlying regulatory network from the "ground up," coupled with microfluidic technology for the introduction of a dynamically changing environment, provides an exciting framework for systematically testing hypotheses regarding the cellular response to changing environmental cues.

Systems biology has grown rapidly in the wake of the human genome project, as it has become clear that an integration of experimental and computational research will be required to quantitatively describe complex biological systems. The utilization of high-throughput technologies has led to the successful reconstruction of gene regulatory networks in many organisms,[58–61] along with the development of quantitative models for many complex and fundamental cellular processes.[62–67] To complement the progress of genome-scale measurement technologies, a primary focus of synthetic biology is to model and construct novel genetic circuits that reproduce the behavior of natural systems and contribute to our understanding of how complex biological functions arise from the connectivity of gene regulatory networks.[68–72] The significant potential of synthetic biology rests on the ability to reliably design and build synthetic gene circuits and to quantitatively monitor the behavior of these novel networks for comparison with computational models. The rapid growth of the field and the impressive studies that have already been accomplished highlight the potential that synthetic biology has to make great contributions to our understanding of natural living systems.

References

1. C. C. Guet, M. B. Elowitz, W. Hsing and S. Leibler, Combinatorial synthesis of genetic networks, *Science* **296**: 1466–1470 (2002).
2. A. Goldbeter, Computational approaches to cellular rhythms, *Nature* **420**: 238–245 (2002).

3. M. B. Elowitz and S. Leibler, A synthetic oscillatory network of transcriptional regulators, *Nature* **403**: 335–338 (2000).
4. E. Fung, W. W. Wong, J. K. Suen, T. Bulter, S. G. Lee and J. C. Liao, Asynthetic gene-metabolic oscillator, *Nature* **435**: 118–122 (2005).
5. M. R. Atkinson, M. A. Savageau, J. T. Myers and A. J. Ninfa, Development of genetic circuitry exhibiting toggle switch or oscillatory behavior in *Escherichia coli*, *Cell* **113**: 597–607 (2003).
6. A. Becskei and L. Serrano, Engineering stability in gene networks by autoregulation, *Nature* **405**: 590–593 (2000).
7. H. H. McAdams and A. Arkin, Stochastic mechanisms in gene expression, *Proc. Natl. Acad. Sci. USA* **94**: 814–819 (1997).
8. A. Arkin, J. Ross and H. H. McAdams, Stochastic kinetic analysis of developmental pathway bifurcation in phage λ-infected *Escherichia coli* cells, *Genetics* **149**: 1633–1648 (1998).
9. T. B. Kepler and T. C. Elston, Stochasticity in transcriptional regulation: origins, consequences, and mathematical representations, *Biophys. J.* **81**: 3116–3136 (2001).
10. J. Paulsson, O. G. Berg and M. Ehrenberg, Stochastic focusing: fluctuation-enhanced sensitivity of intracellular regulation, *Proc. Natl. Acad. Sci. USA* **97**: 7148–7153 (2000).
11. A. M. Arias and P. Hayward, Filtering transcriptional noise during development: concepts and mechanisms, *Nat. Rev. Genet.* **7**: 34–44 (2006).
12. M. Thattai and A. van Oudenaarden, Intrinsic noise in gene regulatory networks, *Proc. Natl. Acad. Sci. USA* **98**: 8614–8619 (2001).
13. E. M. Ozbudak, M. Thattai, I. Kurtser, A. D. Grossman and A. van Oudenaarden, Regulation of noise in the expression of a single gene, *Nat. Genet.* **31**: 69–73 (2002).
14. M. B. Elowitz, A. J. Levine, E. D. Siggia and P. S. Swain, Stochastic gene expression in a single cell, *Science* **297**: 1183–1186 (2002).
15. J. M. Raser and E. K. O'Shea, Control of stochasticity in eukaryotic gene expression, *Science*, **304**: 1811–1814 (2004).
16. J. M. Pedraza and A. van Oudenaarden, Noise propagation in gene networks, *Science* **307**: 1965–1969 (2005).
17. N. Rosenfeld, J. W. Young, U. Alon, P. S. Swain and M. B. Elowitz, Gene regulation at the single-cell level, *Science* **307**: 1962–1965 (2005).
18. D. Volfson, J. Marciniak, N. Ostroff, W. Blake, L. S. Tsimring and J. Hasty, Origins of extrinsic variability in eukaryotic gene expression, *Nature* **439**: 861–864 (2006).
19. R. Ashton, C. Padala and R. S. Kane, Microfluidic separation of DNA, *Curr. Opin. Biotechnol.* **14**: 497–504 (2003).
20. B. Paegel, R. Blazej and R. Mathies, Microfluidic devices for DNA sequencing: sample preparation and electrophoretic analysis, *Curr. Opin. Biotechnol.* **14**: 42–50 (2003).
21. N. Lion, T. Rohner, L. Dayon, I. Arnaud, E. Damoc, N. Youhnovski, Z. Wu, C. Roussel, J. Josserand, H. Jensen, J. S. Rossier, M. Przybylski and H. H. Girault, Microfluidic systems in proteomics, *Electrophoresis* **24**: 3533–3562 (2003).
22. A. Groisman, M. Enzelberger and S. Quake, Microfluidic memory and control devices, *Science* **300**: 955–958 (2003).
23. D. Huh, W. Gu, Y. Kamotani, J. B. Grotberg and S. Takayama. Microfluidics for flow cytometric analysis of cells and particles, *Physiol. Meas.* **26**: R73–R98 (2005).
24. J. M. Ng, I. Gitlin, A. D. Stroock and G. M. Whitesides, Components for integrated poly(dimethylsiloxane) microfluidic systems, *Electrophoresis* **23**: 3461–3673 (2002).
25. H. Mao, T. Yang and P. S. Cremer, A microfluidic device with a linear temperature gradient for parallel and combinatorial measurements, *J. Am. Chem. Soc.* **124**: 4432–4435 (2002).
26. S. K. W. Dertinger, D. T. Chiu, N. L. Jeon and G. M. Whitesides, Generation of gradients having complex shapes using microfluidic networks, *Anal. Chem.* **73**: 1240–1246 (2001).

27. F. Lin, W. Saadi, S. W. Rhee, S. J. Wang, S. Mittal and N. L. Jeon, Generation of dynamic temporal and spatial concentration gradients using microfluidic devices, *Lab Chip* **4**: 164–167 (2004).
28. J. W. Hong, V. Studer, G. Hang, W. F. Anderson and S. Quake, A nanoliter-scale nucleic acid processor with parallel architecture, *Nature Biotechnol.* **22**: 435–439 (2004).
29. A. Fu, H. Chou, C. Spence, F. Arnold and S. Quake, An integrated microfabricated cell sorter, *Anal. Chem.* **74**: 2451–2457 (2002).
30. N. Q. Balaban, J. Merrin, R. Chait, L. Kowalik and S. Leibler, Bacterial persistence as a phenotypic switch, *Science* **305**: 1622–1625 (2004).
31. A. R. Wheeler, W. R. Throndset, R. J. Whelan, A. M. Leach, R. N. Zare, Y. H. Liao, K. Farrell, I. D. Manger and A. Daridon, Microfluidic device for single-cell analysis, *Anal. Chem.* **75**: 3581–3586 (2003).
32. A. Tourovskaia, X. Figueroa-Masot and A. Folch, Differentiation-on-a-chip: a microfluidic platform for long-term cell culture studies, *Lab Chip* **5**: 14–19 (2005).
33. A. Groisman, C. Lobo, H. Cho, J. K. Campbell, Y. S. Dufour, A. M. Stevens and A. Levchenko, A microfluidic chemostat for experiments with bacterial and yeast cells, *Nat. Methods* **2**: 685–689 (2005).
34. F. K. Balagadde, L. You, C. L. Hansen, F. H. Arnold and S. R. Quake, Long-term monitoring of bacteria undergoing programmed population control in a microchemostat, *Science* **309**: 137–140 (2005).
35. S. Cookson, N. Ostroff, W. L. Pang, D. Volfson and J. Hasty, Monitoring dynamics of single-cell gene expression over multiple cell cycles, *Mol. Syst. Biol.* **1**: 2005.0024 (2005).
36. N. Tesla. U.S. Patent No. 1,329,559, 3 Feb 1920.
37. D. C. Duffy, O. J. A. Schueller, S. T. Brittain and G. M. Whitesides, Rapid prototyping of microfluidic switches in poly(dimethyl siloxane) and their actuation by electro-osmotic flow, *J. Micromech. Microeng.* **9**: 211–217 (1999).
38. S. Bendib and O. Français, Analytical study of microchannel and passive microvalve "application to micropump simulator." In: *Design, Characterisation, and Packaging for MEMS and Microelectronics 2001* (Adelaide, Australia, 2001), pp. 283–291.
39. J. Hasty, D. McMillen, F. Isaacs and J. J. Collins, Computational studies of gene regulatory networks: *in numero* molecular biology, *Nat. Rev. Genet.* **2**: 268–279 (2001).
40. L. Glass and S. A. Kauffman, The logical analysis of continuous, non-linear biochemical control networks, *J. Theor. Biol.* **39**: 103–129 (1973).
41. M. A. Savageau, Comparison of classical and autogenous systems of regulation in inducible operons, *Nature* **252**: 546–549 (1974).
42. J. Reinitz and J. R. Vaisnys, Theoretical and experimental analysis of the phage lambda genetic switch implies missing levels of co-operativity, *J. Theor. Biol.* **145**: 295–318 (1990).
43. A. D. Keller, Model genetic circuits encoding autoregulatory transcription factors, *J. Theor. Biol.* **172**: 169–185 (1995).
44. P. Smolen, D. A. Baxter and J. H. Byrne, Frequency selectivity, multistability, and oscillations emerge from models of genetic regulatory systems, *Am. J. Physiol.* **274**: C531–C542 (1998).
45. D. M. Wolf and F. H. Eeckman, On the relationship between genomic regulatory element organization and gene regulatory dynamics, *J. Theor. Biol.* **195**: 167–816 (1998).
46. J. Hasty, J. Pradines, M. Dolnik and J. J. Collins, Noise-based switches and amplifiers for gene expression, *Proc. Natl. Acad. Sci. USA* **97**: 2075–2080 (2000).
47. W. Bialek, *Advances in Neural Information Processing Systems 13* (The MIT Press, Cambridge, MA, 2001).
48. D. T. Gillespie, Exact stochastic simulation of coupled chemical-reactions, *J. Phys. Chem.* **81**: 2340–2361 (1977).
49. J. Hasty, F. Isaacs, M. Dolnik, D. McMillen and J. J. Collins, Designer gene networks: towards fundamental cellular control, *Chaos* **11**: 207–220 (2001).

50. F. J. Isaacs, J. Hasty, C. R. Cantor and J. J. Collins, Prediction and measurement of an autoregulatory genetic module, *Proc. Natl. Acad. Sci. USA* **100**: 7714–7719 (2003).
51. M. Ptashne, A. Jeffrey, A. D. Johnson, R. Maurer, B. J. Meyer, C. O. Pabo, T. M. Roberts and R. T. Sauer, How the λ repressor and cro work, *Cell* **19**: 1–11 (1980).
52. J. O. Nehlin, M. Carlberg and H. Ronne, Yeast galactose permease is related to yeast and mammalian glucose transporters, *Gene* **85**: 313–319 (1989).
53. K. Szkutnicka, J. F. Tschopp, L. Andrews and V. P. Cirillo, Sequence and structure of the yeast galactose transporter, *J. Bacteriol* **171**: 4486–4493 (1989).
54. K. Lai and M. I. Klapa, Alternative pathways of galactose assimilation: could inverse metabolic engineering provide an alternative to galactosemic patients?, *Metab. Eng.* **6**: 239–244 (2004).
55. K. Lai and L. J. Elsas, Overexpression of human udp-glucose pyrophosphorylase rescues galactose-1–phosphate uridyltransferase-deficient yeast, *Biochem. Biophys. Res. Commun.* **271**: 392–400 (2000).
56. D. E. Bassett, Jr., M. A. Basrai, C. Connelly, K. M. Hyland, K. Kitagawa, M. L. Mayer, D. M. Morrow, A. M. Page, V. A. Resto, R. V. Skibbens and P. Hieter, Exploiting the complete yeast genome sequence, *Curr. Opin. Genet. Dev.* **6**: 763–766 (1996).
57. C. M. Coughlan and J. L. Brodsky, Use of yeast as a model system to investigate protein conformational diseases, *Mol. Biotechnol.* **30**: 171–180 (2005).
58. S. Tavazoie, J. D. Hughes, M. J. Campbell, R. J. Cho and G. M. Church, Systematic determination of genetic network architecture, *Nat. Genet.* **22**: 281–285 (1999).
59. R. U. Ibarra, J. S. Edwards and B. O. Palsson, *Escherichia coli* K-12 undergoes adaptive evolution to achieve *in silico* predicted optimal growth, *Nature* **420**: 186–189 (2002).
60. T. Ideker, V. Thorsson, J. A. Ranish, R. Christmas, J. Buhler, J. K. Eng, R. Bumgarner, D. R. Goodlett, R. Aebersold and L. Hood, Integrated genomic and proteomic analyses of a systematically perturbed metabolic network, *Science* **292**: 929–934 (2001).
61. T. S. Gardner, D. di Bernardo, D. Lorenz and J. J. Collins, Inferring genetic networks and identifying compound mode of action via expression profiling, *Science* **301**: 102–105 (2003).
62. L. L. Breeden, Periodic transcription: a cycle within a cycle, *Curr. Biol.* **13**: R31–R38 (2003).
63. I. Simon, J. Barnett, N. Hannett, C. T. Harbison, N. J. Rinaldi, T. L. Volkert, J. J. Wyrick, J. Zeitlinger, D. K. Gifford, T. S. Jaakkola and R. A. Young, Serial regulation of transcriptional regulators in the yeast cell cycle, *Cell* **106**: 697–708 (2001).
64. B. Vogelstein, D. Lane and A. J. Levine, Surfing the p53 network, *Nature* **408**: 307–310 (2000).
65. K. W. Kohn and Y. Pommier, Molecular interaction map of the p53 and Mdm2 logic elements, which control the Off-On switch of p53 in response to DNA damage, *Biochem. Biophys. Res. Commun.* **331**: 816–827 (2005).
66. T. J. Begley and L. D. Samson, Network responses to DNA damaging agents, *DNA Repair (Amst.)* **3**: 1123–1132 (2004).
67. J. Bartek, C. Lukas and J. Lukas, Checking on DNA damage in S phase, *Nat. Rev. Mol. Cell Biol.* **5**: 792–804 (2004).
68. J. Hasty, D. McMillen and J. J. Collins, Engineered gene circuits, *Nature* **420**: 224–230 (2002).
69. S. Basu, Y. Gerchman, C. H. Collins, F. H. Arnold and R. Weiss, A synthetic multicellular system for programmed pattern formation, *Nature* **434**: 1130–1134 (2005).
70. S. Mangan, A. Zaslaver and U. Alon, The coherent feedforward loop serves as a sign-sensitive delay element in transcription networks, *J. Mol. Biol.* **334**: 197–204 (2003).
71. N. Rosenfeld, J. W. Young, U. Alon, P. S. Swain and M. B. Elowitz, Gene regulation at the single-cell level, *Science* **307**: 1962–1965 (2005).
72. J. M. Pedraza and A. van Oudenaarden, Noise propagation in gene networks, *Science* **307**: 1965–1966 (2005).

CHAPTER 26

NETWORK GENOMICS

Trey Ideker

Abstract

Network Genomics is an emerging area of bioengineering which models the influence of genes (hence, genomics) in the context of a larger biomolecular system or network. A *biomolecular network* is a comprehensive collection of molecules and molecular interactions that regulate cellular function. Molecular interactions include physical binding events between proteins and proteins, proteins and DNA, or proteins and drugs, as well as genetic relationships dictating how genes combine to cause particular phenotypes. *Thinking about* biological systems as networks goes hand-in-hand with our ability to experimentally *measure and define* biomolecular interactions at large scale. Once we have catalogued all of the interactions present in a network, we may begin to ask questions such as: "How many different molecules are bound by a typical protein?"; "What is the topological structure of the network?"; "How are signals transmitted through the network in response to internal and external events?"; "Which parts of the network are evolutionarily conserved across species, and which parts differ?" Perhaps most importantly, we can begin to use the interaction network as a storehouse of information from which to extract and construct computer-based models of cellular processes and disease.

1. Introduction

The idea that the best way to learn about a biological system is by comprehensively mapping all of its components has been most successful in the context of the Human Genome Project.[1,2] Throughout the late 1980s and 1990s, a large team of researchers worked to determine the complete DNA sequences of all genes in the human genome and deposited this information into public databases such as GenBank.[3] Nowadays, whenever biologists are confronted with a DNA sequence corresponding to an unknown gene, they use software programs such as BLAST[4] to query GenBank for sequences that are similar. Chances are that some of these similar sequences will correspond to genes or proteins with functions that have been at least partially characterized, and by association, it can be inferred that the function of the novel sequence is related. Starting from an initial query sequence, searching a genome database efficiently yields information about how this sequence is positioned in a greater functional context (Fig. 1, left).

In our "post-genomic" era, focus is shifting from the function of individual proteins to understanding how many proteins interact together to govern signaling, regulatory, structural, and metabolic processes. In contrast to the systematic methods of genome sequencing, however, efforts to identify and characterize cellular processes have typically proceeded in a molecule-directed fashion, beginning with an initial protein of interest and trying to establish other proteins that are involved in the same process. These approaches implicate additional proteins that are possibly involved in the same pathway, which can themselves be used as a basis for future genetic and biochemical experiments.

Although molecule-directed approaches have been successful in assembling most knowledge we have about cellular function to date, they are associated with several

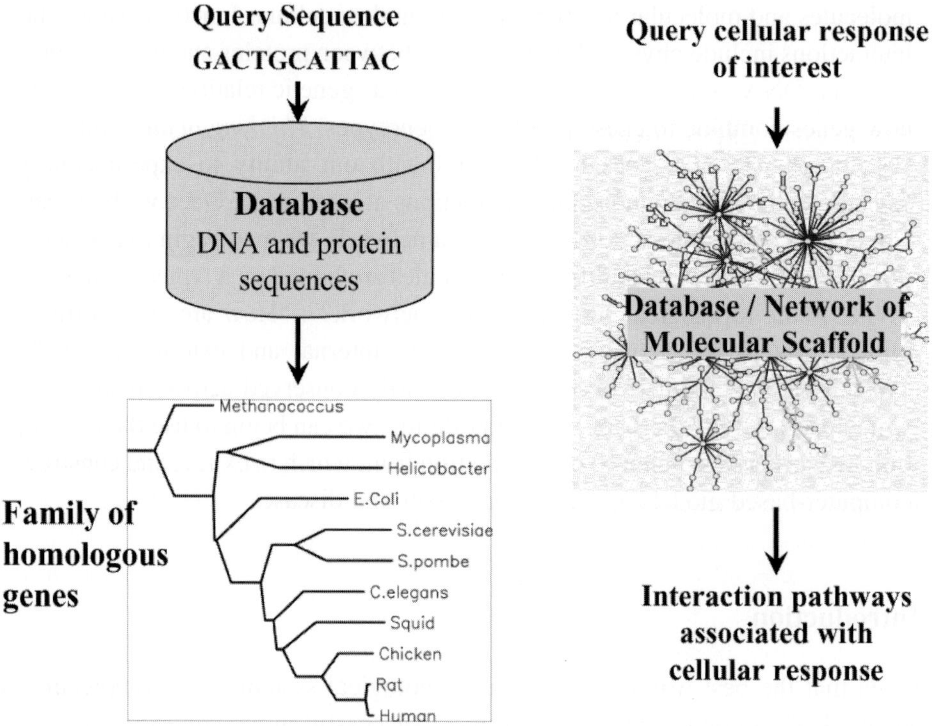

Fig. 1. From protein sequences to networks. Much of bioinformatics involves methods to search large databases of biomolecular information. Such methods have been used widely in the context of the Human Genome Project (left) to search databases of DNA and protein sequences for identical or near matches versus a query. Network genomics methods (right) also perform a database search, in this case of a molecular interaction network to extract the particular interaction pathways and complexes associated with a cellular response or disease of interest.

inherent difficulties. First is the time required: accurate models of cellular processes emerge only after evidence has been accumulated over many years, researchers and laboratories. Second, these approaches do not directly reveal how different cellular processes influence each other or may reveal this crosstalk only accidentally. Third, despite encouraging efforts to construct consolidated functional databases[5] (http://stke.sciencemag.org; http://www.afcs.org), the vast amount of information on the various cellular processes remains fairly decentralized, being buried in the primary literature.

To circumvent some of these problems, an increasing number of researchers are adopting a systematic approach to mapping cellular processes and disease (Fig. 1, right). The key preliminary step involves building a comprehensive database of molecular interactions that broadly covers many aspects of cellular function and physiological responses. Although this step constitutes a sizable initial investment, the molecular interaction database provides a broad foundation for more directed studies to follow. Just as we may use BLAST to query the human genome for sequences of interest, network discovery and search tools enable us to query the molecular interaction network to identify and map cellular processes of interest in a systematic fashion.

2. Systematic Experiments for Characterizing Networks and States

Biologists shed light on molecular interaction networks using two complementary approaches (Table 1).

First, it is possible to systematically measure the molecular interactions themselves, by screening for protein-protein, protein-DNA and small molecular interactions. Several methods are available for measuring *protein-protein interactions* at large scale — two of the most popular being the yeast two-hybrid system[6,7] and protein co-immunoprecipitation (coIP) in conjunction with tandem mass spectrometry.[8,9] Although the vast majority of protein interactions have been generated for the budding yeast *Saccharomyces cerevisiae*, protein interactions are becoming available for a variety of other species, including *Helicobacter pylori*[10] and *Caenorhabditis elegans*,[11] and are catalogued in public databases such as BIND[12] and DIP™.[13] One drawback of these high-throughput measurements is an associated high error rate.[14] As discussed further below, one approach for addressing this problem may be to integrate several complementary data sets together (e.g. two-hybrid interactions with coIP data or gene expression profiles) to reinforce the common signal. Figure 2a shows one example of a large protein-protein interaction network for yeast that has been experimentally defined using the two-hybrid approach,[6,15] while the basic two-hybrid methodology is illustrated in Fig. 3a.

Protein-DNA interactions, which commonly occur between transcription factors and their DNA binding sites, constitute another interaction type that can be measured at high-throughput. Lee *et al.*[16] used the technique of chromatin immunoprecipitation to characterize the complete set of promoter regions bound under nominal conditions for

Table 1. Large scale interaction datasets and sources (see text).

	1) *Directly observe the interactions*	2) *Observe states induced by the interactions*
	Protein→DNA interactions	**Gene expression**
METHOD	Chromatin immunoprecipitation followed by microarray analysis	DNA microarrays; SAGE
DATABASE	• **TRANSFAC** transfac.gbf.de/TRANSFAC/ • **BIND** www.bind.ca/	• **GEO** www.ncbi.nlm.nih.gov/geo/ • **ArrayExpress** www.ebi.ac.uk/microarray/ArrayExpress/
	Protein–protein interactions	**Protein levels, locations, modifications**
METHOD	Two hybrid system; Co-immunoprecipitation followed by mass spectrometry	Mass spectrometry; 2D PAGE; Protein tagging followed by fluorescence microscopy; Protein arrays
DATABASE	• **BIND** www.bind.ca/ • **DIP** dip.doe-mbi.ucla.edu/ • **BRITE** www.genome.ad.jp/brite/ • **MIPS** mips.gsf.de	• **SWISS-2DPAGE** us.expasy.org/ch2d/ • **TRIPLES** ygac.med.yale.edu/triples/ • **Scansite** scansite.mit.edu/
	Metabolic interactions and reactions	**Metabolite and drug levels**
METHOD	No truly systematic measurements, although protein arrays show promise	Mass spectrometry; 2D NMR Current challenge is to determine the molecular identity of all distinct compounds detected
DATABASE	• **MetaCyc** biocyc.org/metacyc/ • **KEGG** www.genome.ad.jp/kegg/ • **Klotho** www.biocheminfo.org/klotho/	Public repositories of metabolic profiles not widely available, although data exchange standards for expression profiles, such as MAGE-ML, may support metabolic data in future.

each of >100 transcription factors in yeast, yielding >5000 novel protein-DNA interactions in that organism. Figure 2b shows an example protein-DNA transcriptional network in sea urchin as defined by Davidson *et al.*,[17] while Fig. 3b illustrates the chromatin immunoprecipitation technique.

A third type of interaction that is often measured at large scale is the so-called *genetic interaction* (Fig. 2c). Unlike protein-protein and protein-DNA interactions, genetic interactions refer not to direct physical relationships between molecules, but rather to how two genes can combine (either directly or indirectly) to cause a particular phenotype such as cell death. Genetic interactions consist of two major varieties: *synthetic lethal* interactions, in which mutations in two non-essential genes are lethal when combined; and *suppressor* interactions, in which one mutation is lethal but combination with a second restores cell viability. Screens for genetic interactions have been used extensively to shed light on pathway organization in model organisms,[2,18–20] while in humans, genetic interactions are critical in linkage analysis of complex diseases[21] and

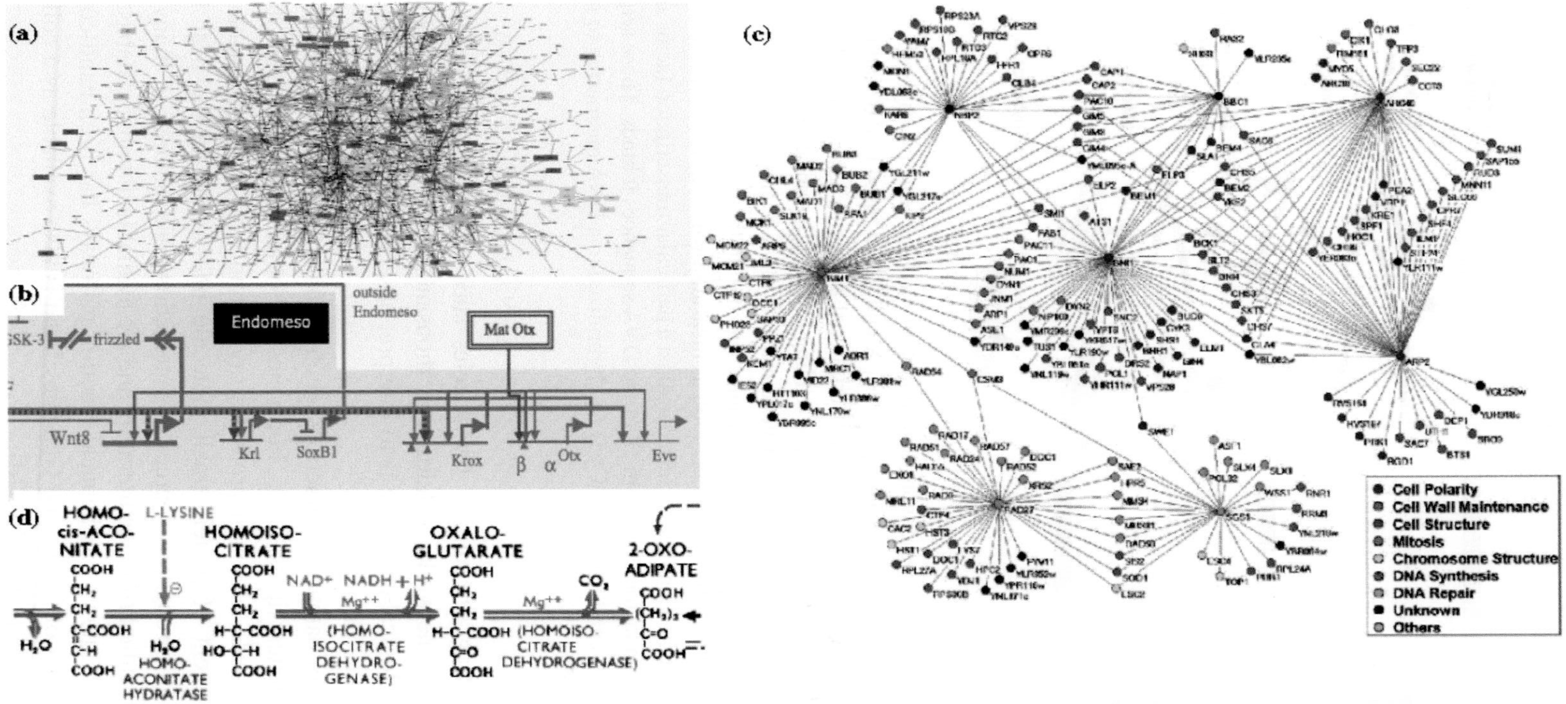

Fig. 2. Different large-scale molecular interaction networks. **(a)** Pairwise protein-protein interactions.[15] **(b)** Protein-DNA interactions between transcription factors and the promoters they bind *in vivo*.[17] **(c)** Genetic interactions in the form of synthetic-lethal relationships.[24] **(d)** Interactions and reactions among metabolites (courtesy of Roche Applied Sciences: see http://www.expasy.org/cgi-bin/search-biochem-index).

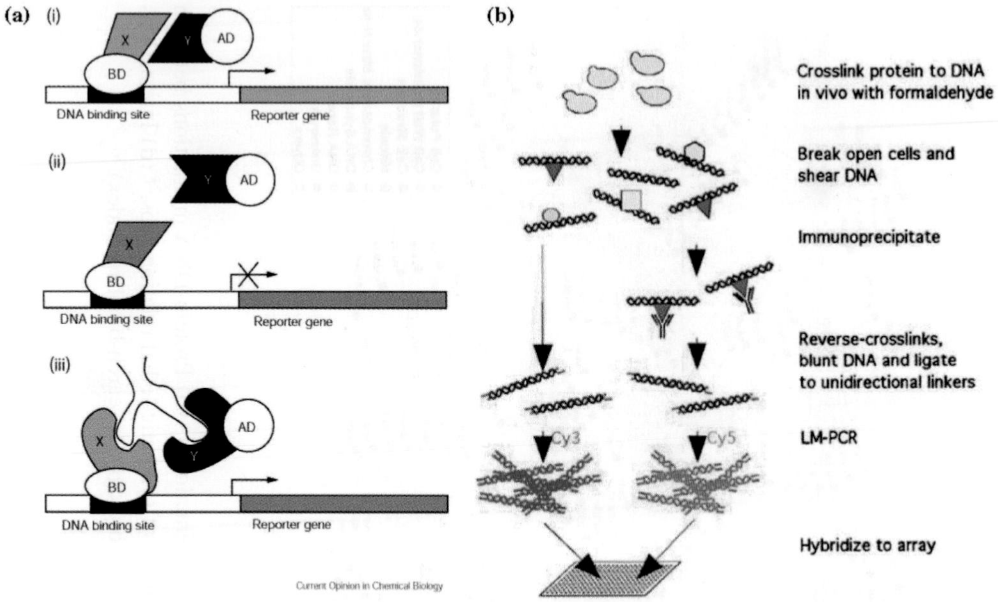

Fig. 3. Experimental methods for network determination. **(a)** The yeast two-hybrid system for measuring protein-protein interactions. Two proteins, X and Y, are fused to the binding domain (BD) and activation domain (AD) of a yeast transcriptional activator, respectively. The interaction between X and Y reconstitutes the activator, leading to transcription of the reporter gene as shown by the arrow in panel (i). The reporter gene remains off if X and Y do not interact (ii), although artifacts can occur when a third molecule has the ability to bridge X and Y (iii). Adapted from Drees.[49] **(b)** Protein-DNA interaction as measured with chromatin immunoprecipitation. The steps illustrate how antibodies raised against a particular transcription factor are used to extract that transcription factor along with all of the DNA sequences it binds. These sequences are labeled with a fluorescent dye (Cy3 or Cy5) and hybridized to a DNA microarray to identify them. From Simon *et al.*[50]

in high-throughput drug screening.[22] For species such as yeast, experiments have defined large genetic networks cataloguing thousands of such interactions.[18,23,24] A major, ongoing biological challenge is to interpret observed genetic interactions in a mechanistic, physical cellular context.[20]

Additional types of pathway interactions, such as those between proteins and small molecules (carbohydrates, lipids, drugs, hormones and other metabolites), are difficult to measure at large scale, although protein array technology[25–27] might enable high-throughput measurement of protein-small molecule interactions in the near future. However, where high-throughput sources of interactions are lacking, in many cases these data have been culled from the literature and are available in public databases. For instance, the set of all known biochemical reaction pathways has been hand-coded in the Kyoto Encyclopedia of Genes and Genomes[28] and is also available as a collection of searchable pictorial diagrams (Fig. 2d).

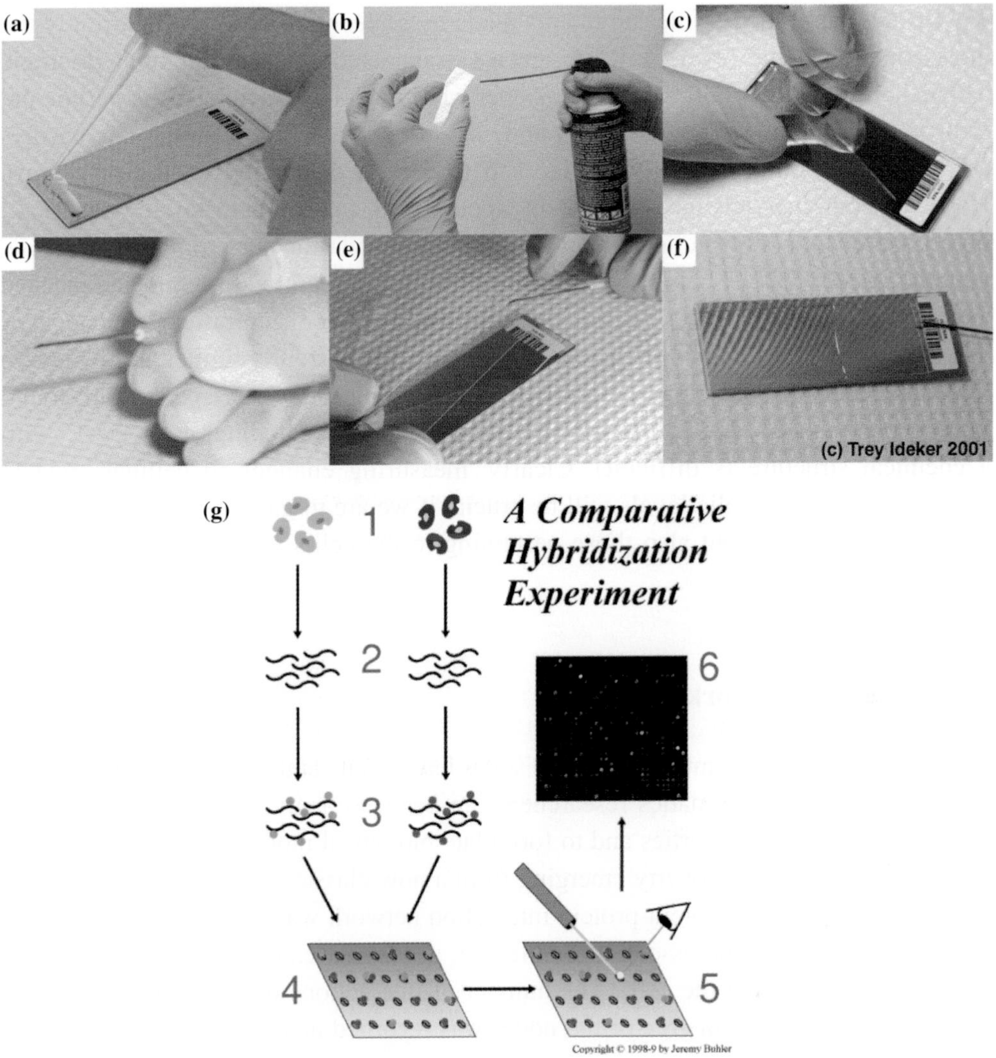

Fig. 4. Network behavior explored through gene expression analysis. Panels **(a)** to **(f)** illustrate a portion of the standard laboratory procedure for measuring gene expression with a DNA microarray experiment, while **(g)** provides an overview of the experimental process. The steps show **(a)** depositing the labeled RNA (or cDNA) on the microarray surface, preparing the coverslip **(b)** and **(c)**, and using a medium-gauge needle **(d)** to slowly lower the coverslip onto the slide **(e)** and **(f)**. From Ideker *et al.*[51] and http://www.cs.wustl.edu/~jbuhler/research/array/.

In addition to characterizing molecular interactions, it is also important to measure the molecular and cellular states induced by the interaction wiring. For example, global changes in gene expression are measured with DNA microarrays (see Fig. 4),[29] whereas changes in protein abundance,[30] protein phosphorylation state[31] and metabolite concentrations[32] can be quantified with mass spectrometry, NMR and other advanced techniques.

Of these approaches, measurements made by DNA microarrays are currently the most comprehensive (every mRNA species is detected), high-throughput (a single technician can assay multiple conditions per week), well-characterized (experimental error is appreciable, but understood), and cost-effective (whole-genome microarrays are purchased commercially for US$50 to US$1000, depending on the organism). However, continued advances in protein labeling and separation technology are making measurements of protein abundance and phosphorylation state almost as feasible, with the primary barrier being the expense and expertise required to set up and manage a mass spectrometry facility. Measurement of metabolite concentrations, an endeavor otherwise known as metabolomics/metabonomics,[33] is currently limited not by detection (thousands of peaks, each representing a different molecular species, are found in a typical NMR spectrum) but by identification (matching each peak with a chemical structure is difficult). Clearly, measuring changes in cellular state at the protein and metabolic levels will be crucial if we are to gain insight into not only regulatory pathways, but also those pertaining to the cell's signaling and metabolic circuitry.

3. Analysis of Network Properties

Once enough molecular interactions and states have been catalogued in publicly accessible databases, bioinformatics researchers can begin to analyze them to characterize their global network properties and to formulate biological models. Figure 5 illustrates an interesting network property emerging from a now classic analysis by Jeong *et al.*[34] In this study, the yeast protein-protein interaction network was combined with a list of yeast genes known to be essential for life (i.e. genes that cannot be deleted from the genome without killing the cell). The number of interactions for each protein node in the network (called the *degree* of the node) was recorded and proteins binned according to their degree.

As shown in Fig. 5, proteins of high degree were essential much more frequently than proteins of low degree. Hence, the global network property: the most highly connected proteins are those that are most essential for life. Observations such as this one are a means to understanding how networks exert control on cellular processes through their signaling, regulatory, and metabolic pathways.

4. Networks as Pathway Models

Numerous review articles and textbooks provide diagrams outlining the signaling, regulatory, or metabolic mechanisms that govern a cellular process, where these mechanisms are often referred to as *pathways*. At a high level, pathways and networks are parallel concepts. Like networks, pathways include some number of molecular components

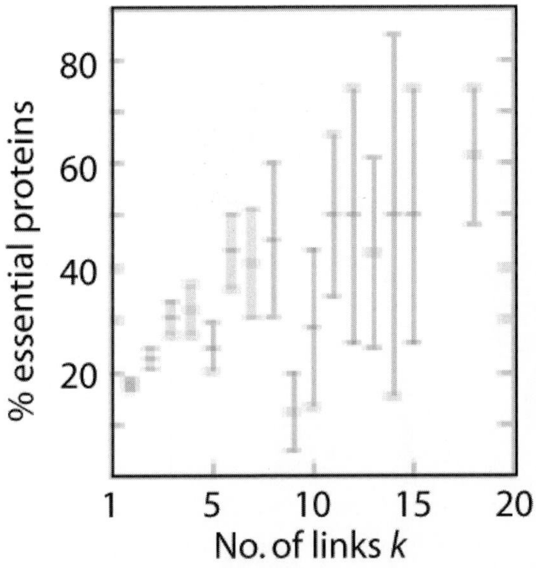

Fig. 5. An example of global network property. Proteins that are highly connected in the protein network tend to also be essential for life. The figure shows that in an analysis of the yeast protein-protein interaction network, the more links a protein has (x-axis), the more likely that deletion of the yeast gene encoding that protein is lethal to the cell (% essential, y-axis). Adapted from Jeong *et al.*[34]

such as genes, proteins and/or small molecules wired together by intermolecular interactions. However, unlike the large-scale and systematic networks of the previous section, pathway diagrams usually invoke a relatively small number of components, are carefully tailored to illustrate a predetermined concept, and rely on accompanying textual descriptions.

5. An Example Network Representation: The Galactose Utilization System in Yeast

Figure 6 gives an example of a pathway diagram and corresponding network representation[35] for the well-studied process of galactose utilization in yeast (see Lohr *et al.*[36] for a review of this pathway).

To arrive at the network representation, we constructed a yeast molecular interaction network from the 2709 protein-protein interactions compiled by Schwikowski *et al.*[15] and 317 protein-DNA interactions present in either of two publicly accessible online databases (as of July, 2000): TRANSFAC[37] or the *Saccharomyces cerevisiae* Promoter Database (SCPD).[5] The Cytoscape network visualization and modeling platform www.cytoscape.org[38] was used to layout and display these physical interactions as a two-dimensional network, focusing on the region containing the *GAL* genes.

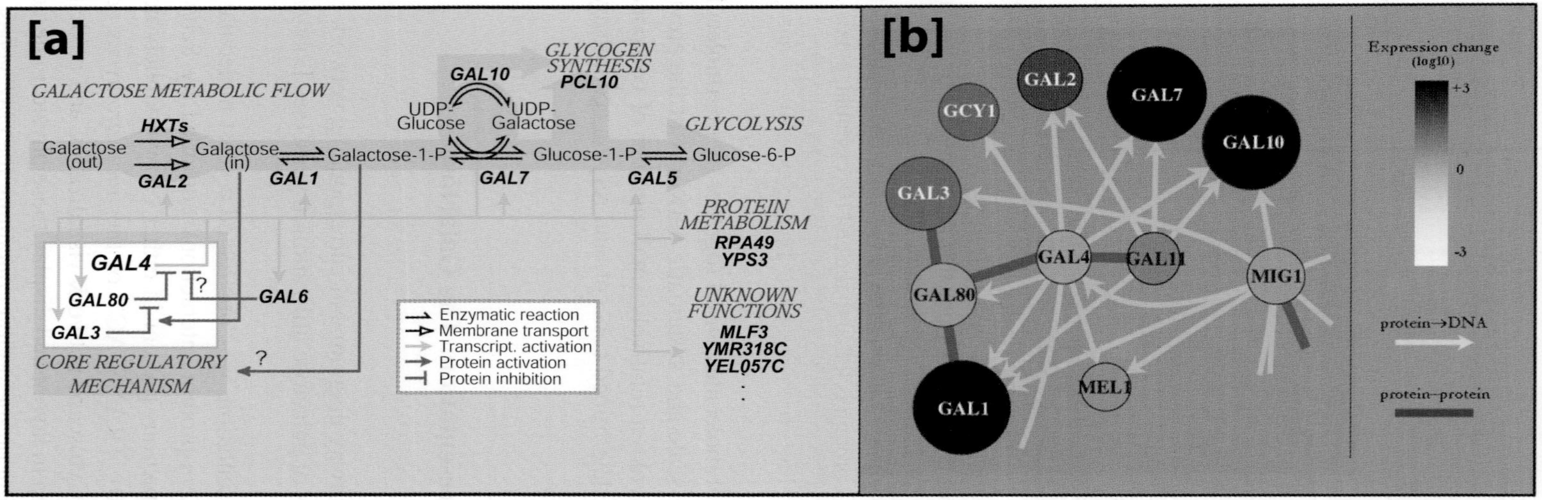

Fig. 6. Networks encode the biochemical "circuitry" responsible for cellular function. **(a)** Classical diagram of the galactose utilization pathway alongside **(b)** its corresponding network representation. The galactose-utilization system employs at least nine genes: four encode the enzymes that catalyze the conversion of galactose to glucose-6-phosphate (*GAL1*, *5*, *7*, and *10*), while a fifth (*GAL2*) encodes a transporter molecule that sets the state of the system. If galactose is present in the yeast cell, the system is turned on; if galactose is absent, the system is turned off. The transcription factors Gal3, Gal4, Gal80, and possibly Gal6 regulate this on/off switch. In the network, nodes represents genes, while edges between nodes signify either protein-protein (blue edges) or protein→DNA (yellow directed edges) interactions. Background gray represents no change in expression; increasing shades of gray represent increasing levels of mRNA expression; and decreasing shades of gray represent decreasing levels of expression in response to a *GAL80* knockout (red node).

The network conveys much of the same high level information as the pathway diagram. According to both the diagram and the network, Gal4 is a transcription factor that binds to the promoters of many other *GAL* genes, thereby regulating their transcription through protein-DNA interactions. Both representations show clearly that the activity of Gal4 may be influenced by protein-protein interactions with Gal80. The correspondence between the pathway and network representations, however, is not exact. For instance, this particular pathway diagram does not include genes like *GAL11* and *MIG1* that appear in the more systematic interaction network, whereas the interaction network misses the regulation of *GAL5* and does not include enzymatic reactions.

Notice that the network specifies only which protein-DNA interactions can take place: unlike the more flexible pathway diagram representation, it does not dictate whether the interactions activate or repress transcription, whether the effect on transcription is rapid or gradual, or in the case that multiple interactions affect a gene, how these interactions should be combined to produce an overall level of mRNA for the gene. Similarly, the network does not specify whether the Gal80p-Gal4p protein-protein interaction, as shown in Fig. 6, results in these proteins forming a functional complex or whether one protein modifies another. Since these levels of information are not encoded in the protein-protein and protein-DNA databases, they are also absent from the network display. Much of this information is known outside the context of the databases: classic genetic and biochemical experiments[6,36] have determined that Gal4p is a strong transcriptional activator, and that Gal80p can bind to Gal4p to repress this function. This information therefore appears in the classic pathway representation but not in the network.

Color is used to integrate information about the perturbation state of the network. In this case, node colors show changes in mRNA expression level in response to a deletion of *GAL80*. When *GAL80* is deleted (indicated in red), we see strong increases in expressions of *GAL1*, *7*, and *10* (indicated by their grayscale intensities). Importantly, we can begin to explain why we see these changes using interactions present in the underlying network. *GAL80* connects to the downstream affected genes through a path of length two: *GAL80*—*GAL4*→*GAL1*, *7*, *10*, and it is a reasonable hypothesis that this is the path by which a *GAL80* deletion evokes the observed changes in expression.

6. Decomposing the Network into Functional Modules

We have seen that in small networks or network regions, such as the one shown in Fig. 6, it is possible to visually inspect the network for interactions that explain changes in expression and other phenomena. However, visual inspection becomes much more difficult in large networks such as those shown in Fig. 2. Instead, researchers have turned to bioinformatic tools able to automatically extract pathways of interest from the global molecular interaction network. Implementing these network search tools is

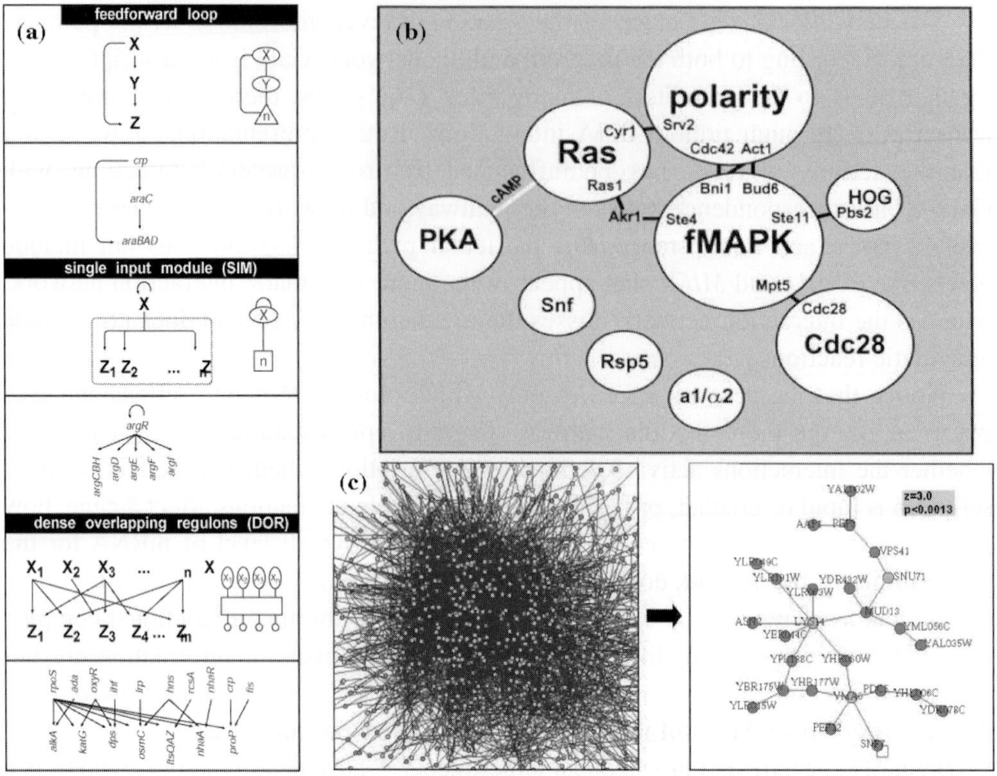

Fig. 7. Decomposing the network into its functional modules. **(a)** Recurrent network motifs;[39] **(b)** modules with significant biological functions;[40] and **(c)** modules activated in response to DNA damage.[44]

analogous to using BLAST to extract matching sequences from a DNA sequence database (see Sec. 1).

For example, in one study, a network of ~600 protein-DNA interactions from *E. coli* was searched to identify small *network motifs.*[39] Network motifs are patterns of interactions (i.e. subnetwork topologies) that recur with higher-than-expected frequencies. Three recurrent network motifs are shown in Fig. 7a: a feed-forward loop, in which a transcription factor X regulates another transcription factor Y and a protein Z that is also regulated by Y; a single input module, in which one transcription factor directly regulates many genes; and a so-called dense overlapping regulon, involving two or more transcription factors regulating a downstream layer of genes in a combinatorial fashion.

Each of these motifs occurs more frequently in the *E. coli* network than in random networks. For instance, the feed-forward loop appeared 34 times in the real network versus. 4.4 ± 3 over 1000 random networks ($P < 0.001$). Searching for motifs reduces the large and complex network of interactions to a small number of "manageable pieces" that, by virtue of their statistical significance, are likely to be biologically meaningful.

A related means of organizing the network is to partition it into a series of *network modules*, in which proteins have many interactions within modules but few interactions between modules. Of course, such an approach presupposes that biological networks are organized in this way, i.e. contain this modular structure. Rives and Galitski[40] identified modular structure in the yeast protein-protein interaction network using a hierarchical clustering technique — similar to techniques that have been used to construct evolutionary trees based on DNA sequences[41] or to cluster microarray profiles.[42] To perform this clustering, they first computed the shortest distance between each pair of proteins in the network. For example, if proteins A and B are directly connected in the network their shortest distance will be one, whereas if A and B do not directly interact but are bridged by a third protein C such that the path A-C-B exists, their shortest distance will be two. These shortest-distance data were then used to identify clusters of proteins for which all shortest distances within the cluster were small, and these were designated as network modules. Figure 7b shows modular structure for the region of the yeast network regulating the process of cellular filamentation. Each large circle encapsulates a specific module of densely connected proteins.

7. Finding Network Modules that are Active Under Specific Biological Conditions

Another method of filtering the network to identify pathways is through determination of *active* modules, i.e. connected regions of the network whose genes are associated with significant coordinated changes in mRNA expression, phenotype, or other molecular state.[43] Identifying active modules reduces network complexity by pinpointing just those regions whose states are perturbed by the conditions of interest, while removing false-positive interactions and interactions not involved in the perturbation response. The remaining subnetworks represent concrete hypotheses regarding the underlying signaling and regulatory mechanisms in the cell.

Recently, Begley *et al.*[44] used such an approach to map genes and pathways required for the cellular response to DNA damage. For each gene-knockout strain in yeast (libraries of all single gene-knockout strains are publicly available), they tested whether the strain was able to grow in the presence of MMS, a powerful DNA-damaging agent. Wild-type cells can, in fact, grow under a moderate concentration of MMS, but many gene-knockout mutants either grow slowly or not at all under these conditions.

How do these "MMS-sensitive genes" map onto the protein-protein and protein-DNA interaction network? Figure 7c shows one network module containing a significant number of MMS-sensitive proteins. In this figure, a node is colored green if deletion of that gene results in slow growth or death in the presence of MMS, red if the deletion has no effect for growth in MMS, and gray if the node has not yet been tested by phenotypic growth assay. Of the gene-knockouts tested, approximately 400 of them were MMS-sensitive. Using the automated screen for pathways, 100 of these were associated with an active network module having many other MMS-sensitive nodes in close proximity.

8. Networks and Evolution

The rapid growth of protein network information raises a host of new questions in evolutionary and comparative biology. Given that protein sequences and structures are conserved in and among species, are networks of protein interactions conserved as well? Is there some minimal set of protein interaction pathways required for all species? Which interactions are present in pathogens but not in their hosts? Can we measure evolutionary distance at the level of network connectivity, rather than at the level of DNA or protein sequence? Mounting evidence suggests that conserved protein interaction pathways indeed exist and may be ubiquitous: for example, proteins in the same pathway are typically present or absent in a genome as a group,[45] and several hundred protein-protein interactions in the yeast network have also been identified for the homologous proteins in worms.[46]

One method to identify conserved interaction pathways involves a search for high-scoring *pathway alignments* using an algorithm called PathBLAST.[47] Pathway alignments consist of two paths, one from each network, in which proteins of the first path are paired with putative orthologs occurring in the same order in the second path (Fig. 8a). Pathway alignments are scored by the degree of protein sequence similarity at each pathway position and by the quality of the protein interactions they contain. PathBLAST implements an efficient search through all possible alignments among two networks to identify the highest scoring pathway alignments overall.

PathBLAST was used in a series of whole-network comparisons among the protein-protein interaction networks of the budding yeast *Saccharomyces cerevisiae* and the bacterial pathogen *Helicobacter pylori*.[47] Both the yeast (14,489 interactions among 4688 proteins, assembled from mass spectrometry and two-hybrid studies) and the *H. pylori* networks (1465 interactions among 732 proteins from a single two-hybrid study[10]) were extracted from the DIP database.[48]

Kelley *et al.* analyzed the yeast and bacterial networks to select the 150 highest scoring pathway alignments of length four (four proteins per path). By combining all overlapping pathway alignments, they found that each of the 150 fell into one of five conserved network regions, two of which are shown in Figs. 8b and 8c. Analysis of these findings reveals many well-characterized interaction pathways as well as many unanticipated pathways whose significance is reinforced by their presence in the networks of both species. PathBLAST has also been used to identify *paralogous* interaction pathways and complexes — that is, pathways whose proteins and interactions

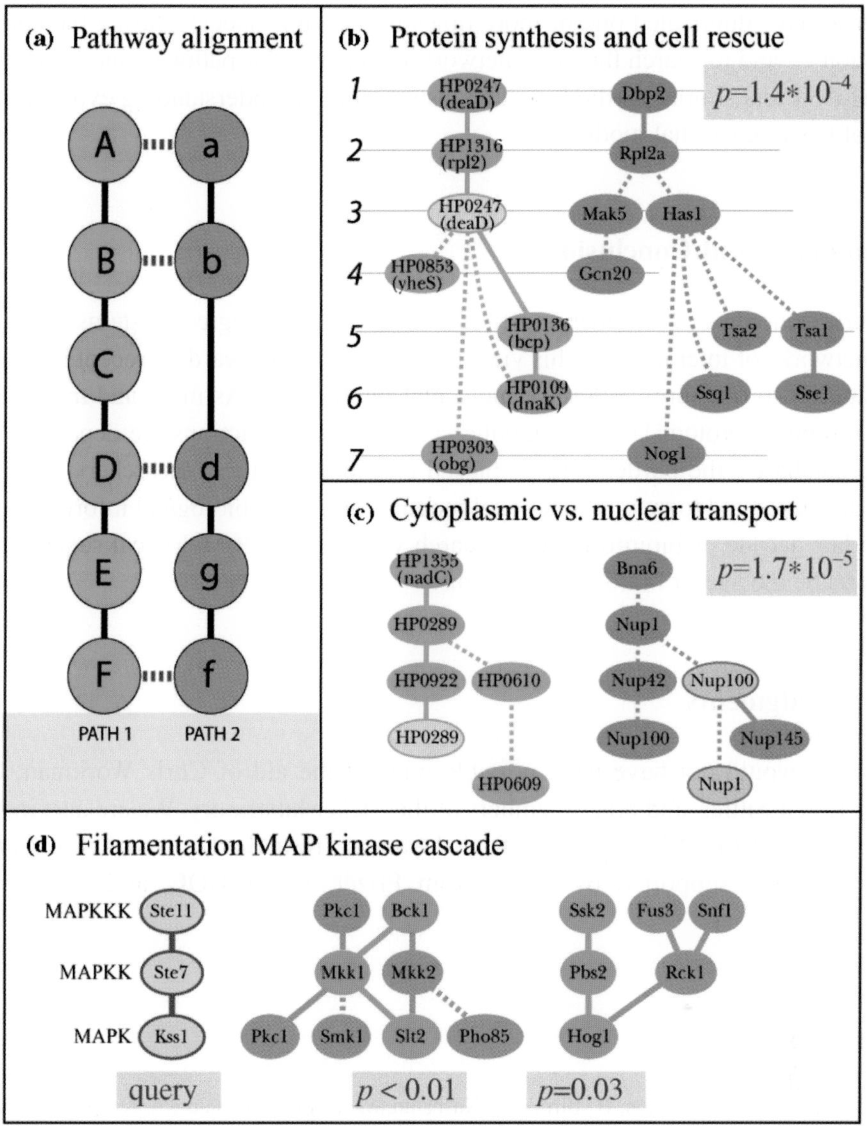

Fig. 8. Evolutionary conservation between protein networks. **(a)** A model pathway alignment between two protein networks, where vertical solid lines indicate direct protein-protein interactions within a single pathway and horizontal dotted lines link proteins with significant sequence similarity. An interaction in one pathway may skip over a protein in the other (protein C) introducing a "gap," while proteins at a particular position that are dissimilar in sequence (proteins E and g) introduce a "mismatch." Panels **(b)** and **(c)** show aligned network regions extracted from the complete protein interaction networks of *H. pylori* (orange; left) vs. *S. cerevisiae* (green; right). Bacterial/yeast protein pairs with significant sequence similarity are placed on the same row [e.g. deaD and Dbp2 in row 1 of **(b)**]. **(d)** Querying the yeast network with a specific MAP kinase pathway involved in the yeast filamentation response. In panels **(b)** to **(d)**, solid links indicate direct protein interactions, whereas dotted links indicate proteins connected at distance two through a common neighbor (i.e. a gap).[47]

appear to have duplicated one or more times in the same species during the course of evolution — and to search the yeast network with particular pathway queries (Fig. 8d). PathBLAST and similar methods will be instrumental in understanding evolution at the level of basic functional modules.

9. Summary and Conclusions

Network Genomics is an emerging field which views biological systems in terms of their networks of interactions. This view has been driven forward by recent increases in our capacity to measure large scale molecular interactions. As more and more data on protein-protein, protein-DNA, metabolic, and genetic interactions are deposited in the public databases, the burden will be placed on bioinformatics researchers to organize these databases and to extract functional modules and other biological information. Just as the last decade of bioinformatics research was the age of biological sequences, the next will be the age of biological networks.

Acknowledgments

This work would not have been possible without the aid of Chris Workman, Owen Ozier, Taylor Sittler, and other members of the Ideker laboratory. We are also indebted to our collaborators Richard Karp, Roded Sharan, Tom Begley, and Leona Samson. This work was supported by grants from Pfizer, PNNL/DOE, and the Whitaker Foundation.

References

1. E. S. Lander, L. M. Linton, B. Birren, C. Nusbaum, M. C. Zody, J. Baldwin, K. Devon, K. Dewar, M. Doyle, W. FitzHugh *et al.*, Initial sequencing and analysis of the human genome, *Nature* **409**: 860–921 (2001).
2. J. C. Venter, M. D. Adams, E. W. Myers, P. W. Li, R. J. Mural, G. G. Sutton, H. O. Smith, M. Yandell, C. A. Evans, R. A. Holt *et al.*, The sequence of the human genome, *Science* **291**: 1304–1351 (2001).
3. D. A. Benson, I. Karsch-Mizrachi, D. J. Lipman, J. Ostell and D. L. Wheeler, GenBank, *Nucleic Acids Res.* **31**: 23–27 (2003).
4. S. F. Altschul, W. Gish, W. Miller, E. W. Myers and D. J. Lipman, Basic local alignment search tool, *J. Mol. Biol.* **215**: 403–410 (1990).
5. P. D. Karp, Pathway databases: a case study in computational symbolic theories, *Science* **293**: 2040–2044 (2001).
6. P. Uetz, L. Giot, G. Cagney, T. A. Mansfield, R. S. Judson, J. R. Knight, D. Lockshon, V. Narayan, M. Srinivasan, P. Pochart *et al.*, A comprehensive analysis of protein-protein interactions in *Saccharomyces cerevisiae*, *Nature* **403**: 623–627 (2000).

7. S. Fields and O. Song, A novel genetic system to detect protein-protein interactions, *Nature* **340**: 245–246 (1989).
8. C. Gavin, M. Bosche, R. Krause, P. Grandi, M. Marzioch, A. Bauer, J. Schultz, J. M. Rick, A. M. Michon, C. M. Cruciat *et al.*, Functional organization of the yeast proteome by systematic analysis of protein complexes, *Nature* **415**: 141–147 (2002).
9. Y. Ho, A. Gruhler, A. Heilbut, G. D. Bader, L. Moore, S. L. Adams, A. Millar, P. Taylor, K. Bennett, K. Boutilier *et al.*, Systematic identification of protein complexes in Saccharomyces cerevisiae by mass spectrometry, *Nature* **415**: 180–183 (2002).
10. J. C. Rain, L. Selig, H. De Reuse, V. Battaglia, C. Reverdy, S. Simon, G. Lenzen, F. Petel, J. Wojcik, V. Schachter *et al.*, The protein-protein interaction map of *Helicobacter pylori*, *Nature* **409**: 211–215 (2001).
11. J. Walhout, R. Sordella, X. Lu, J. L. Hartley, G. F. Temple, M. A. Brasch, N. Thierry-Mieg and M. Vidal, Protein interaction mapping in *C. elegans* using proteins involved in vulval development, *Science* **287**: 116–122 (2000).
12. G. D. Bader, I. Donaldson, C. Wolting, B. F. Ouellette, T. Pawson and C. W. Hogue, BIND — The Biomolecular Interaction Network Database, *Nucleic Acids Res.* **29**: 242–245 (2001).
13. I. Xenarios and D. Eisenberg, Protein interaction databases, *Curr. Opin. Biotechnol.* **12**: 334–339 (2001).
14. C. M. Deane, L. Salwinski, I. Xenarios and D. Eisenberg, Protein interactions: two methods for assessment of the reliability of high throughput observations, *Mol. Cell Proteomics* **1**: 349–356 (2002).
15. B. Schwikowski, P. Uetz and S. Fields, A network of protein-protein interactions in yeast, *Nat. Biotechnol.* **18**: 1257–1261 (2000).
16. T. I. Lee, N. J. Rinaldi, F. Robert, D. T. Odom, Z. Bar-Joseph, G. K. Gerber, N. M. Hannett, C. T. Harbison, C. M. Thompson, I. Simon *et al.*, Transcriptional regulatory networks in *Saccharomyces cerevisiae*, *Science* **298**: 799–804 (2002).
17. E. H. Davidson, J. P. Rast, P. Oliveri, A. Ransick, C. Calestani, C. H. Yuh, T. Minokawa, G. Amore, V. Hinman, C. Arenas-Mena *et al.*, A genomic regulatory network for development, *Science* **295**: 1669–1678 (2002).
18. J. L. Hartman, B. Garvik and L. Hartwell, Principles for the buffering of genetic variation, *Science* **291**: 1001–1004 (2001).
19. L. Avery and S. Wasserman, Ordering gene function: the interpretation of epistasis in regulatory hierarchies, *Trends Genet.* **8**: 312–316 (1992).
20. L. Guarente, Synthetic enhancement in gene interaction: a genetic tool come of age, *Trends Genet.* **9**: 362–366 (1993).
21. P. Sham, Shifting paradigms in gene-mapping methodology for complex traits, *Pharmacogenomics* **2**: 195–202 (2001).
22. S. Dolma, S. L. Lessnick, W. C. Hahn and B. R. Stockwell, Identification of genotype-selective antitumor agents using synthetic lethal chemical screening in engineered human tumor cells, *Cancer Cell* **3**: 285–296 (2003).
23. L. S. Huang and P. W. Sternberg, *Methods in Cell Biology*, Vol. 48, eds. H. F. Epstein and D. C. Shakes (Academic Press, San Diego, 1995), pp. 99–122.
24. H. Tong, M. Evangelista, A. B. Parsons, H. Xu, G. D. Bader, N. Page, M. Robinson, S. Raghibizadeh, C. W. Hogue, H. Bussey *et al.*, Systematic genetic analysis with ordered arrays of yeast deletion mutants, *Science* **294**: 2364–2368 (2001).
25. G. MacBeath and S. L. Schreiber, Printing proteins as microarrays for high-throughput function determination, *Science* **289**: 1760–1763 (2000).
26. H. Zhu, M. Bilgin, R. Bangham, D. Hall, A. Casamayor, P. Bertone, N. Lan, R. Jansen, S. Bidlingmaier, T. Houfek *et al.*, Global analysis of protein activities using proteome chips, *Science* **293**: 2101–2105 (2001).

27. B. B. Haab, M. J. Dunham and P. O. Brown, Protein microarrays for highly parallel detection and quantitation of specific proteins and antibodies in complex solutions, *Genome Biol.* **2**: research0004.1–research0004.13 (2001).
28. M. Kanehisa, S. Goto, S. Kawashima and A. Nakaya, The KEGG databases at GenomeNet, *Nucleic Acids Res.* **30**: 42–46 (2002).
29. J. L. DeRisi, V. R. Iyer and P. O. Brown, Exploring the metabolic and genetic control of gene expression on a genomic scale, *Science* **278**: 680–686 (1997).
30. S. P. Gygi, B. Rist, S. A. Gerber, F. Turecek, M. H. Gelb and R. Aebersold, Quantitative analysis of complex protein mixtures using isotope-coded affinity tags, *Nat. Biotechnol.* **17**: 994–999 (1999).
31. H. Zhou, J. D. Watts and R. Aebersold, A systematic approach to the analysis of protein phosphorylation, *Nat. Biotechnol.* **19**: 375–378 (2001).
32. J. L. Griffin, C. J. Mann, J. Scott, C. C. Shoulders and J. K. Nicholson, Choline containing metabolites during cell transfection: an insight into magnetic resonance spectroscopy detectable changes, *FEBS Lett.* **509**: 263–266 (2001).
33. J. K. Nicholson, J. Connelly, J. C. Lindon and E. Holmes, Metabonomics: a platform for studying drug toxicity and gene function, *Nat. Rev. Drug Discov.* **1**: 153–161 (2002).
34. H. Jeong, S. P. Mason, A. L. Barabasi and Z. N. Oltvai, Lethality and centrality in protein networks, *Nature* **411**: 41–42 (2001).
35. T. Ideker, V. Thorsson, J. A. Ranish, R. Christmas, J. Buhler, J. K. Eng, R. Bumgarner, D. R. Goodlett, R. Aebersold and L. Hood, Integrated genomic and proteomic analyses of a systematically perturbed metabolic network, *Science* **292**: 929–934 (2001).
36. D. Lohr, P. Venkov and J. Zlatanova, Transcriptional regulation in the yeast GAL gene family: a complex genetic network, *FASEB J.* **9**: 777–787 (1995).
37. E. Wingender, X. Chen, E. Fricke, R. Geffers, R. Hehl, I. Liebich, M. Krull, V. Matys, H. Michael, R. Ohnhauser *et al.*, The TRANSFAC system on gene expression regulation, *Nucleic Acids Res.* **29**: 281–283 (2001).
38. P. Shannon, A. Markiel, O. Ozier, N. S. Baliga, J. T. Wang, D. Ramage, N. Amin, B. Schwikowski and T. Ideker, Cytoscape: a software environment for integrated models of biomolecular interaction networks, *Genome Res.* **13**: 2498–2504 (2003).
39. S. S. Shen-Orr, R. Milo, S. Mangan and U. Alon, Network motifs in the transcriptional regulation network of *Escherichia coli*, *Nat. Genet.* **31**: 64–68 (2002).
40. A. W. Rives and T. Galitski, Modular organization of cellular networks, *Proc. Natl. Acad. Sci. USA* **100**: 1128–1133 (2003).
41. W.-H. Li, *Molecular Evolution* (Sinauer Associates, Inc., Sunderland, 1997).
42. B. Eisen, P. T. Spellman, P. O. Brown and D. Botstein, Cluster analysis and display of genome-wide expression patterns, *Proc. Natl. Acad. Sci. USA* **95**: 14863–14868 (1998).
43. T. Ideker, O. Ozier, B. Schwikowski and A. F. Siegel, Discovering regulatory and signalling circuits in molecular interaction networks, *Bioinformatics* **18**(Suppl. 1): S233–240 (2002).
44. T. J. Begley, A. S. Rosenbach, T. Ideker and L. D. Samson, Damage recovery pathways in *Saccharomyces cerevisiae* revealed by genomic phenotyping and interactome mapping, *Mol. Cancer Res.* **1**: 103–112 (2002).
45. D. Karp, M. Riley, S. M. Paley and A. Pellegrini-Toole, The MetaCyc Database, *Nucleic Acids Res.* **30**: 59–61 (2002).
46. L. R. Matthews, P. Vaglio, J. Reboul, H. Ge, B. P. Davis, J. Garrels, S. Vincent and M. Vidal, Identification of potential interaction networks using sequence-based searches for conserved protein-protein interactions or "interologs", *Genome Res.* **11**: 2120–2126 (2001).
47. B. P. Kelley, R. Sharan, R. M. Karp, T. Sittler, D. E. Root, B. R. Stockwell and T. Ideker, Conserved pathways within bacteria and yeast as revealed by global protein network alignment, *Proc. Natl. Acad. Sci. USA* **100**: 11394–11399 (2003).

48. I. Xenarios, L. Salwinski, X. J. Duan, P. Higney, S. M. Kim and D. Eisenberg, DIP, the Database of Interacting Proteins: a research tool for studying cellular networks of protein interactions, *Nucleic Acids Res.* **30**: 303–305 (2002).
49. B. L. Drees, Progress and variations in two-hybrid and three-hybrid technologies, *Curr. Opin. Chem. Biol.* **3**: 64–70 (1999).
50. I. Simon, J. Barnett, N. Hannett, C. T. Harbison, N. J. Rinaldi, T. L. Volkert, J. J. Wyrick, J. Zeitlinger, D. K. Gifford, T. S. Jaakkola *et al.*, Serial regulation of transcriptional regulators in the yeast cell cycle, *Cell* **106**: 697–708 (2001).
51. T. Ideker, S. Ybarra and S. Grimmond, *Hybridization and Post-Hybridization Washing* (CSHL Press, Cold Spring Harbor, 2003).

CHAPTER 27

GENOMES, GENOMIC TECHNOLOGIES AND MEDICINE

Xiaohua Huang

Abstract

The genome is the blueprint of life for all organisms. It encodes in digital form all the hereditary instructions for building, running, maintaining and reproducing an organism. The sequencing of the human genome was one of the greatest breakthroughs in biology and medicine in the last century. With the human genome sequence, scientists can now attempt to enumerate the genes encoding the proteins that form and operate the molecular circuitries in the cells. Genome sequencing with currently available technologies, however, remains slow and expensive for the routine sequencing of individual human genomes for many biomedical applications. For example, sequencing the genomes of a large number of individuals would allow us to examine all the genetic differences in the human populations to search for the genetic basis of complex traits, to perform association studies to identify the molecular etiology of a variety of diseases, and to study human evolution. Identifying the causal genes and variants would represent a significant step towards improved diagnosis, prevention and treatment of diseases. This chapter describes our strategies and recent progresses in engineering the next generation technologies for genome sequencing and for digital enumerations of the molecular components in the cells. In the new technological paradigm we are creating, micro- and nano-technologies are used to engineer fully automated miniaturized "lab-on-a-chip" devices to enable massive parallel manipulations and analyses of biological molecules on an unprecedented scale so that each individual human genome can be sequenced for as little as US$1000.

1. Introduction

The diploid human genome consists of 22 pairs of autosomes and two sex chromosomes. Each chromosome is a linear deoxyribonucleic acid (DNA) molecule. The diploid genome is enormous, containing about six billion base pairs. In nature, there are only four kinds of bases in the DNA molecule: A (adenine), C (cytosine), G (guanine) and T (thymine). The order of these bases in the chromosomal DNA molecules contains in digital form the genetic operation system or the hereditary instructions upon which human genetics and physiology are based. It is thus quite obvious that the genome sequence holds the secret to our full understanding of human genetics and physiology at the molecular level. The ambitious international Human Genome Project was initiated in the early 1990s to obtain a copy of the human genome sequence. To sequence the genome is to decode the order of the bases in the linear DNA molecules. The sequencing of the human genome was made possible by many technological innovations over the past several decades. As a result of that *tour de force* landmark effort, a consensus copy of the human genome sequence has been obtained at very high accuracy.[1,2] The availability of the human genome sequence is revolutionizing biomedical research and medicine.[1–3]

The ability to sequence the genomes of normal, neoplastic and cancer cells from many individuals will enable comparative genomics and association studies to dissect the genetic basis of cancer[4] and complex traits/diseases,[5] and to carry out pharmacogenomic studies to tailor medicine and treatment for each individual (often referred to as personalized medicine).[6] But genome sequencing with current technologies remains a very slow and expensive process (months to years at a cost of ten to 50 million dollars per mammalian-sized genome[7,8]), requiring very expensive instruments and factory-style facilities. The on-going effort to develop the next generation technologies for routine sequencing of individual human genomes at extreme low cost has been referred to as the "Personal Genome Project" (PGP) or the US$1000 genome sequencing project.[8,9] Such revolutionary technologies will have a profound impact on biomedical research and medicine.[3]

2. Genome Sequencing Technologies

The sequencing of the human genome was one of the greatest triumphs in science, engineering and medicine in human history. Engineering and technology development played very important roles in the sequencing of the human genome. It was made possible by the numerous technological advances, including the invention of dideoxy termination DNA sequencing method by Sanger and colleagues,[10] the invention of the polymerase chain reaction (PCR) technique for DNA amplification,[11] the development of fluorescent dyes, enzymes, capillary gel electrophoresis and automated sequencers[12] for DNA sequencing, and the advances in automation and computation.

The Sanger dideoxy sequencing method developed by Fredrick Sanger and co-workers in the 1970s remains the mainstay of major genome sequencing efforts.[9,10] The vast majority of the available DNA sequences have been obtained with automated 96- and 384-capillary DNA sequencers using the Sanger method.[9] In the Sanger DNA sequencing method, a ladder of fluorescently-labeled DNA fragments with single base resolution is generated from the DNA template to be sequenced by DNA synthesis using a DNA polymerase in the presence of a small percentage of dideoxyribonucleotides. The dideoxy-terminated DNA fragments are resolved by electrophoresis in a thin slab gel or gel-filled capillary array. A laser-based fluorescence detection method is used to determine the order of DNA fragments and thus the sequence of the DNA molecule. About 500–1000 bases can be obtained from each sequencing reaction and sequencing run.

To determine the sequence of the six billion base-pair human genome consisting of DNA molecules that are millions of bases long, the random shotgun sequencing strategy developed by Sanger and co-workers in the 1980s is commonly used.[13] The genome is randomly fragmented into smaller pieces and the DNA fragments are cloned and sequenced piece by piece. The genome sequence is then assembled from those DNA fragments through sequence alignment based on the overlaps between those sequences. The human genome contains numerous highly repetitive nearly identical sequences of various sizes. Sophisticated mathematic algorithms and powerful computers are required for assembling the genome sequence. Many expensive automated sequencers and large amounts of costly reagents are required for the multi-step manipulations of hundreds of millions of individual DNA samples to provide sufficient coverage of a human genome.[1,2] Therefore, it is not surprising that genome sequencing with currently available technologies remains slow and prohibitively expensive for routine sequencing of individual human genomes.

Many technologies aiming at dramatically reducing sequencing cost are being developed.[8,9] The emerging technologies include three general categories: gel-based microelectrophoretic, non-gel based, and single molecule methods. Most gel-based microelectrophoretic methods focus on further improving multiplexing and miniaturization by microfabricating capillary arrays on glass wafers and integrating amplification and sequencing onto a single device.[14,15] Efforts for improving integration and throughput to reduce cost by two orders of magnitude have been recently reported.[8,16] Non-gel based methods include sequencing by hybridization,[17,18] pyrosequencing,[19,20] and the massive parallel signature sequencing (MPSS) technology invented by Sydney Brenner and co-workers.[21] In single molecule methods, sequence information is directly read out from a single DNA molecule. These include nanopore and other single molecule techniques.[22–25] Recently Quake's group has demonstrated that sequence information can be obtained from single DNA molecules by synthesis and fluorescence detection using total internal reflection fluorescence (TIRF) microscopy.[24] However, only a few nucleotides can be sequenced with this technology.

Except the single molecule methods, all other methods rely on some sort of efficient technique for preparing the millions of DNA samples for sequencing. Several recently reported PCR-based methods aim to achieve the parallelism and throughput required for preparing DNA samples for genome sequencing: the amplification of single molecules by PCR in water-in-oil micro-emulsions (emulsion PCR),[26] in a thin layer of polyacrylamide gel (polony PCR),[27,28] on a solid surface tethered with primers (surface PCR),[29] or in pico-liter wells of microfabricated titer plates (PTPCR).[30] Although the powerful emulsion PCR method is utilized almost exclusively in many high throughput DNA sequencing methods that are being developed,[16,31] the amplification of long DNA template is highly inefficient by emulsion PCR.[31] The gel layer used in polony PCR method may not be compatible for downstream sequencing procedure and therefore limits its application for genome sequencing. Surface PCR method seems to be quite straightforward, but it is highly inefficient for DNA amplification.[29] PTPCR cannot be easily scaled up for amplifying millions of samples.[30] In addition, all PCR-based methods have many intrinsic limitations due to the repetitive cycles of denaturation and hybridization used in PCR mechanism. For example, due to the exponential nature of the PCR amplification, a significant fraction of the amplified products may contain one or more errors, and compartmentalization of the reaction vessels and thermal cycling are required. The methods we are developing can overcome these limitations, and can potentially be used for massive parallel amplification of several copies of a human genome on a single microscope glass slide.

3. A New Paradigm and Technological Platform for Genomic Analyses

As described above, the major bottleneck in current genome sequencing technologies can be attributed to the requirements for large facilities, expensive instruments and reagents for manipulating the hundreds of millions of individual DNA samples to sequence the six billion base-pairs of the diploid human genome. To be able to sequence a human genome with one single instrument at extremely low cost and high speed (e.g. ~US$1000, within hours), we need to create a new paradigm to achieve unprecedented multiplexing, parallelization and miniaturization so that hundreds of millions of DNA samples can be manipulated in parallel in a single miniature device. One of the major thrust of our research and engineering efforts is to create such a new paradigm to engineer a next generation integrated "lab-on-a-chip" system for high-throughput genomic analyses. The strategy is illustrated in Fig. 1.

The cornerstone of our strategy is the development of methods for massive parallel manipulations and analyses of single DNA molecules. We have developed a new method, called rolling circle amplification (RCA), for efficient amplification of single DNA molecules. Micro- and nano-fabrication tools commonly used in the semiconductor industries are used to fabricate microfluidic devices and nano-arrays for parallel cloning of single DNA molecules and downstream processing to enable genome-scale DNA amplification and sequencing on a single chip. We are developing novel DNA sequencing methods and high-throughput imaging system for high-throughput on-chip DNA sequencing. These multi-disciplinary research and engineering efforts will lay down a solid conceptual and technological framework for the development of a future generation of tools for genomic analyses with single-molecule and single-cell sensitivities, moving us one step closer to the revolutionary US$1000 genome sequencing technology.[8]

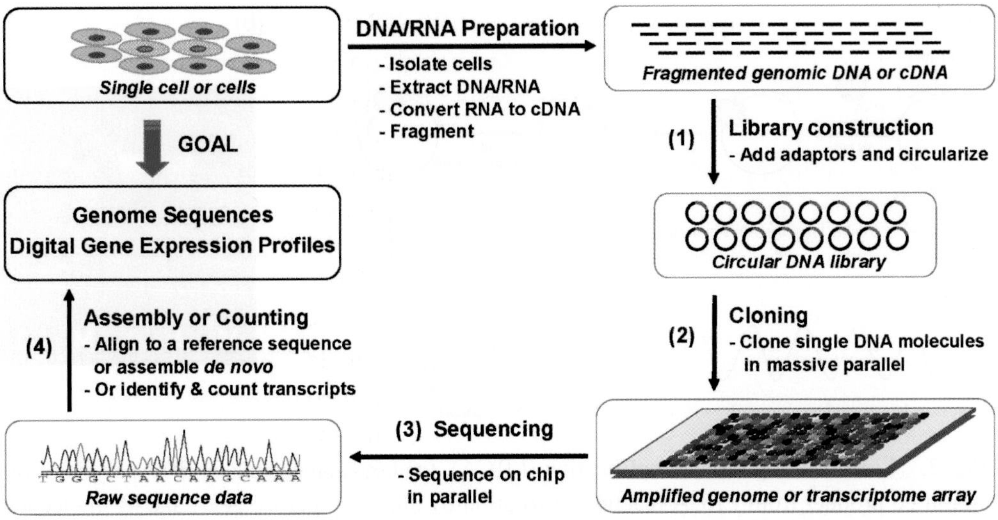

Fig. 1. Systems engineering approach to genome sequencing and gene expression profiling.

4. Rolling Circle Amplification — A Powerful Method for DNA Amplification

In collaborations with others, we have developed the powerful rolling circle amplification (RCA) technology for isothermal DNA amplification and single molecule detection.[32] In linear rolling circle amplification (LRCA), a circular DNA template is amplified by the continuous synthesis of a complementary strand of the circular template by a DNA polymerase through a rolling circle strand-displacement mechanism. The overall mechanism is illustrated in Fig. 2a.

Briefly, in LRCA DNA synthesis is initiated from a primer hybridized to the circular template. Once the complete complementary strand is synthesized, the polymerase reaches the 5′ end of the primer. If the DNA polymerase (e.g. Bst and ϕ29 DNA polymerases) has strand-displacement capability, the complementary strand is continuously displaced from the 5′ end and DNA synthesis continues through a rolling circle mechanism. Since the template is circular, a linear amplification ensues. The product consists of linear tandem repeats of the complementary strand of the original template. One noticeable feature of LRCA is that the linear amplified product remains attached to the primer and is co-localized with the primer if the primer is attached to an object (e.g. a glass surface). This is a very useful property when it is desirable to prevent the amplified product from diffusing away from the location of amplification. Amplification by RCA is usually rapid and carried out isothermally. A typical result of LRCA amplification is shown in Fig. 2b. The template used was a 78-base circle and the amplification reaction was performed with Bst DNA polymerase at 55°C for 15 minutes. As can be observed, the template was amplified more than 1000-fold in 15 minutes.

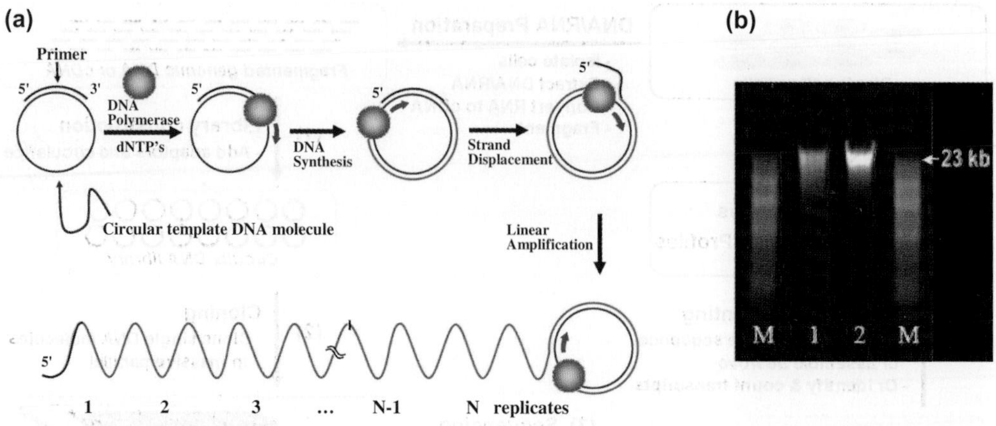

Fig. 2. Linear rolling circle amplification (LRCA). **(a)** LRCA mechanism. **(b)** Analysis of the amplified product from LRCA reaction in solution by alkaline agarose gel. Lane M: molecular weight markers (Hind III digested λ DNA); Lanes 1 and 2: LRCA products.

Another variant of the RCA is called hyperbranched rolling circle amplification (HRCA). HRCA can be used for even greater amplification of DNA. In HRCA, two primers are used. One primer hybridizes to the circular template and initiates the LRCA, and the other primer hybridizes to the complementary strand of the template. The mechanism of a HRCA reaction is schematically illustrated in Fig. 3a.

The first primer initiates the linear amplification from the circular template. The sequence of the second primer is designed to be identical to one part of the template. Therefore, the second primer will hybridize to the newly synthesized single-stranded product initiated from the first primer. DNA synthesis is initiated from this primer and the same strand-displacement mechanism can also take place. The first primer then hybridizes to the newly synthesized single-stranded product initiated from the second primer, and the same mechanism of DNA synthesis and strand-displacement takes place and so on. A hyperbranched mechanism of amplification ensues. The hyperbranched geometric amplification in HRCA results in tremendous amplification. The final amplified products are linear double-stranded DNA molecules consisting of concatenated tandem repeats of the template sequence with various unit lengths. This can be confirmed by the appearance of the ladder of bands in Fig. 3b which shows the results of HRCA amplification reactions with various numbers of input circular templates. With HRCA, tremendous amplification (up to 10^8 fold or more) can be obtained from a single DNA molecule.

5. Massive Parallel Separation and Amplification of Single DNA Molecules

We have successfully demonstrated that single DNA molecules can be cloned in a massive parallel manner with the powerful RCA technology. In one experiment, two synthetic

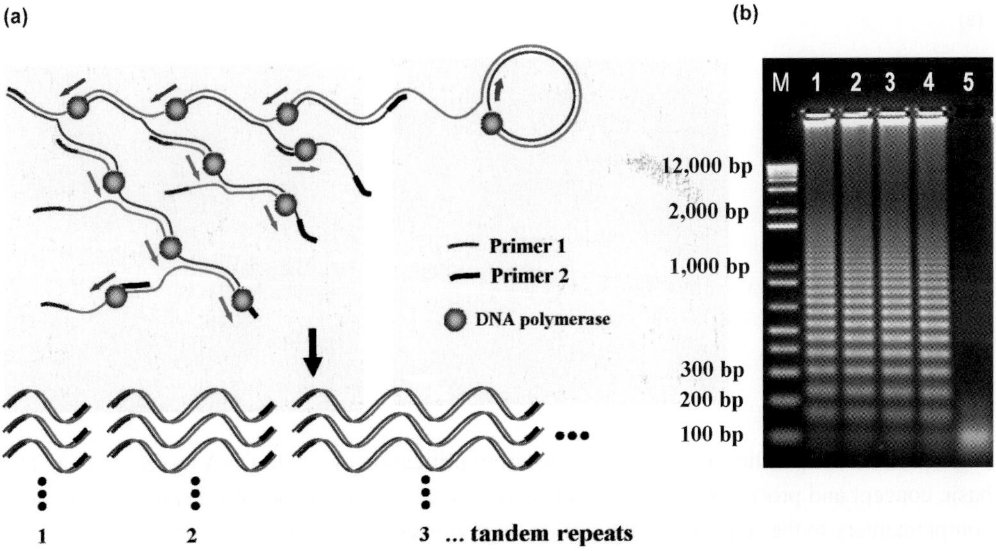

Fig. 3. Hyperbranched rolling circle amplification (HRCA). **(a)** HRCA mechanism. The arrows indicate the directions of DNA synthesis. Green arrow: synthesis from primer 1 initiating the LRCA; red arrows: synthesis from other primer 1; blue arrows: synthesis from primer 2. **(b)** Agarose gel electrophoresis analysis of the products from HRCA reactions in solution. Lane M: DNA molecular weight marker. Lanes 1 to 5: amplified products from HRCA reaction with input of 10,000, 1000, 100 and 1 circular template molecules, respectively.

circular DNA templates were used as a model system to demonstrate the feasibility of massive parallel separation and cloning of single DNA molecules on a monolayer of primers tethered to a glass surface. Figure 4a illustrates the basic concept and the experimental procedures.

The individual DNA molecules were separated and amplified by linear RCA on the surface. The amplified products were labeled either directly by performing the amplification in the presence of a fluorescently labeled nucleotide triphosphate or indirectly by hybridizing a fluorescently labeled oligonucleotide probe onto the products as shown here. After the molecules were compacted with streptavidin through the tetravalent binding of streptavidin to the biotin molecules on the probes, the amplified products from each individual DNA molecule on the surface could easily be detected by fluorescent measurement with an epifluorescence microscope. A typical result is shown in Fig. 4b. The spots in general are very small (<1 μm) and well separated from one another. Each red or green spot very likely represents the amplified product(s) from an individual DNA molecule. Each yellow spot, however, probably represents the amplified products from two different molecules in close proximity.

In an alternative method, the single DNA molecules were first amplified in solution with biotinylated primers. The amplified molecular clones were then captured on a glass surface via biotin-avidin affinity binding. A typical image is shown in Fig. 5c. The density

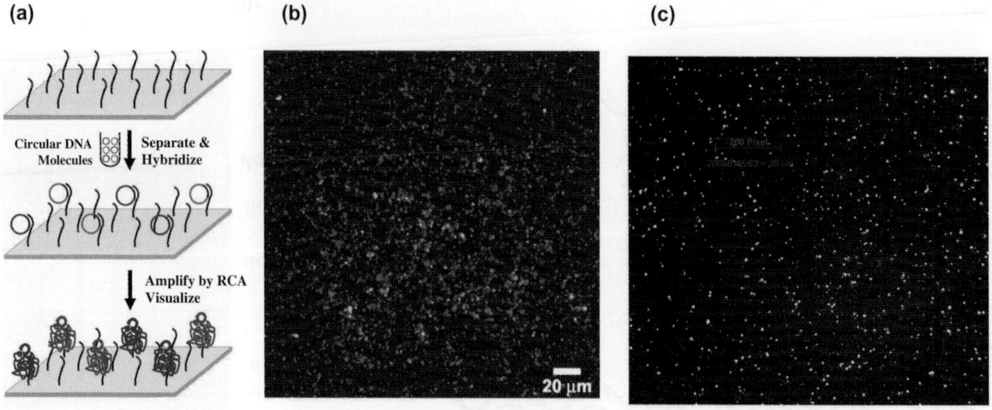

Fig. 4. Massive parallel separation and linear amplification of single DNA molecules. **(a)** The basic concept and procedure. The DNA molecules, each of which contains a *common* sequence complementary to the sequence of the primer immobilized on a solid surface, are spread onto the surface so that the individual DNA molecules are separated from one another and hybridized to the immobilized primers. These individual DNA molecules are amplified by LRCA and the amplified products are visualized by fluorescent detection. **(b)** Fluorescent image of rolling circle amplification (LRCA) products from single DNA molecules of two different circles on a small portion of a microscope slide. The amplified products were detected with two fluorescent probes (red and green). **(c)** Fluorescent image of a small portion of the cover glass. The DNA molecules were amplified in solution with a biotinylated primer. The amplified products were hybridized to a Cy5-labeled oligonucleotide probes and then captured onto the NeutrAvidin-coated cover glass via biotin-avidin affinity binding.

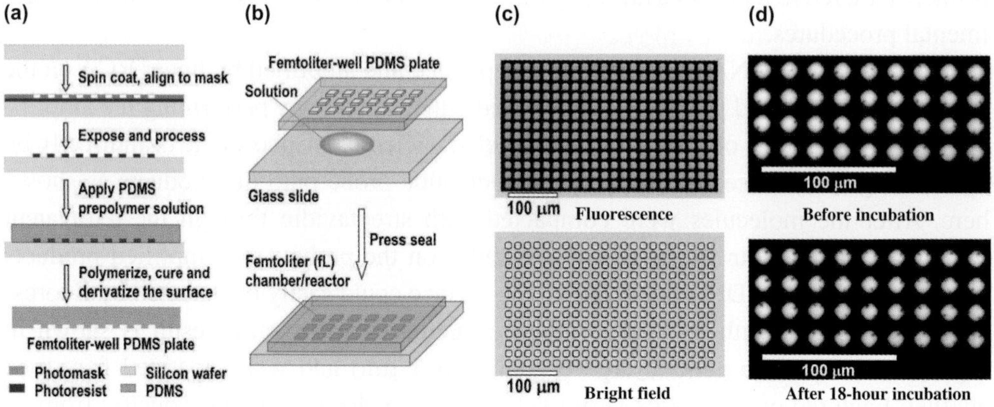

Fig. 5. High density arrays of femtoliter wells on PDMS plates. **(a)** Microfabrication procedure. **(b)** Filling and sealing. **(c)** Images of wells filled with a fluorescent solution. **(d)** Fluorescent images of wells before and after heating at 60°C for 18 hours.

of the molecular clones seen in the image corresponds well with the expected density of input molecules if most of the molecular clones are captured onto the surface.

This implies that most of input circular DNA molecules have been successfully amplified in solution and captured onto the surface. It is estimated that each fluorescent spot contains about 300–600 Cy5 molecules. Since the Cy5-labeled probe hybridizes to only one sequence in the circle, the number of the Cy5 molecules in each spot reflects the number of copies of the amplified template. Surprisingly, even though the molecules were not specifically condensed through biotin-streptavidin binding as used in the two-color experiment, the long linear amplified DNA molecules assume a very compact structure (<1 μm).

Although varying in sizes, those single molecular clones are only a few hundred nanometers in diameter. Hundreds of millions of these molecular clones can be accommodated on an area of a single microscope slide. So we have essentially demonstrated that DNA fragments from several copies of a human genome can be cloned and separated on the surface of a single conventional microscope glass slide to a sufficient quantity to be detected with conventional fluorescence microscopy in one simple process with a very small amount of reagent (25 cm $\times$ 75 cm).

6. Engineering Microfluidic Devices and Nano-Arrays for Genomic Analyses

As can be readily observed in Figs. 5b and 5c, the amplified molecular clones are randomly distributed and each molecule is amplified to a different degree as reflected by the different sizes of the spots or molecular clones. In practice, downstream applications (e.g. DNA sequencing and high-throughput fluorescence imaging) would be greatly simplified if the individual molecules are amplified to an equal or similar molar quantity and ordered in an array. Our strategy is to use microfluidic devices and nano-arrays to order these single molecular clones.[33]

We have demonstrated that PDMS plates with high-density arrays of wells with volumes from femtoliters to attoliters can be fabricated using standard photolithographic methods. The basic procedure is illustrated in Fig. 5a. We have shown that the PDMS surface is very hydrophilic after the surface is treated by O_2 plasma oxidation and derivatized with polyethylene glycol (PEG). The wells can be filled with aqueous solutions and the contents of the wells can also be replicated onto a glass slide (Fig. 5b).

High-density arrays of one-femtoliter wells (1 fL, $1 \times 1 \times 1$ μm^3, with 1 μm spacing) can be fabricated. With this density, up to 500 million wells can be fabricated on a $1'' \times 3''$ area, which are sufficient for amplifying and processing several copies of a whole human genome. In addition, we are using conventional and phase-shifting photolithographic methods to fabricate very-high-density arrays of functionalized

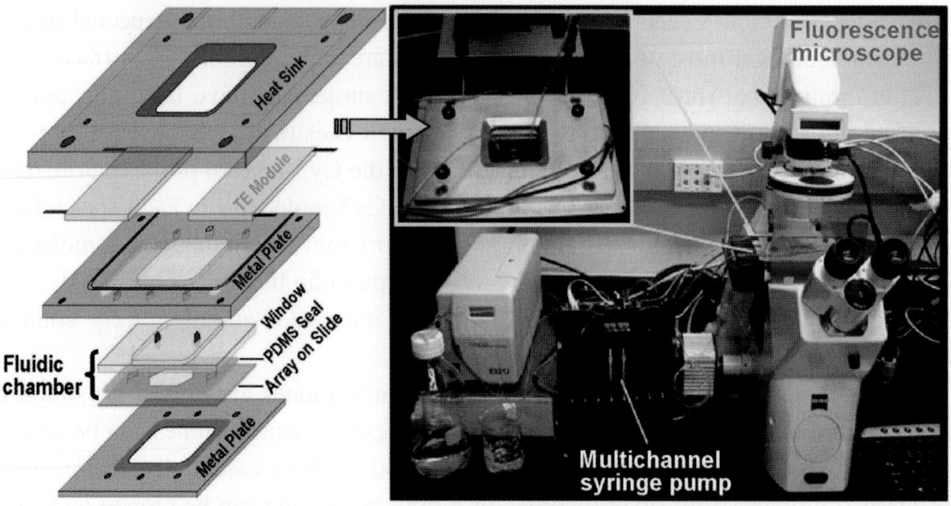

Fig. 6. An automated setup for high-throughput genomic analysis. The center piece of the instrument is a fully automated microscope (Axiovert 200M, Carl Zeiss) equipped with a high density CCD camera (Hamamatsu). The left panel shows the design of the custom-built reaction and imaging chamber. The microscope slide or cover glass with DNA samples immobilized on the glass surface is assembled into the chamber. The temperature is controlled by a Peltier thermoelectric device and the fluid flow is controlled with a multi-channel syringe pump. All instruments are controlled via a computer. The entire setup is shown in the right panel.

nano-scale features on glass or silicon wafers to order the single molecular clones for even greater improvements in throughput and miniaturization, and reduction in reagent cost.

7. New Methods for DNA Sequencing and High-Throughput Imaging System

We have demonstrated the feasibility of massive parallel separation and cloning of hundreds of millions of single DNA molecule using the rolling circle amplification technology. Several copies of a human genome could potentially be cloned on a single chip of the size of a conventional microscope slide. Each of the molecular clone containing hundreds to thousands of identical copies of a DNA molecular species can serve as an addressable nano-reactor for extremely high-throughput and rapid DNA sequencing. Towards this end, we are developing new methods for decoding the sequences of the single DNA molecular clones on the whole genome chips in a highly parallel manner. Two methods are being developed. One is called sequencing by denaturation[34] and another is called sequencing by ligation.

Another major effort in our lab is to integrate the microfluidic system, micro- and nano-devices into a high-speed and high-throughput fluorescence imaging system to enable the rapid sequencing of the large number of single DNA molecular clones *en masse*. A prototype system is shown in Fig. 6. Our long-term goal is to engineer such an integrated and fully automated bench-top system that will enable whole genome sequencing at extremely low cost and at high speed with a single miniaturized device.

8. Summary

The sequencing of the human genome was made possible by the numerous technological innovations over the last several decades and the *tour de force* heroic efforts of thousands of scientists and engineers involved in the international Human Genome Project. The sequencing of the human genome and many other model organisms has ushered in a new exciting era of genomic research and medicine. A copy of the consensus sequence of the human genome, however, provides only the first level of understanding of our genome. Many applications require sequencing the genomes of many human individuals. Technologies for dramatically increasing the efficiency of DNA sequencing to reduce sequencing cost will help unleash the full power of genomics. Such technologies will undoubtedly transform biomedical research and medicine.

The multi-disciplinary research and engineering efforts in our laboratory aim to create a new paradigm to achieve unprecedented multiplexing, parallelization and miniaturization for high-throughput genome sequencing and digital enumerations of biomolecular species in the cells. We have developed technologies for massive parallel amplification cloning of single DNA molecules. New technologies for DNA sequencing and high-throughput imaging are being developed. One of our long-term goals is to engineer a fully automated miniaturized "lab-on-a-chip" bench-top system that will enable whole genome sequencing at high speed and at extremely low cost.

Acknowledgments

This work was supported in part by a grant from NIH/National Human Genome Research Institute (grant no. HG003587). The author thanks Ying-Ja Chen and Kristopher Barbee for their contributions to the work described here.

References

1. E. S. Lander, L. M. Linton, B. Birren, C. Nusbaum, M. C. Zody, J. Baldwin, K. Devon, K. Dewar, M. Doyle, W. FitzHugh *et al.*, Initial sequencing and analysis of the human genome, *Nature* **409**: 860–921 (2001).
2. J. C. Venter, M. D. Adams, E. W. Myers, P. W. Li, R. J. Mural, G. G. Sutton, H. O. Smith, M. Yandell, C. A. Evans, R. A. Holt *et al.*, The sequence of the human genome, *Science* **291**: 1304–1351 (2001).

3. F. S. Collins, E. D. Green, A. E. Guttmacher and M. S. Guyer, A vision for the future of genomics research, *Nature* **422**: 835–847 (2003).
4. P. A. Futreal, A. Kasprzyk, E. Birney, J. C. Mullikin, R. Wooster and M. R. Stratton, Cancer and genomics, *Nature* **409**: 850–852 (2001).
5. L. Kruglyak, Prospects for whole-genome linkage disequilibrium mapping of common disease genes, *Nat. Genet.* **22**: 139–144 (1999).
6. D. Roses, Pharmacogenetics and the practice of medicine, *Nature* **405**: 857–865 (2000).
7. R. H. Waterston, K. Lindblad-Toh, E. Birney, J. Rogers, J. F. Abril, P. Agarwal, R. Agarwala, R. Ainscough, M. Alexandersson, P. An *et al.*, Initial sequencing and comparative analysis of the mouse genome, *Nature* **420**: 520–562 (2002).
8. National Human Genome Research Institute, Revolutionary genome sequencing technologies: the US$1000 genome, http://grants.nih.gov/grants/guide/rfa-files/RFA-HG-04-003.html (2004).
9. J. Shendure, R. D. Mitra, C. Varma and G. M. Church, Advanced sequencing technologies: methods and goals, *Nat. Rev. Genet.* **5**: 335–344 (2004).
10. F. Sanger, S. Nicklen and A. R. Coulson, DNA sequencing with chain-terminating inhibitors, *Proc. Natl. Acad. Sci. USA* **74**: 5463–5467 (1977).
11. K. Mullis, F. Faloona, S. Scharf, R. Saiki, G. Horn and H. Erlich, Specific enzymatic amplification of DNA *in vitro*: the polymerase chain reaction, *Cold Spring Harb. Symp. Quant. Biol.* **51**(Pt. 1): 263–273 (1986).
12. L. M. Smith, Automated DNA sequencing and the analysis of the human genome, *Genome* **31**: 929–937 (1989).
13. F. Sanger, A. R. Coulson, G. F. Hong, D. F. Hill and G. B. Petersen, Nucleotide sequence of bacteriophage lambda DNA, *J. Mol. Biol.* **162**: 729–773 (1982).
14. B. M. Paegel, R. G. Blazej and R. A. Mathies, Microfluidic devices for DNA sequencing: sample preparation and electrophoretic analysis, *Curr. Opin. Biotechnol.* **14**: 42–50 (2003).
15. R. H. Liu, J. Yang, R. Lenigk, J. Bonanno and P. Grodzinski, Self-contained, fully integrated biochip for sample preparation, polymerase chain reaction amplification, and DNA microarray detection, *Anal. Chem.* **76**: 1824–1831 (2004).
16. M. Margulies, M. Egholm, W. E. Altman, S. Attiya, J. S. Bader, L. A. Bemben, J. Berka, M. S. Braverman, Y. J. Chen, Z. Chen *et al.*, Genome sequencing in microfabricated high-density picolitre reactors, *Nature* **437**: 376–380 (2005).
17. M. Chee, R. Yang, E. Hubbell, A. Berno, X. C. Huang, D. Stern, J. Winkler, D. J. Lockhart, M. S. Morris and S. P. Fodor, Accessing genetic information with high-density DNA arrays, *Science* **274**: 610–614 (1996).
18. N. Patil, A. J. Berno, D. A. Hinds, W. A. Barrett, J. M. Doshi, C. R. Hacker, C. R. Kautzer, D. H. Lee, C. Marjoribanks, D. P. McDonough *et al.*, Blocks of limited haplotype diversity revealed by high-resolution scanning of human chromosome 21, *Science* **294**: 1719–1723 (2001).
19. M. Ronaghi, Improved performance of pyrosequencing using single-stranded DNA-binding protein, *Anal. Biochem.* **286**: 282–288 (2000).
20. J. M. Lage, J. H. Leamon, T. Pejovic, S. Hamann, M. Lacey, D. Dillon, R. Segraves, B. Vossbrinck, A. Gonzalez, D. Pinkel *et al.*, Whole genome analysis of genetic alterations in small DNA samples using hyperbranched strand displacement amplification and array-CGH, *Genome Res.* **13**: 294–307 (2003).
21. S. Brenner, M. Johnson, J. Bridgham, G. Golda, D. H. Lloyd, D. Johnson, S. Luo, S. McCurdy, M. Foy, M. Ewan *et al.*, Gene expression analysis by massively parallel signature sequencing (MPSS) on microbead arrays, *Nat. Biotechnol.* **18**: 630–634 (2000).
22. D. W. Deamer and D. Branton, Characterization of nucleic acids by nanopore analysis, *Acc. Chem. Res.* **35**: 817–825 (2002).

23. A. Meller and D. Branton, Single molecule measurements of DNA transport through a nanopore, *Electrophoresis* **23**: 2583–2591 (2002).
24. I. Braslavsky, B. Hebert, E. Kartalov and S. R. Quake, Sequence information can be obtained from single DNA molecules, *Proc. Natl. Acad. Sci. USA* **100**: 3960–3964 (2003).
25. M. J. Levene, J. Korlach, S. W. Turner, M. Foquet, H. G. Craighead and W. W. Webb, Zero-mode waveguides for single-molecule analysis at high concentrations, *Science* **299**: 682–686 (2003).
26. D. Dressman, H. Yan, G. Traverso, K. W. Kinzler and B. Vogelstein, Transforming single DNA molecules into fluorescent magnetic particles for detection and enumeration of genetic variations, *Proc. Natl. Acad. Sci. USA* **100**: 8817–8822 (2003).
27. R. D. Mitra and G. M. Church, *In situ* localized amplification and contact replication of many individual DNA molecules, *Nucleic Acids Res.* **27**: e34 (1999).
28. R. D. Mitra, V. L. Butty, J. Shendure, B. R. Williams, D. E. Housman and G. M. Church, Digital genotyping and haplotyping with polymerase colonies, *Proc. Natl. Acad. Sci. USA* **100**: 5926–5931 (2003).
29. C. Adessi, G. Matton, G. Ayala, G. Turcatti, J. J. Mermod, P. Mayer and E. Kawashima, Solid phase DNA amplification: characterisation of primer attachment and amplification mechanisms, *Nucleic Acids Res.* **28**: e87 (2000).
30. H. Leamon, W. L. Lee, K. R. Tartaro, J. R. Lanza, G. J. Sarkis, A. D. deWinter, J. Berka, M. Weiner, J. M. Rothberg and K. L. Lohman, A massively parallel PicoTiterPlate based platform for discrete picoliter-scale polymerase chain reactions, *Electrophoresis* **24**: 3769–3777 (2003).
31. J. Shendure, G. J. Porreca, N. B. Reppas, X. Lin, J. P. McCutcheon, A. M. Rosenbaum, M. D. Wang, K. Zhang, R. D. Mitra and G. M. Church, Accurate multiplex polony sequencing of an evolved bacterial genome, *Science* **309**: 1728–1732 (2005).
32. P. M. Lizardi, X. Huang, Z. Zhu, P. Bray-Ward, D. C. Thomas and D. C. Ward, Mutation detection and single-molecule counting using isothermal rolling-circle amplification, *Nat. Genet.* **19**: 225–232 (1998).
33. K. Barbee and X. Huang, A new method for fabricating high-density biomolecular arrays (submitted).
34. Y.-J. Chen and X. Huang, DNA sequencing by denaturation: principle and thermodynamic simulations (submitted).

SECTION VIII

SOCIO-ECONOMICAL ASPECTS OF BIOENGINEERING

CHAPTER 28

ETHICS FOR BIOENGINEERS

Michael Kalichman

Abstract

It seems that the news media are filled daily with examples of "ethical violations," "misconduct," and "shirking of responsibility." Although these reports are typically in high profile areas such as politics, sports, and business, there is no reason to assume that scientists and engineers are immune from lapses of good judgment. The goal of much of science and engineering is to generate new knowledge, but this work is typically done behind closed doors. Therefore, the risk of being caught is low and the temptation for misrepresentation is great. However, precisely because of the benefits of this new knowledge, it should be expected that scientists and engineers will be particularly concerned about the integrity of their disciplines. The challenges are to identify the ethical dimensions of the work that we do, to be aware of the tools and resources necessary to avoid the ethical pitfalls, to develop the skills for ethical decision-making, and to be clear about the obligation to act responsibly. Integrity is not just an option or an afterthought, but central to what it means to be an outstanding bioengineer.

1. Ethics and Ethical Decision-Making

We tend to think of ourselves as ethical people. Although we all have failings, it is more appropriate to say that we sometimes do bad things than that we are bad people. We might occasionally speed on the freeway or sometimes accept a little more credit than we deserve, but most of us are essentially good people and we wish to do the right thing. Whether or not this is true of every individual, our moral compasses are typically well-established by the time we are ready for college. So, if that's true, then what is the point of learning about ethics in a bioengineering course?

To understand the relevance of ethics training to scientists and engineers, it is important to be clear about what such training does and does not cover. First, there is little need to argue that criminal behavior, lying, cheating, and stealing are wrong; if adults do not already accept this, it is not likely that their world view will change because of a single lecture or even an entire course. Second, ethics is not just about specific laws or principles; explicit rules simply cannot anticipate all possible ethical challenges. Much of the definition for responsible conduct in science is based on standards that are not written down, not obvious, continuing to change, and variable among different areas of science, different research groups, and even different individuals. Third, being ethical is something more than slavishly following a single ethical principle; there are many admirable ethical principles[1] (e.g. "do unto others as you would have them do unto you," tell the truth, do no harm, or maximize the benefit for the greatest number), which frequently come into conflict. For example, what if you could save the lives of ten people by the killing of one member of your class? Fourth, ethics is not about choosing the one "right" path; it is typically a process by which we use all available information and our critical thinking skills to come up with the best possible solution to a problem. In short, ethical decision-making is not unlike the process of scientific inquiry. We may never find the absolute truth, but we do our best and learn from experience. Finally, ethics for scientists and engineers requires much more of a focus on the elements and boundaries of what we do as researchers, rather than on law or ethical theory.

The following sections review some of the key aspects of our roles as researchers: Data Management (how do we deal with the information we collect?); Human and Animal Subjects (what are our obligations to the subjects of our research?); Managing Conflicts of Interest (how do we minimize the risks of bias in our research?); Publication, Authorship, and Credit (what are our responsibilities in reporting our findings?); Mentoring (what are our obligations to ensure that the next generation of scientists is sufficiently prepared?); Whistleblowing (what are our roles and responsibilities when something goes wrong?); and Social Responsibility (what are our responsibilities to those outside of the research community?).

2. Data Management

The defining element of all research is the data (information) collected during the process of experimentation and observation. To ensure that research is conducted responsibly it is important to consider not only the definition of data, but also how records should be kept, the nature of data ownership, the advantages, disadvantages, and requirements for sharing data, and the obligation and requirements for storing the data.

2.1 *What are data?*

Although it might seem easy to define data, it does not take long to realize that data take many different forms.[2] For example, in bioengineering, primary data could include stress measurements in studies of heart papillary muscle, blood levels of insulin, electrophysiological recordings from a peripheral nerve, images showing fluorescent markers for activation of cell signaling pathways, or digitized movies of a contracting muscle. All of these examples are clearly the essential basis for any conclusions that would be published in a research paper.

The primary raw data are important, but many other kinds of "data" might be needed to verify what was found or, in the case of an allegation of research misconduct, whether the reported research findings were not made up or misrepresented. In this case, the concept of data might be extended to include, for example, complex analyses of raw data, custom-designed software, a transgenic animal, or a newly derived stem cell line. Because data can have so many different forms, the nature of recordkeeping is more difficult than simply noting numbers in a laboratory notebook.

2.2 *Recordkeeping*

Adequate records are essential for many reasons. First, much of research involves trial and error. If records of such work are not kept, then a researcher risks having to waste time in repeating what was already done. Second, when it is time to describe what was done for a paper or a presentation, complete and accurate records are a prerequisite. Third, once work has been reported, records are needed to allow for answering new questions that may arise. Finally, if any question should arise about the integrity of the research or of parallel work done by other researchers, then adequate records are the gold standard for verifying what actually occurred.

The criteria for adequate records are not easily defined for the many different forms of data described above. However, a case can be made that there is a minimum that is appropriate for all kinds of research. For relatively little cost in money or time, it should be possible to keep a bound notebook with numbered pages in which nominal records are kept[3,4] on each day in which research is conducted, minimal entries should include the items listed in Table 1. Good records are important, but it is also important to consider who actually owns those records.

Table 1. Recommended elements of research records.

Element	Description
Page Number	Keep records in a bound notebook, with numbered pages, to increase confidence that nothing has been added or removed.
Date	Date records sequentially for easy verification of what was done and in what order.
Title	Title each entry to make it easier to find what was done.
Name(s)	List name(s) for future reference as to who was involved.
Description	Briefly describe goal(s), what was done, and what was found.
Other records	Note where related research materials are stored and can be identified.

2.3 *Ownership*

Our intuition might suggest that because our research involves the discovery of new knowledge, we are the owners of the data that we generate. However, from a legal perspective, this is not normally the case. Research is usually conducted in an institution (whether for profit or not) and the rights to the ownership of work funded either by the institution or through grants and contracts to individuals at that institution are retained by the institution.

Although the institution typically owns the data, decisions about what to do with the data are nearly always ceded to the principal investigator who received the funding for the project. There are two principal exceptions to this rule. One exception is the case of intellectual property concerns (e.g. copyright, patents, licensing), which would be negotiated between the researcher and the institution. The second exception is when the institution needs access to records because of an investigation into possible misconduct or in responding to a legitimate request under federal or state freedom-of-information laws. Otherwise, all rights of ownership tend to remain with the researchers. One of the primary dimensions of that ownership is the question of what and when data should be shared with others.

2.4 *Sharing*

Science is fueled by the rapid exchange of information. If the goal is progress, then it makes sense that early and complete sharing of data would help to ensure that the pace of new discoveries be as fast as possible. However, there are many reasons that scientists might choose not to share with other scientists either before or even after publication. Some of these reasons are most clearly in the interest of science. For example, if a study is only partially complete and analyses are not yet definitive, then it could be misleading to release data or results that are inaccurate. Before publication, a researcher might also choose not to share his/her data because of a fear that other researchers will make use of that information to publish the same findings first or to obtain an advantage in developing the same or a competing marketable product. Even after publication, some researchers might be reluctant to share raw data because of a desire to make use of that data for further analyses of their own or to avoid being targeted for harassment by competitors or detractors. It is apparent that sharing data may have disadvantages, but it is important to note that there are several advantages of data sharing that are not merely possible, but are probable.

Sharing of data with others, even before publication, is not only of value to scientific progress, but can also be of tremendous benefit to those who share. By sharing data with others, a door is opened for insights gained by reciprocal sharing of data, for new collaborations or friendships, and for increased awareness of your collegial place in

Table 2. Considerations for record retention.

Consideration	Description
Regulations	Are there government or institutional regulations that stipulate how long records should be kept?
Reconstruction	What records are necessary to reconstruct what was done?
Ongoing study	Is this area of research ongoing and of continuing interest?
Confidentiality	If human subjects are involved, will confidentiality be difficult to maintain?
Record location	How easy will it be to relocate individual records?
Storage	What are the costs in space and expense for adequate record storage?

your research community. This kind of networking can yield valuable returns as others are aware of your reputation when it comes time to write letters of recommendation, or to review your grants, papers, or promotion. The one qualification to be added to these advantages is to note that caution should be exercised to avoid sharing with others who are known to take unscrupulous advantage of what they have learned in confidence. In most fields, this is rarely the case.

2.5 *Retention*

Research records are important, but they can only be of use as long as they are retained. There are some federal regulations regarding record retention, but these provide only nominal guidance, typically requiring that records be kept for three to five years after the completion of a grant. Responsible conduct in research depends on more than specifying a particular number of years. Important considerations for record retention include both which records need to be retained and the importance of keeping those records. Some important considerations are summarized in Table 2.

3. Human Subjects

Research in much of science, and particularly in bioengineering, is directed towards improving the human condition. Because such improvements are by definition new, it is not known in advance whether they will in fact be both effective and safe. Therefore, before new developments can be confidently made available, it is first necessary to find volunteers willing to accept the risks of participating in a research study. To understand modern considerations for research with human subjects, it is helpful to be aware both of past abuses of the privilege to conduct research with human subjects and the protections that are now in place.

3.1 *Abuses of human subjects*

In the name of science, some physicians and scientists have withheld effective treatments from people who are suffering, recruited subjects with misleading information, and even tortured and killed the subjects of their research. These incidents are very rare, but the need to prevent them is clear. One of the most extreme examples of the abuse of human subjects is the experiments conducted in Nazi Germany for the purpose of supporting the German war effort and promoting the "Aryan race." A characteristic example of this research was one study in which prisoners from concentration camps were immersed in near freezing water to determine how long it would take for them to die.[5] While such extremes did not occur in experiments in the United States, many studies were performed that failed to respect the rights of individuals to decide whether or not to participate in research. For example, to study the risks of exposure to radiation, plutonium was injected into hospital patients without their knowledge.[6] At the conclusion of World War II, trials of those responsible for the worst atrocities in Germany resulted in the creation of the Nuremberg Code, a set of ten guidelines for ethical research with human subjects.[7] Unfortunately, many of these guidelines were violated for a study conducted in Tuskegee, Alabama. Beginning in 1932, over 400 African-American men with syphilis were enlisted in a study to monitor the course of their disease. It was not until 1972 that it became widely known that these men were not told that they were participating in a research study and that they were not treated for their disease when an effective treatment (penicillin) became available in the 1940s.[8] Although such abuses of research subjects may be infrequent, even one is too many.

3.2 *Responsible research with human subjects*

During the 1960s and 1970s, the Tuskegee case was only one of several examples of research that was clearly inconsistent with high ethical standards. As a result, a commission was appointed to identify those principles that should be considered in the prospective review of all future research involving human subjects. In 1979, the Belmont Commission[9] proposed three principles to serve as a standard for all human research (Table 3). Those principles are now the centerpiece for federally required Institutional Review Boards (IRBs). At nearly all US research institutions, any study involving human subjects is prohibited until reviewed and approved by the IRB.

Table 3. Principles for research with human subjects.[9]

Principle	Description
Autonomy	• Individuals should have the right to decide whether or not to participate in research based on knowing what will or may be a consequence of participation. • Individuals who do not have the ability to make big independent judgments should have special protection against possible exploitation.
Beneficence	• Any possible harm should be minimized, if not eliminated. • The possible benefits of a research project should outweigh the risks to participants.
Justice	• Those groups that bear the burdens should reasonably expect that they will also receive the benefits of a research project.

4. Animal Subjects

Modern biomedical research has brought numerous advances to the quality and duration of human life. The speed with which these advances were achieved is due largely to the ability to learn about the fundamentals of biology in non-human animal species before testing in humans. The choice to study animals before putting human subjects at risk is one of the ten guidelines included in the Nuremberg Code:[7] "The experiment should be so designed and based on the results of animal experimentation and a knowledge of the natural history of the disease or other problem under study, that the anticipated results will justify the performance of the experiment." The premise, accepted by nearly all biomedical scientists, is that the complex molecular, biochemical, and physiological characteristics of animal life can only readily be studied in the intact, living animal, and that these characteristics are sufficiently preserved across many different species. Despite the importance of cell culture or computer modeling, such alternatives are only interpretable based on studies that have been completed in the whole animal.

4.1 *Abuses of animal subjects*

The benefits and privilege of working with animal subjects are not readily accepted by everyone. A large portion of the public is convinced in their opposition. The seeds of this animal rights movement are based on two philosophical arguments. The first is that the relief of some human suffering cannot be used to justify widespread animal suffering for any purpose, including experimentation. This utilitarian argument is most frequently attributed to Peter Singer.[10] The second argument emphasizes the inherent rights of animals to be free from suffering and to live a full life.[11] Unfortunately, these arguments have been successful, in part, because of cases in which some researchers have either abused the privilege of working with animal subjects, or have left themselves open to that perception.

One particularly well known example was a series of studies designed to better understand the nature of head injury in humans by studies in conscious baboons. Videotapes of these experiments were stolen by the Animal Liberation Front (ALF) and then edited and released by the People for the Ethical Treatment of Animals (PETA).[12] Whether or not this particular experiment was justifiable, its value would not be readily apparent to the general public. Therefore, the perception was that researchers were willfully causing pain and distress to animal subjects for no good reason. These perceptions helped to support the view that researchers abuse animal subjects for no other purpose than additional grant funding, rather than that researchers select appropriate animal models to better understand human physiology and pathophysiology. These very different views seem at first glance irreconcilable. However, it is important to note that most individuals are not so easily categorized as being either opposed to any use of animals or openly accepting of any use.

4.2 *Responsible research with animal subjects*

The privilege of conducting research with animals places a special burden on scientists and engineers to balance the potential benefits of research against appropriate respect for the subjects of research. Although each researcher may have personal criteria for what is and is not acceptable, the research community has adopted three specific principles to foster appropriate protections for the animal subjects of research. These principles, published by Russell and Burch,[13] are summarized in Table 4. Respect for these principles is ensured because they form the basis for institutional review of animal subjects research, which is now required for all federally funded research institutions. These reviews are commonly carried out by the Institutional Animal Care and Use Committees (IACUCs).

Table 4. Principles for research with animal subjects.[13]

Principle	Description
	If the goals of the research project can still be met, then the project should be:
Replacement	• Conducted without the use of living animals, or with less sentient animals.
Reduction	• Conducted with fewer animals.
Refinement	• Refined to decrease pain or suffering.

5. Managing Conflicts of Interest

Conflicts of interest are frequent, inevitable, and potentially damaging. Based on the media, and on some governmental regulations, we might assume that conflicts of interest are only financial. However, a cursory examination of any of our lives will remind us that we are daily faced with conflicting interests. With limited time, is it better to study for the upcoming examination in organic chemistry or in calculus? Choosing one interest over the other means that one of our responsibilities will receive less, and possibly too little, attention. That choice does not have to mean that we will do something unethical. Conflicts of interest are not in themselves unethical. It is what we do with those conflicts or what we do because of them that might get us in trouble.

5.1 *Risks of conflicts of interest*

There are two kinds of risks that we face because of conflicts of interest. The first, intentional misconduct, is the more obvious. If I have a financial interest in developing a new prosthetic device, what will I do if the data are not quite good enough to ensure continued and necessary support from venture capitalists? If I choose to fabricate or falsify the data, then I will have intentionally chosen to maximize the interest of my short-term gain (financial) while sacrificing the competing long-term interests of science, technology transfer, and my personal integrity. A second risk of conflicts of interests, unintentional bias, is more subtle. If I have a personal interest in improving treatment of juvenile onset diabetes because my daughter is a diabetic, is it possible that my hope for rapid success will cause me to unintentionally conclude that my results are better than they are? Although scientists and engineers often view themselves as being highly objective, we are all still human beings with hopes and biases. If we conduct an experiment with a treatment and control group, and there is any room for subjective decision-making about the endpoints, then we risk making our results look better than they actually are.

5.2 *Minimizing the risks of conflicts of interest*

Because we know that bias might result from any kind of conflict of interest, not just financial, it is important to consider how we might minimize that risk. The answer has three parts. First, it is a good idea to eliminate conflicts of interest whenever practical. It is not essential that someone be both a researcher and an entrepreneur, and certainly not for the same product. Second, if a conflict of interest is unavoidable, then find ways to ensure that those who have something to gain from positive (or negative) results are not in a position to bias the results either intentionally or unintentionally. This can often be done simply, for example, by blinding (hiding the identity) of the members of the control and experimental groups until it is time to conduct final analyses of the data. Another useful measure is to have people without a conflict of interest responsible for monitoring the project and/or for all subjective determinations and final interpretations of the data. Finally, even when these precautions are taken, it is useful for the community of science to take one further step, which is to disclose to others the nature of your conflict of interest (and what you have done to minimize the risks of bias).

6. Publication, Authorship, and Credit

For all practical purposes, results from experiments do not exist unless they are communicated to other researchers. The primary method for communication of scientific discoveries is through publication in scientific journals. As science has continued to explode, so has the number of journals. The result is that there may now be on the order of over 20,000 peer-reviewed journals (i.e. articles are published only after review and acceptance by peers competent to judge the quality of the work), and at least ten times that number of all types of journals. With many papers published in each of these journals every year, the literature is vast. It is important both that editors of journals be selective about what is published and that readers have the tools (e.g. appropriate search engines) to find relevant publications. Given this situation, what are the responsibilities of authors for producing a valuable contribution to the literature?

6.1 *Publication*

Published manuscripts should not only have significant merit, but should also be accurate and truthful. Most manuscripts, particularly those in peer-reviewed journals, are bound in printed issues that are intended to be kept indefinitely. While the quantity of publications will make this goal increasingly impractical, many journals are now turning to electronic storage and access through Web-based resources. For both print and electronic publications, it is important that authors should never take their responsibilities lightly. Once a paper has been accepted and published, it effectively becomes a permanent record or report attributed to its authors. As a minimum, this means that authors have an obligation to avoid any intentional misrepresentation of what was found. However, a higher and appropriate standard calls for the highest levels of integrity in the reporting of what was found. Some examples of responsibilities for authors are listed in Table 5. Clearly, publishing a manuscript implies a considerable obligation for those listed as authors.

6.2 *Authorship and credit*

Authorship of published, peer-reviewed articles is one of the most significant measures used to judge the merits of an academic career. When well-known researchers are

Table 5. Responsibilities for authors.

Responsibility	Description
Ideas	A study originates with asking the right questions or defining hypotheses.
Preparation	Before data collection can begin, the study must be planned and designed.
Performance	Data collection and recordkeeping are the defining elements of a research study.
Analysis	Once data have been collected, they must be analyzed and interpreted before communication to other researchers.
Report	The results of a research study are typically reported through a manuscript (paper), which must be drafted and revised before publication.
Assurance	One or more members of a research team must be in a position to offer assurance that the study and all findings are presented truthfully.

Table 6. Guidelines for requirements for authorship.[15]

Requirements	Description
1. Work	"Substantial contributions to **conception and design**, or **acquisition of data**, or **analysis and interpretation** of data."
2. Writing	"**Drafting** the article **or revising** it critically for important intellectual content."
3. Approval	"**Final approval** of the version to be published."

introduced for public presentations, it is frequent that the introduction will include a comment about how many papers they have published. Reviews for academic advancement, grant applications, and even for publication of new papers are often based in part on the researcher's success in publishing papers. Because authorship is so highly valued, it may be surprising to know that it is not always a good thing to be an author on a paper. In more than one case, co-authors have discovered only after a paper was published that one of the other contributors had falsified or fabricated his or her contribution.[14] The result is that co-authors will always share some blame in the published literature for a manuscript known to be — at least in part — fraudulent.

As much as it is desirable to have the credit of being an author on a paper, it is worth considering what criteria are expected for authorship. On the one hand, it is fair to say that criteria within different disciplines and even in different research groups vary widely. On the other hand, some general guidelines do exist. The guidelines of the International Committee of Medical Journal Editors[15] are summarized in Table 6. While these guidelines are helpful, it is also useful to think more generally about the responsibilities of authors.

A proposed minimal set of guidelines is summarized in Table 7.

Table 7. Minimal guidelines for authorship.

Contributors	Contributions and credit
Each author is responsible for	• A substantial and new contribution. • A review of the corresponding raw data.
All authors must have	• Read the final manuscript before publication. • Agreed to be named as an author.
Everyone who meets the criteria for authorship should be	• Included as an author.
Contributors who do not qualify for authorship should be	• Acknowledged.

7. Mentoring

As employees, students, and even researchers begin a new endeavor, there is much that they do not know about standards of conduct, recognizing and overcoming obstacles, and taking full advantage of opportunities. This information cannot typically be found in a book, nor is it something that they will have learned in the course of their general education. Because each career trajectory is so specific, it should be apparent that the best source of information will be someone who has recently traveled down the same career path. For example, an undergraduate student interested in graduate education in bioengineering might turn to current or recent graduates from graduate programs in bioengineering. When one of these more experienced individuals is willing and able to help someone less experienced, then they are serving in the role of a mentor. The term *mentor* comes from the Greek story of Odysseus who, when departing for the Trojan War, left the care of his son Telemachus in the hands of a trusted friend, Mentor.

7.1 *Roles and responsibilities of mentors*

Ideally, mentors are experienced advisors, not supervisors or employers. The distinction is important because decisions should be in the domain of the trainee and will hopefully not be tainted by the risk of coercion or bias on the part of a mentor. The kinds of advice and support that might be provided by such a mentor are wide-ranging.[16] Some examples of roles and responsibilities are summarized in Table 8. In reviewing the nature of mentors, it should be clear that we all might find ourselves in the role of mentor for someone less experienced, whether we are a beginning student or a senior researcher approaching the end of his or her career.

7.2 *Roles and responsibilities of trainees*

Mentors can be an invaluable resource to someone less experienced, but it is not the responsibility of the mentor to seek out the relationship. It is up to each of us to recognize

Table 8. Examples of roles and responsibilities of mentors.

Roles and responsibilities	Description
Develop mentoring relationship with trainee	• Be available. • Allow for differences in personalities. • Let trainees make their own decisions. • Teach by words and example.
Promote learning of research knowledge and skills	• Critical thinking and creativity. • Methods. • Communication skills.
Promote career development	• Job market. • Networking. • Varied career options.
Promote socialization	• Ethical development. • Teaching skills. • Working as part of a team. • Administration and planning.
Address special circumstances	• Gender. • Race. • Disabilities.
Promote effective mentoring	• Keep learning about mentoring relationships.

that the paths ahead of us are largely unknown and that our chances for success will be greatly increased if we can gain help from those that have already succeeded. However, it should also be clear that the goal is rarely to find just one mentor. Not only will we need different kinds of help at different stages of our life, but it is unlikely that our unique background and ambitions will be precisely matched to one individual. Therefore, it is most appropriate that we think about finding multiple mentors to address our particular circumstances (e.g. female, underrepresented minority, older student) and aspirations (e.g. teaching job in a liberal arts college). A list of recommended roles and responsibilities for trainees is summarized in Table 9. And, as noted for mentors, our mentoring relationships are relevant throughout our lives.

8. Whistleblowing

8.1 *What is whistleblowing?*

Although we will sometimes admire a whistleblower for bringing to light the hidden misdeeds of others, our society more often views whistleblowers with disdain and skepticism.[17] Certainly, if you look for synonyms of the word "whistleblower," you will find terms that are clearly pejorative (e.g. blabbermouth, busybody, rat, or stool pigeon).

Table 9. Examples of roles and responsibilities of trainees.

Roles and responsibilities	Description
Clarify goals	• Career plans. • Needs. • Expectations.
Locate prospective mentors	• Knowledgeable individuals. • Compatible personalities.
Distinguish between mentors and supervisors	• Not all supervisors are mentors. • Not all mentors are supervisors.
Promote effective mentoring	• Keep learning about mentoring relationships.

This negative view begins on the playground where the child who "tells" on another child is called a tattletale. This model values loyalty to our peers over revealing bad behavior, but it also helps to protect the bully. If the bad behavior is minor (returning the volleyball to the wrong locker room), it seems easy to excuse this view. However, if the behavior is more serious (bringing guns to campus), then surely the welfare of our community depends more on whistleblowing than on protecting the individual who may be preparing to do serious harm to others.

In a research environment it is very unlikely, although not impossible, that lives will be at stake if a potential whistleblower remains silent. On the other hand, it is worth considering the importance of whistleblowing to the integrity of the research enterprise. Research is an esoteric endeavor, generating new knowledge for which there is typically no clear way to judge whether the supporting data have been reported truthfully. The only way to know what really happened behind the closed doors of the research environment is for one or more members of that research group to speak out when they see something that is wrong. This act of whistleblowing is the most certain way to identify waste of research dollars, generation of misleading results, or even abuses to animal or human subjects of research.

8.2 *Roles and responsibilities for whistleblowing*

Whistleblowers often suffer adverse and sometimes severe consequences for having reported wrongdoing by others.[17] Many state and federal regulations are now in place with the intention of protecting whistleblowers, but the risks are still very real. Despite these risks, it should be clear that if you witness misconduct, then you at least have an ethical obligation to act. For most of us, it would be difficult to live with ourselves knowing that we had failed to speak up about something that we knew to be wrong. Separate from our ethical obligations, it is also important to keep in mind that if we do

Table 10. Roles and responsibilities for whistleblowers.

Roles and responsibilities	Description
Obligation	A witness to misconduct has an obligation to act.
Perspective	The perspective of a respected colleague or someone with more experience can help identify mistakes in judgment and options for how best to respond to perceived misconduct.
Conflict resolution	Disputes about potential misconduct or disagreements about responsible conduct can often be resolved by good conflict resolution skills or turning to a trusted third party for arbitration or mediation.
Proper channels	Institutional channels for addressing misconduct will typically be preferable to the unpredictable consequences of making an allegation public.
Allegation	The whistleblower's role is solely to bring an allegation forward for proper investigation. Whistleblowing is not a substitute for other mechanisms to address a grievance or perceived sleights.

not speak up, but someone else does, then our silence about the misconduct may be suspect in itself.

How can you best balance this ethical obligation with the potential practical consequences?[18] To begin with, it is worth knowing that you have many actions that might be taken before or instead of publicly naming someone as guilty of some form of misconduct. If you find that you are faced with concerns about the conduct of others, then you should consider the roles and responsibilities as outlined in Table 10. Some of these steps will protect you from making false or inappropriate charges and others will help to insulate you from the possibility of repercussions. The end result in all cases will be to improve the research enterprise, and your environment in particular.

9. Social Responsibility

9.1 *The privilege of conducting research*

Any discussion about the responsible conduct of research would be incomplete without an emphasis on the privilege of conducting research. Certainly all researchers working in public research institutions have a direct obligation to the public funders of those institutions. Even many private research institutes and companies receive at least some public support. Further, all research or use of animal or human subjects is potentially subject to municipal, state, or federal regulatory oversight. The implication of these connections between the public and research is that just as research is now supported, the public could as easily decrease funding or increase restrictions on research. In short, it is of both ethical and practical importance that researchers foster increased public understanding and engagement.

9.2 *Public understanding of science*

Communication between researchers and the public could be greatly improved. Many outside of science and engineering lack even a basic understanding of what science can and cannot do, much less how science proceeds. Ideally, successful communication about the nature of science should begin with K-12 education programs. A second important source of information for both children and their parents should be the media, including newspapers, magazine, radio, television, and now the Internet. Unfortunately, the completeness and accuracy of what is taught in schools or publicized in the media depends on the scientific training of the teachers and journalists. As the pace and complexity of science continue to increase, the likelihood that people trained as teachers or journalists will have the necessary understanding of science is decreasing. The unavoidable conclusion is that researchers must spend some time outside the walls of their research institutions so as to engage the public. This means taking the time to visit classrooms and talk about science, working with schools to develop high quality science and science fair programs, and effectively communicating with the media.

9.3 *Public approval of science*

Science and engineering are creating new challenges for society. The derivation of stem cells from human embryos raises questions about the moral status of the fertilized human egg. Continuing improvements in prosthetic body parts, artificial intelligence, and interfaces between humans and machines will challenge us to decide just what it means to be human. The development of new technologies to prolong life, without always maintaining the quality of life, prompts concerns about the morality of keeping a body alive, long after the mind is gone. These questions are largely matters of public debate and policy. If scientists and engineers fail to participate in these discussions, then decision-making will occur in the absence of a complete understanding of what science has shown, what is possible, and what is improbable. Researchers should consider outreach and dialogue with the public as an essential part of the responsible conduct of research.

10. Summary and Conclusions

The practice of science and engineering is filled with challenges for responsible and ethical conduct. One of the primary responsibilities of all researchers is to recognize this characteristic of their profession and be sensitive to the consequences of their actions for themselves, for their profession, and for the community. Because consequences lie in the future and are rarely certain, the secondary responsibility of researchers is to use their critical thinking skills and experience to make the best possible decisions about their conduct. Taking these steps will sometimes result in mistakes, but as long as we learn from our mistakes, we will continue to do better. The only clear mistake would be to fail to engage in the process of ethical decision-making. In short, we should apply the same analytical approach we use in the practice of science and engineering to the ethical dimensions of our work.

Acknowledgments

This work was supported in part by grant AI01591 from the National Institutes of Health.

References

1. L. M. Hinman, *Ethics: A Pluralistic Approach to Moral Theory*, 4th ed. (Wadsworth/Thomson, Belmont, CA, 2007).
2. Department of Health and Human Services, *Data Management in Biomedical Research, Report of a Workshop*, April 1990 (Chevy Chase, Maryland, 1990).
3. H. M. Kanare, *Writing the Laboratory Notebook* (American Chemical Society, Washington, DC, 1985).
4. F. L. Macrina, Scientific recordkeeping. In: *Scientific Integrity: Text and Cases in Responsible Conduct of Research*, 3rd ed. (American Society of Microbiology Press, Washington, DC, 2005), pp. 269–296.
5. R. L. Berger, Nazi science: the Dachau hypothermia experiments, *N. Engl. J. Med.* **322**(20): 1435–1440 (1990).
6. Advisory Committee on Human Radiation Experiments, *Final Report* (Oxford University Press, New York, NY, 1996).
7. *Trials of War Criminals Before the Nuremberg Military Tribunals*, Under Control Council Law No. 10, Vol. 2 (US Government Printing Office, Washington, DC, 1949), pp. 181–182.
8. J. H. Jones, *Bad Blood: The Tuskegee Syphilis Experiment* (Free Press, New York, NY, 1981).
9. National Commission for the Protection of Human Subjects of Biomedical and Behavioral Research, *The Belmont Report, Ethical Principles and Guidelines for the Protection of Human Subjects of Research* (Department of Health, Education, and Welfare, US Government Printing Office, Washington, DC, 1979).
10. P. Singer, *Animal Liberation* (Random House, New York, NY, 1975).
11. T. Regan, *The Case for Animal Rights* (University of California Press, Berkeley, CA, 1983).
12. C. R. McCarthy, *Part 2: The Historical Background of OPRR's Responsibilities for Humane Care and Use of Laboratory Animals*. Available at: http://onlineethics.org/reseth/nbac/hmccarthy.html (accessed November 1, 2005) (2001).
13. W. M. S. Russell and R. L. Burch, *Principles of Humane Animal Experimentation* (Charles C. Thomas, Springfield, IL, 1959).

14. R. L. Engler, J. W. Covell, P. J. Friedman, P. S. Kitcher and R. M. Peters, Misrepresentation and responsibility in medical research, *N. Engl. J. Med.* **317**: 1383–1389 (1987).
15. International Committee of Medical Journal Editors, *Uniform Requirements for Manuscripts Submitted to Biomedical Journals: Writing and Editing for Biomedical Publication.* Available at: http://www.icmje.org (accessed November 1, 2005) (2005).
16. National Academy of Sciences, National Academy of Engineering, and Institute of Medicine, *Adviser, Teacher, Role Model, Friend: On Being a Mentor to Students in Science and Engineering* (National Academy Press, Washington, DC, 1997).
17. Research Triangle Institute, *Consequences of Whistleblowing for the Whistleblower in Misconduct in Science Cases*, report submitted to Office of Research Integrity. Available at: http://ori.hhs.gov/documents/consequences.pdf (accessed November 1, 2005) (1995).
18. C. K. Gunsalus, How to blow the whistle and still have a career afterwards, *Sci. Eng. Ethics* **4**: 51–64 (1999).

CHAPTER 29

OPPORTUNITIES AND CHALLENGES IN BIOENGINEERING ENTREPRENEURSHIP

Jen-Shih Lee

Abstract

The contributions made by biomedical engineering entrepreneurs have led to not only better understanding in biology and medicine, but also better health care for people. In this chapter we first review some of the innovations made by the biomedical engineering industry and viewed by physicians as significance in improving the health of their patients. Two innovations, cardiac pacemakers and hemodialysis, are highlighted to elaborate on their entrepreneurial growth into billion dollar industries. Pointers are offered for readers to evaluate the chance of success for their inventions and to gain insights on the commitment required for entrepreneurship. It is concluded with the heading "Biomedical Engineers Mean Business" to encourage bioengineering students to consider entrepreneurs as their career option when opportunities arise.

1. Introduction

The biomedical industry has become a major economic factor of the United States. It contributes greatly to the advancement of medicine and biology, the betterment of health care, and the enhancement of food production worldwide. This US$200 billion per year industry employs millions of professionals and production workers in the US. It is one of the few US industries with export that far exceeds import. *"Our nation's economic competitiveness has come to rely more and more on this industry's success.... The challenge today, particularly here in America, where the field has made its greatest strides — is to keep the stream of creativity and innovations flowing and bring more of these advancements into use,"* as noted by Don P. Giddens, the President of the American Institute for Medical and Biological Engineering (2004/2005) and the Dean of College of Engineering, Georgia Institute of Technology.[1]

Many biomedical companies build their success on innovative devices, products, and drugs. Section 2 of this chapter first lists the 24 key medical and biological innovations voted by the fellows of AIMBE in 2005 to the AIMBE Hall of Fame (www.aimbe.org). The same section also describes the view of many renowned primary-care physicians on how these innovations have significantly improved the health of their patients. In Sec. 3, two innovations, cardiac pacemakers and hemodialysis, are chosen to elaborate on their development from conception decades ago to their subsequent growth into a multibillion industry.

Ninety-five percent of the biomedical industries in the US are small business, an entity full of entrepreneurial spirit. Their success records have induced many of our biomedical engineering graduates wanting to become entrepreneurs after they have gained some work experience in academia or industry. They may already have an invention or product that they want to see medical application for the benefit of patients. Sections 4 and 5 provide some pointers for readers to evaluate the chance that their invention will succeed and the commitment required for entrepreneurship.

I taught a course on "Biomedical Engineering Entrepreneurship" at the University of Virginia and a shortened one at the University of California San Diego. Many topics included here originated from the lectures of these two courses. In particular, the pointers given in Sec. 5 are from a lecture delivered by Wendell E. Dunn III, professor of the Darden Business School, University of Virginia. In this course, the students were asked to select an invention to build their entrepreneurship, to write a business plan, and to present the plan for critics. To conclude this chapter, I use the heading "Biomedical Engineers Mean Business" for Sec. 6 to highlight a proposition that biomedical engineering undergraduates and graduate students are to learn entrepreneurship in their course work so that they are prepared in taking an entrepreneurial career when opportunities arise. It is such entrepreneurships that will contribute to the growth of the nation.

2. Innovations in Biomedical Engineering

"Innovation will be the single most important factor in determining America's success through the 21st century... America's challenge is to unleash its innovation capacity to drive productivity, standard of living and leadership in global markets... For the past 25 years, we have optimized our organizations for efficiency and quality. Over the next quarter century, we must optimize our society for innovation." This is the opening resolution of an Innovate America report presented by Samuel J. Palmisano, chairman and chief executive officer of IBM Corp and G. Wayne Clough, president of Georgia Institute of Technology at the 2004 National Innovation Initiative Summit in Washington DC.[2]

One of AIMBE's goals is to accelerate the economy of the nation and the improvement of health care through innovations in medical and biological engineering. To promote public awareness of the contributions made by biomedical engineers and to assure a public policy for a healthy environment for medical and biological engineering innovation lead to the establishment by AIMBE of an Advocacy Committee chaired by John T. Watson, Professor of Bioengineering at UCSD. As one outcome of their deliberation, 24 key innovations of medical and biological engineering are inducted into the AIMBE Hall of Fame at the 2005 AIMBE Annual Meeting. Grouped according to the decades that the innovations first have wide usage, they are:

- 1950s and Earlier
  - Artificial kidney
  - X-ray
  - Electrocardiogram
  - Cardiac pacemaker
  - Cardiopulmonary bypass
  - Antibiotic production technology
  - Defibrillator
- 1960s
  - Heart valve replacement
  - Intraocular lens
  - Ultrasound
  - Vascular grafts
  - Blood analysis and processing
- 1970s
  - Computer assisted tomography (CT)
  - Artificial hip and knee replacement
  - Balloon catheter
  - Endoscopy
  - Biological plant/food engineering
- 1980s
  - Magnetic resonance imaging (MRI)
  - Laser surgery
  - Vascular stents
  - Recombinant therapeutics
- 1990s Until Today
  - Genomic sequencing and micro-arrays
  - Positron emission tomography
  - Image-guided surgery

The Advisory Committee formulated the following three criteria for the fellows of AIMBE to select from a list of 60 innovations into the AIMBE Hall of Fame:

- The innovations represent significant engineering achievement
- They are in general use
- Most importantly, the innovations save lives and improve the quality of life for a large number of people.

How primary care physicians view the importance of medical innovations is the question addressed by an article co-authored by Victor R. Fuchs and Harold C. Sox, Jr. Thirty innovations were chosen based on the emphasis in articles published in the *Journal of the American Medical Association* and the *New England Journal of Medicine* in the past 25 years.[3] Two hundred and twenty-five physicians with distinction in their profession and actively involved in patient care were asked to select five to seven that would have the most adverse effect on their patients if the innovations did not exist. The ranking of the top ten medical innovations in the study are:

(1) MRI and CT (magnetic resonance imaging and computed tomography)
(2) ACE inhibitors — for treatment of high blood pressure
(3) Balloon angioplasty — procedure to open blocked blood vessels of the heart
(4) Statins — drugs used for lowering cholesterol and preventing coronary heart disease
(5) Mammography
(6) Coronary artery bypass graft
(7) Proton pump inhibitors and H2 blockers — used to treat gastroesophageal reflux disease
(8) SSRIs (selective serotonin reuptake inhibitors) and new non-SSRI anti-depressants
(9) Cataract extraction and lens implant
(10) Hip and knee replacement.

Six of these ten innovations, other than the four innovative drugs, are in the AIMBE's Hall of Fame. Fuchs and Sox said the most surprising finding of their study was "the extent to which the leading innovations were an outgrowth of physical sciences (physics, engineering, and computer science) rather than disciplines traditionally associated with the 'biomedical sciences'."

Many medical and biological engineering innovations in the AIMBE's Hall of Fame are produced from the collaboration among engineers, physicists, computer scientists and medical professionals. The challenges and opportunities are for biomedical engineers to use their diverse knowledge in engineering, biology and medicine for leading the collaborative effort in innovations and to employ their entrepreneurial skill to translating the innovations into biological and clinical applications and in commercializing them for the benefit of people.

3. Innovators and Entrepreneurs

Since the time of these innovations, stunning progress for more functional systems and better patient care has been made. This section highlights the progress by first examining how cardiac pacemakers[4] and hemodialysis[5] were invented and then a discussion on the current state of cardiac rhythm management and the growth of hemodialysis service.

3.1 *Cardiac pacemakers*

In 1954, Dr. C. Walton Lillehei (who is acknowledged as the "Father of Open-Heart Surgery") of the University of Minnesota Hospital began operating on infants with congenital heart disorders that robbed their blood of oxygen and caused a bluish, or "cyanotic," cast to their skin — hence the use of the term "blue babies" to describe them. The operation, while effective, often interfered with the ability of the heart to conduct electrical impulses to produce a regular heart beat, resulting in a condition known as "heart blocks."

To remedy this problem, Lillehei used AC-powered external pacemakers to stimulate the tiny hearts to beat after surgery. Built on vacuum-tube technology, the pacemaker was bulky but could be wheeled around and plugged into the wall. On October 31, 1957, a three-hour electrical-power outage in Minneapolis threatened the lives of the babies who had undergone surgery and required the use of the pacemaker. Tragically, one baby died that night. The next day, Lillehei asked Earl E. Bakken if his company Medtronic could come up with something better.

Bakken then recalled seeing a circuit for an electronic transistorized metronome in *Popular Electronics* magazine that transmitted clicks through a loud speaker with an adjustable rate to fit the music. He dug up the back issue and built that circuit into a four-inch square, one-and-a-half-inch thick metal box with terminals for connection to the battery and with wires to carry the pulses to the heart. In four weeks, he had designed, constructed, and tested in his garage shop the first small, self-contained, battery-powered, transistorized, external pacemaker that could be taped to a patient's chest wall. When pacing was no longer needed, the wires could be withdrawn without having to reopen the chest.

The next day, after the functional demonstration of the device at the university's animal lab on a dog, the device was already in use by a little girl. Because of the response of the medical community to the success of Lillehei's operations on blue babies, soon the general press had picked up the story of the little box that kept children's hearts beating after surgery. People were calling it a "miracle." By the end of 1958, some 60 orders for the pacemaker came into Medtronic from all over the country.

The confluence of transistors, battery, materials, and cardiac surgery created a fast-moving stream of new therapeutic possibilities. Better electrodes, implantable cardiac pacemakers (fixed rate and on demand), heart-failure-targeting stimulators, implantable defibrillators, and external pacemaker and defibrillation products are coming on to the market. To contrast with the first sale figure of Medtronic cardiac pacemakers, the world now has a multibillion-dollar industry producing some ten million cardiac rhythm management systems a year.

3.2 *Hemodialysis*

Today 400,000 patients with chronic kidney failure, which is known as the end stage renal disease (ESRD), in the US and two million worldwide are undergoing chronic dialysis. Left untreated, both acute renal failure and end-stage renal disease produce uremia and death. Here are excerpts from the speech on the 2002 Lasker Award for Clinical Research celebrating the achievements of the two scientists who made hemodialysis possible: Willem Kolff (who is acknowledged as the "Pioneer of Artificial Organs") and Belding Scribner.

> *"Our story begins in 1938 at a small medical ward at the University of Groningen Hospital in the Netherlands. The physician in charge was Willem Kolff, who had just graduated from medical school. One of his first patients was a 22-year-old man in uremic coma. The young Dr. Kolff, then only 28 years old, watched helplessly for four days as the young man died in front of his eyes. He had no treatment to offer — if only he could find a way to remove the toxic metabolic wastes that accumulated in blood when the kidney failed...Despite the difficult circumstances of Nazi-occupied Netherlands, Kolff miraculously cajoled an enamel manufacturing company to help him obtain scarce materials in order to construct the first artificial kidney. This machine, which came to be known as the "rotating-drum hemodializer," consisted of 130 feet of cellophane tubing made from sausage casing, wrapped 30 times around a horizontal drum made out of aluminum strips. As the drum rotated through a bath of salt solution contained in an enamel tank, the patient's blood was exposed to the dialysis bath, allowing rapid and efficient removal of the toxic wastes.*
>
> *When World War II ended, Kolff donated all five of his artificial kidneys to hospitals in London, Poland, The Hague, Montreal, and Mount Sinai Hospital here in New York City. This extraordinary act of generosity enabled physicians throughout the world to become familiar with the new technique of dialysis. He also provided blueprints of his "rotating-drum hemodializer" to George Thorn at the Peter Bent Brigham Hospital in Boston. This led to the manufacture of the Kolff-Brigham kidney, which was an improved stainless steel version of the original...*
>
> *The Kolff kidney solved the problem of acute renal failure, but what about the hundreds of thousands of patients with chronic end-stage renal disease for whom prolongation of life requires repeated dialysis three times a week forever? In the late 1950s, the conventional wisdom among kidney experts was that chronic intermittent dialysis would never be possible because of two insurmountable problems, one technical*

and one psychological. The technical problem was one of circulatory access; every time a patient was hooked up to a dialysis machine veins and arteries were damaged, and after six or seven treatments, physicians would run out of places to connect the machine. The psychological problem stemmed from the widely held mystical belief that a cellophane dialyzer outside the body could never permanently replace the complex functions of a normal organ. After all, according to the experts, the kidney was a sacred organ. Above and beyond its excretory function, it produces three essential hormones: erythropoietin for forming red blood cells, renin for maintaining blood volume and blood pressure, and hydroxylated vitamin D for preventing breakdown of the bones.

In 1960, the impossible suddenly became possible. The psychological and technical barriers to chronic dialysis came crashing down through the research of Belding Scribner, a young professor of medicine at the University of Washington in Seattle...His idea was elegant in its simplicity: sew plastic tubes into an artery and a vein in the patient's arm for connection to the artificial kidney. When the dialysis treatment was over, keep the access to the circulation open by hooking the two tubes together outside the patient's body via a small U-shaped device, made of Teflon. This U-shaped Teflon device, which came to be known as the Scribner Shunt, served as a permanently installed extension of the patient's own circulatory system, shunting the blood from the tube in the artery back to the tube in the vein. Whenever the patient needed to be dialyzed again, no new incisions in the blood vessels had to be made. The Shunt was simply disconnected from the tubes in the patient's arm, and the patient was hooked up again to the machine...

The contributions of Willem Kolff and Belding Scribner revolutionized the treatment of kidney disease, saving and prolonging the useful lives of millions of people...."

As summarized by the US Renal Data System, the patient population increases by 8% a year; the health care cost incurred by these US patients comes to US$20 billion.[6] The annual mortality rate of hemodialysis patients is about 18%. Thirty percent of the patients often experience symptoms such as cramps, headaches, nausea or dizziness. These symptoms usually develop before the detection of measurable decrease in blood pressure. In severe cases, shock or death can happen.

These symptoms in close association with the development of hypotension may indicate their interrelationship to low blood flow to organs, in particular the brain. Low cardiac output is one key factor leading to hypotension. Since the patient's heart still functions normally, the Starling's principle characterizes the development of low cardiac output as a result of low venous return. The low blood volume (i.e. hypovolemia) and shifting of blood to fill an expanding microcirculation (i.e. microvascular dilatation) are two factors leading to low venous return.[7] If hypotension results from hypovolemia, the correct treatment is to replenish the low blood volume with fluid infusion to the circulation, for example. On the other hand, if the blood is shifted to fill the microcirculation of the liver, then the fluid infusion may not be effective in increasing the venous return.

Currently the physicians do not have a technique capable of differentiating whether the symptoms and hypotension result from hypovolemia or microvascular dilatation. Most often, the counter-measure chosen is not effective in alleviating the hypotensive symptoms.

In the US, about 0.4% of patients perform hemodialysis at home while 14% of the patients in New Zealand do so. Because patients can take home hemodialysis more often and over a period shorter than that available in the hospital, the patient experiences less fluctuation in blood pressure, and the hypotensive symptoms are less likely to develop. An increase in use of home hemodialysis requires better assurance of patient safety. There remain challenges and opportunities for biomedical engineers to further improve the monitoring technique capable of alleviating hypotensive symptoms, assuring a safe hemodialysis for the patient, and reducing mortality.

4. Your Venture

4.1 *Entrepreneurship*

Webster's dictionary defines an entrepreneur as someone who organizes, operates, and takes risks in a business venture in expectation of gaining profits. One popular conception of the entrepreneur is the lone-wolf model: the person who goes out and does everything required to start a business. This model of a business in relative isolation is simply not practical just because no one can be a master of all the skills required to establish and operate a high-tech business in this 21st century.

Many promising ventures in high tech start out as a small business in research and development. During this launching phase, a new product is invented, an improvement of existing technology is developed, or a service is improved. Depending on the complexity of the product, improvement, or service some businesses can be launched with a capital investment in the range of US$100,000 to US$1 million. For this phase, federal funding through the Small Business Innovation Research (SBIR) program has a record in getting many small business entities into an excellent start. When the potential of the product is demonstrated over the course of the launching phase, the company may then need to raise additional capital of US$1 million to US$10 million to build up the product's manufacturing capability, as well as to sufficiently market the product so that it can generate an annual revenue of US$100 million to US$1 billion for the new venture.

If you want to become a successful entrepreneur, you must possess a critical attribute: perseverance. The ability to stick with a task for as long as it takes to complete it is essential if you hope to raise venture capital, move your product to the manufacturing and marketing phase, and make the first few sales. You will seldom get a chance to present your business plan to a venture capitalist as the result of a single phone call or an introduction from an acquaintance. In fact, the venture capitalist may intentionally put you off for a while just to see how well you persevere in the face of delays or an initially lukewarm reception. Perseverance is a trait that the venture capitalist will want to see you demonstrate before he or she grants you an appointment to make a ten-minute presentation of your business plan.

Making the first sale of your product, which is an equally challenging task, can only be done with perseverance. The perseverance of the whole entrepreneurial team, committed to developing and selling a high-quality product, is essential if you want your product to penetrate the market significantly, particularly in the face of competition.

The entrepreneurial team needs a leader who can

- build up the team with qualified individuals,
- inspire everyone with the vision of the company,
- take calculated risks as the company develops and markets its product,
- make changes in the company direction when warranted, and
- take a leadership role in managing the other members of the team and their time.

A partnership arrangement, which operates by consensus, seldom works for a developing company.

One popular myth about entrepreneurs is that they are basically risk-takers. This is not true. Entrepreneurs are risk *managers*. They are very good at figuring out the minimum essential risks they will have to take to get the company off the ground, and they devise ways to share these risks with other people.

For example, entrepreneurs may start a company with their own money, but as soon as they can, they will borrow money from banks and investors so that they do not have to use their own money more than necessary. They will develop contracts and enter into business arrangements that require customers, vendors, and other people to bear some of the risks.

Entrepreneurs who see themselves as risk-takers are actually gamblers, and in the long term, the house always wins.

Something else that entrepreneurs do well is figuring out what might appear to be a risk that really is not. They will dig around to discover that the venture is really quite different than what one might have perceived from the outside. By cutting right to the heart of the venture, they can figure out the minimum essential work and the minimum essential risks that have to be taken in order to make the venture a success.

Entrepreneurship is an art that can be learned. Think about it this way: in medical school, professors teach their students the best scientific practices for healing sick patients, as well as more subjective things such as appropriate bedside manner. But there are limits to what any professor can teach. To develop into a capable physician and provide the best possible patient care, a medical student must develop the art of getting to the heart of a problem.

The same is true for the entrepreneur: an entrepreneur can be taught best business practices and develop more subjective attributes such as strong interpersonal skills, but they, too, must develop the art of getting to the heart of a problem.

Since entrepreneurship is a team effort, you have two options in an entrepreneur team: lead or follow. If you are inclined to lead, you can build a venture and do whatever is in your mind. If you are inclined to follow, you can join an entrepreneur team and support the growth of the new venture.

For example, a technically oriented engineer with an inclination to follow will serve well in the technology group, which follows the mission of the company by producing a product that is significantly superior to that of the competitors.

There are also opportunities for biomedical engineers who wish to work in a new venture's business group, which will include sales associates, accountants, and lawyers, all of whom support the entrepreneurial leader in managing the operation of the company.

Regardless of the role you might choose to play, the goal of the entire entrepreneur team is to carry out the R&D of the product, to develop the business plan, to pursue the mission set for the company, and to manufacture and market the products of the company. As a factor for consideration, the entrepreneur can outsource some of these action items to more experienced companies to achieve cost saving, timely completion of the project, and a larger sales volume.

4.2 *Your product or invention*

To gain a better understanding of the potential viability of your venture idea before you go to the expense of patenting the technology and launching your venture, you can start by answering the following questions:

(1) *Does your device solve an important medical problem*? What is the disease being addressed by your product? How many patients will be using your product? What cannot be accomplished clinically without your product? The importance of the medical problem may be closely related to the size of the market and patient demand for your device. Inventing a device or developing a new drug that can improve or treat many clinical conditions or diseases may imply that you have a potentially big market for your product.

(2) *Do physicians and patients want your device*? It is important to meet the needs of customers, i.e. the physicians, patients, biomedical engineers, nurses, or other health professionals who would be most likely to use your product. Engineering a device and then looking for its medical applications will not, in most cases, produce a product that can precisely fit the needs of the physician and patient. Once a device is developed, these customers should be continuously solicited for feedback on ways to improve your product.

(3) *How will the Food and Drug Administration (FDA) regulate your device*? In the opinion of many biomedical engineering entrepreneurs, working out a favorable FDA classification is a task that requires your utmost attention. You need to

know the complexity and uncertainty involved in navigating FDA rules and regulations. These rules and regulations will dictate the level of work that faces you in getting your product to market. The more the FDA demands justification for your product, the more capital you will need to get your device or drug into the marketplace, if ever. The related questions that need to be answered are whether government or insurance companies will reimburse for the use of your product or could you convince them to pay the reimbursement like that of hemodialysis care?

(4) *What is the technological advantage of your invention over that of your competitor*? In the case of device development, an improvement in sensitivity by a factor of ten in clinical power will be extremely desirable for you to successively market your device. Likewise, you will have an overwhelming edge if your device is non-invasive while the products of your competitor are invasive. Cost will also be a factor, but it is much less important than the clinical power of the device and the quality of the product and its service. In addition, you should also be able to answer the following questions:

- Who are your industrial competitors?
- What is the intellectual property position of your invention?
- Will your product be more powerful over those of your competitors?
- Is your product a hot area for development?

(5) *Is the timing right*? The success with the transistorized external pacemaker is an example of good timing. First, transistors had already been invented. The transistors are what enabled Bakken to miniaturize the pacemaker that previously had been constructed with vacuum-tube technology. Secondly, open-heart surgery had progressed to the point that it opened up the opportunity to develop pacing devices that would help return the heart function to normal. Similarly, the new material Teflon transforms the Scribner Shunt to success.

The answers to these questions will also require research, but of a different sort: interviewing people about the prospect of your product; checking with literature on the science and engineering of your invention; surfing Web sites on disease and patient needs; and attending scientific, engineering, and medical meetings to identify your competitors and to learn about development trends and state-of-the-art research already underway.

When you are ready to solicit investments from venture capitalists to develop, manufacture, and market your product, you will need to be able to provide them with positive answers to the following three questions:

(1) Is the market for your product big?
(2) Can you gain a significant percentage of the market over your competitors?
(3) Can your entrepreneurial team do the job?

Good market research will help you provide objective answers to the first two of these three questions. By the time you carry out this research on your venture and its marketability, you will have obtained a great deal of information. Integrate them into a meaningful market study and development strategy for you to understand the risks and their management, to assess the size of the market, and to compete effectively. Once you have demonstrated the feasibility of your invention and complete its initial development, you may obtain some more marketing information for you to develop a business plan and then raise funding for manufacturing and marketing your product.

As a guideline for the research and market analysis, Sheridan Snyder, the founder of Genzyme, recommends a minimum of 50 interviews with relevant experts — half conducted by telephone and half in person. Such extensive interviewing helps to ensure that accurate assumptions are being made about the competitive environment for a particular technology. *"No matter what you think your science is, it's going to have to move out into the world and compete against others,"* he said. *"The sooner you know what the competition is, the better off your company will be."*

5. Entrepreneurial Career

The best entrepreneurs come in one of two forms. The first type of entrepreneur has a feeling and an affinity for people; he or she recognizes that there are many opportunities to be exploited and seeks those opportunities out. The second type of entrepreneur is more technically oriented; he or she has a way of solving a problem or making use of a platform (or enabling) technology. By solving a whole class of problems with a certain technical approach, this second type of entrepreneur seeks out ways of implementing this technical approach commercially.

While business schools attract and graduate numerous entrepreneurs, they are not the largest source of entrepreneurs. Business schools provide them with extensive knowledge about how to run a business with great efficiency. Consequently, business schools are great places to find chief financial officers, marketing experts, operation managers and strategy planners for your entrepreneurial team, but not necessarily people who would take or manage risks.

What are necessary to be an entrepreneur are courage, perseverance, and leadership — attributes that are widely distributed in society. Schooling is not necessarily a requirement. Anyone with these attributes has the potential to become a successful entrepreneur, as long as he or she also has a great idea, a way to raise the operational capital, and a means to sell the product.

Students can learn about the entrepreneurial game while they are still in school by participating in fund raising activities for their colleges and universities, or by helping existing firms market their product lines. Hands-on experiences like these are great ways of picking up basic skills and gaining credibility within a specific organization — credibility that could lead to an invitation to continue on full time after graduation.

An enterprising student can always find an alumnus or alumna who owns his or her own business and ask to be put to work. Using such old-school ties is a great way to get one's foot in the door and gain industry experience.

Most cities have local technical councils or organizations actively promoting entrepreneurship and helping people develop and make business-plan presentations. Becoming involved with these councils or organizations is another good way for students to study new ventures or participate on a small three- or four-person technical team and learn how to write a business plan or business concept.

Another entry point to entrepreneurship is by gaining employment with a family business. If you have no prior business experience, look around for business people who know you, or who know someone you know, because these are the people who are most likely to give you the greatest break. If you like the business they are in, you could offer to work for them or volunteer over the summer, or part-time during a school term, or for a year or two after you graduate.

Big companies may have semi-independent entrepreneur entities in operation. Working in these entities will gain you valuable entrepreneurial skills. Students wanting to gain some business experience can also seek out academic entrepreneurial ventures — new businesses that have been built on the research of one or more professors. These days, a growing number of professors have businesses on the side that are sanctioned and supported by the university where the professor works. Most of these business activities are licensed from the university. However, very few professors have much excess time on their hands to fully develop their ventures into commercial successes, and would welcome any additional help they could get. Students specifically could look for opportunities to work on the development of a new or improved technology, while at the same time learning something about the business aspects of the venture.

Resourceful students may find other career entries to entrepreneurship beyond those just mentioned. The important thing to remember is that there are many ways for would-be entrepreneurs to gain business experience without actually having to attend business school or commit to working in a particular industry.

6. Biomedical Engineers Mean Business

This is the headline of an article written by reporter Charlotte Crystal in *Inside UVA*.[8] *The article noted that "Students will learn (from the course on 'Biomedical Engineering Entrepreneurship') how to conduct a patent search and determine whether competitive products are available; how to submit a medical device application to the Food and Drug Administration; how to apply for a federal Small Business Innovation Research grant; how to develop financial projections for the new product; how to write a formal business plan; and how to make a presentation to venture capitalists and negotiate a business deal that will protect their interest in their intellectual property."* This is one challenge and opportunity in bioengineering education for our students to know something about business, so that they will be prepared, later when they want to build companies around their technology.

Recognizing the emerging field of biomedical engineering, many engineers, scientists and physicians gathered to found the Biomedical Engineering Society in 1968 with the purpose to promote the increase of biomedical engineering knowledge and its utilization (www.bmes.org). Today 108 universities in the US have education programs with emphasis in biomedical engineering or bioengineering (www.Whitaker.org). These programs attract the best and brightest university students to work on the new discipline. Only when these biomedical engineers mean business, will their creativities and innovations be transformed to entrepreneurial ventures destined to expand biomedical industry, create jobs, enhance the nation's economy, save lives, and improve people's health.

Acknowledgments

The author acknowledges the support from NIH SBIR grants HL 57136 and DK 55423 and a Challenge Award from the Virginia's Center for Innovative Technology.

References

1. News Release on "AIMBE Hall of Fame Salutes Achievements in Medical, Biological Engineering That Have Saved and Improved Lives Worldwide." Available at www.aimbe.org (February 17, 2005).
2. J. Palmisano and G. W. Cough, Innovate America, *National Innovation Initiative Summit and Report*, Council on Competitiveness (www.compete.org) (2004), p. 7.
3. V. R. Fuchs and H. C. Sox, Jr., Physicians' views of the relative importance of thirty medical innovations, *Health Affairs* **20**: 30–42 (2001).
4. E. E. Bakken, *One Man's Full Life* (Medtronic Inc., Minneapolis, 1999).
5. J. L. Goldstein, Comments and Acceptance Speech of 2002 Albert Lasker Award, appeared in www.laskerfoundation.org and in *J. Am. Soc. Nephrol.* **13**: 3027–3030 (2002). Excerpted with permission from the Albert Lasker Foundation.
6. US Renal Data System, *USRDS 2003 Annual Data Report*, National Institute of Diabetes and Digestive and Kidney Diseases, Bethesda, MD. Available at www.usrds.org (2003).
7. J. S. Lee, 1998 Distinguished lecture: biomechanics of the microcirculation, an integrative and therapeutic perspective, *Ann. Biomed. Eng.* **28**: 1–13 (2000). Reprinted with permission.
8. C. Crystal, Biomedical engineers mean business, *Inside UVA* (March 28, 2003) p. 10.

CHAPTER 30

HOW TO MOVE MEDICAL DEVICES FROM BENCH TO BEDSIDE

Paul Citron

Abstract

Although there are similarities in how any new technology in any industry migrates from Research and Development (R&D) to the ultimate customer, the medical device industry has certain elements that are unique to it. Understanding these factors and accommodating them as an integral part of the business development plan can make the difference between a mere laboratory curiosity and an innovation that serves the needs of seriously ill patients, and at the same time produces financial returns for the industry. We will examine some of these factors in this chapter and address how they relate to the success of the innovation process, as well as the regulatory process and FDA's role and its requirements.

1. Some Background Considerations

The following observations apply to a medical device company that is already in business and has one or more commercialized products.

It is an immutable fact that yesterday's technological breakthrough will soon be an obsolete product. One need only look at the rapid flow of evolutionary improvements that occur in telecommunications, transportation, personal computers, and other consumer products to appreciate this. Engineers have a predisposition to not be satisfied with the state-of-the-art and seek to continually advance product capabilities. In addition, competitive pressure makes the process of continual *relevant* advancements an imperative. Companies that are content to maintain the status quo and rest on past technological innovations often fail. It is often only a question of time.

Clayton Christenson in his book, *The Innovators Dilemma*, addresses the crucial importance of *sustaining technologies* to the continued success of a company. Sustaining technologies are contained within next-generational product iterations. They are the innovations that keep the product fresh in the minds of customers and set the product apart from competitive offerings and give it an advantage in the marketplace.

A key point to note is that a technological innovation, absent perceived value by the customer, likely will not result in a sustaining technology. So, especially in the medical device field, adding new features and capabilities without due consideration to the relevant unmet needs of the eventual user may waste R&D resources. It can also have serious negative consequences for market share of the product. The most elegant design can and often does lose out to the more useful product.

What are some of the characteristics of the next-generational innovations of medical device? First, the next-generational product is firmly rooted within the context of the current generation. It builds on what is already out there, and great caution must be invoked to make certain the new innovations do not lead to confusion in how the device is used. This is particularly true in instances where some new aspect of the product may create a safety risk for the patient if it is not correctly configured or applied by the physician. Even though the labeling (e.g. instruction manual), as required by the FDA, may be clear, one simply cannot assume it is read by those using the device or whether the departures from previous practice are duly noted and understood. Changes to software and user interfaces are particularly a potential trap. Think about the frustrations, miscues, even disasters, you have experienced when moving from one generation of PowerPoint to the next. Unexpected consequences in patients that result from changes in the user interface can be dire.

The requirements for next-generational innovations are more or less obvious in most cases. In the case of implants, physicians and patients want them to be "smaller, lighter, more capable, cheaper, and more reliable." Your competitors know this too. While the outcome may be obvious, the methods to achieve them may be very complex and not obvious. That is where engineering expertise comes into play.

As an aside, this is a good time to bring up the notion of technology-driven versus market-driven organizations. As a general rule, the industry pioneer is usually technology-driven. After all, the first product generations are on a steep learning curve related mostly to making the product work in a heretofore unexploited application. The first physician-users are highly invested in the process of refining how the product is applied to patients, as well as helping the engineers make the technology more practicable. The first competitors are often more market-driven. They have the advantage of entering the market when many of the initial vagaries have been addressed. Such organizations tend to use marketing methodologies to better define what aspects of the technology need further refinement and also help set the priorities for selection among many possible feature improvements. Market-driven organizations set priorities as viewed from the perspective of potential users. Technology-driven companies have a difficult time making the transition to market-driven. Many are eventually eclipsed by later entrants if they do not make the transition of their operating culture to one that better balances these distinct cultures.

Some medical devices are truly stand-alone. Others are part of a system. Let us take the cardiac pacemaker as an example. The pacemaker system consists of an implanted stimulation module (most often referred to as the pacemaker) that contains a battery and sophisticated electronics for proper stimulation, information storage, and communication. The pacemaker is connected to a flexible lead that interconnects the module, implanted just under the skin, to heart tissue. When the battery is depleted after several years, perhaps as long as a decade, the physician makes a small incision, removes the pacemaker from its "pocket" beneath the skin, disconnects the lead, and reconnects a new pacemaker in its place. The new device may be several model generations newer and more sophisticated than the one it replaces. Another component of the system is the external programmer (the programming device) which is used by the physician during and after implantation of the stimulator module to interrogate the implanted pacemaker and to configure certain device settings to tailor its performance to the patient's condition. The programmers are essentially highly complex, specialized, and expensive proprietary computer systems. The purpose of this brief tutorial on pacemakers is to introduce the topic of *legacy* requirements for next-generational products. The replacement pacemaker needs to be able to connect properly to the old lead. The programmer needs to be able to "talk" to the old unit as well as the new one.

Unlike many other product markets, respect for legacy requirements is important and unfortunately can be a constraining factor in the innovation process for medical devices. Here are two examples where legacy considerations are important. Since a previously implanted lead may not be easily removed from the heart, physicians expect new pacemaker generations to be compatible with leads already implanted. In the case of programmers, the seemingly simple solution of offering physicians a new one to operate alongside the old one is often not a desirable option. First, programmers are relatively expensive and are most often given to the customer as an accessory, rather than sold. In addition, floor space in the procedure room and in storage rooms is very limited. Most customers use more than one vendor of pacemaker, so the issue of storage is multiplied since physicians need to find room for several programmers. Also, it is difficult to remember which programmer works with which device, and the user interfaces are likely very different. Having only one programmer per manufacturer causes less confusion and reduces sharply the possibility of user error. These examples highlight the importance of legacy constraints as well as forward compatibility considerations.

It is interesting to note that pioneering companies have a much longer tail of legacy products than newer entrants. This is yet another factor that can give newer companies an advantage since they can implement radically different designs with less regard to accommodating previous generations.

2. Where Do Innovations Come From? Next-Generational Ideas

Next-generational ideas can come from anywhere. This is the trite answer. In general practice, they come from internal R&D staff, marketing personnel, and physician-customers. There is an interplay among these groups that is crucial to a pipeline of relevant, competitively distinguishable new products.

The role of the R&D staff is to be the undisputed experts in understanding how the product works and how it performs in actual use. Especially for life-supporting, life-improving technologies, it is the responsibility of R&D to study carefully all aspects of product performance after the product is sold, not just predictive testing and simulations of anticipated performance. An important aspect of this is analysis of returned products for signs of unanticipated wear and other signs of impending failure or performance compromise. As an example, implanted electro-mechanical devices such as pacemakers are designed to maintain a hermetic atmosphere for certain components such as the microelectronics module and battery. Since the device must operate in a highly corrosive "wet" environment, it is important to know whether moisture is slowly finding its way into the hermetic enclosure. Were this to happen, sudden catastrophic failures could occur, endangering the patient. There could well be a situation where the current generation of devices perform admirably with a battery that has a life-time of, let us say, only three years, but where incorporation of an improved, higher-capacity battery that has a life-expectancy of six years would result in moisture-related failures as critical moisture levels develop inside the device. Paradoxically, the introduction of a major design improvement, notably extended service lives of the device through advancements in battery technology, would result in a spike of unanticipated device failures in later years.

The R&D staff should understand the implications of new technological advancements in the field. They should be able to assess how current performance limitations could be addressed by the incorporation of newer technology. At the same time, caution must be exercised regarding adopting the new technology before some of its latent performance characteristics are teased out. What may be suitable for a consumer entertainment product or automotive application may prove not to be robust enough for a medical device that a patient relies upon. The device literature is, unfortunately, rich with examples where promising new component technologies, which had proven themselves in other applications, have failed miserably when used for implant applications.

R&D staff must also have a firm grasp of how their product performs in the context of other medical devices and interventions, as well as with other everyday technologies that can affect performance. As examples, how will the product perform when exposed to modern imaging technologies such as MRI or to therapeutic diathermy? Will it be affected by electromagnetic interference from microwave ovens or cell phones? Conversely, will it adversely affect the performance of other technologies the patient may need in the future? A device does not have to be highly sophisticated (a relative term as used in this example) in order to raise issues of safe interactions with other technologies. Consider a breast implant. How will it affect the physician's ability to visualize early, subtle tissue abnormalities when conducting mammograms for breast cancer?

Marketing personnel's role is to have a "pulse" of the market. They should be conversant with technological capabilities, present and emerging, and align them with market needs and also with the ability of the market to properly apply the technology. With respect to the latter point, marketing personnel must take responsibility for training and education of the customer and patient, not just promotion of the product. This may take the form of seminars, workshops, and literature. Marketing personnel also has the responsibility for being evenhanded in the approach it takes to the education process. This means they must disclose the potential risks and complications associated with the product, not just the positive aspects. It means picking representative case examples of outcomes, not just the exceptionally good examples. It means setting realistic expectations.

Marketing is the source for competitive intelligence. It must be conscious of competitive offerings and what is the customers' perception of them. From this knowledge should come a working understanding of relative strengths and weaknesses of the respective products and strategies to close gaps and leapfrog the competition. In order to leapfrog, though, marketing must also have an understanding of competitors' trends relative to where the company is headed. The competitive intelligence process is not devious, unethical, or illegal. Public information sources abound ranging from research presentations at medical and scientific meetings, published reports in the medical literature and by the company (e.g. annual reports), executive presentations to investment groups, etc.

Perhaps most important, marketing personnel should have ultimate responsibility for setting priorities for next-generational product features and product positioning. This, of course, is done with active consultation with R&D staff, particularly as it relates to the relative risk (to schedule and cost) of each feature. Consequently, developing the specification is a shared responsibility with R&D. Stated another way, marketing should own the product specification. R&D should view it as an achievable description and is responsible for its execution, on time and on budget.

An often underappreciated partner in the design of next-generational products is the customer-user, or the physician. Note that this physician is often different from the research physician. These are the practitioners "in the trenches," not at the lab bench. They know what works and does not in actual practice. They know what performance gaps exist, even though they do not know how, technologically, to solve them. But they can inform the company what the issues are. The customer can place your technology in the context of what it really does in practice. Companies often misperceive the relative importance of their products to the other activities performed on a day-to-day basis. Companies also misperceive what really goes on in the customer's setting. To this latter point, having a program that exposes key marketing and R&D staff to the environment where the technology is used can have a profound impact on the success of next-generational products.

This section began with the question where do next-generational ideas come from? If there were a balance scale, it would tilt heavily to the role of R&D working with Marketing, collaborating with actual customers, all as colleagues in the innovation process. A seminal point to be made again here is that the hands-on experience of the physician who routinely uses the product is invaluable. It is why *listening to the customer* is central to an effective next-generational innovation process.

3. Where Do Innovations Come From? Breakthroughs

Breakthrough products are those that, for the first time, (1) address a long-standing unmet clinical need creating a new market or (2) are discontinuous enhancements that cause a sudden shift in the share in existing markets because of their newly-found singular importance in the clinical setting. Whereas next-generational innovations are usually predictable and part of a strategic plan, breakthroughs in existing product lines often start as unplanned events. They emanate from one or a few individuals in the organization who have a "vision." Often, breakthrough improvements go against the flow of conventional thinking. Consequently, breakthrough ideas in their formative period are often ridiculed by management and customers, because they are unobvious and perhaps even heretical.

There are many reasons for the hostile reception given to major innovations in their formative period. The first is that most "great" ideas eventually prove not to work. Another source of skepticism in a company setting is the fact that the new idea must compete for scarce R&D resources with existing products or advancements already on the product plan. It is nearly always easy to make an argument for why pouring more money into existing product lines that have proven markets should take priority over highly speculative investments in new, unproven areas. In addition, timelines for breakthrough products are normally much longer than those for next-generational innovations. Since markets may not yet exist for the pioneering breakthrough, it is very difficult to accurately estimate the business opportunity. Nevertheless, progressive and successful companies do undertake projects that eventually are considered breakthroughs and provide engines of vigorous financial growth.

Interestingly, current customers may not be a good source of guidance for major advances and innovations. Let us examine some of the reasons. Existing customers are highly vested in their current ways of doing things. It is very difficult to get them to change treatment protocols that are viewed as effective and that are a standard of care within the medical community. If physicians perceive the current method works well enough, there is a natural bias against untested new methods. There is a fear of unanticipated, latent side effects and complications. In those instances where the new innovation could steer medical practice to another medical specialist, the current customer consciously or subconsciously comes up with reasons and seemingly reasonable arguments why the innovation is flawed and why the level of care may suffer.

In the previous section, the notion of listening to the customer regarding next-generational enhancements was introduced. In the case of breakthroughs, customer input is not as reliable for the reasons just stated and also because of something called "functional fixedness." It is *the human tendency to fixate on the way products and services are normally used, making people unable to imagine alternative functions* (from "Turning customer input into innovation," *Harvard Business Review*, Vol. 80, No. 1, January 2002). The very customers who can provide excellent input on how to incrementally improve a product may be highly unreliable for innovations that are major departures from current practice. Functional fixedness thus further complicates the process of selecting from a large number of projects when allocating resources. Proposals that are easy to visualize or are variations on current practice and ways of doing things have a much easier chance of garnering support than those ideas that are still unproven but have the potential of providing very high levels of value.

Let us look at some factors that facilitate successful breakthroughs.

There really is no reliable formula for picking which potential breakthrough investments to make. One characteristic that seems to increase the likelihood of success is who the idea champions are. If they have had previous breakthrough successes, this is a strong indicator that their concept merits serious consideration. Other rules of thumb include staging investments so projects start small and grow only as milestones are met; having a high-level management champion who protects the project from cost-cutting moves that would reallocate budget to seemingly safer projects; forming a small R&D team that is passionate about the project and that possesses above-average technical and organizational skills to deal with setbacks and also to scavenge resources.

4. The Regulatory Process and Innovation

Medical technologies are regulated by the Food and Drug Administration (FDA) in the United States and equivalent bodies in other countries. Even though there is a general awareness among start-up founders and managers that the FDA's requirements must be met, there is often a misperception of the specifics involved and how to prepare and stage activities so that new products can enter the marketplace as soon as possible. All too often, FDA considerations are put off until far too late in the R&D process. When this happens, schedules can be seriously affected and there may be substantial unplanned expenses in order to comport to requirements set by the FDA.

At this point, a brief primer of the FDA's role and its requirements may be useful. Although the federal government had some authority to regulate the flow of medical devices in commerce since the early 1900s, it was not until 1976 and the passage of the Medical Device Amendments that the FDA gained sweeping authority to regulate essentially all aspects of devices. The 1976 act gave the agency authority to grant approval for the conduct of clinical studies to secure safety and effectiveness data; to approve or "clear" new products for commercial release as well as deny approval; to define and enforce good manufacturing practices; and, to require companies to provide proper notification of defective products.

It is interesting to note that drugs came under FDA regulatory control much earlier, in 1938, with the passage of the Food, Drug and Cosmetic Act. This act gave the agency the explicit authority to review new drugs prior to commercial release for safety. This act also had provisions to allow the agency to intercede in the event of misbranded or adulterated devices, but not the *a priori* authority of the 1976 Act.

The failure to grant regulatory authority for devices in the 1938 Act stems from the relatively low impact therapeutic devices had prior to the 1960s. It was not until the clinical application of highly sophisticated devices such as prosthetic heart valves, implanted cardiac pacemakers, and orthopedic implants that safety concerns relative to devices required regulatory oversight. The inclusion of devices as the responsibility of the FDA was driven by Congress. Congress wisely recognized that medical devices were very different from drugs. Consequently, the provisions of the Medical Device Amendments of 1976 created a regulatory pathway that was distinct from that of drugs. It also led to the creation of a new branch of the FDA, the Center for Devices and Radiological Health (CDRH) that had administrative responsibility over devices. CDRH joined the Center for Drug Evaluation and Research (CDER) and Center for Biologics Evaluation and Research (CBER) as more or less parallel and quasi-independent regulatory groups that had as their mission protection of the American public from unsafe and ineffective technologies.

Table 1 summarizes the major differences between drugs and devices that have relevance for how they are regulated and why a common approach for drugs and devices within CDER is highly inappropriate. These intrinsic differences require a very different approach, operational mindset, and technical skills in performing the various steps and the formulation of relevant requirements for the regulatory process. The pronounced difference between the indirect[a] therapeutic action of drugs in comparison to

Table 1. Summary of major differences between drugs and devices.

Devices	Drugs
Direct mechanism of action — Readily apparent response	Indirect mechanism of action — Metabolites, GI and liver inactivation
Site/organ-specific therapy	Systemic treatment
Uniform patient response to therapy (generally)	Variable patient response — Dosing — Side-effects, toxicity
High initial cost	Costs accumulate over treatment
Automatic therapy	Dependence on patient compliance
Progressive efficacy improvement (new features, fewer complications)	Efficacy static
Next-generational cost-effectiveness improvement	Cost-effectiveness relatively constant

[a]Drugs, whether administered by mouth or via other methods, generally circulate throughout the body and can affect multiple tissues in addition to the intended target tissue or organ. In addition, drug compound pharmacokinetics are often modified as they are processed by the liver, leading to highly variable inter-patient results. Devices, on the other hand, are in most instances site specific in their location and action.

the direct action of devices is responsible for a very different set of clinical evaluation models and imperatives for drugs versus devices. For instance, concerns about latent effects of drugs on non-target tissue or organs such as liver toxicity or carcinogenicity suggest that drug trials must enroll many more patients and go on for substantially longer periods of time than devices in most instances.

Three device classifications were established by the 1976 Act, based in large measure on the risk posed by the device. Class I devices are those that are perceived to pose the least risk to patients. They include items such as tongue depressors, elastic bandages, hand-held surgical instruments, and examination gloves. Such devices do not require FDA approval to be marketed. Class II devices pose intermediate risk to patients. Agency "clearance" is required in order to market the product. Products that are eligible for Class II designation include items such as electric wheelchairs and external infusion pumps. Class III devices are those that have the potential to pose substantial risk for patients. They often are life-support or life-enhancing devices for which safety and effectiveness must be demonstrated and assured. Examples of Class III devices include prosthetic heart valves, implanted drug pumps, brain stimulators, and ventricular assist devices. Class III devices require extensive pre-clinical and clinical testing as a prerequisite to securing FDA pre-market approval (PMA). A provision of the 1976 Act, the so-called 510k exemption, permits a somewhat abbreviated regulatory process for those otherwise Class III devices and some Class II devices that are "equivalent" to devices that were on the market prior to May 28, 1976. An example of a device category that falls into the 510k category is the cardiac pacemaker. It should be noted that the determination of whether a specific device qualifies for 510k status is not automatic and often requires discussion and negotiation with the agency.

From time to time, companies that come up with breakthrough products are tempted to use the complex and costly FDA approval process as a barrier to entry for others who follow. The thinking is that setting high regulatory hurdles may discourage fast followers from entering the market. At a minimum it is hoped that the regulatory process as influenced by the pioneer will cause delays and give the pioneer a greater period of marketplace exclusivity. This is not a sound strategy for a number of reasons. Self-imposed requirements that seem easy today (but are viewed to be difficult for the competitor, such as an unnecessarily large clinical study or taking the Pre-Market Approval (PMA) route when a 510k route was possible) may prove to be complex and expensive tomorrow — they may also become unnecessary but the agency will be reluctant to remove them, thereby adding cost and delay to the process. As an independent and highly fluid oversight and enforcement organization, the FDA is not and should not be viewed as an ally and business partner. It is not and should not be. Finally, overly prescriptive design "rules" that are suggested by the first study sponsor may become obsolete as more is learned about the technology, but may be difficult for the agency to eliminate in the future. As an example, the commercial pioneer in the cochlear implant field "taught" the FDA that implanted electrodes that were inserted too deep in the cochlea were dangerous. They did this in part because competitors were evaluating longer electrode systems that went deeper into the canal. This false impression became a problem when it was shown that deeper placement produced markedly superior results, but the safety concern was embedded in the FDA's thinking and had to be disproved.

As biomedical engineering advances, the lines between a device and a biologic or device and drug begin to blur. Let us look at some examples. Wire stents used to prop open a coronary artery are easily classified as devices. But what about a wire stent that elutes a drug that further improves long-term efficacy? Metal "cages" are often used to immobilize spinal vertebra as a treatment for severe back pain. These are devices in the classical sense. How does one classify a cage that contains bone morphogenic protein that produces a more reliable and robust fusion? What about tissue-engineered products that may have a mechanical component that provides short-term functional performance but is designed so that the artificial elements are replaced or enhanced by what appears to be natural tissue? Such "combination" products are becoming increasingly common. As has already been noted, the regulatory pathway and requirements for each of the traditional branches are markedly different from the others for good reason. The involvement of two, and perhaps three regulatory branches poses a distinct regulatory challenge for the study sponsor. Issues such as consistency, jurisdictional authority, least burdensome requirements, and clinical and technological expertise emerge. They further complicate an already difficult process. The confusion can negatively affect the innovation process because it adds substantial opportunity-risk (time and money) to an already difficult circumstance. The FDA is developing protocols that define which branch of the FDA takes the lead in a combination product approval and what constitutes a reasonable set of requirements in order to demonstrate safety and effectiveness.

One step the FDA has taken was the creation of the Office of Combination Products (OCP) at the end of 2002. According to the FDA, OCP duties include:

- *assigning an FDA Center to have primary jurisdiction for review of a combination product;*
- *ensuring timely and effective premarket review of combination products by overseeing reviews involving more than one agency center;*
- *ensuring consistency and appropriateness of postmarket regulation of combination products;*
- *resolving disputes regarding the timeliness of premarket review of combination products;*
- *updating agreements, guidance documents or practices specific to the assignment of combination products;*
- *submitting annual reports to Congress on the Office's activities and impact;*
- *working with FDA Centers to develop guidance or regulations to clarify the agency regulation of combination products; and*
- *serving as a focal point for combination products issues for internal and external stakeholders.*

5. Some Closing Observations

The growth and even survival of a medical device company depend on a continuous flow of new innovations. Clearly, next-generational iterations and improvements are expected by customers to address shortcomings and to harness emerging technological advancements that can make the product more reliable, more effective, and provide greater value in its intended applications. Breakthrough products that offer fundamentally new approaches for solving clinical problems are important ingredients in allowing a company to broaden its presence in the marketplace and more fully serve the needs of its customers. Although there is not a precise formula for where next-generational versus breakthrough innovations come from, this chapter has provided some generalizations to consider and exploit.

This chapter has also examined a number of factors that influence the success of a new idea in the medical device field. A key point is that a new idea, even a seemingly great idea, is an unfulfilled promise unless it becomes widely used in the marketplace. Only after it is adopted by physicians who have the ability to choose from among alternatives does the idea become an innovation.

Finally, a number of external forces greatly influence the innovation process for better or worse. Key among these is the regulatory process as has been noted. Other factors not discussed in this chapter that can affect the movement of new ideas from bench to bedside include: the ability to secure adequate levels of reimbursement in a timely manner; the degree of difficulty in mastering implantation techniques; whether an adequate market size exists to financially support the infrastructure needed to sustain the technology and still secure adequate levels of revenue and profit; and, whether patients believe the technology offers them a desirable treatment option. This last consideration often does not receive enough attention from engineers and scientists as they develop what they hope will be significant innovations. We must always keep in mind that medical device technologies serve patients, not the other way around.

INDEX

3-D model 121, 126–128
3-D nano-mechanics 117, 127

acetylene 197
acridine orange 359
actin 45, 46, 235–238
  cytoskeleton 85
  protofilament 119, 120, 123, 125–128
actin-disrupting agent cytochalasin D 85
action potential 39, 42, 46, 47, 49
activator protein-1 (AP-1) 81, 83
active module 465
adapter molecule 82, 86
adenosine 103
adenosine triphosphate (ATP) 103, 412–416
adenovirus 93, 95
airflow resistance 189
airway 181, 183–185, 188–191
  peripheral 190
  proximal 190
alternative nuclear transfer (ANT) 306
alveolar
  epithelial cell 190, 192
  instability 190
  interstitial region 192
alveoli 183–187, 190, 202, 205, 206
Alzheimer's disease 298
American Institute for Medical and Biological Engineering (AIMBE) Hall of Fame 508–510
amyotrophic lateral sclerosis (Lou Gehrig's disease) 300, 301
anatomy 225, 226, 228, 229, 239
anesthetic gas 194, 197
angina 111
angiogenesis 87, 315–323
animal
  model 300
  subject 490, 495, 496
anisotropy 44, 45
aorta 38, 40
aortic valve 38–40
apoptosis 81, 86
architecture 225, 232
array 475, 476, 480, 481
arterial
  hypoxemia and hypercapnia, causes of 205
  network 100
  pressure 102–107, 110, 111
  vascular tree 184
arteriole 99–107, 109–111
asthma 189
astrocyte 294, 295
atherogenesis 80, 87, 89, 93, 94
atheroma 80
atherosclerosis 80, 81, 90, 92, 94, 99, 110
atrio-ventricular node 42
auto-digestion 131, 144
autoregulation of blood flow 99, 100, 103, 110, 111

bacterial flagella 416, 418
balloon angioplasty 92–95
baroreceptor 40

benzyl alcohol 85
Bessis, Marcel 359
beta cell 298
beta-glucuronidase 363
bicarbonate ion 192
binding 212
bioengineering 81, 453
bioinformatics 307, 401, 403, 408, 410, 411
biomechanics 13, 14, 20, 28, 30, 31
Biomedical Engineering Society 520
biophotonics 353
birth defect 297, 300
BLAST algorithm 407, 408
blastocyst 291–293, 302, 303, 305–307
blood
  loss 149–151, 154
  pressure analysis 24
  substitute 149, 153, 154
  viscosity 170, 171, 177, 178
blood : gas partition coefficient 201
body fluid 209–213, 217
  regulation 216
bone marrow transplant 300
boundary condition 196, 200, 201
bovine aortic endothelial cell (BAEC) 82, 85, 87, 88
brain 102, 103, 106, 111
branch point 80, 81, 87–91, 94
BrdU incorporation 87, 88
bundle of His 42
bypass surgery 92, 95

calcium (Ca^{2+}) 37, 45–49, 108–110
  channel 46
California Institute of Regenerative Medicine (CIRM) 304
cancer 298, 300, 353, 357, 362
capillary 181, 184–186, 192, 205, 206
  endothelial cell 192
  network 181, 182, 184, 185
  sheet 163, 170, 171, 173, 174
  transit time 196
carbon
  dioxide 103
  monoxide 198
cardiac
  arrhythmia 49
  cycle 38, 44
  electromechanics 37, 43
  function 37
  hypertrophy 64, 65
  output 186, 191, 196, 197, 199, 205
cardiac myofibres sheet 40, 41, 43, 44, 49
cardiovascular
  disease 132, 140, 142, 145
  system 99, 100, 102
career 507, 508, 518, 519
cataract surgery 359
cavitation bubble 359
Cbl 85, 86
cDNA cloning 121
cell 37, 40–42, 45–47, 49
  activation 137, 139–145
  cycle 87
  division 292, 298, 300
  membrane adhesion 139
cell-based therapy 295, 297, 300
cellular
  contraction 55
  function 453–455, 460, 462
  signaling 421, 425
central nervous system (CNS) 327–333, 335, 337, 342, 343, 345–347
centrin 361, 362
centriole 361, 362
cFos 81, 83, 86
chemotactic agent 81
chemotherapy 300
chest wall 186, 187, 192
cholesterol 85, 92
chromosome 297, 306
chronic bronchitis 189
circulatory system 37
circumferential stress 80
cis-element 81
cJun 81, 83, 86
c-Jun N-terminal kinase (JNK) 82–87, 94
cloning 363
coated stent 95

compliance 40, 49
computational
  biology 428, 439, 441, 448
  model 43, 45, 49
computer simulation 42
concentrating mechanism (urinary) 216, 218, 219
confined compression 245–250, 254, 256
conflict
  of interest 490, 497
  resolution 503
confocal microscope 85
connective tissue cell 294
constitutive
  equation 27, 28, 31
  model 45
contractility 40
convection equation 199
cornea 357, 358
coronary artery disease 92, 93, 95, 111
coronary vessel 54
Coulomb force 211
countercurrent 209, 215, 219, 220
crossbridge 46, 49
cyclin dependent kinase (cdk) 87
cytoskeleton 62, 63
  protein 84

data management 490
deadspace 188, 191, 199
degenerative 292
degenerative disease 292
degree (number of interactions for each protein node) 460, 466
dense body 109
  denudation 93
depolarization 42, 44, 46
derivation process 302
diabetes 298–300
diastole 38, 45, 47, 54, 57, 59
differentiated cell 294, 295, 297
differentiation 291–295, 297, 298, 300, 301, 307
diffusion 181, 182, 186, 190, 192–199, 201–203, 205, 206, 211, 212, 219, 418, 421, 422
  coefficient (D) 85
  equation 194, 195
  limitation 199, 201–203, 205
diffusion-limited 197, 198
DiI 85
disturbed flow 88, 89, 92
DNA
  hybridization 369, 370, 372–375, 382
  microarray 459, 460
  microarray technology 86
  plasmid 363
  sequence 453, 458, 464, 465
dopamine 298, 299
drug delivery 320

ejection 38, 40
elastic
  deformation 118, 120
  recoil 186, 187
electric field 353, 359
electrical activation 55, 59, 60
electrocardiogram (ECG) 39, 42, 43
electron microscopic 91
electrophysiology 37, 43–45, 47, 49
electrotonic coupling 47
embryonic
  development 297, 298
  fibroblast 296
  stem cell line 296, 297, 305
emphysema 189
endocardium 38, 41, 42
endothelial cell (EC) 79–88, 90, 91, 94, 103
  apoptosis 81, 86
  death 91, 92
  denudation 93
  mitosis 81, 91, 92
  survival 86, 87, 90
  turnover 79, 80, 82, 91, 92, 94
endothelium 134, 136, 137, 139, 140, 143
energy conversion 410, 411
engraftment 301

entrepreneurship 507, 508, 514, 515, 519
epicardium 38, 41–43, 45, 410, 411
equilibrium 410, 411
ERK 82, 83, 85, 86, 88, 94
erythrocyte 119–121, 123, 124, 128
  tropomodulin (E-Tmod) 119–127
ethical decision-making 489, 490, 505
ethics 301, 302
Evans blue albumin (EBA) 91, 92
evolution 100
extracellular matrix (ECM) 80, 83–86

feeder layer 296
fertilized egg 292, 293, 305
fibronectin 84
Fick's law of diffusion 190, 193
filament 45–47, 49
finite element method 42, 44
finite strain 59
Flk-1: VEGF receptor 2, 79, 84–87, 94
flow 79–82, 85, 87–95
  pattern 81, 87, 88, 91, 95
  volume 106
flow chamber/channel 79, 81, 82, 85, 87, 88, 93
  step flow channel 79, 87, 88, 93
fluorescence 353, 361, 362, 364
fluorescence correlation spectroscopy (FCS) 364
fluorescence recovery after photobleaching (FRAP) 85, 364
fluorescent resonance energy transfer (FRET) 364
fluorocarbon 153
flux 211, 212, 216–219
foam cell 80
focal adhesion kinase (FAK) 82
force-velocity
  relationship 48, 238
Frank-Starling mechanism 40, 44, 47
free energy 212, 215, 219
functional
  fixedness 526, 527
  lung unit 182–185, 188, 189, 191

gap junction 41
gas
  exchange equation 194
  kinetic theory 196
  molecular weight 196
gene expression 25, 26, 81–83, 86–88, 90, 95
  analysis 459
genetic disease 300
genomics 474, 483
  cloning 122, 128
  sequencing 473–477, 483
geometrical
  factor 307
  mismatch 93, 95
glaucoma 359
glia 338, 340
global cardiac function 56
glomerulus 213, 214, 218, 220
  filtration 214, 217, 218, 221
glycocalyx 79, 86
G-protein coupled receptor 79, 86
gravity 211
Grb2 82
green fluorescent protein (GFP) 353, 361, 362
growth 292, 307
  factor 80, 307
growth arrest and DNA damage inducible protein 45 (GADD45) 87

H^+ ion 103
heart 37, 38, 40–45, 47, 49
  anatomy/structure 54
  attack 298
  disease 299, 300
  failure 54, 62–65, 69–71, 73–75
  rate 40
HeLa cell 362
hemodialysis 507, 508, 511–514, 517
hemodilution 149, 154, 157
hemodynamic force 80, 81, 94
hemoglobin 149, 150, 152–158, 186, 192–194, 196
  carbamino 193
  vesicle 157

hemolytic anemia 118, 120, 123, 124
hemorrhage 111, 149, 157
Henry's law 195, 198, 201
heterogeneity 45, 47, 191
high-throughput drug screening 307, 458
homeostasis 212, 213
  function 87
human
  embryo 302
  feeder cell 296
  fibrosarcoma cell 363
  genome 402–404, 408
  subject 493–495, 502, 503
human aortic endothelial cell (HAEC) 86, 87
human subjects
  autonomy 495
  beneficence 495
  justice 495
human umbilical vein endothelial cell (HUVEC) 81, 82, 90
hypercapnia 204–206
hyperelasticity 45
hyperopia 357
hypotension 513, 514
hypoventilation 205
hypovolemia 513, 514
hypoxemia 204–206
hypoxic pulmonary vasoconstriction 191

immune
  deficiency 300
  system 296, 300
immunosuppressive drug 296
impedance 163, 164, 172, 173, 175–178
implantable glucose sensor 279
incompressible 45
indentation 245–251, 255, 256
inert gas 194, 195, 197, 198, 201, 202
inflammation 87, 132, 133, 142, 144, 145
inner cell mass 292, 293, 296, 305–307
innovation 507–511, 520
insertion 226
Institutional Animal Care and Use Committee (IACUC) 496
Institutional Review Board (IRB) 494
insulin 298–300
integrin 79, 83–86, 94, 99, 109
  $\alpha_6\beta_1$ 84
  $\alpha_v\beta_3$ 84, 85
integrin-blocking antibody 85
  6S6 85
  LM609 85
intellectual property 492, 517, 519
intercalated disk 41
intercellular junction protein 79, 86
intercostal orifice 91, 92
interdisciplinary research 307
internal environment of the body 100
Internet 353, 359, 364, 365
invention 507, 508, 516–518
in vitro fertilization (IVF) 292, 305, 306
ion channel 79, 86
isolated small artery 107
isotropy 44, 45
isovolumic
  contraction 38, 39
  relaxation 38, 40

junctional complex (JC) 117, 119, 120, 123, 125–128
juxtaglomerular apparatus 110

kanamycin 363
keratinocyte 294
kidney physiology 209–213, 216, 221
knee
  arthroplasty 261, 263, 266, 267, 270, 274–276
  force 261, 266
  kinematics 266
knockout (KO) mouse/mice 117, 120–123, 128
Krüppel-like factor-2 (KLF-2) 90

lactic acid 103
LacZ 93
laminar 79, 86, 87, 94
laminin 84

laser
argon ion 359
femtosecond 360
scissors 353, 359, 361, 363, 364
trapping 353
tweezers 364
laser assisted *in situ* keratomileusis (LASIK) 357
Law of Laplace 108, 109
relationship 58
left atrium 37, 38
left ventricle 38–41, 43, 49
legacy requirement 523
length-tension relationship 22, 236, 238
leukocyte 133, 136, 137, 139, 140, 142
ligand 80
living system 401, 402, 409–412, 418, 424
LM609 85
low density lipoprotein (LDL) 80–82, 89, 91, 92, 94, 95
oxidized 80
macrophage 80, 81

macrophage 80, 81
macula densa 110
macular degeneration 356, 357
magnetic resonance imaging (MRI) 41, 59
diffusion weighted 41
maldistribution 188, 191, 192
MAP kinase 467
market-driven 522
mass-conservation equation 199
mathematical model 411, 419
matrix metalloproteinase-1 87
matrix protein 307
mechanical circulatory support system (MCSS) 69–75
mechanical
force 298, 307
interdependence 190, 206
mismatch 93
property 244–246, 249, 256
weakness 124
mechanics 37, 44, 45, 49
mechano-chemical transduction 79, 81, 83, 95
mechano-electrical feedback 37, 43, 49
mechanosensing 81, 84, 86
mechanosensor 109
mechanotransduction 60–62, 65, 79, 86, 95
medical imaging 307
membrane 117–128
lipid bilayer 84, 85
lipid fluidity 85
mechanics 117–120, 125, 128
sensor 79, 84
skeletal network 118–120
thickness 193
mentoring 490, 500–502
metabolic
engineering 395
hypothesis, the 99, 102–106
metabolism 388, 390, 395, 410, 411, 413, 424
oxidative 103
microarray 388, 390, 392, 399, 421
microcirculation 133, 136, 137, 140–143, 150, 151, 154, 155
microelectronic array 369–376, 380, 381
micro-environment 307
microfabrication 480
microfluidics 435, 436, 438, 439
microkeratome 358
micropipette aspiration 120, 123, 124
microplasma 363
microtubule 360, 361
microvascular dilatation 513, 514
midmyocardial 47, 49
migration 81, 86
minute ventilation 185
mitogen activated protein kinase (MAPK) 81, 82, 85, 88
mitosis 81, 91, 92, 360–362
metaphase 361
mitotic cell 361
mitral valve 38, 39
mixing entropy 219, 220

MMS (methyl-methane-sulfonate) 465, 466
molecular
- biology 117, 128
- configuration tensor 28
- network 453–455, 457, 460, 461, 463, 465–467
- ruler 119, 123, 125, 126

monocyte 80, 81, 82, 95
monocyte chemotactic protein-1 (MCP-1) 79–83, 88, 89, 93–95
mortality 513, 514
motion 209, 211, 212
mouse feeder cell 296
mRNA expression 462, 463, 465
muco-ciliary transport system 189
multiphoton absorption 360
multi-photonic 353
multipotent 291, 295, 296
muscle LIM protein (MLP) 62–64
myocardial ischemia 63, 65
myocardium 38, 40, 42–45, 47–49, 102, 103, 111
myocyte 40–46
myofiber 39–43, 45, 48, 49
myofibril 235, 236
myogenic (autoregulation) 220, 221
myopia 357
myosin 45, 46, 225, 235–238
- head 46

nanoengineering 327, 335, 337, 345
nanofabrication 369, 371, 375, 381, 476
nanoparticle 314, 322, 323
nanotechnology 327, 328, 336–343, 345–347
negative feedback 209, 213, 219, 221
nephron 213–215, 217, 218, 220, 221
- anatomy 214

nervous system 99, 102, 105, 111
network 387–396, 399
- genomics 453, 454, 468
- module 465
- motif 464

neurobiology 328, 347
neurodegenerative disease 300
neuron 294, 295, 299, 327, 330–336, 338–341, 343, 344
neuroscience 327, 328, 337–342, 344, 347
nitrous oxide 197
notebook 491
nuclear transfer 306, 307
nutrient 100

Oct-4 297
oligodendrocyte 294, 295
opening angle 15–19
optical breakdown 358
optoinjection 363
optoporation 363
organ level 37
orthotropy 43, 44
osmotic gradient (renal medullary) 212, 214, 215, 219
osteoarthritis 268, 300
oxidized 80
oxygen 103
oxygen carrying capacity 149, 150, 152–155
oxygen free radical formation 139, 140

$p21^{cip1}$ 87
p53 87
p60Src 82
pacemaker 507, 508, 511, 512
pancreatic digestive enzyme 144
paralysis 300
parenchymal cell 100, 102
Parkinson's disease 299, 300
particle deposition 189
pathBLAST 466, 468
pathway (regulatory or signaling) 454, 458, 460, 463, 465, 466
- alignment 466
- paralogous interaction 466

patient
- privacy 304
- right 304
- safety 304

PD98059: inhibitor for ERK 88

perfusion 182, 194, 197, 198, 200–206
perfusion-limited 197
pericardium 37, 45
permeability 79–82, 89–92, 94
perseverance 514, 515, 518
personalized medicine 474
perspective 492, 503
phosphorylation cascade 82, 83
photoablation 353, 355, 357, 359
photochemical 353, 356, 357
photodisruptive 353
photodynamic therapy (PDT) 353, 357
photothermal 353
plaque 93
plasma
    expander 149, 151, 152, 154, 157
    formation 359
plasticity 295, 296
platelet 133–135, 137, 139, 142, 143
pluripotent 291, 293, 295–297, 301, 304–308
pneumothorax 187, 188
pore theory 218
porphyrin 356
positive pressure mechanical ventilation 187
potassium 45, 47, 49
potential difference 211, 215
P-R interval 42
preimplantation genetic diagnosis (PGD) 305, 306
preload 40, 48
pre-market approval (PMA) 529
pressure 80, 93, 94
    drop 100–102, 111
pressure-flow relationship 163, 167, 171, 177, 178
prestretch 60
proliferation 81, 86–89, 93, 95, 295, 297, 298
Proposition 71 on stem cells 303, 304
protein kinase 82
protein kinase C 110
protein-protein interaction 455–458, 461, 463, 464, 466, 467
pseudopod formation 137–139, 142
public policy 301
publication 490, 492, 498–500
pulmonary
    artery 163, 165, 167–169, 171, 172, 176, 177
    circulation 191, 192
    valve 38
    vascular resistance 192
    vasculature 165, 172
    vein 165, 168, 169
pulsatile 90
Purkinje system 42
P-V loop 57, 58
P-wave 42

QRS complex 42

Ras 79, 82–86, 93, 94
RasN17, a negative mutant of Ras 82, 93–95
ratio of ventilation to blood flow (V/Q ratio) 191, 192, 201, 204
re-attachment region 88
receptor tyrosine kinase 84
reciprocating 90
reconstruction 387–389, 392, 394, 396, 399
recordkeeping 491, 499
regenerating cardiac function 75
regenerative
    medicine 291, 299, 301, 303, 304
    therapy 296
regional myocardial function 59
remodeling 86, 87
renal
    autoregulation 219
    tubule 213, 214
repolarization 42, 49
resistance 40, 42
respiratory
    distress of the newborn 190
    muscle 186, 187
re-stenosis 93–95
resuscitation 149, 153, 157, 158

retinoblastoma gene product (Rb) 87
rheumatoid arthritis 300
right atrium 37, 38, 40, 42
right ventricle 38, 40, 42
"RoboLase" 359, 364
ryanodine receptor 46

salt 210, 215–217, 219, 220
saphenous vein 92, 93
sarcomere 45–49, 56, 61, 62, 65, 225, 229–232, 234–238
sarcoplasmatic reticulum 46
secondary prevention 74, 75
self-assembly 369–371, 375–377, 380–382
self-renewal 297, 307
septum 38, 42, 45
sequence alignment 404–408
shape index 91
Shc 82, 84, 85
shear 298, 307
shear stress 79–87, 90, 94, 298
    laminar 79, 86, 87, 94
    pulsatile 90
    reciprocating 90
shock 131, 132, 141, 142, 144, 145
    wave 359, 363
shunting 206
signal transduction 81, 94
signaling pathway 79, 81–84, 86, 88, 95
single blastomere cell 305
single molecule DNA amplification 474, 476, 477
sinoatrial node 42
skeletal muscle 225, 226, 229–239
skeletal muscle origin 226, 238
smooth muscle 100, 102, 104, 108, 109, 111
sodium 45, 46
    pump 215, 219
solubility 195–198, 200–202
Sos 82
specific marker 297
spectrin (Sp) 119, 120, 123–128
spinal nerve 300
stem cell 291–308
    bone marrow 293
    embryonic 121–123, 291–293, 295–306
    epithelial 294
    mesenchymal 294
    neural 294
    skin 293, 294
    somatic 293
    tissue 292, 294
Strahler model 165, 166
strain 41, 44, 45, 49
    energy 45
stress 39, 40, 44, 45, 47–49, 118, 120, 123, 125, 126
    normal 80
    oxidative 90
    sensor 61
stretch 307
stretch-activated channel 49
stretch-sensitive ion channel 108
stroke 300
SU-1498: Flk-1 blocking agent 85
subcultured 296, 297
    passage 296
surface
    area 181, 190, 193, 195, 198
    tension 190
surfactant 190, 192, 206
survival 86, 87, 90
suspension complex 120, 122, 126, 127
sympathetic nervous system 111
synthetic biology 427–429, 431, 432, 439, 440, 448
systemic circulation 191
systems biology 307, 387–389, 392, 396, 401, 411, 419, 421
systole 38, 40, 57–60

targeted delivery 307
technology-driven 522
telemetry 261, 263–266
teratoma 291, 297
thermal
    mechanism 356
    photocoagulation 356

Tie2 87
tissue 37, 40, 42, 44, 45, 47
  pressure hypothesis 106
  remodeling 17, 18, 20, 21, 25–28, 30
topology 117, 119, 125, 127
total artificial heart 69–72
TPA responsive element (TRE) 81–83
trachea 183, 184, 187
traction force microscopy 56, 57
transcription factor 81, 83, 90, 297, 455–458, 462–464
trans-differentiation 295
transfection 363
transformation 212
transmural 41, 47, 49
transplantation therapy 300
tricuspid valve 38
trophectoderm 306
tropomyosin (TM) 46, 120, 125, 126
troponin C 46, 47
Tschachotin 359
tubular transport 218, 219
tubulin 361
tubuloglomerular feedback 110, 111, 220
tumor vasculature 316, 322, 323
turnover 79, 80, 82, 91, 92, 94
T-wave 42
tyrosine phosphorylation 84, 85

ultraviolet excimer laser 357, 358
umbilical cord blood 294
undifferentiated state 295
urea 210, 216, 219, 220
utilitarian 495

valve 54, 57
vascular
  graft 92, 93
  smooth muscle 100, 102, 108, 109, 111
  smooth muscle cell (SMC) 80, 93
vascular endothelial growth factor (VEGF) 79, 84
vasodilator metabolite 102–104
venous vascular tree 184
ventilation 181, 182, 185–188, 191, 192, 194, 195, 199–206
ventilation/perfusion ratio 200–203
ventricular assist device 69, 70, 71, 73
ventricular filling 38, 40
vessel compliance 169
virus 296
vitronectin 84
voltage gated Ca^{2+} channel 110

wall tension 108–110
waste product 100
Weibel model 165, 166

X-ray imaging 59

yeast two-hybrid system 455, 458

Z-disc 56
zero-stress state 14–21, 26